SYMMETRIES IN SCIENCE VIII

SYMMETRIES IN SCIENCE VIII

Edited by

Bruno Gruber
Southern Illinois University at Carbondale in Niigata
Niigata, Japan

PLENUM PRESS · NEW YORK AND LONDON

Library of Congress Cataloging-in-Publication Data

Symmetries in science VIII / edited by Bruno Gruber.
        p.   cm.
    Includes bibliographical references and index.
    ISBN 0-306-45119-0
    1. Symmetry--Congresses. 2. Symmetry (Physics)--Congresses.
  I. Gruber, Bruno, 1936-   . II. Symposium on Symmetries in Science
  (8th : 1994 : Bregenz, Austria)
  Q172.5.S95S922   1995
  530.1'522--dc20   95-36857
                                                           CIP

Proceedings of a Symposium on Symmetries in Science VIII,
held August 8–12, 1994, in Bregenz, Austria

ISBN 0-306-45119-0

© 1995 Plenum Press, New York
A Division of Plenum Publishing Corporation
233 Spring Street, New York, N. Y. 10013

10 9 8 7 6 5 4 3 2 1

Foreword

The symposium "Symmetries in Science VIII" was held in August of 1994 at the Cloister Mehrerau in Bregenz, Austria. The symposium was supported by Southern Illinois University at Carbondale, the Land Vorarlberg, and the Landeshaupstadt Bregenz. I wish to thank Dr. John C. Guyon, President of Southern Illinois University at Carbondale; Dr. Hubert Regner, Amt der Vorarlberger Landesregierung; and Dipl. Vw. Siegfried Gasser, Buergermeister der Landeshauptstadt Bregenz and Lantagsabgeordneter, for their generous support of the symposium.

Finally I wish to thank Frater Albin of the Cloister Mehrerau for his support and cooperation in this endeavor, which made for a successful meeting in a most pleasant environment.

Bruno Gruber

CONTENTS

ON $O_m \times GL_n$ HIGHEST WEIGHT VECTORS

Helmer Aslaksen, Eng-Chye Tan[1], and Chen-bo Zhu

Department of Mathematics
National University of Singapore
Kent Ridge, Singapore 0511
Singapore

INTRODUCTION

Let $\mathbb{C}^{m,n}$ be the vector space of $m \times n$ complex matrices and $\mathcal{P}(\mathbb{C}^{m,n})$ be the algebra of complex-valued polynomials on $\mathbb{C}^{m,n}$. Let $GL_m \times GL_n$ act on $\mathcal{P}(\mathbb{C}^{m,n})$ by pre- and post-multiplication as follows:

$$(g_1, g_2)f(x) = f(g_1^{-1} x g_2),$$

where $x \in \mathbb{C}^{m,n}$, $(g_1, g_2) \in GL_m \times GL_n$, $f \in \mathcal{P}(\mathbb{C}^{m,n})$. We choose a system of coordinates on $\mathbb{C}^{m,n}$ as follows:

$$\begin{bmatrix} x_{11} & x_{12} & \cdots & x_{1n} \\ x_{21} & x_{22} & \cdots & x_{2n} \\ \vdots & \vdots & \cdots & \vdots \\ x_{m1} & x_{m2} & \cdots & x_{mn} \end{bmatrix}.$$

It is easy to describe the infinitesimal action of the Lie algebra $\mathfrak{gl}_m \oplus \mathfrak{gl}_n$ of $GL_m \times GL_n$ on $\mathcal{P}(\mathbb{C}^{m,n})$. Let

$$\begin{aligned}
L_{jk} &= \sum_{s=1}^{n} x_{js} \frac{\partial}{\partial x_{ks}}, \quad 1 \le j, k \le m, \\
R_{jk} &= \sum_{s=1}^{m} x_{sj} \frac{\partial}{\partial x_{sk}}, \quad 1 \le j, k \le n.
\end{aligned} \tag{1}$$

Then

$$\begin{aligned}
\mathfrak{gl}_m &= \operatorname{span}\{L_{jk}, 1 \le j, k \le m\}, \\
\mathfrak{gl}_n &= \operatorname{span}\{R_{jk}, 1 \le j, k \le n\}.
\end{aligned}$$

---

[1]Presenter of the paper at the conference

Symmetries in Science VII, Edited by
B. Gruber, Plenum Press, New York, 1995

We have the Cartan decomposition

$$\mathfrak{gl}_m = \mathfrak{o}_m \oplus \mathfrak{p},$$

where $\mathfrak{o}_m$ is the Lie algebra of O_m sitting in GL_m, and

$$\mathfrak{o}_m = \mathrm{span}\{L_{jk} - L_{kj}, 1 \leq j < k \leq m\},$$
$$\mathfrak{p} = \mathrm{span}\{L_{jk} + L_{kj}, 1 \leq j \leq k \leq m\}. \tag{2}$$

Fix a reductive group G and a maximal unipotent subgroup U of G. For any rational representation V of G, let $\mathcal{P}(V)$ be the algebra of polynomials functions on V. The ring of U invariants in $\mathcal{P}(V)$ will be called the ring of G highest weight vectors in $\mathcal{P}(V)$. Our problem can be stated simply as follows:

Problem: Describe the ring $R_{m,n}$ of $O_m \times GL_n$ highest weight vectors in $\mathcal{P}(\mathbb{C}^{m,n})$.

For small m, this is easy.

Proposition 1 *(a) $R_{1,n}$ is the polynomial algebra in x_{11}.*

(b) $R_{2,1}$ is the polynomial algebra in x_{11} and x_{21}.

(c) If $n \geq 2$, $R_{2,n}$ is freely generated by

$$x_{11}, \qquad x_{21}, \; and \qquad \begin{vmatrix} x_{11} & x_{12} \\ x_{21} & x_{22} \end{vmatrix}.$$

Let

$$\Delta_{jk} = \sum_{s=1}^{m} \frac{\partial^2}{\partial x_{sj} \partial x_{sk}}, \quad 1 \leq j \leq k \leq n, \; \text{and}$$
$$r_{jk}^2 = \sum_{s=1}^{m} x_{sj} x_{sk}, \quad 1 \leq j \leq k \leq n. \tag{3}$$

Define the space of harmonics

$$\mathcal{H} = \{f \in \mathcal{P}(\mathbb{C}^{m,n}) \mid \Delta_{jk} f = 0, \; 1 \leq j \leq k \leq n\}, \tag{4}$$

and let

$$\mathcal{I} = \mathcal{P}(\mathbb{C}^{m,n})^{O_m}$$

be the ring of O_m invariants. It is known [6] that

$$\mathcal{P}(\mathbb{C}^{m,n}) = \mathcal{I} \cdot \mathcal{H}. \tag{5}$$

The description of $R_{m,1}$ follow easily from the theory of spherical harmonics. It says that $\mathcal{H}$ is O_m stable and the O_m highest weight vectors in $\mathcal{H}$ (with respect to an appropriate choice of a positive system of roots, which we will elaborate in the next section) are z_1^k for $k = 0, 1, \ldots$, where

$$z_1 = x_{11} - i x_{21}.$$

Furthermore, $\mathcal{I}$ is generated by r_{11}^2 as an algebra. The ring of O_m highest weight vectors in $\mathcal{P}(\mathbb{C}^{m,1})$ is thus

$$\mathbb{C}[z_1, r_{11}^2].$$

For $n > 1$, the GL_n-stucture of $\mathcal{I}$ and in particular its $O_m \times GL_n$ highest weight vectors are well-known from classical invariant theory (see Theorem 1 in Section 3). In [6], Kashiwara and Vergne generalised the theory of spherical harmonics to give the set of $O_m \times GL_n$ highest weight vectors in the space $\mathcal{H}$. These give partial information on $R_{m,n}$, and to get a complete picture, one has to understand how the highest weight vectors multiply as in (5) above.

We had determined the ring $R_{m,2}$ and also found a set of generators for $R_{m,3}$, along with some relations. Beyond $n = 3$, the situation gets too complicated. This is understandable, because $R_{m,n}$ contains in particular information on the restriction of GL_m representations (with depth at most $\min(m,n)$) to its subgroup O_m, which is known to be a difficult problem.

HARMONIC POLYNOMIALS

We recall some results from Kashiwara and Vergne [6]. We choose O_m so that its Lie algebra $\mathfrak{o}_m$ is described by (2). Now choose a Cartan subalgebra (different from the one in Section 2) for $\mathfrak{o}_m$ as follows:

$$\mathfrak{h}_L = \left\{ H = \begin{bmatrix} A_1 & & \\ & A_2 & 0 \\ & 0 & \ddots \end{bmatrix} \middle| A_j = \begin{bmatrix} 0 & ih_j \\ -ih_j & 0 \end{bmatrix}, h_j \in \mathbb{C}, 1 \le j \le \left[\frac{m}{2}\right] \right\} \tag{6}$$

and also a Cartan subalgebra for $\mathfrak{gl}_n$:

$$\mathfrak{h}_R = \left\{ \begin{bmatrix} d_1 & & \\ 0 & \ddots & 0 \\ & & d_n \end{bmatrix} \middle| d_j \in \mathbb{C} \right\}. \tag{7}$$

Define the following linear functionals on $\mathfrak{h}_L$:

$$e_j(\text{above } H) = h_j, \quad 1 \le j \le \left[\frac{m}{2}\right]. \tag{8}$$

The roots of $\mathfrak{o}_m$ are

$$\Delta = \begin{cases} \{\pm e_j \pm e_k, j \ne k\} \cup \{\pm e_k\} & \text{if } m \text{ is odd,} \\ \{\pm e_j \pm e_k, j \ne k\} & \text{if } m \text{ is even.} \end{cases} \tag{9}$$

We choose a positive system as follows:

$$\Delta^+ = \begin{cases} \{e_j \pm e_k, j < k\} \cup \{e_k\} & \text{if } m \text{ is odd,} \\ \{e_j \pm e_k, j < k\} & \text{if } m \text{ is even.} \end{cases} \tag{10}$$

For GL_n, we choose the Borel subalgebra of upper triangular matrices to fix a positive root system for $\mathfrak{gl}_n$.

We could use the theory of highest weights to parametrize irreducible finite-dimensional representations of SO_m by tuples of integers $(\alpha_1, \alpha_2, \ldots, \alpha_{\left[\frac{m}{2}\right]})$ satisfying

$$\begin{cases} \alpha_1 \ge \alpha_2 \ge \ldots \ge \alpha_{\left[\frac{m}{2}\right]} \ge 0 & \text{if } m \text{ is odd,} \\ \alpha_1 \ge \alpha_2 \ge \ldots \ge |\alpha_{\left[\frac{m}{2}\right]}| & \text{if } m \text{ is even,} \end{cases}$$

and likewise an irreducible finite-dimensional representation of GL_n by an n-tuple of integers $(\alpha_1, \alpha_2, \ldots, \alpha_n)$ with $\alpha_1 \geq \alpha_2 \geq \ldots \geq \alpha_n$.

If $m = 2l+1$, O_{2l+1} is isomorphic to a direct product $SO_{2l+1} \times \mathbb{Z}_2$. A representation of O_{2l+1} can then be parametrized by

$$(\alpha_1, \ldots, \alpha_l; \varepsilon) \simeq (\alpha_1, \ldots, \alpha_l) \otimes \varepsilon,$$

where $(\alpha_1, \ldots, \alpha_l) \in \widehat{SO_{2l+1}}$ and ε is a one-dimensional representation of $\mathbb{Z}_2$ trivial or nontrivial according to $\varepsilon = 1$ or -1.

If $m = 2l$, O_{2l} is isomorphic to a semi-direct product $SO_{2l} \rtimes \mathbb{Z}_2$. If $(\alpha_1, \ldots, \alpha_l) \in \widehat{SO_{2l}}$ and $\alpha_l \neq 0$, then

$$\mathrm{Ind}_{SO_{2l}}^{O_{2l}}(\alpha_1, \ldots, \alpha_l)$$

is irreducible and we denote this by $(\alpha_1, \ldots, \alpha_l; 0)$. If $\alpha_l = 0$, then there are two ways to extend $(\alpha_1, \ldots, \alpha_l)$ to a representation of O_{2l}, namely

$$(\alpha_1, \ldots, \alpha_l; \varepsilon) \simeq (\alpha_1, \ldots, \alpha_l) \otimes \varepsilon,$$

where $\varepsilon = \pm 1$. A representation of O_{2l} can then be parametrized by

$$(\alpha_1, \ldots, \alpha_l; 0) \qquad \text{if } \alpha_l \neq 0,$$

or

$$(\alpha_1, \ldots, \alpha_l; \pm 1) \qquad \text{if } \alpha_l = 0.$$

Now define

$$z_{jk} = x_{2j-1,k} - i x_{2j,k} \tag{11}$$

for $1 \leq j \leq \left[\frac{m}{2}\right], 1 \leq k \leq n$, and the following $t \times t$ determinants for $1 \leq t \leq \min\left(\left[\frac{m}{2}\right], n\right)$:

$$\alpha_t = \begin{vmatrix} z_{11} & z_{12} & \cdots & z_{1t} \\ \vdots & \vdots & \vdots & \vdots \\ z_{t1} & z_{t2} & \cdots & z_{tt} \end{vmatrix}, \tag{12}$$

$$\gamma_t = \begin{vmatrix} r_{11}^2 & r_{12}^2 & \cdots & r_{1t}^2 \\ \vdots & \vdots & \vdots & \vdots \\ r_{t1}^2 & r_{t2}^2 & \cdots & r_{tt}^2 \end{vmatrix}. \tag{13}$$

It is not difficult to check that with respect to the positive systems for $\mathfrak{o}_m$ and $\mathfrak{gl}_n$, each α_t is an $O_m \times GL_n$ highest weight vector of weight

$$(\underbrace{1, \ldots, 1}_{t \text{ copies}}, 0, \ldots, 0; (-1)^t) \otimes (\underbrace{1, \ldots, 1}_{t \text{ copies}}, 0, \ldots, 0), \tag{14}$$

and each γ_t is an $O_m \times GL_n$ highest weight vector of weight

$$(0, \ldots, 0; 1) \otimes (\underbrace{2, \ldots, 2}_{t \text{ copies}}, 0, \ldots, 0). \tag{15}$$

The following results are well known, and can be found in [6] and [8]:

Theorem 1 (a) *The space of O_m-invariant polynomials in $\mathcal{P}(\mathbb{C}^{m,n})$ is generated by r_{jk}^2, $1 \le j \le k \le n$. The subring of $O_m \times GL_n$ highest weight vectors in $\mathcal{I}$ is generated freely by*

$$\gamma_1, \dots, \gamma_{\min(\left[\frac{m}{2}\right], n)}.$$

(b) *Assume that $m \ge 2n$. The ring of $O_m \times GL_n$ highest weight vectors in $\mathcal{H}$ is generated freely by*

$$\alpha_1, \dots, \alpha_n.$$

(c) $\mathcal{P}(\mathbb{C}^{m,n}) = \mathcal{H} \cdot \mathcal{I}$.

Remarks. Analogous results hold for the cases where $m < 2n$ and they can be found in [6]. It turns out that there are more generators in $\mathcal{H}$ for $m < 2n$. In this paper, we will only discuss the cases when $m \ge 2n$ to facilitate exposition. In fact, the cases where $m < 2n$ could be treated in a similar fashion, but it is more messy, and so we will only present the results for $m \ge 2n$.

STRUCTURE OF $R_{m,2}$ AND APPLICATIONS

We have already described $R_{1,2}$ and $R_{2,2}$. We treat the case when $n = 2$ and $m \ge 4$, and simply state the result for $R_{3,2}$. First, we have a simple result about tensor products of GL_2 representations. Let

$$\mathfrak{gl}_2 = \operatorname{span}\{R_{11}, R_{22}, R_{12}, R_{22}\},$$

with the usual commutation relations

$$[R_{ij}, R_{st}] = \delta_{js}R_{it} - \delta_{ti}R_{sj}.$$

We will select R_{12} to be the positive root vector in $\mathfrak{gl}_2$. As usual, if V is a representation of GL_2, we say that $v \in V$ is a GL_2 highest weight vector if $R_{12}v = 0$ and v is an eigenvector for R_{11} and R_{22}. The following lemma is easy.

Lemma 1 *Let $V_{(\lambda_1, \lambda_2)}$ be an irreducible finite-dimensional representation of GL_2 with highest weight (λ_1, λ_2). Then*

$$V_{(\lambda_1, \lambda_2)} \otimes V_{(1,0)} \simeq \begin{cases} V_{(\lambda_1+1, \lambda_2)} \oplus V_{(\lambda_1, \lambda_2+1)} & \text{if } \lambda_1 \ne \lambda_2, \\ V_{(\lambda_1+1, \lambda_2)} & \text{if } \lambda_1 = \lambda_2, \end{cases}$$

and if $\lambda_1 \ne \lambda_2$, the highest weight vector of $V_{(\lambda_1, \lambda_2+1)}$ is given (up to a scalar) by

$$v = R_{21}v_{(\lambda_1, \lambda_2)} \otimes v_{(1,0)} - (\lambda_1 - \lambda_2)v_{(\lambda_1, \lambda_2)} \otimes R_{21}v_{(1,0)},$$

where $v_{(\lambda_1, \lambda_2)}$ and $v_{(1,0)}$ are the highest weight vectors in $V_{(\lambda_1, \lambda_2)}$ and $V_{(1,0)}$, respectively.

From now on, we restrict our discussion to vectors in $\mathcal{P}(\mathbb{C}^{m,n})$. Consider the map

$$\pi : \mathcal{I} \otimes \mathcal{H} \longrightarrow \mathcal{P}(\mathbb{C}^{m,n})$$

by polynomial multiplication. Theorem 1(c) says that the map is onto. To find the $O_m \times GL_n$ representations appearing in $\mathcal{P}(\mathbb{C}^{m,n})$, consider $1 \otimes \sigma_1$ appearing in $\mathcal{I}$ and $\rho \otimes \sigma_2$ appearing in $\mathcal{H}$. Since

$$(1 \otimes \sigma_1) \otimes (\rho \otimes \sigma_2) \simeq \rho \otimes (\sigma_1 \otimes \sigma_2)$$

as $O_m \times GL_n$ representations, it suffices to understand the projection of $\rho \otimes \tilde{\sigma}$ under π, where $\tilde{\sigma}$ is a GL_n subrepresentation of $\sigma_1 \otimes \sigma_2$.

Let $< f >$ denote the GL_2 module in $\mathcal{P}(\mathbb{C}^{m,2})$ generated by the polynomial f. From the discussion above, we are to find the GL_2 highest weight vectors in the tensor product

$$< \alpha_1^a \alpha_2^b > \otimes < \gamma_1^c \gamma_2^d > .$$

Since γ_2 and α_2 each generates a one-dimesional representation of GL_2, it suffices to consider

$$< \alpha_1^a > \otimes < \gamma_1^c > .$$

Observe from Lemma 1 that if v_1 and v_2 are two GL_2 highest weight vectors of weights (λ_1', λ_2') and $(\lambda_1'', \lambda_2'')$, then the highest weight vectors of

$$< \alpha_1 > \otimes < v_1 v_2 >$$

are $\alpha_1 v_1 v_2$ and

$$\begin{aligned}
(R_{21}(v_1 v_2))\alpha_1 &- (\lambda_1' + \lambda_1'' - \lambda_2' - \lambda_2'')v_1 v_2 (R_{21}\alpha_1) \\
&= [(R_{21}v_1)\alpha_1 - (\lambda_1' - \lambda_2')v_1(R_{21}\alpha_1)]v_2 \\
&\quad + [(R_{21}v_2)\alpha_1 - (\lambda_1'' - \lambda_2'')v_2(R_{21}\alpha_1)]v_1.
\end{aligned} \tag{16}$$

Now using Lemma 1 on $< \alpha_1 > \otimes < \gamma_1 >$ we get

$$\beta_1 = \begin{vmatrix} z_{11} & z_{12} \\ r_{11}^2 & r_{12}^2 \end{vmatrix}$$

of $O_m \times GL_2$ weight $(1, 0, \ldots, 0; -1) \otimes (2, 1)$. Another application of Lemma 1 on $< \alpha_1 > \otimes < \beta_1 >$ gives

$$\beta_2 = \begin{vmatrix} 0 & z_{11} & z_{12} \\ z_{11} & r_{11}^2 & r_{12}^2 \\ z_{12} & r_{21}^2 & r_{22}^2 \end{vmatrix}$$

of $O_m \times GL_2$ weight $(2, 0, \ldots, 0; 1) \otimes (2, 2)$. Since β_2 transforms under GL_2 by the determinant representation,

$$< \alpha_1 > \otimes < \beta_2 > = < \alpha_1 \beta_2 > .$$

We also have the relation

$$\alpha_1^2 \gamma_2 + \beta_1^2 + \gamma_1 \beta_2 = 0$$

by a direct computation.

6

Theorem 2 *(Roger Howe)*

(a) The ring $R_{1,2}$ is the polynomial algebra $\mathbb{C}[x_{11}]$ while $R_{2,2}$ is freely generated by

$$\alpha_1, \qquad \overline{\alpha_1}, \quad and \quad \begin{vmatrix} z_{11} & \overline{z_{11}} \\ z_{12} & \overline{z_{12}} \end{vmatrix}.$$

(b) The ring $R_{3,2}$ is freely generated by α_1, γ_1 and δ where

$$\delta = \begin{vmatrix} x_{11} & x_{12} \\ x_{21} & x_{22} \end{vmatrix}.$$

(c) If $m \geq 4$, $R_{m,2}$ is generated by α_1, α_2, β_1, β_2, γ_1 and γ_2 with one relation:

$$\alpha_1^2 \gamma_2 + \beta_1^2 + \gamma_1 \beta_2 = 0.$$

Proof. We have already shown (a). We will prove (c). Because

$$< \alpha_1^a > \otimes < \gamma_1^c > \hookrightarrow < \alpha_1 > \otimes (< \alpha_1^{a-1} > \otimes < \gamma_1^c >),$$

we can use equation (16) and a straightforward induction to conclude that α_1, α_2, β_1, β_2, γ_1 and γ_2 generate $R_{m,2}$. It is not difficult to compute that the Krull dimension of $R_{m,2}$ is 5. It is also simple to see that α_1, α_2, β_2, γ_1 and γ_2 are algebraically independent. The ideal of relations among α_1, α_2, β_1, β_2, γ_1 and γ_2 will therefore have to be a principal ideal. Since the polynomial

$$\alpha_1^2 \gamma_2 + \beta_1^2 + \gamma_1 \beta_2$$

is irreducible, it generates this ideal. $\square$

A consequence of the above result is a branching law for O_m representations appearing in a GL_m representation with depth at most two.

Corollary 1 *We have the following formula for the restriction of a representation $(a, b, 0, \ldots, 0)$ of GL_m to O_m:*

$$(a, b, 0, \ldots, 0)\Big|_{O_m} = \sum_{\delta \in S} (x_\delta, y_\delta, 0, \ldots, 0; (-1)^\varepsilon)$$

where

$$S = \{\delta = (a_1, a_2, b_1, b_2, c_1, c_2) \in \mathbb{Z}_+^6 \mid x_\delta = a_1 + a_2 + b_1 + 2b_2, \quad y_\delta = a_2,$$
$$a_1 + a_2 + 2b_1 + 2b_2 + 2c_1 + 2c_2 = a, \quad \varepsilon = a_1 + b_1,$$
$$a_2 + b_1 + 2b_2 + 2c_2 = b, \quad b_1 = 0, 1\}$$

GENERATORS FOR $R_{m,3}$

We proceed to study the case $n = 3$ and $m \geq 6$. As in the previous section we use Theorem 1 to reduce our study to tensor products of the form

$$< \alpha_1^a \alpha_2^b \alpha_3^e > \otimes < \gamma_1^c \gamma_2^d \gamma_3^f > .$$

Since $< \alpha_3 >$ and $< \gamma_3 >$ each generates a one-dimensional representation of GL_3, it suffices to study

$$< \alpha_1^a \alpha_2^b > \otimes < \gamma_1^c \gamma_2^d > .$$

Because

$$< \alpha_1^a \alpha_2^b > \otimes < \gamma_1^c \gamma_2^d > \hookrightarrow < \alpha_1 > \otimes (< \alpha_1^{a-1} \alpha_2^b > \otimes < \gamma_1^c \gamma_2^d >)$$

if $a \neq 0$, and

$$< \alpha_1^a \alpha_2^b > \otimes < \gamma_1^c \gamma_2^d > \hookrightarrow < \alpha_2 > \otimes (< \alpha_1^a \alpha_2^{b-1} > \otimes < \gamma_1^c \gamma_2^d >)$$

if $b \neq 0$, we shall first consider decompositions of the GL_3 modules

$$< \alpha_1 > \otimes V \qquad \text{and} \qquad < \alpha_2 > \otimes V$$

for an arbitrary irreducible GL_3 module V.

Let us introduce some notations. Recall our selection of the set of diagonal matrices

$$\mathfrak{h}_R = \left\{ \begin{bmatrix} d_1 & 0 & 0 \\ 0 & d_2 & 0 \\ 0 & 0 & d_3 \end{bmatrix} \middle| d_j \in \mathbb{C} \right\}$$

as the Cartan subalgebra in $\mathfrak{gl}_3$. Define the functionals

$$\varepsilon_j \left(\begin{bmatrix} d_1 & 0 & 0 \\ 0 & d_2 & 0 \\ 0 & 0 & d_3 \end{bmatrix} \right) = d_j, \quad j = 1, 2, 3.$$

The non-zero roots of $\mathfrak{gl}_3$ are

$$\Delta = \{ \pm(\varepsilon_i - \varepsilon_j) \mid 1 \leq i < j \leq 3 \},$$

where the root vectors of $\varepsilon_i - \varepsilon_j$ and $-(\varepsilon_i - \varepsilon_j)$ are the matrix units E_{ij} and E_{ji} respectively. These root vectors act on $\mathcal{P}(\mathbb{C}^{m,3})$ as R_{ij} described by (1.1). Select a positive system as follows

$$\Xi = \{ e_1 = \varepsilon_1 - \varepsilon_2, \ e_2 = \varepsilon_2 - \varepsilon_3 \}.$$

If h is a GL_3 highest weight vector in $\mathcal{P}(\mathbb{C}^{m,3})$, we shall denote by $< h >$ the GL_3 module generated by h. If V_1 and V_2 are two GL_3 sub-modules in $\mathcal{P}(\mathbb{C}^{m,3})$, we have the multiplication map

$$\Phi : V_1 \otimes V_2 \longrightarrow \mathcal{P}(\mathbb{C}^{m,3}),$$
$$\Phi(v_1 \otimes v_2) = v_1 v_2, \qquad v_1 \in V_1, v_2 \in V_2.$$

If V_{λ_1} and V_{λ_2} are GL_3 modules of highest weight λ_1 and λ_2, respectively, standard representation theory says that

$$V_{\lambda_1} \otimes V_{\lambda_2} = \sum_{\lambda \in P} m_\lambda V_{\lambda_1 + \lambda_2 - \lambda} = \sum_{\lambda \in P} W_{\lambda_1 + \lambda_2 - \lambda},$$

where P is the set of sums of positive roots (including the zero sum), m_λ is the multiplicity of the representation $V_{\lambda_1 + \lambda_2 - \lambda}$ in the tensor product, and $W_{\lambda_1 + \lambda_2 - \lambda}$ is the $V_{\lambda_1 + \lambda_2 - \lambda}$-isotypic component. For $\lambda \in P$, let

$$\pi_\lambda : V_{\lambda_1} \otimes V_{\lambda_2} \longrightarrow W_{\lambda_1 + \lambda_2 - \lambda}$$

be the projection onto the $V_{\lambda_1 + \lambda_2 - \lambda}$-isotypic component, and define another map

$$F_\lambda : V_{\lambda_1} \otimes V_{\lambda_2} \longrightarrow \Phi(W_{\lambda_1 + \lambda_2 - \lambda}),$$
$$F_\lambda = \Phi \circ \pi_\lambda.$$

Thus, if $V_1 = <v_1>$ and $V_2 = <v_2>$, then

$$F_0(v_1 \otimes v_2) = v_1 v_2,$$

and so

$$F_0(V_1 \otimes V_2) = <v_1 v_2>.$$

The following lemma is straightforward.

Lemma 2 *Let $V_1 = <v_1>$ and $V_2 = <v_2>$ be irreducible GL_3 modules generated by highest weight vectors v_1 and v_2 of weight $(\lambda_1, \lambda_2, \lambda_3)$ and (μ_1, μ_2, μ_3) respectively. We have the following formulae.*

(a) $\quad F_{e_1}(v_1 \otimes v_2) = (\lambda_1 - \lambda_2)v_1(R_{21}v_2) - (\mu_1 - \mu_2)(R_{21}v_1)v_2,$

(b) $\quad F_{e_2}(v_1 \otimes v_2) = (\lambda_2 - \lambda_3)v_1(R_{32}v_2) - (\mu_2 - \mu_3)(R_{32}v_1)v_2,$

(c) $\quad F_{e_1+e_2}(\alpha_1 \otimes v_1) = \alpha_1(R_{21}R_{32}v_1) - (\lambda_2 - \lambda_3)\alpha_1(R_{31}v_1)$
$\qquad\qquad - (\lambda_1 - \lambda_3 + 1)(R_{21}\alpha_1)(R_{32}v_1) + (\lambda_2 - \lambda_3)(\lambda_1 - \lambda_3 + 1)(R_{31}\alpha_1)v_1,$

(d) $\quad F_{e_1+e_2}(\alpha_2 \otimes v_1) = \alpha_2(R_{21}R_{32}v_1) - (\lambda_1 - \lambda_2 + 1)\alpha_2(R_{31}v_1)$
$\qquad\qquad - (\lambda_1 - \lambda_3 + 1)(R_{32}\alpha_2)(R_{21}v_1) + (\lambda_1 - \lambda_2)(\lambda_1 - \lambda_3 + 1)(R_{31}\alpha_2)v_1.$

Proposition 2 *Let $V_1 = <v_1>$ and $V_2 = <v_2>$ be irreducible GL_3 modules generated by highest weight vectors v_1 and v_2 of weight $(\lambda_1, \lambda_2, \lambda_3)$ and (μ_1, μ_2, μ_3) respectively. We have the following formulae.*

(a) $\quad F_{e_1}(\alpha_1 \otimes (v_1 v_2)) = v_1 F_{e_1}(\alpha_1 \otimes v_2) + v_2 F_{e_1}(\alpha_1 \otimes v_1),$

(a) $\quad F_{e_2}(\alpha_2 \otimes (v_1 v_2)) = v_1 F_{e_2}(\alpha_2 \otimes v_2) + v_2 F_{e_2}(\alpha_2 \otimes v_1),$

(c) $\quad F_{e_1+e_2}(\alpha_1 \otimes (v_1 v_2))$
$$= \frac{[F_{e_2}(F_{e_1}(\alpha_1 \otimes v_2) \otimes v_1) + (\lambda_2 + \mu_2 - \lambda_3 - \mu_3 + 1)v_1 F_{e_1+e_2}(\alpha_1 \otimes v_2)]}{(\mu_2 - \mu_3 + 1)}$$
$$+ \frac{[F_{e_2}(F_{e_1}(\alpha_1 \otimes v_1) \otimes v_2) + (\lambda_2 + \mu_2 - \lambda_3 - \mu_3 + 1)v_2 F_{e_1+e_2}(\alpha_1 \otimes v_1)]}{(\lambda_2 - \lambda_3 + 1)},$$

(d) $\quad F_{e_1+e_2}(\alpha_2 \otimes (v_1 v_2))$
$$= \frac{[F_{e_1}(F_{e_2}(\alpha_2 \otimes v_1) \otimes v_2) + (\lambda_1 + \mu_1 - \lambda_2 - \mu_2 + 1)v_2 F_{e_1+e_2}(\alpha_2 \otimes v_1)]}{(\mu_1 - \mu_2 + 1)}$$
$$+ \frac{[F_{e_1}(F_{e_2}(\alpha_2 \otimes v_2) \otimes v_1) + (\lambda_1 + \mu_1 - \lambda_2 - \mu_2 + 1)v_1 F_{e_1+e_2}(\alpha_2 \otimes v_2)]}{(\lambda_1 - \lambda_2 + 1)}.$$

Proof. The first two formulae follow directly from Lemma 2. It is not difficult to verify (c) and (d). The difficult part is to believe such a formula holds and then to find it. For full details, please refer to [1]. $\square$

Define

$$\beta_1 = \begin{vmatrix} z_{11} & z_{12} \\ r_{11}^2 & r_{12}^2 \end{vmatrix}$$

$$\beta_2 = \begin{vmatrix} 0 & z_{11} & z_{12} \\ z_{11} & r_{11}^2 & r_{12}^2 \\ z_{12} & r_{21}^2 & r_{22}^2 \end{vmatrix} \qquad \beta_3 = \begin{vmatrix} z_{11} & z_{12} & z_{13} \\ z_{21} & z_{22} & z_{23} \\ r_{11}^2 & r_{12}^2 & r_{13}^2 \end{vmatrix} \qquad \beta_4 = \begin{vmatrix} z_{11} & z_{12} & z_{13} \\ r_{11}^2 & r_{12}^2 & r_{13}^2 \\ r_{21}^2 & r_{22}^2 & r_{23}^2 \end{vmatrix}$$

$$\beta_5 = \begin{vmatrix} z_{11} & 0 & z_{12} & z_{13} \\ z_{21} & 0 & z_{22} & z_{23} \\ 0 & r_{11}^2 & r_{12}^2 & r_{13}^2 \\ 0 & r_{21}^2 & r_{22}^2 & r_{23}^2 \end{vmatrix} \qquad \beta_6 = \begin{vmatrix} 0 & z_{11} & z_{12} & z_{13} \\ z_{11} & r_{11}^2 & r_{12}^2 & r_{13}^2 \\ z_{12} & r_{21}^2 & r_{22}^2 & r_{23}^2 \\ z_{13} & r_{31}^2 & r_{32}^2 & r_{33}^2 \end{vmatrix} \qquad \beta_7 = \begin{vmatrix} 0 & z_{11} & z_{12} & z_{13} \\ 0 & z_{21} & z_{22} & z_{23} \\ z_{11} & r_{11}^2 & r_{12}^2 & r_{13}^2 \\ z_{12} & r_{21}^2 & r_{22}^2 & r_{23}^2 \end{vmatrix}.$$

$$\beta_8 = \begin{vmatrix} z_{11} & z_{21} & 0 & 0 & 0 \\ 0 & 0 & z_{11} & z_{12} & z_{13} \\ 0 & 0 & r_{11}^2 & r_{12}^2 & r_{13}^2 \\ z_{12} & z_{22} & r_{21}^2 & r_{22}^2 & r_{23}^2 \\ z_{13} & z_{23} & r_{31}^2 & r_{32}^2 & r_{33}^2 \end{vmatrix} \qquad \beta_9 = \begin{vmatrix} 0 & 0 & z_{11} & z_{21} & z_{13} \\ 0 & 0 & z_{21} & z_{22} & z_{23} \\ z_{11} & z_{22} & r_{11}^2 & r_{12}^2 & r_{13}^2 \\ z_{12} & z_{22} & r_{21}^2 & r_{22}^2 & r_{23}^2 \\ z_{13} & z_{23} & r_{31}^2 & r_{32}^2 & r_{33}^2 \end{vmatrix}.$$

The $O_m \times GL_3$ weights of the above polynomials are given in the following table.

	O_m weight	GL_3 weight
α_1	$(1,0,\dots,0;-1)$	$(1,0,0)$
α_2	$(1,1,0,\dots,0;1)$	$(1,1,0)$
α_3	$(1,1,1,0,\dots,0;-1)$	$(1,1,1)$
γ_1	$(0,\dots,0;1)$	$(2,0,0)$
γ_2	$(0,\dots,0;1)$	$(2,2,0)$
γ_3	$(0,\dots,0;1)$	$(2,2,2)$
β_1	$(1,0,\dots,0;-1)$	$(2,1,0)$
β_2	$(2,0,\dots,0;1)$	$(2,2,0)$
β_3	$(1,1,0,\dots,0;1)$	$(2,1,1)$
β_4	$(1,0,\dots,0;-1)$	$(2,2,1)$
β_5	$(1,1,0,\dots,0;1)$	$(3,2,1)$
β_6	$(2,0,\dots,0;1)$	$(2,2,2)$
β_7	$(2,1,0,\dots,0;-1)$	$(2,2,1)$
β_8	$(2,1,0,\dots,0;-1)$	$(3,2,2)$
β_9	$(2,2,0,\dots,0;1)$	$(2,2,2)$

Theorem 3 *If $m \geq 6$, the ring of $O_m \times GL_3$ highest weight vectors in $\mathcal{P}(\mathbb{C}^{m,3})$ is generated by*

$$\alpha_i, \ i=1,2,3, \qquad \beta_j, \ j=1,\dots,9, \qquad \gamma_k, \ k=1,2,3.$$

Remarks. We also have an (incomplete) set of relations:

$$(1) \qquad \beta_1^2 + \beta_2\gamma_1 + \alpha_1^2\gamma_2 = 0,$$
$$(2) \qquad \beta_4^2 + \beta_6\gamma_2 - \beta_2\gamma_3 = 0,$$
$$(3) \qquad \beta_5^2 + \alpha_2^2\gamma_1\gamma_3 + \beta_3^2\gamma_2 - \beta_9\gamma_1\gamma_2 = 0,$$
$$(4) \qquad \beta_7^2 + \beta_2\beta_9 - \alpha_2^2\beta_6 = 0,$$
$$(5) \qquad \beta_8^2 + \alpha_1^2\beta_9\gamma_3 + \beta_6\beta_9\gamma_1 - \beta_3^2\beta_6 = 0,$$
$$(6) \qquad \beta_1\beta_3 - \beta_7\gamma_1 - \alpha_1\beta_5 = 0.$$

These relations are some of those that were found using computer algebra software [2] and [9] and the methods of [7].

Proof. By Proposition 2, it suffices to find a set of highest weights S that are closed under F_{e_1} and F_{e_2}, i.e., $x, y \in S$ implies that $F_{e_i}(x \otimes y)$ is generated by elements in S. This can be done, after some tedious computations (see [1]). $\square$

References

[1] H. Aslaksen, E-C. Tan and C. Zhu, On certain rings of highest weight vectors, to appear in *Journal of Algebra*.

[2] D. Bayer and M. Stillman, Macaulay: A system for computation in algebraic geometry and commutative algebra. Source and object code available for Unix and Macintosh computers. Contact the authors, or download from `zariski.harvard.edu` via anonymous ftp.

[3] R. Howe, Remarks on classical invariant theory, *Trans. Amer. Math. Soc.* **313** (1989), 539 – 570.

[4] R. Howe, Appendix to remarks on classical invariant theory, preprint.

[5] R. Howe, Oral communication.

[6] M. Kashiwara and M. Vergne, On the Segal-Shale-Weil representations and harmonic polynomials, *Invent. Math.* **44** (1978), 1 – 47.

[7] M. Stillman, Methods for computing in algebraic geometry and commutative algebra, *Acta Appl. Math.* **21** (1990), 77 – 103.

[8] H. Weyl, "The Classical Groups", Princeton Univ. Press, Princeton, New Jersey, 1946.

[9] Wolfram Research, "Mathematica", Version 1.2, Champaign, Illinois, 1990.

DEPARTMENT OF MATHEMATICS, NATIONAL UNIVERSITY OF SINGAPORE, SINGAPORE 0511, REPUBLIC OF SINGAPORE
e-mail: `mathelmr, mattanec, matzhucb@nus.sg`

INVARIANT THEORY OF MATRICES

Helmer Aslaksen[1], Eng-Chye Tan and Chen-bo Zhu

Department of Mathematics
National University of Singapore
Singapore 0511
Republic of Singapore
e-mail: `mathelmr(mattanec, matzhucb)@nus.sg`

INTRODUCTION

Let F be a field of characteristic 0, let $M(n,m) = M(n,m,F)$ denote the set of $n \times m$ matrices over F and let $W = W(n,m,F)$ be the vector space of m-tuples of $n \times n$ matrices over F. Let $V \subset W$ be a vector space on which a group $G \subset GL(n,F)$ acts by simultaneous conjugation. We will denote the polynomial functions on V by $P(V)$ and the G invariants by $P(V)^G$.

We will first consider affine invariants, i.e., invariants under the general linear group. Expressions of the form $\operatorname{tr} A^2 B^2 AB$ are invariant, since

$$\operatorname{tr}(gAg^{-1})^2(gBg^{-1})^2 gAg^{-1}gBg^{-1} = \operatorname{tr} gA^2B^2ABg^{-1} = \operatorname{tr} A^2B^2AB.$$

The following theorem is due to Gurevich [6], Sibirskii [14] and Procesi [12].

Theorem 1 *Let $G = GL(n,F)$ act on $W(n,m,F)$ by conjugation. The invariants are of the form $\operatorname{tr} P(A_1,\dots,A_m)$, where P is a noncommutative polynomial in m matrix variables.*

We will next consider invariants under the orthogonal group. The conjugation action is now the same as the congruence action $(gMg^{-1} = gMg^t)$, and we see that in additions to the affine invariants, expressions of the form $\operatorname{tr} A^t B$ are now also invariants. Namely,

$$\operatorname{tr}(gAg^t)^t gBg^t = \operatorname{tr} gA^tg^tgBg^t = \operatorname{tr} A^tB.$$

The next theorem is due to Sibirskii [14] and Procesi [12].

Theorem 2 *Let $G = O(n,F)$ act on $W(n,m,F)$ by conjugation. The invariants are of the form $\operatorname{tr} P(A_1,\dots,A_m,A_1^t,\dots,A_m^t)$, where P is a noncommutative polynomial in $2m$ matrix variables.*

---

[1]Presenter of the paper at the conference.

Symmetries in Science VII, Edited by
B. Gruber, Plenum Press, New York, 1995

We can similarly consider invariants under the unitary group. If we set $M^* = \overline{M}^t$, then we get the following theorem due to Sibirskii [14] and Procesi [12].

Theorem 3 *Let $G = U(n)$ act on $W(n, m, \mathbb{C})$ by conjugation. The invariants are of the form* $\operatorname{tr} P(A_1, \ldots, A_m, A_1^*, \ldots, A_m^*)$, *where P is a noncommutative polynomial in $2m$ matrix variables.*

We can also consider symplectic invariants. If $n = 2k$, we let

$$J_n = \begin{pmatrix} 0 & -I_k \\ I_k & 0 \end{pmatrix},$$

and define the symplectic involution by

$$M^* = J_n^{-1} M^t J_n = -J_n M^t J_n,$$

then we can state the following theorem due to Procesi [12].

Theorem 4 *Let $G = Sp(k, F)$ act on $W(2k, m, F)$ by conjugation. The invariants are of the form* $\operatorname{tr} P(A_1, \ldots, A_m, A_1^*, \ldots, A_m^*)$, *where P is a noncommutative polynomial in $2m$ matrix variables.*

Unfortunately, these theorems give infinite sets of generators for the invariants. We will outline how to find finite sets of generators.

For $n = 2$ the Cayley-Hamilton Theorem says that

$$A^2 - A \operatorname{tr} A + 1/2I[(\operatorname{tr} A)^2 - \operatorname{tr} A^2] = 0$$

or in its polarized version

$$AB + BA - A \operatorname{tr} B - B \operatorname{tr} A + I[\operatorname{tr} A \operatorname{tr} B - \operatorname{tr} AB] = 0.$$

It is well known (see for example [5] or [11]) that the polarized Cayley-Hamilton Theorem implies that any product of three 2×2 matrices is reducible. That is, ABC can be written as a linear combination of matrix products with fewer factors and coefficients expressible in terms of traces. Writing

$$\operatorname{tr}(A, B) = \operatorname{tr} AB - \operatorname{tr} A \operatorname{tr} B,$$

we have

$$\begin{aligned}
2ABC &= A(BC + CB) + (AB + BA)C - [B(AC) + (AC)B] \\
&= AB \operatorname{tr} C + AC \operatorname{tr} B + A \operatorname{tr}(B, C) + AC \operatorname{tr} B + BC \operatorname{tr} A \\
&\quad + C \operatorname{tr}(A, B) - AC \operatorname{tr} B - B \operatorname{tr}(AC) - I \operatorname{tr}(AC, B) \\
&= AB \operatorname{tr} C + AC \operatorname{tr} B + BC \operatorname{tr} A \\
&\quad + A \operatorname{tr}(B, C) - B \operatorname{tr}(AC) + C \operatorname{tr}(A, B) - I \operatorname{tr}(AC, B).
\end{aligned}$$

More generally, it follows from the work of Procesi [12] and Razmyslov [13] that the product of $n^2 - 1$ matrices of order n is reducible. For 3×3 matrices the product of 6 matrices is reducible [5]. Using these results, we get finite sets of generators for the algebras of invariants.

SPECIAL ORTHOGONAL GROUP

We will now try to generalize these results to other classical groups. We will start with $SO(n)$, where $n = 2k$. Recall that if M is a skewsymmetric matrix, then $\det M$ is a square and is 0 if the size of M is odd (see for example [8]). We can thus define the Pfaffian of an $n \times n$ skewsymmetric matrix by

$$(\mathrm{pf}\, M)^2 = \det M, \qquad \mathrm{pf}\, J_n = 1.$$

The Pfaffian satisfies the relation

$$\mathrm{pf}(gMg^t) = \mathrm{pf}\, M.$$

For an arbitrary matrix, we let $\widetilde{\mathrm{pf}}\, M$ denote the Pfaffian of the skewsymmetric projection,

$$\widetilde{\mathrm{pf}}\, M = \mathrm{pf}(M - M^t).$$

This is clearly an $SO(2k, F)$ invariant. By abuse of notation we will refer to $\widetilde{\mathrm{pf}}$ as the Pfaffian, too. We have

$$\mathrm{pf} \begin{pmatrix} a & b \\ c & d \end{pmatrix} = b - c.$$

For $G = SO(2, F)$, the invariants $P[W(2, m, F)]^G$ were determined in [1]. They are generated by the invariants $\operatorname{tr} P(A, A^t)$ and $\widetilde{\mathrm{pf}}\, P(A, A^t)$ where $A \in W(2, m, F)$ and P is a noncommutative polynomial.

We will see that for n odd we do not get any more invariants when we restrict $O(n, F)$ to $SO(n, F)$. In the even case we have the following crucial Lemma that indicates why the Pfaffian appears.

Lemma 5 *Let $x_1, \dots, x_{2k} \in F^{2k}$ and let $[x_1, \dots, x_{2k}]$ denote the determinant of the matrix with columns $x_1, \dots, x_{2k}$. Then*

$$[x_1, \dots, x_{2k}] = \widetilde{\mathrm{pf}}(x_1 x_2^t + \cdots + x_{2k-1} x_{2k}^t).$$

Proof. Let X be the matrix with columns $x_1, \dots, x_{2k}$ and set

$$\tilde{J} = \begin{pmatrix} 0 & 1 & & & \\ -1 & 0 & & & \\ & & \ddots & & \\ & & & 0 & 1 \\ & & & -1 & 0 \end{pmatrix}.$$

Then

$$\widetilde{\mathrm{pf}}(x_1 x_2^t + \cdots + x_{2k-1} x_{2k}^t) = \mathrm{pf}(X \tilde{J} X^t) = \det X \, \mathrm{pf}\, \tilde{J} = \det X.$$

$\square$

When $k > 1$ the Pfaffian will no longer be linear. Instead we will consider the polarized Pfaffian, which we will denote by pl. Thus $\mathrm{pl}(A_1, \dots, A_k)$ is the coefficient of $t_1 \cdots t_k$ in the expansion of $\widetilde{\mathrm{pf}}(t_1 A_1 + \cdots + t_k A_k)$. It is a symmetric, multilinear function of k matrices and satisfies $\mathrm{pl}(A, \dots, A) = k!\, \widetilde{\mathrm{pf}}(A)$.

When proving our main theorem about the invariants of $SO(2k, F)$ we will use classical invariant theory and consider decomposable matrices of the form $A_i = u_i v_i^t$. Notice that rank $u_i v_i^t \leq 1$. We will need the following simple Lemma about the polarized Pfaffian.

Lemma 6 *Let $A_1, \ldots, A_k$ be of the form $A_i = u_i v_i^t$. The polarized Pfaffian is alternating in the A_i, i.e.,*

$$\mathrm{pl}(A_{i_1}, \ldots, A_{i_k}) = 0$$

if two of the A_{i_j} are equal. Hence

$$\widetilde{\mathrm{pf}}(A_1 + \cdots + A_k) = \mathrm{pl}(A_1, \ldots, A_k).$$

Proof. Set $A = t_1 A_{i_1} + \cdots + t_k A_{i_k}$. Then $\mathrm{pl}(A_{i_1}, \ldots, A_{i_k})$ is the coefficient of $t_1 \cdots t_k$ in the expansion of $\widetilde{\mathrm{pf}}(A)$. But if the A_{i_j} are not all distinct, the rank of A will be at most $k - 1$. But then the rank of $A - A^t$ will be at most $2k - 2$, so $\det(A - A^t) = 0$. It follows that

$$\widetilde{\mathrm{pf}}(A) = \mathrm{pf}(A - A^t) = 0,$$

and hence $\mathrm{pl}(A_{i_1}, \ldots, A_{i_k}) = 0$ if two of the A_{i_j} are equal. Combining this with the multilinearity of pl, we get

$$\widetilde{\mathrm{pf}}(A_1 + \cdots + A_k) = \mathrm{pl}(A_1 + \cdots + A_k, \ldots, A_1 + \cdots + A_k)/k! = \mathrm{pl}(A_1, \ldots, A_k).$$

$$\square$$

We can now prove our main theorem.

Theorem 7 *When $G = SO(2k+1, F)$, the invariants $P[W(2k+1, m, F)]^G$ are the same as the $O(2k+1, F)$ invariants. When $G = SO(2k, F)$, the invariants $P[W(2k, m, F)]^G$ are generated by traces and polarized Pfaffians of the A_i and A_i^t, $i = 1, \ldots, m$, i.e.,*

$$\mathrm{tr}\, P(A, A^t) \quad and \quad \mathrm{pl}\big(P_1(A, A^t), \ldots, P_k(A, A^t)\big),$$

where $P, P_1, \ldots, P_k$ are noncommutative polynomials and $A \in W(2k, m, F)$.

Proof. The proof will follow from classical invariant theory. We can first reduce the problems to finding the multihomogeneous invariants of order $(d_1, \ldots, d_r)$ in m matrices, and then reduce further to studying multilinear invariants of dm matrices where $d = \sum_{i=1}^{r} d_i$. We will identify $F^n \otimes F^n$ and $M(n, F)$ using

$$u \otimes v \to uv^t.$$

We can assume that $A_i = u_i \otimes v_i$ (the symbolic method). The invariants of $(F^n \otimes F^n)^{\otimes dm}$ are generated by inner products and determinants, i.e., invariants of the form

$$\phi(x_1 \otimes \cdots \otimes x_{2dm})$$
$$= [x_1, \ldots, x_n] \cdots [x_{n(l-1)+1}, \ldots, x_{nl}] \langle x_{nl+1}, x_{nl+2} \rangle \cdots \langle x_{2dm-1}, x_{2dm} \rangle$$

where $\langle x_i, x_j \rangle = x_i^t x_j$ and $[x_1, \ldots, x_n]$ denotes the determinant of the matrix with columns $x_1, \ldots, x_n$.

We first observe that if n is odd, we must have an even number of determinants. But then the whole expression will be an $O(n, F)$ invariant. This proves the first part of the theorem.

We can therefore assume that $n = 2k$. We observe that

$$\langle x_i, x_j \rangle = x_i^t x_j = \mathrm{tr}\, x_i x_j^t = \mathrm{tr}\, x_i \otimes x_j.$$

If w_i is u_i or v_i and w_i' the other one, then a product of the form

$$\langle w_{i_1}, w_{i_2}' \rangle \langle w_{i_2}, w_{i_3}' \rangle \cdots \langle w_{i_r}, w_{i_1}' \rangle$$

is equal to

$$\operatorname{tr} B_{i_1} \cdots B_{i_r},$$

where B_i equals A_i or A_i^t depending on whether w_i is u_i or v_i.

We will now show that ϕ can be written in terms of traces and polarized Pfaffians of the A_i and the A_i^t.

Since the product of two determinants is an $O(2k, F)$ invariant and hence expressible in terms of scalar products, we can assume that we have only one determinant. By reordering the A_i and replacing A_i by A_i^t, i.e., interchanging u_i and v_i, we can assume that

$$\phi = \pm [u_1, v_1, \dots, u_l, v_l, u_{l+1}, \dots, u_{2k-l}]$$
$$\langle v_{l+1}, u_{2k-l+1}\rangle \langle v_{2k-l+1}, u_{2k-l+2}\rangle \cdots \langle v_{2k-l+d_1}, v_{l+2}\rangle$$
$$\langle v_{l+3}, u_{2k-l+d_1+1}\rangle \cdots \langle v_{2k-l+d_1+d_2}, v_{l+4}\rangle$$
$$\cdots \langle v_{2k-l-1}, u_{2k-l+d_1+\cdots+d_{k-l-1}+1}\rangle \cdots \langle v_{2k-l+d_1+\cdots+d_{k-l}}, v_{2k-l}\rangle Q.$$

where Q only involves scalar products of u_i and v_i for $i > 2k-l+d_1+\cdots+d_{k-l}$. But this can be expressed as the trace of a polynomial in A_i and A_i^t for $i > 2k-l+d_1+\cdots+d_{k-l}$. We can now apply Lemma 6 and we get that

$$[u_1, v_1, \dots, u_l, v_l, u_{l+1}, \dots, u_{2k-l}] = \widetilde{\operatorname{pf}}(A_1 + \cdots A_l + u_{l+1}u_{l+2}^t + \cdots u_{2k-l-1}u_{2k-l}^t)$$
$$= \operatorname{pl}(A_1, \dots, A_l, u_{l+1}u_{l+2}^t, \cdots, u_{2k-l-1}u_{2k-l}^t). \tag{1}$$

But since the function pl is multilinear, we can combine

$$\langle v_{l+1}, u_{2k-l+1}\rangle \langle v_{2k-l+1}, u_{2k-l+2}\rangle \cdots \langle v_{2k-l+d_1}, v_{l+2}\rangle$$

with $u_{l+1}u_{l+2}^t$ to get

$$u_{l+1}u_{l+2}^t v_{l+1}^t u_{2k-l+1} v_{2k-l+1}^t u_{2k-l+2} \cdots v_{2k-l+d_1}^t v_{l+2}$$
$$= u_{l+1} v_{l+1}^t u_{2k-l+1} v_{2k-l+1}^t u_{2k-l+2} \cdots v_{2k-l+d_1}^t v_{l+2} u_{l+2}^t$$
$$= A_{l+1} A_{2k-l+1} \cdots A_{2k-l+d_1} A_{l+2}^t.$$

Repeating this for each of the last $k - l$ terms in (1), we get that

$$\phi/Q = \operatorname{pl}(A_1, \dots, A_l, A_{l+1}A_{2k-l+1}\cdots A_{2k-l+d_1}A_{l+2}^t,$$
$$\cdots, A_{2k-l-1}A_{2k-l+d_1+\cdots+d_{k-l-1}+1}\cdots A_{2k-l+d_1+\cdots+d_{k-l}}A_{2k-l})^t.$$

This shows that ϕ is of the required form. $\qquad\square$

For similar results for the case of $SO(p, q)$, see [4].

CLASSICAL GROUPS

It turns out that we can use our results about $SO(n)$ invariants to deduce similar results for all the other classical groups. Let us first define what we mean by a classical group. Let F be either $\mathbb{R}$ or $\mathbb{C}$ and let D be a division algebra over F with involution $a \mapsto a^\#$. Let V be a vector space over D and let $(\,,)$ be a nondegenerate sesquilinear form on V that is Hermitian or skew-Hermitian relative to $^\#$. (If the involution is the identity we get bilinear symmetric and skewsymmetric forms.) Let G be the isometry group of $(\,,)$. We will call such groups classical isometry groups.

Table 1

	$\mathbb{R}$	$\mathbb{C}$	$\mathbb{H}$
bilinear and symmetric	$O(p,q)$	$O(n,\mathbb{C})$	$-$
bilinear and skewsymmetric	$Sp(k,\mathbb{R})$	$Sp(k,\mathbb{C})$	$-$
sesquilinear and Hermitian	$-$	$U(p,q)$	$Sp(p,q)$
sesquilinear and skew-Hermitian	$-$	$U(p,q)$	$Sp(k,\mathbb{H})$

If the involution is either the identity or complex or quaternionic conjugation, we can describe the classical isometry groups by Table 1. (We are assuming that $n = p + q$ or $n = 2k$ when appropriate.) A "$-$" means that there are no such forms. Helgason [7] writes $SO^*(2k)$ for the group we have called $Sp(k,\mathbb{H})$.

The remaining classical groups are the general and special linear groups $GL(n,F)$ and $SL(n,F)$ and the special isometry groups, i.e., the intersections of $SL(n,F)$ and the classical isometry groups.

As explained in the introduction, Sibirskii [14] gave a set of generators of $P[W]^G$ when G equals $GL(n,F)$, $O(n,F)$ or $U(n)$. His proof for the case of $U(n)$ is essentially Weyl's unitary trick, which can easily be adapted to work for $U(p,q)$. These results were also proved independently by Procesi [12], who also solved the problem for $Sp(k,F)$. His proof for the $O(n,\mathbb{R})$ case can also easily be modified for the case of $O(p,q)$. For the special isometry groups, all the cases are trivial except for the special orthogonal groups.

It therefore remains to discuss the quaternionic groups $Sp(k,\mathbb{H})$ and $Sp(p,q)$. This is discussed in [4].

QUIVERS

We are often interested in more general situations. We might be interested in invariants under block diagonal subgroups like $GL(m) \times GL(n)$ or $O(m) \times O(n)$. It turns out that again the invariants can be expressed in terms of traces of certain products of matrices. In order to express which products, we can form a quiver, and the allowed products will correspond to cycles in the quiver. For details, see [2, 9, 10], but we will outline one example here.

Consider the action of $GL(m) \times GL(n)$ on $M(n+m)$ given by

$$\begin{pmatrix} g_1 & 0 \\ 0 & g_2 \end{pmatrix} \begin{pmatrix} A & B \\ C & D \end{pmatrix} \begin{pmatrix} g_1 & 0 \\ 0 & g_2 \end{pmatrix}^{-1} = \begin{pmatrix} g_1 A g_2^{-1} & g_1 B g_2^{-1} \\ g_2 C g_1^{-1} & g_2 D g_2^{-1} \end{pmatrix}.$$

We can think of this as an action on

$$V = M(m,m) \oplus M(m,n) \oplus M(n,m) \oplus M(n,n) = M_{1,1} \oplus M_{1,2} \oplus M_{2,1} \oplus M_{2,2}.$$

We will represent this action by the following quiver.

1   2

The invariants are now given by

$$\operatorname{tr} A_{i_1,i_2} A_{i_2,i_3} \cdots A_{i_{l-1},i_l},$$

where $i_1, \ldots, i_l$ is a cycle in the quiver and $A_{i,j} \in M_{i,j}$.

For a classical group, its Levi subgroups are usually products of general linear, orthogonal or symplectic groups. Their invariant theory is useful in understanding the invariant theory in universal enveloping algebras [2].

References

[1] H. Aslaksen, *SO(2) invariants of a set of 2×2 matrices*, Math. Scand. **65** (1989), 59–66.

[2] H. Aslaksen, E.-C. Tan and C.-B. Zhu, *Quivers and the invariant theory of Levi subgroups*, J. Funct. Anal. **120** (1994), 163–187.

[3] H. Aslaksen, E.-C. Tan and C.-B. Zhu, *Generators and relations of invariants of 2×2 matrices*, Comm. Algebra **22** (1994), 1821–1832.

[4] H. Aslaksen, E.-C. Tan and C.-B. Zhu, *Invariant theory of special orthogonal groups*, Pacific J. Math. (to appear).

[5] J. Dubnov and V. Ivanov, *Sur l'abaissement du degré des polynômes en affineurs*, Dokl. Akad. Nauk SSSR **41** (1943), 95–98.

[6] G. B. Gurevich, *Foundations of the theory of algebraic invariants*, P. Noordhoff Ltd, Groningen, The Netherlands, 1964.

[7] S. Helgason, *Differential Geometry, Lie Groups, and Symmetric Spaces*, Academic Press, New York, USA, 1978.

[8] N. Jacobson, *Basic Algebra I*, W. H. Freeman and Co., New York, USA, 1985.

[9] W. H. Klink and T. Ton-That, *Invariant theory of the block diagonal subgroups of $GL(n,\mathbb{C})$ and generalised Casimir operators*, J. of Algebra **145** (1992), 187 – 203.

[10] H. Kraft, *Invariants of quivers (following Procesi)*, private communication.

[11] L. Le Bruyn, *Trace rings of generic 2 by 2 matrices*, Memoirs of the AMS **66** (1987), no. 363.

[12] C. Procesi, *The invariant theory of $n \times n$ matrices*, Adv. Math. **19** (1976), 306–381.

[13] Ju. P. Razmyslov, *Trace identities of full matrix algebras over a field of characteristic zero*, Math. USSR Izv. **8** (1974), 727–760.

[14] K. S. Sibirskii, *Algebraic invariants for a set of matrices*, Siberian Math. J. **9** (1968), 115–124.

SYMMETRIES OF ELEMENTARY PARTICLES REVISITED

A. O. Barut

International Center of Physics and Applied Mathematics
ICPAM, P.K. 126
Edirne, Turkey
and
University of Colorado
Boulder, CO 80309

INTRODUCTION

Many of us became interested in symmetries of particles and in group theory, a major theme of these series of conferences Symmetries in Science, in the sixties because of the very remarkable symmetry properties of the newly found many fundamental particles at that time. One was very curious about the origin of these symmetries. What have we learned about the origin of these symmetries. What have we learned about these mysterious symmetries after some thirty years? The symmetries were mainly the occurrence of octets and decouplets in the states of mesons baryons and heavy leptons which where attributed to the representations of the internal symmetry group $SU(3)$. Why $SU(3)$, what does it mean? Well, we have not solved it, but pushed the problem a bit further under the carpet, we did not explain the symmetries, nor the group $SU(3)$, but we assigned the same symmetries, or the same quantum numbers to hypothetical particles called quarkes, without really explaining them.

Quarks are the fundamental representations of the algebra of $SU(3)$. Every representation of $SU(3)$ can be constructed out of tensor product of fundamental representations. Hence this is a mathematical description of observed symmetries of hadrons, it does not explain them. By explaining we mean to attempt to reduce the symmetry to the structure of the particles. In fact this so-called flavor—$SU(3)$ has now been abandonned as a basic symmetry, one now believes that it should follow from the QCD Lagrangian as an approximative result. The group that gave rise to the notion of quarks as fundamental representations is gone. Instead we have three repetitions of doublet of quarks and there is no reason for these three repetitions, or families.

The search for the origin of particle symmetries stopped when the quarks hastily were elevated to be the real physical constituents rather than being just mathematical fundamental representations. The analogy to H-atom is quite striking: The fundamental representation of the symmetry of the states of the H-atom ($SO(4)$) are four dimensional and the states of the H-atom can be constructed out of these four "quarks", which however do not exist in free space, they are like normal modes of the full Coulomb Hamiltonian with e and p interacting via the Coulomb potential.

The purpose of this essay is to discuss a phenomenon in electrodynamics that possibly could unravel the real constituents of hadrons (like e and p in H-atom) and explain the $SU(3)$-mystery.

Symmetries in Science VII, Edited by
B. Gruber, Plenum Press, New York, 1995

II. The Real Physical Constituents of Hadrons

We assume two, only two, elementary constituents: electron and neutrino. They are absolutely stable, indestructable and not modifiable. All states of hardons, mesons, leptons, vector mesons, heavy leptons, can be constructed out of these two. No hypotehtical quarks, Higgs, gluons would be needed. We further assume that an unstable particle decays because of its instability as a composite object, i.e. it is a resonance short lived or sometimes quite long lived. No new particles and forces will be introduced.

This idea has to be developed along two lines, first a kinematical construction of all other states from e and ν, secondly one must have a dynamical theory and show the existence of resonances; One must try this conservative approach, otherwise physics would become too speculative with the introduction of many new hypothetical particles and new forces for every need.

The Aufbau-principle is rather simple. Next to e and ν_e the most stable particles are p and ν_μ which can be constructed as $(e^+e^+e^-)$ and $(\nu_e\bar{\nu}_e\nu_e)$-states. Heaving constructed p and ν_μ which are rather stable, they can be used as building blocks for other states. Thus, the next two states are $n = (pe^-\nu_e)$ and $\mu = (\nu_\mu e^-\nu_e)$. These are the most stable among all hadrons. The phase space shows that $n - p$ mass difference and the rate of decay is much smaller for neutron than for μ. Every such composite state provides a quantum number, number of states, which is conserved for processes shorter than its lifetime. So we have a proton number (baryon number) a μ-number ("strangeness") a n-number (N) and a ν_μ-number ("charm") which are conserved for processes shorter than the lifetime of p, μ, n, and ν_μ, respectively. Then mesons are essentially constructed as $\ell\bar{\ell}$ states where ℓ stands for leptons, and baryons as $b\ell\bar{\ell}$ where b stands for basic baryons such as proton and neutron. Such a scheme can be brought in correspondence with the quark model, but the constituents here are the observed real physical particles which are seen in their decay.

III. The Dynamical Problem

In standard particle physics the strong interactions are attributed to quarks exchanging gluons, and the weak interactions to the exchange of heavy vector mesons. These models are formulated in the framework of gauge theories, but the gauge symetry is so broken that no trace of it is left. The framework of gauge theory may be accidental. Thus it is generally assumed that electrons and neutrinos cannot have strong and weak interactions. However, the electromagnetic interactions can manifest themselves as strong and weak interactions. These are due to magnetic interactions between e^+e^- and e^- and ν_e. First of all, a tiny anomalous magnetic moment of the neutrino of the order of $10^{-10}\mu_0$ results in $e - \nu_e$ scattering cross section of the same order of magnitude as in Weinberg-Salam theory. This shows how weak the weak interactions are. This is also about the experimental upper limit for the anomalous magnetic moment of the neutrino. Secondly, the anomalous magnetic moment interactions become very strong at short distances.

A good way to visualize these effects is that of a deep potential well at short distances between the leptons, which can support a positive energy resonance (formation of hadrons) followed by a tunnelling (decay of the resonance). A nonperturbative covariant two (or many) body equation has been developed which allows one to extrapolate electrodynamics to short distances. This is not possible with the usual perturbative QED, where one needs to be able to sum infinitely many Feynman graphs. Short of the complete solution of the relativistic two-body problem with magnetic interactions, many approximate models have been studied with the property of a potential barier mentioned above.

This is a simple and economical approach to particle physics. It does not introduce new hypothetical forces and particles. Particle physics becomes a continuation of molecular physics to short distances.

In this scheme the internal quantum numbers are just labeling different species of stable states, they are discrete quantum numbers. The iso-spin multiplets differ by the exchange of $(e^-\bar{\nu}_e)$-quanta, they have similar mass because the mass of $(e^-\bar{\nu}_e)$ is small. In the decouplet of baryons the masses of the levels $\Delta, \Gamma, \Sigma, \Xi$ and Ω^- differ by about 140 MeV, which is the mass of an additional μ^- at each level. There are no new internal coordinates on which the internal symmetries or gauge symmetries act. Actually, we use only algebras and not groups, and only the low-dimensional representations of the internal algebras which coincide with the representations of discrete finite subalgebras. Ultimately, however, all masses and mass-differences (like the important $n-p$ mass difference) must be calculated dynamically. Thus, the origin of internal quantum numbers and symmetries is in principle no mystery.

The importance of magnetic spin-dependent interactions is also shown by very large polarization effects observed in high energy collision which QCD cannot explain.

The so-called standard model of electroweak interactions can be derived from magnetic interactions. With electron and neutrino as fundamental particles there are four basic magnetic currents: $e^+e^-, e^-\bar{\nu}_e, e^+\nu_e, \nu_e\bar{\nu}_e$ which are the four currents in the electroweak theory. In fact, it is possible to establish rigorously from the magnetic interactions the Fermi or the Weinberg-Salam from the Lagrangian.

This brings us to a fundamental question in physics; namely what basic entities we use at every stage of physics. We follow the scientific credo: Why invent new hypothetical (unseen) forces, when the traditional, time honored physics can explain the same phenomena, even if sometimes in a subtle way. For example, we do not need new forces to expalin the chemical binding of neutral atoms (but it is still electromagnetic), or α-decay. But is is possible to introduce new phenomenological entities and give them appropriate properties and write effective Lagrangians fullfilling general symmetry principles.

I think it is best to view the standard model as an effective theory. Although it is formulated in the framework of a fundamental gauge theory, the gauge symmetry is so broken that there is no trace of it left. In pactice in fact one uses the old Fermi-theory, the vector mesons are not explicitly seen except by their decay products. In analogy again it is conceivable to write chemical forces, or α-decay, as an effective new Lagrangian theory with new coupling constants and new exchange particles.

IV. Conclusions

The secret to the internal symmetries of elementary particles is stability. This is the conclusion of the model outlined here. Internal symmetries are about the type of species of observed particles. Why we have not just one type of particle, but a whole set of them, differeing in charge and other properties. Starting from just e and ν and building successively all other states the different internal quantum numbers arise as follows: If we form first a rather stable state A which itself is a building block for other less stable states. This particular state A defines a species, and an internal quantum number S_A, namely the number of constituents of type A. It is conserved for all processes shorter than the lifetime of A. For example, the meson $\bar{\mu} = (\bar{e}\bar{\nu}_e\nu_e)$ has a lifetime of atoms 10^{-6}sec. For processes shorter than $10^{-6}, \mu$ is stable. We can identify μ-number with strangeness. Indeed the so-called strange particles can be built with a muon or with 1,2,3 muons in the decouplet of baryons $\Sigma^*\Xi^*$ and Ω^- coressponding to strangeness 1,2,3. The mass diferences in the decouplet are roughly one muon mass.

Proton is almost absolutely stable, so proton number (= baryon number) is conserved. Neutron number is conserved for processes shorter than about 10^3s, ν_μ-number can be identified with charm. Similarly we have τ- and ν_τ-numbers.

In standard model there are six leptons as fundamental particles, and in addition six quarks of three different colours. Here we have simply two fundamental leptons (e, ν_e) and composite higher mass leptons $(\mu, \nu_\mu, \tau, \nu_\tau, \ldots)$. The ground states of the composite states of the same spin constitute the original $SU(3)$-flavor multiplets, at the same time, excited states of the same composite structures will give excited states of hadrons, if they are stable. Thus both species of ground states and excited states come from the same dynamics and are governed by the stability of bound states.

References

For more technical details see:
"Stable Particles as Building Blocks of Matter" *Surveys in High Energy Physics* **1**, 113 (1980).
"Unification based on Electromagnetism," *Ann. der Physik*, **43**, 83 (1986).
"Connection Between Stable Particle Model and the Integrally Charged Quark Model" *Lett. N. C.* **35**, 200 (1982).
"Foundations of Electrodynamics and Particle Physics" in Frontiers in Fundamental Physics, edit. by F. Selleri (Plenum Press, 1994).

PERTURBATIVE SU(1,1)

Haluk Beker

Physics Department
Boğaziçi University
Bebek , İstanbul , Turkey

INTRODUCTION

The dynamical group $SO(4,2)$ was first used to gain insight into the H - atom problem. Later it was applied to other systems, such as hadrons, where only scant information exists about the constituents and / or the binding interaction. This is appropriate , since group theoretical treatment of such systems emphasizes global quantum numbers while ignoring internal coordinates that are not accessible to experiment.

A concrete realisation of $SO(4,2)$ generators using only position $\bar{r}$ and its conjugate momentum $\bar{p}$, tailored exclusively for the H - atom problem , was constructed nearly a quarter of a century ago[1]. The question : " What are the most general generators of $SO(4,2)$ that may be constructed using only $\bar{r}$ and $\bar{p}$? " was answered fifteen years later[2]. Subsequently , this representation was further generalized to include arbitrary spin. Thus , the group $SO(4,2)$ was formally extended to $SU(2,2)$[3].

Of the two commuting subgroups of $SU(2,2)$:
$SU(2,2) \supset SU(2) \otimes SU(1,1)$, the generators of $SU(1,1)$ are necessarily scalars since they commute with the $SU(2)$ sector. The non-compact subalgebra of $SU(1,1)$ is to be used for spectrum generation. The generators of the $SU(1,1)$ algebra are given by :

$$\Lambda_1 = \frac{G}{G'^2} K^2 + \frac{Q}{G} - \frac{G\,G'''}{2\,G'^3} + \frac{3}{4}\frac{G\,G''^2}{G'^4} - \frac{G}{4}$$

Symmetries in Science VII, Edited by
B. Gruber, Plenum Press, New York, 1995

$$\Lambda_2 = \frac{G}{G'}K - i\frac{G\,G''}{2\,G'^2}$$

$$\Lambda_3 = \frac{G}{G'^2}K^2 + \frac{Q}{G} - \frac{G\,G'''}{2\,G'^3} + \frac{3}{4}\frac{G\,G''^2}{G'^4} + \frac{G}{4}$$

where $G = G(r)$ is an arbitrary function, K is the hermitian radial momentum conjugate to r with the representation :

$$K \to - i\left(\frac{d}{dr} + \frac{1}{r}\right)$$

and Q is the Casimir operator[4].

Λ_3 has the discrete spectrum D_+ :

$$\text{Spectrum}\,(\Lambda_3) \;=\; \text{Spectrum}\left(\frac{G}{G'^2}K^2 + \frac{Q}{G} - \frac{G\,G'''}{2\,G'^3} + \frac{3}{4}\frac{G\,G''^2}{G'^4} + \frac{G}{4}\right)$$

$$= v + \frac{1}{2} + \sqrt{Q + \frac{1}{4}} \qquad (v = 0, 1, 2, \dots)$$

The above formalism is related to at least two other familiar topics. One is the intimate relationship between the SU(1,1) group and the confluent hypergeometric differential equation (CHGDE). The other topic is the basic property of pure stationary bound states. Both relations deserve brief sketches.

RELATION TO CHGDE

Let us agree to represent $\dfrac{d^2 y}{dx^2} + F_1(x)\dfrac{dy}{dx} + F_0(x)\,y = 0$, the most general second order, homogeneous linear differential equation in standard form, i.e. $F_2(x) = 1$, as $[1, F_1, F_0]\,y = 0$. By transforming the dependent variable, the DE is put into invariant form $[1, 0, I]\,\Psi = 0$, where the invariant function $I = I(x)$ is given by : $I = F_0 - \left(\dfrac{F_1}{2}\right)' - \left(\dfrac{F_1}{2}\right)^2$ and

$$\Psi = \left[e^{\int \frac{F_1}{2}\,dx}\right] y$$

is the new dependent variable. Now let us consider the

CHGDE with G as its independent variable : $[G, \gamma - G, -\alpha]\, {}_1F_1(\alpha, \gamma; G) = 0$.

To guarantee square integrable solutions, the choice $\alpha = -\nu$, $(\nu = 0, 1, 2, \ldots)$ is necessary. Applying the procedure of putting a DE into invariant form, we

obtain
$$\left[1, 0, \frac{\frac{\gamma}{2} - \frac{\gamma^2}{4}}{G^2} + \frac{\nu + \frac{\gamma}{2}}{G} - \frac{1}{4} \right] e^{-G/2} \, G^{\gamma/2} \, {}_1F_1(-\nu, \gamma; G) = 0.$$

At this point let us acknowledge that G is indeed a function; $G = G(r)$ and let us convert the above DE to another one with derivatives with respect to r.

The relations
$$\frac{d}{dG} = \frac{1}{G'} \frac{d}{dr} \quad , \quad \frac{d^2}{dG^2} = \frac{1}{G'^2} \frac{d^2}{dr^2} - \frac{G''}{G'^3} \frac{d}{dr}$$
lead to

$$\left[\frac{1}{G'^2} , -\frac{G''}{G'^3} , \frac{\frac{\gamma}{2} - \frac{\gamma^2}{4}}{G^2} + \frac{\nu + \frac{\gamma}{2}}{G} - \frac{1}{4} \right] e^{-G/2} \, G^{\gamma/2} \, {}_1F_1(-\nu, \gamma; G) = 0$$

which is in neither invariant nor standard form. Once more going through the procedure of putting a DE into invariant form, we obtain

$$\left[1, 0, \frac{1}{2}\frac{G'''}{G'} - \frac{3}{4}\frac{G''^2}{G'^2} + \left(\frac{\gamma}{2} + \frac{\gamma^2}{4}\right)\frac{G'^2}{G^2} + \left(\nu + \frac{\gamma}{2}\right)\frac{G'^2}{G} - \frac{1}{4}G'^2 \right] \frac{e^{-G/2} \, G^{\gamma/2}}{\sqrt{G'}} \, {}_1F_1(-\nu, \gamma; G) = 0.$$

Remembering $K \rightarrow -i\left(\dfrac{d}{dr} + \dfrac{1}{r}\right)$, therefore $K^2 \rightarrow \dfrac{1}{r^2}\dfrac{d}{dr} r^2 \dfrac{d}{dr}$ and using the

identity: $-r K^2 \dfrac{f}{r} \rightarrow \dfrac{1}{r}\dfrac{d}{dr} r^2 \dfrac{d}{dr}\dfrac{f}{r} = \dfrac{d^2 f}{dr^2}$, the above DE reduces to the operator equation

$$\left[K^2 - \frac{1}{2}\frac{G'''}{G'} + \frac{3}{4}\frac{G''^2}{G'^2} - \left(\frac{\gamma}{2} - \frac{\gamma^2}{4}\right)\frac{G'^2}{G^2} - \left(\nu + \frac{\gamma}{2}\right)\frac{G'^2}{G} + \frac{1}{4}G'^2 \right] \frac{e^{-G/2} \, G^{\gamma/2}}{r \sqrt{G'}} \, {}_1F_1(-\nu, \gamma; G) = 0.$$

Rearranged, it becomes the already familiar spectral relation :

$$\left[\frac{G}{G'^2} K^2 + \frac{\left(\frac{\gamma}{2} - \frac{\gamma^2}{4}\right)}{G} - \frac{G\,G'''}{2\,G'^3} + \frac{3}{4}\frac{G\,G''^2}{G'^4} + \frac{G}{4} \right] R = \left(\nu + \frac{\gamma}{2}\right) R$$

with $\quad Q = \dfrac{\gamma^2}{4} - \dfrac{\gamma}{2} \quad$ and $\quad R = \dfrac{e^{-G/2} \, G^{\gamma/2}}{r \sqrt{G'}} \, {}_1F_1(-\nu, \gamma; G) \qquad 4$

RELATION TO BOUND STATE PROPERTIES

The form of SU(1,1) generators is also related to the fact that for a stationary bound state there is no explosion or implosion of probability. This can be expressed mathematically as $<K> = 0$; the expectation value of radial momentum being zero. The demonstration is straightforward : The radial wave function of a pure state can always be chosen as real, since the radial Schrödinger equation is

$$\left[\frac{1}{r^2} \frac{d}{dr} r^2 \frac{d}{dr} - \frac{l\,(l+1)}{r^2} - \frac{2\,m\,V(r)}{\hbar^2} + \frac{2\,m\,E}{\hbar^2} \right] R = 0 \quad \text{and if } R \text{ is a solution so}$$

is R^* or their sum which is real. Then the expectation value

$$<K> = -i \int_0^\infty 4\pi\, r^2 dr\; R \left(\frac{d}{dr} + \frac{1}{r} \right) R \quad \text{may easily be shown to be zero, provided}$$

rR vanishes at $r = 0$ and $r = \infty$. The relation $<K> = 0$ can be generalised to $<\sqrt{f}\, K\, \sqrt{f}> = 0$ where $f = f(r)$ is an arbitrary function, as long as $\sqrt{f}\, R$ is also square integrable. Since K is an even operator, the reason behind this result is not parity as it is in the 1-D analog. The underlying reason can only be the dynamical symmetry of bound state systems, namely the

SU(1,1) algebra . $\sqrt{f}\, K\, \sqrt{f}$ or the identical expression $f\,K - i\,\dfrac{f'}{2}$ must be a linear combination of the step up and step down operators of the algebra. Since these ladder operators themselves are linear combinations of Λ_1 and Λ_2 , we may as well say that $f\,K - i\,\dfrac{f'}{2}$ is a linear combination of Λ_1 and Λ_2.

The operator $f\,K - i\,\dfrac{f'}{2}$ may tentatively be set equal to Λ_2 , without loss of generality because it is possible to similarity transform generators into each other. Once Λ_2 is determined within a similarity transform, the SU(1,1) generators may be determined by requiring :

1. The generators are at most quadratic in K ,
2. Λ_3 does not contain a linear K term ,
3. SU(1,1) commutation relations are satisfied ,
4. The Casimir operator is constant .

The results are :

$$\Lambda_1 = \frac{G}{G'^2} K^2 + \frac{Q}{G} - \frac{G\,G'''}{2\,G'^3} + \frac{3}{4} \frac{G\,G''^2}{G'^4} - \frac{G}{4}$$

$$\Lambda_2 = \frac{G}{G'} K - i \frac{G\,G''}{2\,G'^2}$$

$$\Lambda_3 = \frac{G}{G'^2} K^2 + \frac{Q}{G} - \frac{G\,G'''}{2\,G'^3} + \frac{3}{4}\frac{G\,G''^2}{G'^4} + \frac{G}{4}$$

as expected [5].

UNPERTURBED PROBLEMS

The method that uses SU(1,1) as a spectrum generating algebra to solve quantum mechanical problems hinges on the comparison between the radial Schrödinger equation :

$$\left[\, 1\,,\,0\,,\, \frac{-l\,(l+1)}{r^2} - \frac{2\,m\,V(r)}{\hbar^2} + \frac{2\,m\,E}{\hbar^2}\,\right]\, r\,R\ =\ 0 \quad \text{and the invariant form of the}$$

generalized CHGDE :

$$\left[\, 1\,,\,0\,,\, \frac{1}{2}\frac{G'''}{G'} - \frac{3}{4}\frac{G''^2}{G'^2} + \left(\frac{\gamma}{2} - \frac{\gamma^2}{4}\right)\frac{G'^2}{G^2} + \left(v + \frac{\gamma}{2}\right)\frac{G'^2}{G} - \frac{1}{4}G'^2 \,\right] \frac{e^{-\frac{G}{2}}\,G^{\frac{\gamma}{2}}}{\sqrt{G'}}\ {}_1F_1\,(\,-v\,,\gamma\,;G\,)\ =\ 0$$

The radial Schrödinger equation contains the constant term $\dfrac{2\,m\,E}{\hbar^2}$. The trivial identifications G'^2 , $\dfrac{G'^2}{G}$ or $\dfrac{G'^2}{G^2}$ with this constant term yield linear , quadratic , and exponential solutions for $G\,(\,r\,)$. These solutions in turn , lead to the solutions of H - atom , SHO and $l = 0$ Morse potential problems , respectively . As an instructive example , consider the H - atom problem with

$$V\,(\,r\,) = \frac{-\,e^2}{r} \quad \text{and} \quad G\,(\,r\,) = A\,r\,.$$ Substituting into the two equations given above and equating yields :

$$-\frac{l\,(l+1)}{r^2} + \frac{2\,m\,e^2}{\hbar^2 r} + \frac{2\,m\,E}{\hbar^2}\ =\ \left(\frac{\gamma}{2} - \frac{\gamma^2}{4}\right)\frac{1}{r^2} + \left(v + \frac{\gamma}{2}\right)\frac{A}{r} - \frac{A^2}{4}\ .$$

The equations

$$l\,(l+1)\ =\ \frac{\gamma^2}{4} - \frac{\gamma}{2} \quad,\qquad \frac{2\,m\,e^2}{\hbar^2}\ =\ A\left(v + \frac{\gamma}{2}\right) \quad,\qquad \frac{2\,m\,E}{\hbar^2}\ =\ -\frac{A^2}{4}$$

and $\ r\,R \propto \dfrac{e^{-\frac{A\,r}{2}}\,(A\,r)^{\frac{\gamma}{2}}}{\sqrt{A}}\ {}_1F_1\,(\,-v\,,\gamma\,;G\,)\ $ lead to $\ E\ =\ -\dfrac{m\,e^4}{2\,\hbar^2\,n^2}\ $ with

$$n = \nu + l + 1 \qquad \text{and} \qquad R(r) \propto e^{-\frac{r}{n a_0}} \; r^l \; {}_1F_1\left(-\nu, 2l+2 \; ; \; \frac{2r}{n a_0}\right) \qquad \text{where}$$

$a_0 = \dfrac{\hbar^2}{m e^2}$ is the Bohr radius. Another important example is the SHO

problem with $V(r) = \dfrac{1}{2} m \omega^2 r^2$ and $G(r) = A r^2$. Substituting :

$$-\frac{l(l+1)}{r^2} - \frac{m^2 \omega^2 r^2}{\hbar^2} + \frac{2 m E}{\hbar^2} \;=\; -\frac{3}{4 r^2} + \left(\frac{\gamma}{2} - \frac{\gamma^2}{4}\right)\frac{4}{r^2} + \left(\nu + \frac{\gamma}{2}\right) 4 A - A^2 r^2 \;,$$

the relations $\quad l(l+1) = 4\left(\dfrac{\gamma^2}{4} - \dfrac{\gamma}{2}\right) + \dfrac{3}{4}$

$$\frac{2 m E}{\hbar^2} = \left(\nu + \frac{\gamma}{2}\right) 4 A$$

$$\frac{m^2 \omega^2}{\hbar^2} = A^2 \qquad\qquad \text{and}$$

$$r R \propto \frac{e^{-\frac{A r^2}{2}} \left(A r^2\right)^{\gamma/2} {}_1F_1(-\nu, \gamma \; ; \; A r^2)}{\sqrt{2 A} \; r} \qquad \text{are obtained. Their}$$

solution yields the familiar $\quad E = \left(2\nu + l + \dfrac{3}{2}\right) \hbar \omega$

$$R(r) \propto e^{-\frac{m \omega}{2 \hbar} r^2} \; r^l \; {}_1F_1\left(-\nu, l + \frac{3}{2} \; ; \; \frac{m \omega}{\hbar} r^2\right) \qquad \text{with} \qquad \frac{\gamma}{2} = \frac{l}{2} + \frac{3}{4} \qquad \text{and}$$

$A = \dfrac{m \omega}{\hbar}$. The Morse potential problem is less interesting since it is limited to

$l = 0$ case.

PERTURBATION METHOD

A new perturbation method based on the deformation of SU(1,1) generators
will be presented next[6]. The method outlined above for the unperturbed problem
was characterized by a single arbitrary function $G(r)$. Once the form of $G(r)$
is specified for a given problem, a small correction $g(r)$ may be added to it,
such that $\quad G(r) \rightarrow G(r) + g(r)$.

This deformation of G, deforms the SU(1,1) generators in turn, but keeps
the Casimir operator invariant, since the Casimir operator does not depend on the

form of G. Let us write the equation for the unperturbed case, with a slightly modified notation :

$$-\frac{l(l+1)}{r^2} - \frac{2\,m\,V_0(r)}{\hbar^2} + \frac{2\,m\,E_0}{\hbar^2} = I(G, G', G'', G''')$$

$$= \frac{1}{2}\frac{G'''}{G'} - \frac{3}{4}\frac{G''^2}{G'^2} + \left(\frac{\gamma}{2} - \frac{\gamma^2}{4}\right)\frac{G'^2}{G^2} + \left(v + \frac{\gamma}{2}\right)\frac{G'^2}{G} - \frac{1}{4}G'^2$$

Transforming $G \to G + g$, we expect $V_0 \to V_0 + V_1$, $E_0 \to E_0 + E_1$ where V_1 is the perturbing potential and E_1 is the energy correction. Further, as $G \to G + g$

$$I(G, G', G'', G''') \to I(G, G', G'', G''') + \frac{\partial I}{\partial G}g + \frac{\partial I}{\partial G'}g' + \frac{\partial I}{\partial G''}g'' + \frac{\partial I}{\partial G'''}g'''$$

Thus we reach our starting point which is a formidable looking but essentially simple equation :

$$-\frac{l(l+1)}{r^2} - \frac{2\,m\,V_0(r)}{\hbar^2} - \frac{2\,m\,V_1(r)}{\hbar^2} + \frac{2\,m\,E_0}{\hbar^2} + \frac{2\,m\,E_1}{\hbar^2} = \frac{1}{2}\frac{G'''}{G'} - \frac{3}{4}\frac{G''^2}{G'^2} + \left(\frac{\gamma}{2} - \frac{\gamma^2}{4}\right)\frac{G'^2}{G^2}$$

$$+ \left(v + \frac{\gamma}{2}\right)\frac{G'^2}{G} - \frac{1}{4}G'^2 + \left[-2\left(\frac{\gamma}{2} - \frac{\gamma^2}{4}\right)\frac{G'^2}{G^3} - \left(v + \frac{\gamma}{2}\right)\frac{G'^2}{G^2}\right]g$$

$$+ \left[-\frac{1}{2}\frac{G'''}{G'^2} + \frac{3}{2}\frac{G''^2}{G'^3} + \left(\frac{\gamma}{2} - \frac{\gamma^2}{4}\right)\frac{2\,G'}{G^2} + \left(v + \frac{\gamma}{2}\right)\frac{2\,G'}{G} - \frac{G'}{2}\right]g' + \left[-\frac{3}{2}\frac{G''}{G'^2}\right]g'' + \left[\frac{1}{2\,G'}\right]g'''$$

The assertion of simplicity is somewhat justified when we apply it to the H-atom problem with $G = \dfrac{2\,r}{n\,a_0}$. Terms that are multiplied by g or its derivatives are already small, therefore in those terms the parameters $\dfrac{\gamma}{2}$ or A can be replaced by $l+1$ and $\dfrac{2}{n\,a_0}$, their respective unperturbed values, since second order corrections may safely be neglected. The H-atom perturbation equation now becomes :

$$-\frac{l\,(l+1)}{r^2} + \frac{2}{a_0 r} - \frac{2\,m\,V_1(r)}{\hbar^2} - \frac{1}{n^2\,a_0^2} + \frac{2\,m\,E_1}{\hbar^2} \; =$$

$$\left(\frac{\gamma}{2} - \frac{\gamma^2}{4}\right)\frac{1}{r^2} + \left(\nu + \frac{\gamma}{2}\right)\frac{A}{r} - \frac{A^2}{4} + \left[\frac{l\,(l+1)\,n\,a_0}{r^3} - \frac{n}{r^2}\right]g$$

$$+\left[\frac{-l\,(l+1)\,n\,a_0}{r^2} + \frac{2\,n}{r} - \frac{1}{n\,a_0}\right]g' + \left[\frac{n\,a_0}{4}\right]g'''$$

A SIMPLE EXAMPLE

An illustrative , yet simple example is provided when we choose $g = \varepsilon_0 = \text{constant}$[6]. The above equation reduces to

$$-\frac{l\,(l+1)}{r^2} + \frac{2}{a_0 r} - \frac{2\,m\,V_1(r)}{\hbar^2} - \frac{1}{n^2\,a_0^2} + \frac{2\,m\,E_1}{\hbar^2} \; =$$

$$= \left(\frac{\gamma}{2} - \frac{\gamma^2}{4}\right)\frac{1}{r^2} + \left(\nu + \frac{\gamma}{2}\right)\frac{A}{r} - \frac{A^2}{4} + \left[\frac{l\,(l+1)\,n\,a_0}{r^3} - \frac{n}{r^2}\right]\varepsilon_0$$

We immediately observe that such a choice of g is useful for inverse cubic perturbations . Defining $V_1(r) = \delta\,\dfrac{e^2\,a_0^{\,2}}{r^3}$, where δ is a dimensionless coupling constant , the expression $-\dfrac{2\,m\,V_1(r)}{\hbar^2}$ reduces to $-\dfrac{2\,\delta\,a_0}{r^3}$. Now equating the coefficients of the like powers of r , we obtain the equations :

$$-\,2\,\delta = l\,(l+1)\,n\,\varepsilon_0$$

$$l\,(l+1) = \left(\frac{\gamma^2}{4} - \frac{\gamma}{2}\right) + n\,\varepsilon_0$$

$$\frac{2}{a_0} = \left(\nu + \frac{\gamma}{2}\right)A$$

$$-\frac{1}{n^2\,a_0^2} + \frac{2\,m\,E_1}{\hbar^2} = -\frac{A^2}{4}$$

Their solution yields $E_1 = \delta \dfrac{e^2}{a_0 \, n^3 \left(l+\dfrac{1}{2}\right) l\,(l+1)}$ for the energy correction.

Borrowing the relation $E_1 = <V_1(r)>$ from ordinary first order perturbation

theory, we observe that $\left\langle \dfrac{1}{r^3} \right\rangle = \dfrac{1}{a_0^3 \, n^3 \left(l+\dfrac{1}{2}\right) l\,(l+1)}$, a well known result.

SUMMARY OF RESULTS

To attack problems of greater complexity , choices more sophisticated than $g = $ constant are necessary .. A choice of $g \propto r^k$ leads to terms with r^{k-3} , r^{k-2} and r^{k-1} dependence . Thus , their linear combination becomes the expression for the perturbing potential . Rather than a simple form for g and such a complicated expression for V_1 , we should prefer a complicated sum for g with carefully selected coefficients that will give us a simple V_1 . A method akin to Frobenius' method starts with such a sum for g and ends up with a simple power function expression for V_1 . The energy correction E_1 is obtained as a byproduct . Again borrowing the ordinary first order perturbation result $E_1 = <V_1>$ we obtain :

$$<r^N> = a_0^N \frac{n^{N-1} \Delta_{N+1}}{(N+1)!} \qquad (N \geq -1)$$

$$\left\langle \frac{1}{r^N} \right\rangle = \frac{\Delta_{N+1}}{a_0^N \, n^{N+1} \, (N-2)! \left(l+\dfrac{1}{2}\right) \displaystyle\prod_{k=0}^{(N-3)} \left[l\,(l+1) - \dfrac{k}{2}\left(\dfrac{k}{2}+1\right)\right]} \qquad (N \geq 3)$$

where Δ_D stands for the determinant of the $D \times D$ submatrix located at the upper left hand corner of the following infinite matrix :

$$\begin{bmatrix} n & -n\,[\,l\,(l+1)\,] & 0 & 0 & \cdots \\[2ex] -\dfrac{1}{n} & 3\,n & -2\,n\,[\,l\,(l+1)-\tfrac{3}{4}\,] & 0 & \cdots \\[2ex] 0 & -\dfrac{2}{n} & 5\,n & -3\,n\,[\,l\,(l+1)-2\,] & \cdots \\[2ex] 0 & 0 & -\dfrac{3}{n} & 7\,n & \cdots \\[2ex] 0 & 0 & 0 & -\dfrac{4}{n} & \cdots \\[2ex] \vdots & \vdots & \vdots & \vdots & \ddots \end{bmatrix}$$

It is further observed, with some degree of surprise, that $\Delta_0 = 1$ and $\Delta_{-1} = 0$ may be used with impunity. Since the above matrix is tridiagonal, certain simplifications are possible. The recursion relation for Δ_N

$$\Delta_N = (\,2N-1\,)\,n\,\Delta_{N-1} - (\,N-1\,)^2 \left[\,l\,(l+1)-\frac{N}{2}\left(\frac{N}{2}-1\right)\right]\Delta_{N-2} \qquad (\,N \geq 1\,)$$

and finally the recursion relation for $<r^N>$ is obtained :

$$\frac{(\,N+2\,)}{n^2\,a_0}\,\langle\,r^{N+1}\,\rangle - (\,2N+3\,)\,\langle\,r^N\,\rangle + a_0\,(\,N+1)\left[\,l\,(l+1)-\frac{N}{2}\left(\frac{N}{2}+1\right)\right]\langle\,r^{N-1}\,\rangle = 0$$

The above recursion relation is valid for all N, positive or negative. For positive N it is sufficient only to know that $<r^0> \,=\, 1$, for negative N one should

also know
$$\left\langle\frac{1}{r^2}\right\rangle = \frac{1}{a_0^{\,2}\,n^3\left(l+\dfrac{1}{2}\right)}$$
as starting points..

CONCLUSION

As was shown above, a new perturbation method has been developed based on the deformation of $SU(1,1)$ generators. Applied to the H - atom problem the expressions $<r^N>$ and $\left\langle\dfrac{1}{r^N}\right\rangle$ are evaluated for all N. $<r^N>$ are well - behaved while $\left\langle\dfrac{1}{r^N}\right\rangle$ expressions are singular for $N > 2\,l+2$ as expected. The new method has a remedy for these cases, however, this is presented elsewhere[6]. The analysis of SHO perturbations is also complete and they will be the subject of a future article. It should be stressed however, that the results are of secondary importance. Of significance is the method itself, by which the evaluation of $< f (r) >$, which ordinarily requires integral calculus, is reduced to the high school algebra of solving simultaneous equations. The power of group theoretical methods is that sometimes the achievements transcend the original objectives and insight is gained into things that have not been considered before. We feel and hope the method outlined above has such a potential.

ACKNOWLEDGEMENTS

I would like to thank Prof. Bruno Gruber, Southern Illinois University at Carbondale and Landeshauptstadt Bregenz for the hospitality shown at Cloister Mehrerau, Bregenz, Austria. Thanks are also due to Boğaziçi University Research Fund for their support and Ms. Sumru Başaran for correcting the manuscript.

REFERENCES

1. A multitude of articles exist on SO(4,2), for an extensive list on earlier work the reader is referred to the references of B. G. Wybourne, Classical Groups for Physicists, John Wiley (1974)
2. A.O. Barut and H. Beker, Phys. Rev. Lett. 50 (1983) 1560
3. A.O. Barut, H. Beker and A.J. Bracken, Proc. XIIIth Intern. Colloq. on Group Theoretical Methods in Physics, World Scientific (1984) 269
4. H. Beker, Found. of Physics 23 (1993) 851
5. P. Cordero and G.C. Ghirardi, Nuovo Cimento 2A (1971) 217
6. A.O. Barut, H. Beker and T. Rador, Phys. Letters (to be published)

A DUAL STRUCTURE FOR THE QUANTAL ROTATION GROUPS, SU (2)

L.C. Biedenharn and M.A. Lohe[*]

Center for Particle Physics
University of Texas at Austin, Austin, Texas 78712

Abstract

The Lie algebra of $SU(2)$ can be extended to the universal enveloping algebra and embedded in a non-commutative, co-commutative Hopf algebra. We demonstrate that this structure, for $SU(2)$, permits a *dual structure* to be defined, which is again a Hopf algebra of the same type but with a *deformed* algebra. In the limit of no deformation, this dual Hopf algebra has the Lie algebra of $SU(2)$.

1 Introduction

For compact *Abelian* groups, the concept of a dual structure is well-known; the dual structure is again a compact Abelian group, the *dual group* [1]. Consider the $U(1)$ group. The set of all unitary irreps (characters) is given by:

$$g = g(\alpha) \rightarrow D^{(n)}(g) = e^{in\alpha} \text{ with } n \in \mathbf{Z}. \tag{1.1}$$

Given any two such unitary irreps for the same group element, $D^{(n)}(g)$ and $D^{(m)}(g)$ one can define their product (as complex numbers) giving another irrep $D^{(n+m)}(g)$. This product defines a distinct group, which is Abelian and isomorphic to the group $\mathbf{Z}$. Thus the dual group to $U(1)$ is $\mathbf{Z}$, a well-known result.

The concept of a *dual group* cannot be generalized to non-Abelian groups. We will demonstrate, however, that if one first extends the notion of the Lie algebra of a compact Lie group to the universal enveloping algebra, and further extends this algebra to a Hopf algebra (an algebra—co-algebra structure), by defining a co-commutative co-product, then this Hopf algebra does possess a dual structure of exactly the same type, but with a deformed algebra. In the limit of no deformation, this dual algebra is the Lie algebra of $SU(2)$.

---

[*]Permanent address: Northern Territory University, PO Box 40146, Casuarina, NT Australia

The Abelian group example of duality, $U(1) \leftrightarrow \mathbf{Z}$, shows that the concept of a complete set of unitary irreducible representations (unitary irreps) is essential for defining the dual structure. The famous Peter-Weyl theorem [1] generalizes this concept to compact Lie groups and from this it follows that the matrix elements of the set of all unitary irreps of $SU(2)$ sets up a *pairing* between points in the discrete space of matrix labels and the points of the group manifold. More precisely, let $\{p\} = \{j, m, m'\}$ be the set of matrix labels (j denotes the irrep, m and m' the two 'magnetic' quantum numbers) and let g denote the set of points in the $SU(2)$ group manifold. Then a matrix-element of an irrep is a pairing: $(p, g) \rightarrow \mathbf{C}$. Note that the number of independent integer parameters p is precisely the dimensionality of the manifold for g.

Just as in the $U(1)$ example, there is a straightforward definition of a "dual product" defined (as complex numbers) for irrep matrix elements of the same group element. This is the Kronecker (or direct) product. This product is, however, unsuitable as a dual group product, since it fails to yield a unique "product irrep." There is a deeper reason, however, for the failure of the Kronecker product to function as the dual product. Although the group elements g are indeed dual to the points p, the group product on g belongs to $SU(2)$, whereas the irreps labelled by p are irreps of a factor group of $SU(2) \times SU(2)$, so that two *different* groups are involved for g and for p. (This reflects the non-Abelian nature of $SU(2)$, and is inherent in the problem.) We must therefore seek a different approach for the desired dual structure.

Nonetheless, the failure of this direct approach suggests a way to proceed. Wigner [2] has shown that for $SU(2)$ this product (which we will hereinafter call the "Wigner product") can be put in the form:

$$\mathbf{U}^{-1}(\mathbf{D}^{[m]}(g) \times \mathbf{D}^{[m']}(g))\mathbf{U} = \sum_l \bigoplus \mathbf{D}^{[l]}(g). \tag{1.2}$$

Here $\times$ denotes the direct (Kronecker) product, with $\mathbf{U}$ a unitary transformation (in direct product space) which brings the direct product to the form of a direct sum of irreps. The matrix elements for $\mathbf{U}$ constitute the Wigner-Clebsch-Gordan coefficients (WCG-coefficients). It is our view that the proper interpretation of the failure of this direct approach in defining a dual structure lies in a re-interpretation of the Wigner product law, (1.2), in algebraic terms, as a co-product on carrier spaces of operator-valued irrep matrices, as we shall explain in the next sections.

2 An Interpretation of the Wigner Product Law and Its Generalizations

The Wigner product law, (1.2), may be put in a more recognizable, algebraic, form by introducing the WCG coefficients. Using standard notation [3], the algebraic form of (1.2) is:

$$D^{(j_1)}_{m_1,m_1'}(g) D^{(j_2)}_{m_2,m_2'}(g) = \sum_{j_3, m_3, m_3'} C \begin{smallmatrix} j_1 & j_2 & j_3 \\ m_1 & m_2 & m_3 \end{smallmatrix} \, C \begin{smallmatrix} j_1 & j_2 & j_3 \\ m_1' & m_2' & m_3' \end{smallmatrix} \, D^{(j_3)}_{m_3,m_3'}(g). \tag{2.1}$$

The primary role of the WCG coefficients is that of coupling coefficients for the carrier space vectors of two $SU(2)$ irreps. An equivalent, though less familiar, view of this role is that the WCG coefficients express the action of a co-product on the basis vectors. For the WCG coefficients of $SU(2)$ the co-product, denoted Δ, is co-commutative since we have:

$$\Delta(\mathbf{J}) = \mathbf{J} \otimes \mathbf{1} + \mathbf{1} \otimes \mathbf{J}. \tag{2.2}$$

To bring out more clearly the meaning of (2.1) let us now invert this equation using the orthonormality of the WCG coefficients. Since there are two WCG coefficients on the RHS of (2.1), we see that there are two (commuting) $SU(2)$ groups involved, from the left and right action on the group element g, namely the factor group[†] $SU(2) \star SU(2)$.

To make the resulting relation more conceptually meaningful, let us denote the WCG coupling by $\times$. We then find symbolically:

$$(\mathbf{D}) \overset{\times}{\underset{\times}{}} (\mathbf{D}) = \mathbf{D}. \tag{2.3}$$

Expressed in words, (2.3) asserts that WCG coefficient coupling the irrep carrier spaces of the left and right action $SU(2)$ groups, results in an irrep carrier space of the *co-product* of $SU(2) \star SU(2)$ generators. We conclude that: *the structural meaning of the Wigner product law is to express the co-product action of the $SU(2)$ Hopf algebra, in precisely the same way that the group product law:*

$$\mathbf{D}(g_1)\mathbf{D}(g_2) = \mathbf{D}(g_1 g_2), \tag{2.4}$$

expresses the product action of the $SU(2)$ Hopf algebra.

We remark next that the WCG coefficents encode not only the co-product action but also determine the commutation relations for the Lie algebra. As we shall demonstrate in Section 3, *the Hopf algebra of $SU(2)$ is uniquely determined by the WCG coefficients.*

It follows from these remarks that in order to identify the dual structure to $SU(2)$ we must first find the proper analog, for the dual structure, to the coupling coefficients of $SU(2)$. To do this, let us consider the Wigner product law and its known analogs [3]:

(a) *The Wigner Product Law*
Explicit Form:

$$D^{(j_1)}_{m_1,m_1'}(g)\, D^{(j_2)}_{m_2,m_2'}(g) = \sum_{j_3,m_3,m_3'} C^{\ j_1\ \ j_2\ \ j_3}_{m_1\,m_2\,m_3}\, C^{\ j_1\ \ j_2\ \ j_3}_{m_1'\,m_2'\,m_3'}\, D^{(j_3)}_{m_3,m_3'}(g).$$

Symbolic Form:
$$DD = \sum CDC \tag{2.5}$$

(b) *The Re-Coupling Law*
Explicit Form:

$$C^{\ a\ \ b\ \ e}_{\alpha\,\beta\,\alpha+\beta}\, C^{\ e\ \ \ \ \ d\ \ c}_{\alpha+\beta\,\delta\,\alpha+\beta+\delta} = \sum_f C^{\ b\ \ d\ \ f}_{\beta\,\delta\,\beta+\delta}\left((2e+1)(2f+1)\right)^{\frac{1}{2}} W(abcd;ef) C^{\ a\ \ f\ \ \ \ \ c}_{\alpha\,\beta+\delta\,\alpha+\beta+\delta}$$

Symbolic Form:
$$CC = \sum CWC, \text{ and} \tag{2.6}$$

<hr>

[†] The $\star$ denotes factoring by the relation relating the two invariant (Casimir) operators.

(c) *The Biedenharn-Elliot Identity*
Explicit Form:

$$W(a\alpha b\beta; c\gamma)W(a'\alpha b'\beta; c'\gamma) = \sum_\lambda (2\lambda + 1)W(a'\lambda ac; ac')W(b\lambda\beta c'; b'c)W(a'\lambda\gamma b; ab')$$

Symbolic Form:

$$WW = \sum WWW. \tag{2.7}$$

These three product laws are not only analogous, but as algebraic identities have a well-defined limit structure [3] such that: $\lim(2.7) = (2.6), \lim(2.6) = (2.5)$.

Just as in obtaining (2.3) from (2.1)—using the orthonormality of the WCG coefficients—(2.5) and (2.7) can be inverted, using the orthonormality of the $(6-j)$ coefficients. We will denote the coupling effected by the $(6-j)$ operators by the symbol $\square$. Inverting the re-coupling law, (2.6), in symbolic form we find:

$$C \, {}^{\square}_{\times} C = C, \tag{2.8}$$

and for the B-E identity, (2.7), the symbolic form:

$$W \, {}^{\square}_{\square} W = W. \tag{2.9}$$

In this symbolic form these two identities, (2.8) and (2.9), are clearly structural analogs to (2.3). Moreover we see that:

The symbolic identities, (2.3), (2.8) and (2.9), identify the coupling $\square$ by $(6-j)$ coefficients to be the analog—for the dual Hopf algebra—of the coupling $\times$ by the WCG coefficients for the $SU(2)$ Hopf algebra.

It follows that we can determine *both* the algebra and the co-algebra structure of the dual Hopf algebra from the $(6-j)$ coupling coefficients.

3 The Dual Algebra

We have stated (in Section 2) that the coupling law for the group (the WCG coefficients) determines the Lie algebra structure. This is not difficult to verify. The coupling law for the adjoint $SU(2)$ irrep is the WCG coefficient $C \, {}^{111}_{\alpha\beta\gamma}$. Applied to the generators J_α, which carry the adjoint irrep, we see that vector coupling of J_α and J_β is, by definition:

$$(J \times J)_\gamma \equiv (2)^{\frac{1}{2}} \sum_{\alpha\beta} C \, {}^{111}_{\alpha\beta\gamma} J_\alpha J_\beta. \tag{3.1}$$

Since $C \, {}^{111}_{\beta\alpha\gamma} = -C \, {}^{111}_{\alpha\beta\gamma}$, one sees immediately that (3.1) is anti-symmetric, and yields a commutator, which is the desired relation.

However, the purpose of our argument is to derive the commutation relations in an algebraic way which will generalize to the dual coupling. To do this, let us recall that the adjoint operator J_α is equivalent to a WCG *operator* acting on a generic state $|jm\rangle$. Thus we have:

$$\langle jm + \alpha|J_\alpha|jm\rangle = (j(j+1))^{\frac{1}{2}} C \, {}^{j\,1\,j}_{m\,\alpha\,m+\alpha}. \tag{3.2}$$

Next we use (3.1), (3.2), and a variant of the re-coupling relation, (2.6):

$$\sum_{\alpha\beta} C^{\;1\;1\;1}_{\;\alpha\,\beta\,\gamma}\; C^{\;j\;\;1\;j}_{\;m+\beta\;\alpha}\; C^{\;j\;1\;j}_{\;m\;\beta} = (3(2j+1))^{\frac{1}{2}} W(j1j1;j1) C^{\;j\;1\;j}_{\;m\;\gamma\;m+\gamma}. \tag{3.3}$$

Evaluating the term $(3(2j+1))^{\frac{1}{2}} W(j1j1;j1)$ to be $(2j(j+1))^{-\frac{1}{2}}$, we find for (3.1)—acting on a generic vector—the result:

$$(J \times J)_\gamma \,|\,jm\,\rangle \equiv (2)^{\frac{1}{2}} \sum_{\alpha\beta} C^{\;1\;1\;1}_{\;\alpha\,\beta\,\gamma}\, J_\alpha J_\beta \,|\,jm\,\rangle = J_\gamma \,|\,jm\,\rangle, \tag{3.4}$$

which is precisely the desired commutation relation. We have determined (in Section 2) that the coupling coefficients for the dual structure are the $(6-j)$ coefficients. It is very useful for the further development of this idea to have an explicit realization of the corresponding $(6-j)$ *operators* as linear operators on a well-defined set of vector spaces which carry irreps of the dual algebra. That such a realization exists [4] is in itself rather surprising.

Consider the unitary group $U(3)$. Then the complete set of (integral) unitary irreps can be labelled by the Young frames denoted by $[m] \equiv [m_{13}m_{23}m_{33}]$, where the m_{i3} are (positive, negative or zero) integers such that $m_{13} \geq m_{23} \geq m_{33}$. The individual vectors carrying a given irrep $[m]$ may be labelled uniquely and canonically (to within phase equivalence) by the Gel'fand-Weyl pattern:

$$(m) \equiv \begin{pmatrix} m_{13} & m_{23} & m_{33} \\ & m_{12} & m_{22} \\ & m_{11} & \end{pmatrix}, \tag{3.5}$$

$$\text{where } m_{ij} \geq m_{ij-1} \geq m_{i+1j}. \tag{3.6}$$

An orthonormalized vector will be denoted by $|(m)\rangle$.

The surprising fact, to which we alluded, is that matrix elements of canonical $U(3)$ unit tensor operators have a well-defined limit when the irrep label $m_{33} \to -\infty$. This limiting result involves the $(6-j)$ coefficient. If we denote the (orthonormal) limiting vectors by the modified Gel'fand-Weyl array: $\left|\left(\begin{smallmatrix} m_{13} & & m_{23} \\ & m_{12} & m_{22} \\ & m_{11} & \end{smallmatrix}\right)\right\rangle\!\!\Big\rangle$, we can then project by the $U(2)$ vectors $\left|\begin{smallmatrix} m_{12} & & m_{22} \\ & m_{11} & \end{smallmatrix}\right\rangle$ to obtain the truncated (or equivalently, factored) space of orthonormal vectors denoted by $\left|\left(\begin{smallmatrix} m_{13} & & m_{23} \\ & m_{12} & m_{22} \end{smallmatrix}\right)\right\rangle\!\!\Big\rangle$. (A more detailed discussion may be found in ref. [4].)

Finally we will denote those $(6-j)$ operators, which correspond to generators in the dual algebra, by the symbol: $\left[\begin{smallmatrix} & 0 & \\ 1 & & -1 \\ & \gamma & \end{smallmatrix}\right]$, where: (a) the $U(2)$ irrep label $[1 \;\; -1]$ is equivalent to $j = 1$ in $SU(2)$, (b) the label 0 denotes zero shift in the $U(2)$ labels $[m_{13}\; m_{23}]$ of the truncated carrier space, and (c) the label γ (which corresponds to the magnetic quantum number index $\gamma = (1,0,-1)$ in J_γ) denotes the $U(2)$ shift: $[m_{12}\; m_{22}] \to [m_{12} + \gamma\; m_{22} - \gamma]$ on the labels of the truncated carrier space.

With these definitions, we can now state explicitly the realization of a *general* $(6\!-\!j)$ operator in the dual algebra [5]:

$$\left[\begin{matrix} & f-b & \\ d & & -d \\ & c-e & \end{matrix}\right] \left|\begin{matrix} b & & -b \\ & e-a & -e-a \end{matrix}\right\rangle$$

$$= ((2e+1)(2f+1))^{\frac{1}{2}}\, W(abcd; ef) \left| \begin{array}{cc} f & -f \\ c-a & -c-a \end{array} \right\rangle$$

$$= \begin{pmatrix} f & d & b \\ a & 0 & a \\ c & d & e \end{pmatrix} \left| \begin{array}{cc} f & -f \\ c-a & -c-a \end{array} \right\rangle. \tag{3.7}$$

(In (3.7) we have given the general result for all $(6-j)$ operators. The middle line gives the Racah form, $W(abcd; ef)$, while the bottom line uses a special case of an *orthonormal* $(9-j)$ coefficient whose rows and columns directly show the four angular momentum triangles. Specializing the labels b,c,d,e,f to the generator labels yields the $(6-j)$ generators.)

To determine the analog of the commutation relations, we will use the analog of (3.3), replacing the WCG coefficients by the $(6-j)$ coupling coefficients. Since (3.3) is determined by the re-coupling identity (2.6), we see that the analog to (3.3) must be determined from the B-E identity, (2.7). Unlike the coupling in (3.1), the operator coupling for the dual algebra is not numerical, but operator valued. We find:

$$\left(\begin{bmatrix} 0 \\ 1 \;-1 \\ \cdot \end{bmatrix} \Box \begin{bmatrix} 0 \\ 1 \;-1 \\ \cdot \end{bmatrix} \right)_{\begin{bmatrix} 0 \\ 1 \; {}_{e''-e} -1 \end{bmatrix}} \left| \begin{array}{cc} b & -b \\ e-a & -e-a \end{array} \right\rangle$$

$$\equiv \sum_{e'} (2)^{\frac{1}{2}} \begin{pmatrix} b & 1 & b \\ a & 0 & a \\ e'' & 1 & e' \end{pmatrix} \begin{pmatrix} 1 & 1 & 1 \\ e'' & 0 & e'' \\ e & 1 & e' \end{pmatrix} \begin{pmatrix} b & 1 & b \\ a & 0 & a \\ e' & 1 & e \end{pmatrix} \left| \begin{array}{cc} b & -b \\ e''-a & -e''-a \end{array} \right\rangle. \tag{3.8}$$

In (3.8) we have indicated on the LHS, symbolically, that one has a coupled pair of $(6-j)$ operators acting on a generic state vector while on the RHS we have given an explicit evaluation of the coupled operators in terms $(6-j)$ coefficients. (The middle $(6-j)$ coefficient (RHS (3.8)) is the analog to the WCG coupling $C^{1\,1\,1}_{\alpha\beta\gamma}$ in (3.4).) Since the RHS of (3.8) can be seen to be precisely the sum in the B-E identity of (2.7) the complicated algebraic relation in (3.8) can be evaluated, yielding a simpler result. Thus one finds for (3.8) the result:

$$\left(\begin{bmatrix} 0 \\ 1 \;-1 \\ \cdot \end{bmatrix} \Box \begin{bmatrix} 0 \\ 1 \;-1 \\ \cdot \end{bmatrix} \right)_{\begin{bmatrix} 0 \\ 1 \; {}_{e''-e} -1 \end{bmatrix}} \left| \begin{array}{cc} b & -b \\ e-a & -e-a \end{array} \right\rangle$$

$$= (2)^{\frac{1}{2}} \begin{pmatrix} b & 1 & b \\ b & 0 & b \\ 1 & 1 & 1 \end{pmatrix} \begin{pmatrix} b & 1 & b \\ a & 0 & a \\ e'' & 1 & e \end{pmatrix} \left| \begin{array}{cc} b & -b \\ e''-a & -e''-a \end{array} \right\rangle$$

$$= (b(b+1))^{-\frac{1}{2}} \begin{bmatrix} 0 \\ 1 \quad -1 \\ e''-e \end{bmatrix} \left| \begin{array}{cc} b & -b \\ e-a & -e-a \end{array} \right\rangle. \tag{3.9}$$

(In obtaining the last form for (3.9), we have used the evaluation of the $(6-j)$ coefficient given immediately after (3.3).) We interpret (3.9) as the analog to the commutation relations of (3.3), so that (3.9) defines an algebraic relation on two generators $\begin{bmatrix} 0 \\ 1 \;-1 \\ \alpha \end{bmatrix}$ and $\begin{bmatrix} 0 \\ 1 \;-1 \\ \beta \end{bmatrix}$ of the dual algebra yielding a generator, $\begin{bmatrix} 0 \\ 1 \;-1 \\ \gamma \end{bmatrix}$. Since the coupling coefficient in (3.9) is not anti-symmetric, and does not vanish for $\alpha = \beta$, we no longer have a Lie algebra. Moreover, the evaluations of the coupling coefficients are complicated algebraic functions dependent on the invariants of the given state vectors. Nonetheless, the result in (3.9) *can* be interpreted as the defining algebraic relations for the generators of the dual algebra, determining explicit algebraic results for the generators acting on every representation of the algebra.

The result we have found in (3.9) can be put in a much more perspicuous form using the symbolic coupling $\Box$. In addition, it is useful to note that the $SU(2)$ generators, J_α are *re-normalized* WCG operators, from (3.2). Since the $(6-j)$ operators in the dual algebra are analogs of the WCG operators in $SU(2)$, this implies that the proper analog (in the dual algebra) of the generators, J_α, is a *re-normalized* $(6-j)$ operator. Denoting a (re-normalized) dual generator by $\mathcal{J}_\alpha$, we have as the analog of (3.2):

$$\mathcal{J}_\alpha \left| \begin{matrix} b & -b \\ e-a & -e-a \end{matrix} \right\rangle = (b(b+1))^{\frac{1}{2}} \left[\begin{matrix} 0 \\ 1 & -1 \\ \alpha \end{matrix} \right] \left| \begin{matrix} b & -b \\ e-a & -e-a \end{matrix} \right\rangle. \tag{3.10}$$

With these notational changes, (3.9) becomes:

$$(\mathcal{J}. \ \Box \ \mathcal{J}.)_\alpha = \mathcal{J}_\alpha. \tag{3.11}$$

This result is clearly the analog of the commutation relations for $SU(2)$, which have the form:

$$[J. \times J.]_\gamma = J_\gamma. \tag{3.12}$$

In the limit that $m_{22} \to -\infty$, we find that $\mathcal{J}_\alpha \to J_\alpha$ and that (3.11) $\to$ (3.12). *Thus the dual algebra is a deformation of $SU(2)$, and smoothly approaches the $SU(2)$ algebra in the limit of no deformation.*

4 The Dual Co-Algebra Structure

The existence of a co-algebra structure for $SU(2)$ is easily verified directly from the definition: $\Delta(\mathbf{J}) = \mathbf{1} \otimes \mathbf{J} + \mathbf{J} \otimes \mathbf{1}$. We have noted, however, in Section 2 that the co-algebra structure is also encoded in the WCG coefficients, and this is the approach that is most easily generalized to the dual case.

To demonstrate, using the WCG coefficients, the $SU(2)$ co-algebra structure consider the tensor product of two independent sets of irrep vectors: $|j_i, m_i\rangle, i = 1, 2$. Then we have:

$$|j_1 j_2; JM\rangle \equiv \sum_{m_1, m_2} C \, {}^{j_1}_{m_1} \, {}^{j_2}_{m_2} \, {}^{J}_{M} \, |j_1 m_1\rangle \otimes |j_2 m_2\rangle. \tag{4.1}$$

The co-algebra structure asserts that the co-product:

$$\mathbf{J}^{\text{total}} = \mathbf{j_1} \otimes \mathbf{1} + \mathbf{1} \otimes \mathbf{j_2} \tag{4.2}$$

realizes the generator matrix elements when acting on the tensor product vectors in (4.1). Algebraically, this assertion is the relation:

$$J_\mu \, |j_1 j_2; JM\rangle = (J(J+1))^{\frac{1}{2}} C \, {}^{J}_{M} \, {}^{1}_{\mu} \, {}^{J}_{M+\mu} \, |j_1 j_2; JM+\mu\rangle$$

$$= \sum_{m_1, m_2} \Big((j_1(j_1+1))^{\frac{1}{2}} C \, {}^{j_1}_{m_1} \, {}^{1}_{\mu} \, {}^{j_1}_{m_1+\mu} \, C \, {}^{j_1}_{m_1} \, {}^{j_2}_{m_2} \, {}^{J}_{M} \, |j_1 m_1+\mu\rangle \otimes |j_2 m_2\rangle$$

$$+ (j_2(j_2+1))^{\frac{1}{2}} C \, {}^{j_2}_{m_2} \, {}^{1}_{\mu} \, {}^{j_2}_{m_2+\mu} \cdot C \, {}^{j_1}_{m_2} \, {}^{j_2}_{m_2} \, {}^{J}_{M} \, |j_1 m_2\rangle \otimes |j_2 m_2+\mu\rangle \Big). \tag{4.3}$$

Using the re-coupling identity, (2.6), this result, (4.3), can be seen to imply the algebraic identity:

$$(J(J+1))^{\frac{1}{2}} C^{\ J\ \ 1\ \ J}_{\ M\ \mu\ M+\mu}$$

$$= \sum_{J'} \left((j_1(j_1+1))^{\frac{1}{2}} \left(\begin{array}{ccc} j_1 & 1 & j_1 \\ j_2 & 0 & j_2 \\ J' & 1 & J \end{array} \right) + (j_2(j_2+1))^{\frac{1}{2}} \left(\begin{array}{ccc} j_2 & 1 & j_2 \\ j_1 & 0 & j_1 \\ J' & 1 & J \end{array} \right) \right) C^{\ J\ \ 1\ \ J'}_{\ M\ \mu\ M+\mu} \ . \quad (4.4)$$

It is rather surprising that such a simple property (co-multiplication), which is so easily verifiable, should lead to such a complicated identity.

Since we know from the existence of the co-product, (4.2), that (4.1) implies (4.4), we can *deduce* from the orthonormality of the WCG coefficient that (4.4) implies the validity of the identity:

$$(j_1(j_1+1))^{\frac{1}{2}} \left(\begin{array}{ccc} j_1 & 1 & j_1 \\ j_2 & 0 & j_2 \\ J' & 1 & J \end{array} \right) + (j_2(j_2+1))^{\frac{1}{2}} \left(\begin{array}{ccc} j_2 & 1 & j_2 \\ j_1 & 0 & j_1 \\ J' & 1 & J \end{array} \right) = \delta^{J'}_{J} (J(J+1))^{\frac{1}{2}} \quad (4.5)$$

Algebraic tables of the $(6-j)$ coefficients [3] verify the identity directly.

Consider now the question as to whether or not there exists a co-product for the dual algebra of $SU(2)$. If we were to proceed by analogy to the construction above, we would use the realization of the $(6-j)$ operator algebra acting on the space of state vectors in the truncated space $U(3) : U(2)$, as discussed in Section 3. *Such a procedure cannot succeed, since a co-product does not descend to a "co-product" on the truncated space.*

There is, however, an alternative way to introduce a co-product in the $SU(2)$ group: instead of using a tensor product over the vector space of states—as done in (4.1) for $SU(2)$—one can introduce instead a tensor product over the vector space of tensor operators. In both cases, the tensor product is effected by the same WCG coefficients. The consistency of the second approach (assuming equivariance) requires that the induced action of the generators on a given tensor operator be realized by the WCG coefficients.

The compatibility requirement—for the tensor product of the vector space of tensor operators—is that the induced action of the generators $\mathbf{J}_\mu$ on the $SU(2)$ tensor operator $(\mathcal{O}^m_j)$—an action denoted by $\mathbf{J}_\mu(\mathcal{O}^m_j)$—must be:

$$J_\mu(\mathcal{O}^m_j) \equiv (j(j+1))^{\frac{1}{2}} C^{\ j\ \ 1\ \ j}_{\ m\ \mu\ m+\mu} \ \mathcal{O}^{m+\mu}_j. \quad (4.6)$$

This result is easily verified, since the induced action for $SU(2)$ is given by commutation:

$$J_\mu(\mathcal{O}^m_j) = [J_\mu, \mathcal{O}^m_j]. \quad (4.7)$$

It follows that the induced action of J_μ on a *tensor product* of tensor operators (effected by the WCG coefficients) is a (co-commutative) co-product by the distributive law for commutators. The algebraic verification of this result, as in (4.3)ff, is precisely the same since the same relations leading from (4.1) to (4.5) follow, *mutatis mutandis*. The key to both the results is the re-coupling identity, (2.6).

It should now be clear as to how to define a co-product for the dual algebra: one uses $(6-j)$ operators and the B-E identity in analogy with the WCG operators and the re-coupling identity.

The induced action *on a* $(6-j)$ *tensor operator,* $\left[\begin{smallmatrix} & \gamma & \\ j & & -j \\ & m & \end{smallmatrix}\right]$, *by the re-normalized* $(6-j)$ generators, $\mathcal{J}_\mu$, is now defined by the relation:

$$\mathcal{J}_\mu\left(\left[\begin{matrix} & \gamma & \\ j & & -j \\ & m & \end{matrix}\right]\right)$$

$$\equiv (j(j+1))^{\frac{1}{2}} \begin{pmatrix} j & 1 & j \\ a & 0 & a \\ a+m+\mu & 1 & a+m \end{pmatrix} \left[\begin{matrix} & \gamma & \\ j & & -j \\ & m+\mu & \end{matrix}\right], \tag{4.8}$$

where the parameter $a = \frac{1}{2}(m_{12} - m_{22})$ is determined from the underlying space of vectors $\left| \begin{smallmatrix} m_{13} & & m_{23} \\ & m_{12} & \\ & & m_{22} \end{smallmatrix} \right\rangle$ on which (4.8) is to act. (This space is suppressed in writing (4.8) as an operator relation; strictly speaking, the parameter a should be written as an eigenvalue of an invariant operator, which can be done.) We remark that it is possible, by enlarging the set of generators to the operators $\left[\begin{smallmatrix} & \gamma & \\ 1 & & -1 \\ & m & \end{smallmatrix}\right]$, $(\gamma, m = 1, 0, -1)$ to realize the induced action as an operator relation acting on the vector space of truncated states yielding (4.8).

Next we give an analog to (4.1), expressed now, however, as a tensor product on the vector space of tensor operators. The required coupled vector spaces are given by:

$$\left(\left[\begin{matrix} & 0 & \\ j_1 & & -j_1 \\ & \cdot & \end{matrix}\right] \otimes \left[\begin{matrix} & 0 & \\ j_2 & & -j_2 \\ & \cdot & \end{matrix}\right]\right)\left[\begin{matrix} & 0 & \\ J & & -J \\ & a''-a & \end{matrix}\right]$$

$$\equiv \sum_{a'} \begin{pmatrix} J & j_2 & j_1 \\ a'' & 0 & a'' \\ a & j_2 & a' \end{pmatrix} \left[\begin{matrix} & 0 & \\ j_1 & & -j_1 \\ & a''-a & \end{matrix}\right] \otimes \left[\begin{matrix} & 0 & \\ j_2 & & -j_2 \\ & a'-a & \end{matrix}\right]. \tag{4.9}$$

Since we wish to verify that the co-product has the induced action:

$$\Delta(\mathcal{J}_\mu(\cdot)) = \mathcal{J}_\mu(\cdot) \otimes \mathbf{1} + \mathbf{1} \otimes \mathcal{J}_\mu(\cdot), \tag{4.10}$$

when acting on the tensor product in (4.9), we therefore apply the RHS of (4.10) to the vectors in (4.9). The action of $\mathcal{J}_\mu(\cdot) \otimes \mathbf{1}$ on (4.9)—using (4.8)—is given by:

$$\sum_{a'} (j_1(j_1+1))^{\frac{1}{2}} \begin{pmatrix} j_1 & 1 & j_1 \\ a' & 0 & a' \\ a''+\mu & 1 & a \end{pmatrix} \begin{pmatrix} J & j_2 & j_1 \\ a'' & 0 & a'' \\ a & j_2 & a' \end{pmatrix}$$

$$\times \left[\begin{matrix} & 0 & \\ j_1 & & -j_1 \\ & a''+\mu-a' & \end{matrix}\right] \otimes \left[\begin{matrix} & 0 & \\ j_2 & & -j_2 \\ & a'-a & \end{matrix}\right]. \tag{4.11}$$

We may simplify this expression by applying the inverse to (4.9)—since the $(6-j)$ coefficients involved all have an inverse—and then applying the B-E product identity.

(This procedure is the direct analog of the technique using the re-coupling identity in (4.4).) This yields:

$$\sum_{a',J'} (j_1(j_1+1))^{\frac{1}{2}} \begin{pmatrix} J & j_2 & j_1 \\ a'' & 0 & a'' \\ a & j_2 & a' \end{pmatrix} \begin{pmatrix} J' & j_2 & j_1 \\ a''+\mu & 0 & a''+\mu \\ a & j_2 & a' \end{pmatrix} \begin{pmatrix} j_1 & 1 & j_1 \\ a' & 0 & a' \\ a''+\mu & 1 & a'' \end{pmatrix}$$

$$\times \left(\begin{bmatrix} & 0 & \\ j_1 & \cdot & -j_1 \end{bmatrix} \otimes \begin{bmatrix} & 0 & \\ j_2 & \cdot & -j_2 \end{bmatrix} \right) \begin{bmatrix} & 0 & \\ J & & -J \\ & a''+\mu-a & \end{bmatrix}$$

$$= \sum_{J'} (j_1(j_1+1))^{\frac{1}{2}} \begin{pmatrix} j_1 & 1 & j_1 \\ j_2 & 0 & j_2 \\ J' & 1 & J \end{pmatrix} \begin{pmatrix} J' & 1 & J \\ a & 0 & a \\ a''+\mu & 1 & a'' \end{pmatrix}$$

$$\times \left(\begin{bmatrix} & 0 & \\ j_1 & \cdot & -j_1 \end{bmatrix} \otimes \begin{bmatrix} & 0 & \\ j_2 & \cdot & -j_2 \end{bmatrix} \right) \begin{bmatrix} & 0 & \\ J' & & -J' \\ & a''+\mu-a & \end{bmatrix} . \quad (4.12)$$

The action of $1 \otimes \mathcal{J}_\mu(\cdot)$ on (4.9) is, in exactly the same way, given by the equation above, except that $j_1 \leftrightarrow j_2$ everywhere *aside from the final tensor product term.* Adding these two terms yields the desired action by $\Delta(\mathcal{J}_\mu(\cdot))$ from (4.9). *This sum is easily seen to involve precisely the identity in (4.5).* Thus we find that the co-product Δ implies that the action of $\Delta(\mathcal{J}_\mu(\cdot))$ on the tensor product is exactly in the required form, that is, obeys (4.8) with coupled labels.

To conclude: *We have proved that the dual algebra possesses a (co-commutative) co-product and thus the dual algebra extends to a Hopf algebra.*

5 Concluding Remarks

Let us summarize our results in the form of an explicit proposition which has now been proved.

Proposition 1 *The quantal rotation group, $SU(2)$, extends to a non-commutative, co-commutative, Hopf algebra. There exists a dual Hopf algebra of the same type that is a deformation of the Lie algebra of $SU(2)$.*

An examination of the essentials of the proof of this proposition shows that it is based on very general principles (the three product laws in (2.5),(2.6) and (2.7)), and an apparently special property (the realization of the $(6-j)$ operator algebra in (3.7)). There exists a generalization of the three product laws to $SU(n)$ and, remarkably, a generalization of the realization of the $(6-j)$ operator algebra for $SU(n)$ as well. Thus we *conjecture* that our proposition is, in fact, valid for $SU(n)$ using the same proof, *mutatis mutandis.*

Acknowledgements

This work was supported in part under contract DOE-ER40757-054 of the Center for Particle Physics and is preprint CPP-94-30.

References

[1] A. A. Kirillov, *Elements of the Theory of Representations,* Springer Verlag (Berlin, 1976).

[2] E. P. Wigner, *On the Matrices Which Reduce the Kronecker Products of Representations of S.R. Groups,* in *Quantum Theory of Angular Momentum,* L. C. Biedenharn and H. Van Dam (Eds.), Academic Press (New York, 1965).

[3] L. C. Biedenharn and J. D. Louck, *Angular Momentum in Quantum Physics,* Cambridge University Press (Cambridge, 1984). (Originally published by Addison-Wesley Reading, MA, 1981).

[4] L. C. Biedenharn and J. D. Louck, *The Racah-Wigner Algebra in Quantum Theory,* Addison-Wesley Publ. Co. (Reading, MA 1981).

[5] J. D. Louck and L. C. Biedenharn, J. Math. Phys. **14,** 1336-1357, (1973). (Cf.(4.11).)

SOME POINTS IN THE QUANTIZATION OF RELATIVISTIC GRASSMANN DEPENDENT INTERACTION SYSTEMS

A. Del Sol[1] Mesa and R. P. Martínez y Romero[2]

[1]Instituto de Física, Universidad Nacional Autónoma de México
Apartado Postal 20-364, 01000 México D.F. Mexico
[2] Departamento de Física, Facultad de Ciencias
Universidad Nacional Autónoma de México
Apartado Postal 20-364, 01000, México D. F. Mexico

INTRODUCTION

The Grassmann variables in the context of a physical theory are now widely accepted. We can think, for instance, in supergravity or in superstring theories as examples of the importance they have now in contemporary Physics. These already ubiquitous examples are not certainly the only ones. Since the Grassmann variables are used as the classical equivalent of quantum spin, they also appear in the description of systems where the spin plays an important role[1]. In all these cases, the central idea is based in Dirac's point of view that we should first try to understand a physical theory, and only then try to quantize it[2]. To illustrate this point, we quote the work by Crater and Van Alstine[3], where they construct two body relativistic wave equations for particles with spin. It is not easy to work at the quantum level with this problem, because one must first try to impose certain conditions of compatibilty on the wave function and in the equations themselves, which restrict the class of available potentials for the problem. Crater and Van Alstine translate the problem to the classical level and then they use the general theory of constraints to analyze it. They find that a supersymmetry condition is needed to impose, in order to obtain consistent equations for the problem. Of course, the final theory is obtained after quantizing their results.

However, it is quite frequently to find some ambiguities in the quantization of such problems, associated with the Grassmann variables. For instance, if we have a potential dependent on the Grassmann variable θ, and we are interested in a series expansion of it, we get a different result depending wether the quantization is done before or after the series expansion. This is understandable, since at classical level $\theta^2 = 0$, but at quantum level $\theta^2 = -1/2$. In some papers this point has already been raised[3], but in spite of the fact that the authors solve their particular problem, they don't give us a systematic answer to this point. The existence of ambiguities in the quantization of physical systems is not a new problem. For instance a term of the form $\mathbf{x} \cdot \mathbf{p}$, when quantized becomes the hermitian expression $\frac{1}{2}(\mathbf{x} \cdot \mathbf{p} + \mathbf{p} \cdot \mathbf{x})$, due to the non conmutative character of the $\mathbf{x}$ and $\mathbf{p}$ variables, but the general treatment for the Grassmann variables is still not well understood. What we want to do in this paper is to mention some points associated with the quantization of relativistic Grassmann dependent potential problems, where such ambiguities frequently occur.

FORMULATION OF THE PROBLEM

Let us start with the relativisitic free particle Dirac equation which reads as[4]

$$(\theta \cdot p + m\theta_5)\psi = 0, \tag{1}$$

where we are working in units in which $\hbar = c = 1$, and

$$\theta^\mu \equiv \frac{i}{\sqrt{2}}\gamma_5\gamma^\mu, \qquad \theta_5 \equiv \frac{i}{\sqrt{2}}\gamma_5. \tag{2}$$

Here the $\gamma's$ are the usual Dirac matrices and γ_5 is given by the expression $\gamma_5 \equiv i\gamma^0\gamma^1\gamma^2\gamma^3$. We also define the metric tensor as $g^{\mu\nu} = \mathrm{diag}(-1,1,1,1)$. We have chosen the definitions of our theory in such a way that the Dirac equation becomes in the classical limit a constraint with a definite parity, in this case odd. Anyway, at the quantum level the $\theta's$ matrices satisfy the algebra

$$\begin{aligned}
[\theta^\mu, \theta^\nu]_+ &= -g^{\mu\nu} \\
[\theta_5, \theta^\mu]_+ &= 0 \\
[\theta_5, \theta_5]_+ &= -1
\end{aligned} \tag{3.a.b.c}$$

where the bracket $[\ ,\ ]_+$ means the anticommutator.

In order to obtain the classical limit, we follow the correspondence principle translating the anticommutators into the generalized Dirac brackets

$$\frac{1}{i}[\ ,\ ]_+ \rightarrow \{\ ,\ \}. \tag{4}$$

In the classical limit, we consider the bosonic operators x^μ, and p_μ as real numbers, while the $\theta's$ operators are replaced by the real Grassmann variables θ^μ and θ_5. The algebra of the Grassmann variables is easily obtained from Eq. 3) and 4) as

$$\begin{aligned}
\{\theta^\mu, \theta^\nu\} &= ig^{\mu\nu} \\
\{\theta_5, \theta_5\} &= i \\
\{\theta_5, \theta^\mu\} &= 0
\end{aligned} \tag{5a, b, c}$$

Now the passage to classical mechanics (or *pseudoclassical* as is also called), is obtained by translating the Dirac equation into a first class constraint.

$$\pounds = \theta \cdot p + m\theta_5 \approx 0, \tag{6}$$

where the simbol $\approx$ means weak equality. However, this is not the only first class constraint that we have, since $\pounds$ is odd we can construct another one, which is in fact the mass shell condition,

$$\mathcal{H} = \frac{1}{i}\{\mathcal{L}, \mathcal{L}\} = p^2 + m^2 \approx 0. \tag{7}$$

Let us now introduce a certain type of Grassmann dependent interaction which covers a quite general class of relativistic quantum mechanical problems.

$$p_\mu \to P_\mu = p_\mu + P_\mu^\alpha(x)\theta_\alpha\theta_5 + P_\mu^{\delta\gamma}(x)\theta_\delta\theta_\gamma. \tag{8}$$

Here the coefficients $P_\mu^\alpha(x)$, and $P_\mu^{\alpha\beta}(x)$ depend only in the space – time variable x. This expression embrace a general class of interactions that appear in the literature. We can quote the Dirac oscillator[5], the Kukulin oscillator[6] and the works of Crater and Van Alstine already mentioned[3], as particular examples . In certain cases, minor modifications are needed to use, but they don't alter the essential part of our results. Notice also that P is again an even first class constraint, so we can construct another first class constraint as we already did before, by simply taking the Dirac bracket of $\mathcal{L}$ with itself,

$$\mathcal{H} = \{\mathcal{L}, \mathcal{L}\} = \frac{1}{i}\{\theta \cdot P, \theta \cdot P\} + \frac{2m}{i}\{\theta \cdot P, \theta_5\} + m^2. \tag{9}$$

It is now possible to show that after some algebra

$$\begin{aligned}
\mathcal{H} = P^2 + m^2 &+ 2i(\partial_{[\nu} P_{\mu]})\theta^\nu\theta^\mu + 2[P_\mu^\nu(x)\theta_5 + P_\mu^{[\nu\gamma]}(x)\theta_\gamma]\theta^\mu P_\nu \\
&+ 2g_{\alpha\beta}\theta^\mu P_\mu^\alpha(x)\theta^\rho P_\rho^{[\beta\sigma]}(x)\theta_\sigma\theta_5 + 2m\theta^\sigma P_\sigma^\alpha(x)\theta_\alpha,
\end{aligned} \tag{10}$$

where $P_\mu^{[\alpha\beta]}(x) \equiv P_\mu^{\alpha\beta}(x) - P_\mu^{\beta\,\alpha}(x)$, is an antisymmetric tensor.

On the other hand, the quantum counterpart of this problem is given by

$$\mathcal{L}_q = (\theta \cdot P + m\theta_5)\psi = 0 \tag{11}$$

where all the variables involved in this expresion are quantum operators, according to the corespondence principle. We can now construct the quantum equivalent of the mass shell condition, or Klein-Gordon equation, given by

$$\mathcal{H}\psi = [\mathcal{L}_q, \mathcal{L}_q]_+\psi = 0 \tag{12}$$

We want to remark that if we proceed naively, and we want at this point to quantize our system, we could develop in 10) the expression for P_μ given by Eq. 8) to obtain an incorrect result, in the sense that it is different from the Klein Gordon equation, Eq.12). How can we avoid the differences between both approaches? It is important to point out that the we can encounter, basically, two types of differences. First, the differences that come from the powers of P_μ. In this case, the correct approach would be to quantize Eq. 10) *before* using equation 8), and only then we can expand the resultant expression, if we need to. In this sense we shall call P_μ the relevant variable to quantize. Second, the differences that arises when two indices of the product of two Grassmann (or more) variables become equal. To illustrate this point, let us take the last term in Eq. 10)

$$2m\theta^\sigma P_\alpha^\sigma(x)\theta_\alpha = 2m \sum_{\alpha \neq \sigma} \theta^\sigma P_\sigma^\alpha(x)\theta_\alpha, \tag{13}$$

If we want to quantize now this term, it becomes

$$2m \sum_{\alpha \neq \sigma} \theta^\sigma P_\sigma^\alpha(x)\theta_\sigma - 2m \operatorname{Tr}(P_\sigma^\alpha). \tag{14}$$

Notice that there is a diference in the last term of equation 14), for the quantum case. We can even call these differences *fermionic anomalies*, because they are originated in the properties of the Grassmann variables. In order to avoid such kind of anomalies, we are going to impose the followig conditions for the coefficients of P_μ

$$P_\alpha^\alpha = P_\alpha^{\alpha\beta} = P_\beta^{\alpha\beta} = 0 \tag{15}$$

We can now resume our propositions. If we don't want to have fermionic anomalies in relativistic quantum mechanical problems with a Grassmann dependence in the potential, we must first construct the relevant variable P_μ, writing the interaction in the form 8), imposing the conditions 15) for the coefficients, and then quantizing before any expansion. This quantizaton procedure can be proved by developing Eq. 12) with the condition 15), and comparing the result with 10), to show that both results are in complete agreement. It is interesting to point out that the conditon of relevant variable is valid even in calculations associated with the gauge transformation of the theory, where the anomalies could also be present. This last point can be proved along the same lines as before, althought the calculations are long.

EXAMPLE

We can illustrate our results with the example of the Dirac oscillator. This model was introduced some years ago to explain the barionic mass spectrum in the context of quark confinement[5]. The central idea is the replacement of the momentum $\mathbf{p}$ of the particle in the Dirac equation by

$$\mathbf{p} \to \mathbf{p} - im\omega\mathbf{r}\beta \tag{16}$$

where m is the mass of the particle, ω the frequency of the oscillator, $\mathbf{r}$ the position vector and β the Dirac beta matrix. The Dirac oscillator can be put in the form 11), where

$$P^\mu \equiv p^\mu - 2im\omega x_\perp^\mu (\hat{u} \cdot \theta)\theta_5 \tag{17.a}$$

$$x_\perp^\mu \equiv x^\mu + (\hat{u} \cdot x)\hat{u}^\mu. \tag{17.b}$$

Here u^μ is a frame dependent vector which in a definite frame of reference (the lab. frame) takes the expression $(1,0,0,0)$. We notice that the conditions 15) are automatically satisfied, because $\hat{u}$ is perpendicular to $x_\perp$. Following the correspondence limit, we can construct the classical limit of the model, in such a way that there are no ambiguities in the process. It is not too hard to verify that we can recover the original quantum model if we consider 17) as the relevant variable to quantize. We can also illustrate that this property is not valid in general for any dynamical variable. For instance, let us calculate the classical gauge transformation for the variable p^μ generated by the first class constraint $\mathcal{L}$,

$$\delta p_\mu \approx \epsilon\{p^\mu, \theta \cdot P + m\theta_5\} = -2\epsilon m\omega\theta^\mu(\hat{u} \cdot \theta_5), \tag{18.a}$$

where P is given by 17.a), and ϵ is an infinitesimal parameter. If we calculate now the quantum counterpart

$$\delta p_\mu = \epsilon[p^\mu, \theta \cdot P + m\theta_5] = -2\epsilon m\omega\theta^\mu(\hat{u} \cdot \theta)\theta_5 - \epsilon m\omega\,\hat{u}^\mu\theta_5. \tag{18.b}$$

we again see the presence of anomalies, that would not be present if we calculate the same as above for P_μ instead that for p_μ. As a final comment, we want to point out that in some cases[7] we need to put part of the interaction not only in P_μ, but also in a term of the form $m \to M(x)$, but as far as our discussion concerns, this point is not relevant, since we are introducing only a function of x, and the anomalies are present in the Grassmann part of the interaction.

AKNOWLEDGMENTS

This work was partially supported by grant CONACYT 4846-E9406 and DGAPA ESP I00191 of UNAM.

REFERENCES

1) C.A.P. Galvao and C. Teitelboim, *J. Math. Phys.* **21**, 1863, (1980); *Quantization of Gauge Systems* M. Henneaux and C. Teitelboim. Princeton University Press, New Jersey, (1992); R. Casalbuoni, *Nuovo Cimento*, A33, 286 (1976); A. Barducci, R. Casalbuoni, and L. Lusanna, *Nuovo Cimento A 35*, 377 (1976); F. A. Berezin and M.S. Marinov, *Ann. Phys.* **104**, 336 (1977); F. A. Berezin, *The method of second Quantization*. Academic, New York, (1966).
2) P.A.M. Dirac, *Can. J. Math.* **2**, 129 (1950); *Proc. Roy. Soc. Sect. A* **246**, 326 (1958); *Lectures on Quantum Mechanics*, Belfer Graduate School of Science, Yeshiva Univ., New York (1964).
3) H. Crater and P. Van Alstine, *Ann. Phys.* **21**, 1863 (1980).
4) C.A.P. Galvao and C. Teitelboim *Op. Cit.*
5) M. Moshinsky and A. Szczepaniak, *J. Phys. a: Math. Gen.* **22** L817 (1989). D. Ito , K. Mori, E. Carriere, *Nuovo Cimento a* **51**,119 (1967); P.A. Cook, *Lettere al Nuovo Cimento* **1**,419 (1971). 6) V.I. Kukulin, G. Loyola and M. Moshinsky *Phys. Lett. A* **158** (1991).
7) H. Crater and P. Van Alstine *Phys. Rev. D,* **36**, No. 10, 3007 (1987).

q - DIFFERENCE INTERTWINING OPERATORS FOR $U_q(sl(4))$ AND q - CONFORMAL INVARIANT EQUATIONS

V.K. Dobrev*

Arnold Sommerfeld Institute for Mathematical Physics
Technical University Clausthal
Leibnizstr. 10
38678 Clausthal-Zellerfeld
Germany

1. INTRODUCTION

1.1. Consider a Lie group G, e.g., the Lorentz, Poincaré, conformal groups, and differential equations

$$\mathcal{I} f = j \tag{1.1}$$

which are G-invariant. These play a very important role in the description of physical symmetries - recall, e.g., the examples of Dirac, Maxwell equations, (for more examples cf., e.g., [1]). It is important to construct systematically such equations for the setting of quantum groups. Such equations there are expected as q-difference equations. The hope is that these equations will have less singular behaviour than the classical counterparts.

The approach to this problem used here relies on the following. In the classical situation the differential operators $\mathcal{I}$ giving the equations above may be described as operators intertwining representations of complex and real semisimple Lie groups [2], [3], [4], [5]. (A notable exception are differential equations originating from Casimir operators, e.g., the Laplace-Beltrami operator, cf., e.g., [1].)

To recall the notions, consider a semisimple Lie group G and two representations T, T' acting in the representation spaces C, C', which may be Hilbert, Fréchet, etc. An *intertwining operator* $\mathcal{I}$ for these two representations is a continuous linear map

$$\mathcal{I} : C \longrightarrow C' \tag{1.2}$$

---

* Permanent address : Bulgarian Academy of Sciences, Institute of Nuclear Research and Nuclear Energy, 72 Tsarigradsko Chaussee, 1784 Sofia, Bulgaria.

Symmetries in Science VII, Edited by
B. Gruber, Plenum Press, New York, 1995

such that

$$\mathcal{I} \circ T(g) \;=\; T'(g) \circ \mathcal{I}\,, \quad \forall g \in G\,. \tag{1.3}$$

This is what precisely is meant when we say that the equation (1.1) is a G - *invariant equation*. Note that $\ker \mathcal{I}$, $\operatorname{im} \mathcal{I}$ are invariant subspaces of C, C', resp.[1] If $\ker \mathcal{I} \neq 0$, this means that the equation (1.1) with $j = 0$ has non-trivial solutions.

Such equations exist also for more general classes of Lie groups. However, if G is semisimple (even reductive) then there exists canonical ways for the construction of all intertwining operators and thus, of all G - invariant equations.[2] These operators are of two types - differential and integral. For the canonical construction of the integral invariant operators (which we shall not consider) we refer to [2].

As stated we are interested in the invariant differential operators. There are many ways to find such operators, cf., e.g., [1], however, most of these rely on constructions which are not yet available for quantum groups. Here we shall apply a procedure [5] which is rather algebraic and can be generalized almost straightforwardly to quantum groups. This procedure is recalled in Subsection 1.3.

1.2.　　Here we shall sketch the procedure of [5] illustrating the general notions with the conformal group $SU(2,2)$. Let G be a real semisimple Lie group. Let $\mathcal{G}$ be the Lie algebra of G. We shall use the so-called Bruhat decompositions of $\mathcal{G}$

$$\mathcal{G} \;=\; \mathcal{N}^+ \oplus \mathcal{M} \oplus \mathcal{A} \oplus \mathcal{N}^-\,, \tag{1.4}$$

(considered as direct sum of linear spaces), where $\mathcal{A}$ is a noncompact abelian subalgebra, $\mathcal{M}$ (a reductive Lie algebra) is the centralizer of $\mathcal{A}$ in $\mathcal{G}$ (mod $\mathcal{A}$), and $\mathcal{N}^+$, $\mathcal{N}^-$, resp., are nilpotent subalgebras forming the positive, negative, resp., root spaces of the restricted root system $(\mathcal{G}, \mathcal{A})$. For the conformal group the subalgebras $\mathcal{N}^+$, $\mathcal{M}$, $\mathcal{A}$, $\mathcal{N}^-$, are the subalgebras of translations, Lorentz transformations, dilatations, special conformal transformations, resp.

In general, a real noncompact Lie algebra $\mathcal{G}$ has more than one Bruhat decomposition. This is standard material, cf., e.g., [6]. (It is explained also in [5], or in [7].) Note that $\tilde{\mathcal{P}}^\pm = \mathcal{N}^\pm \oplus \mathcal{M} \oplus \mathcal{A}$ are subalgebras of $\mathcal{G}$, the so-called *parabolic subalgebras*. The parabolic subalgebras with minimal, resp., maximal, dimension are called *minimal*, resp., *maximal parabolic subalgebras* of G.[3] The group $SU(2,2)$ has three non-trivial non-conjugate parabolic subalgebras of dimensions $9, 10, 11$. With the above identification $\tilde{\mathcal{P}}^\pm$ are maximal conjugate parabolic subalgebras; otherwise $\tilde{\mathcal{P}}^+$ is called the Weyl algebra (comprising the Poincaré algebra and the dilatations).

Let us now introduce the corresponding subgroups of G. Let K denote the maximal compact subgroup of G, and let $\mathcal{K}$ denote the Lie algebra of K. Then we have the simply connected subgroups $A = \exp(\mathcal{A})$, $N^\pm = \exp(\mathcal{N}^\pm)$. Further, M is the centralizer of A in G (mod A). (M has the structure $M = M_d M_r$, where M_d is a finite group, M_r is reductive with the same Lie algebra $\mathcal{M}$ as M.) Then $P^\pm = MAN^\pm$ are called parabolic subgroups of G.

---

[1]　If $\ker \mathcal{I} = 0$ and $\operatorname{im} \mathcal{I} = C'$, then the representations T and T' are called *equivalent*, otherwise T and T' are called *partially equivalent*.

[2]　For simplicity we consider mostly semisimple Lie groups, though the same results are valid for reductive Lie groups, since only their semisimple subgroups are essential for these considerations.

[3]　The subalgebra $\mathcal{M}$ is compact only if it is a subalgebra of a minimal parabolic subalgebra. Thus, if $\mathcal{G}$ is compact the Bruhat decomposition is trivialized becoming $\mathcal{G} = \mathcal{M}$.

The importance of the parabolic subgroups stems from the fact that the representations induced from them generate all (admissible) irreducible representations of G. In fact, it is enough to use only the so-called *cuspidal* parabolic subgroups, singled out by the condition that $\operatorname{rank} M = \operatorname{rank} M \cap K$; thus M has discrete series representations.

Let P be a cuspidal parabolic subgroup, ($P = P^+$, $N = N^+$ or $P = P^-$, $N = N^-$ is specified by convenience). Let μ fix a discrete series representation D^μ on the Hilbert space V_μ or the so-called limit of a discrete series representation (cf. [8]). Let ν be a (non-unitary) character of A, $\nu \in \mathcal{A}^*$.

We call the induced representation $\chi = \operatorname{Ind}_P^G(\mu \otimes \nu \otimes 1)$ an *elementary representation* of G. (These are called *generalized principal series representations* (or limits thereof) in [8].)

Consider the space of functions

$$\mathcal{C}_\chi \;=\; \{\mathcal{F} \in C^\infty(G, V_\mu) \mid \mathcal{F}(gman) = e^{\nu(H)} \cdot D^\mu(m^{-1})\mathcal{F}(g)\}\;, \tag{1.5}$$

where $a = \exp(H)$, $H \in \mathcal{A}$. The special property of the functions of $\mathcal{C}_\chi$ is called *right covariance*.

Then the elementary representation (ER) $\mathcal{T}^\chi$ acts in $\mathcal{C}_\chi$, as the left regular representation (LRR), by:

$$(\mathcal{T}^\chi(g)\mathcal{F})(g') \;=\; \mathcal{F}(g^{-1}g')\;, \quad g, g' \in G\;. \tag{1.6}$$

(In practice, the same induction is used with non-discrete series representations of M and also with non-cuspidal parabolic subgroups.) One can introduce in $\mathcal{C}_\chi$ a Fréchet space topology or complete it to a Hilbert space (cf. [8]). Finally, note that in order to obtain the invariant differential operators one may consider the infinitesimal versions of (1.5) and (1.6) (cf. the end of this subsection).

The ERs differ from the LRR (which is highly reducible) by the specific representation spaces $\mathcal{C}_\chi$. In contrast, the ERs are generically irreducible. The reducible ERs form a measure zero set in the space of the representation parameters μ, ν. (Reducibility here is topological in the sense that there exist nontrivial (closed) invariant subspace.) The irreducible components of the ERs (including the irreducible ERs) are called *subrepresentations*.

The importance of the elementary representations stems from the following result:

Theorem. [9], [10] Every irreducible admissible representation of a real connected semisimple Lie group G with finite centre is equivalent to a subrepresentation of an elementary representation of G.

Remark: Admissibility is a technical condition which is usually fulfilled in the physically interesting examples.

The other feature of the ERs which makes them important for our considerations is a highest, resp., lowest weight module structure associated with them. For this we introduce the right action of $\mathcal{G}^{\mathbb{C}}$ (the complexification of $\mathcal{G}$) by the standard formula:

$$(\hat{X}\mathcal{F})(g) \;=\; \frac{d}{dt}\mathcal{F}(g\exp(tX))|_{t=0}\;, \quad X \in \mathcal{G}^{\mathbb{C}}\;, \quad \mathcal{F} \in \mathcal{C}_\chi\;, \quad g \in G\;, \tag{1.7}$$

which is defined first for $X \in \mathcal{G}$ and then is extended to $\mathcal{G}^{\mathbb{C}}$ by linearity. Note that this action takes $\mathcal{F}$ out of $\mathcal{C}_\chi$ for some X but that is exactly why it is used for the construction of the intertwining differential operators.

We illustrate this special property in the case of the minimal parabolic subalgebra. In that case M is compact and V_μ is finite dimensional. Then it has a lowest weight vector v_0. Using this we introduce $\mathbb{C}$-valued realization $\tilde{C}_\chi$ of the space C_χ by the formula:

$$\varphi(g) \equiv \langle v_0, \mathcal{F}(g) \rangle , \qquad (1.8)$$

where $\langle , \rangle$ is the M-invariant scalar product in V_μ. On these functions the right action of $\mathcal{G}^{\mathbb{C}}$ is defined by:

$$(\hat{X}\varphi)(g) \equiv \langle v_0, (\hat{X}\mathcal{F})(g) \rangle . \qquad (1.9)$$

Part of the main result of [5] is:

Proposition. The functions of the $\mathbb{C}$-valued realization $\tilde{C}_\chi$ of the ER C_χ satisfy :

$$\hat{X}\varphi = \Lambda(X) \cdot \varphi , \quad X \in \mathcal{H}^{\mathbb{C}} , \quad \Lambda \in (\mathcal{H}^{\mathbb{C}})^* \qquad (1.10a)$$
$$\hat{X}\varphi = 0 , \quad X \in \mathcal{G}_-^{\mathbb{C}} , \qquad (1.10b)$$

where $\Lambda = \Lambda(\chi)$ is built canonically from χ, [4] $\mathcal{G}_+^{\mathbb{C}}, \mathcal{G}_-^{\mathbb{C}}$, are the positive, negative root spaces of $\mathcal{G}^{\mathbb{C}}$, i.e., we use the standard triangular decomposition $\mathcal{G}^{\mathbb{C}} = \mathcal{G}_+^{\mathbb{C}} \oplus \mathcal{H}^{\mathbb{C}} \oplus \mathcal{G}_-^{\mathbb{C}}$.

Now we note that conditions (1.10) are the defining conditions for the lowest weight vector of a lowest weight module (LWM) over $\mathcal{G}^{\mathbb{C}}$ with lowest weight Λ. Moreover, special properties of a class of lowest weight modules, namely, Verma modules, are immediately related with the construction of invariant differential operators.

To be more specific let us recall that a Verma module is a lowest weight module V^Λ with lowest weight L, such that $V^\Lambda \cong U(\mathcal{G}_+^{\mathbb{C}})v_0$, where v_0 is the lowest weight vector, $U(\mathcal{G}_+^{\mathbb{C}})$ is the universal enveloping algebra of $\mathcal{G}_+^{\mathbb{C}}$.[5] Verma modules are universal in the following sense : every irreducible LWM is ismorphic to a factor-module of the Verma module with the same lowest weight.

Generically, Verma modules are irreducible, however, we shall be mostly interested in the reducible ones since these are relevant for the construction of differential equations. We recall the Bernstein-Gel'fand-Gel'fand [13] criterion according to which the Verma module V^Λ is reducible iff

$$2\langle \Lambda - \rho, \beta \rangle + m\langle \beta, \beta \rangle = 0 , \qquad (1.11)$$

holds for some $\beta \in \Delta^+$, $m \in \mathbb{N}$, where Δ^+ denotes the positive roots of the root system $(\mathcal{G}^{\mathbb{C}}, \mathcal{H}^{\mathbb{C}})$, ρ is half the sum of the positive roots Δ^+.[6] Whenever (1.11) is fulfilled there exists [11] in V^Λ a unique vector v_s, called *singular vector*, such that $v_s \neq v_0$ and it has the properties (1.10) of a lowest weight vector with shifted weight $\Lambda + m\beta$:

$$\hat{X}v_s = (\Lambda + m\beta)(X) \cdot v_s , \quad X \in \mathcal{H}^{\mathbb{C}} , \qquad (1.12a)$$
$$\hat{X}v_s = 0 , \quad X \in \mathcal{G}_-^{\mathbb{C}} , \qquad (1.12b)$$

---

[4] It contains all the information from χ, except about the character ϵ of the finite group M_d.

[5] For more mathematically precise definition, cf. [11], [12].

[6] Note that we are working here with lowest weights shifted by ρ with respect to the notation of [5].

The general structure of a singular vector is [5] :

$$v_s \;=\; P_{m\beta}(X_1^+,\ldots,X_\ell^+)v_0 \;, \tag{1.13}$$

where $P_{m\beta}$ is a homogeneous polynomial in its variables of degrees mk_i, where $k_i \in \mathbb{Z}_+$ come from the decomposition of β into simple roots: $\beta = \sum k_i\alpha_i$, $\alpha_i \in \Delta_S$, the system of simple roots, X_j^+ are the simple root vectors, $\ell = \operatorname{rank}\mathcal{G}^{\mathbb{C}}$.[7]

It is obvious that (1.13) satisfies (1.12a), while conditions (1.12b) fix the coefficients of $P_{m\beta}$ up to an overall multiplicative nonzero constant.

Now we are in a position to define the differential intertwining operators, corresponding to the singular vectors.

Let the signature χ of an ER be such that the corresponding $\Lambda = \Lambda(\chi)$ satisfies (1.11) for some $\beta \in \Delta^+$ and some $m \in \mathbb{N}$.[8] Then there exists an intertwining differential operator [5]

$$\mathcal{D}_{m\beta} \;:\; \tilde{\mathcal{C}}_\chi \;\longrightarrow\; \tilde{\mathcal{C}}_{\chi'} \;, \tag{1.14}$$

where χ' is such that $\Lambda' = \Lambda'(\chi') = \Lambda + m\beta$.

The important fact is that (1.14) is explicitly given by [5]

$$\mathcal{D}_{m\beta}\varphi(g) \;=\; P_{m\beta}(\hat{X}_1^+,\ldots,\hat{X}_\ell^+)\varphi(g) \;, \tag{1.15}$$

where $P_{m\beta}$ is the same polynomial as in (1.13) and $\hat{X}_j^+$ denotes the action (1.7).

One important simplification is that in order to check the intertwining properies of the operator in (1.15) it is enough to work with the infinitesimal versions of (1.5) and (1.6), i.e., work with representations of the Lie algebra. Thus, also in the quantum group setting we work with representations of quantum algebras.

This finishes the sketch of the classical results.

1.3. Now we comment on the application to quantum groups of the approach presented above. The analogue of (1.7) was alreay known [15]. The infinitesimal left regular representation (in [15] the anti-representation was used) and the right covariance (1.5) were introduced for $U_q(sl(2))$ [16], for a Lorentz quantum algebra [17] and for $U_q(sl(n))$ [18]. Here Sections 2. and partly 3. review part of the exposition of [18] for $U_q(sl(n))$ with general n. Then in Sections 4. and 5. we consider in detail the case $n = 4$ (the case $n = 3$ was considered in detail in [18]).

As an immediate application of these developments, presented here in Sections 2.-5., and of the results for the q-deformations, in particular, of the conformal algebra, we are able to use q - conformal invariance to propose new q - Minkowski space-time and q - Maxwell equations. The new q - Maxwell equations are q-conformal invariant and are the first members of an infinite new hierarchy of q - difference equations. We are using an indexless formulation in which all indices are traded for two conjugate variables, $z, \bar{z}$. The proposed new q - Minkowski coordinates together with $z, \bar{z}$ can be interpreted as the six local coordinates of a $SU_q(2,2)$ flag manifold. This was announced in [19] and is presented here in Section 6.

---

[7] A singular vector may also be written in terms of the full Cartan-Weyl basis of $\mathcal{G}_+^{\mathbb{C}}$.

[8] If β is a real root, (i.e., $\beta|_{\mathcal{H}_m^{\mathbb{C}}} = 0$, where $\mathcal{H}_m$ is the Cartan subalgebra of $\mathcal{M}$), then some conditions are imposed on the character ϵ representing the finite group M_d.[14]

2. THE MATRIX QUANTUM GROUP $GL_q(n)$ AND THE DUAL QUANTUM ALGEBRA

Let us consider an $n \times n$ quantum matrix M with non-commuting matrix elements a_{ij}, $1 \le i, j \le n$. The matrix quantum group $\mathcal{A}_g = GL_q(n)$, $q \in \mathbb{C}$, is generated by the matrix elements a_{ij} with the following commutation relations [20] $(\lambda = q - q^{-1})$:

$$a_{i\ell} a_{ij} = q a_{ij} a_{i\ell} , \quad \text{for} \ \ell > j , \tag{2.1a}$$

$$a_{kj} a_{ij} = q a_{ij} a_{kj} , \quad \text{for} \ k > i , \tag{2.1b}$$

$$a_{kj} a_{i\ell} = a_{i\ell} a_{kj} , \quad \text{for} \ k > i , \ \ell > j , \tag{2.1c}$$

$$a_{ij} a_{k\ell} = a_{k\ell} a_{ij} - \lambda a_{i\ell} a_{kj} , \quad \text{for} \ k > i , \ \ell > j . \tag{2.1d}$$

Considered as a bialgebra, it has the following comultiplication $\delta_{\mathcal{A}}$ and counit $\varepsilon_{\mathcal{A}}$:

$$\delta_{\mathcal{A}}(a_{ij}) = \sum_{k=1}^{n} a_{ik} \otimes a_{kj} , \quad \varepsilon_{\mathcal{A}}(a_{ij}) = \delta_{ij} . \tag{2.2}$$

This algebra has determinant D given by [20]:

$$D = \sum_{\rho \in S_n} \epsilon(\rho) \, a_{1,\rho(1)} \ldots a_{n,\rho(n)} = \sum_{\rho \in S_n} \epsilon(\rho) \, a_{\rho(1),1} \ldots a_{\rho(n),n} , \tag{2.3}$$

where summations are over all permutations ρ of $\{1, \ldots, n\}$ and the quantum signature is:

$$\epsilon(\rho) = \prod_{\substack{j < k \\ \rho(j) > \rho(k)}} (-q^{-1}) . \tag{2.4}$$

The determinant obeys [20]:

$$\delta_{\mathcal{A}}(D) = D \otimes D , \quad \varepsilon_{\mathcal{A}}(D) = 1 . \tag{2.5}$$

The determinant is central, i.e., it commutes with the elements a_{ik} [20]:

$$a_{ik} \, D = D \, a_{ik} . \tag{2.6}$$

Further, if $D \neq 0$ one extends the algebra by an element D^{-1} which obeys [20]:

$$DD^{-1} = D^{-1}D = 1_{\mathcal{A}} . \tag{2.7}$$

Next one defines the left and right quantum cofactor matrix A_{ij} [20]:

$$
\begin{aligned}
A_{ij} &= \sum_{\rho(i)=j} \frac{\epsilon(\rho \circ \sigma_i)}{\epsilon(\sigma_i)} \, a_{1,\rho(1)} \ldots \widehat{a}_{ij} \ldots a_{n,\rho(n)} = \\
&= \sum_{\rho(j)=i} \frac{\epsilon(\rho \circ \sigma'_j)}{\epsilon(\sigma'_j)} \, a_{\rho(1),1} \ldots \widehat{a}_{ij} \ldots a_{\rho(n),n} ,
\end{aligned}
\tag{2.8}
$$

where σ_i and σ'_j denote the cyclic permutations:

$$\sigma_i = \{i, \ldots, 1\} , \quad \sigma'_j = \{j, \ldots, n\} , \tag{2.9}$$

and the notation $\hat{x}$ indicates that x is to be omited. Now one can show that [20]:

$$\sum_j a_{ij}\, A_{\ell j} \;=\; \sum_j A_{ji}\, a_{j\ell} \;=\; \delta_{i\ell}\, D \;, \tag{2.10}$$

and obtain the left and right inverse [20]:

$$M^{-1} \;=\; D^{-1}\, A \;=\; A\, D^{-1} \;. \tag{2.11}$$

Thus, one can introduce the antipode in $GL_q(n)$ [20] :

$$\gamma_A(a_{ij}) \;=\; D^{-1}\, A_{ji} \;=\; A_{ji}\, D^{-1} \;. \tag{2.12}$$

Next we introduce a basis of $GL_q(n)$ which consists of monomials

$$f \;=\; (a_{21})^{p_{21}} \ldots (a_{n,n-1})^{p_{n,n-1}} (a_{11})^{\ell_1} \ldots (a_{nn})^{\ell_n} (a_{n-1,n})^{n_{n-1,n}} \ldots (a_{12})^{n_{12}} \;=$$

$$=\; f_{\bar{\ell},\bar{p},\bar{n}} \;, \tag{2.13}$$

where $\bar{\ell}, \bar{p}, \bar{n}$ denote the sets $\{\ell_i\}$, $\{p_{ij}\}$, $\{n_{ij}\}$, resp., $\ell_i, p_{ij}, n_{ij} \in \mathbb{Z}_+$ and we have used the so-called normal ordering of the elements a_{ij}. Namely, we first put the elements a_{ij} with $i > j$ in lexicographic order, i.e., if $i < k$ then $a_{ij}\ (i > j)$ is before $a_{k\ell}\ (k > \ell)$ and $a_{ti}\ (t > i)$ is before $a_{tk}\ (t > k)$; then we put the elements a_{ii}; finally we put the elements a_{ij} with $i < j$ in antilexicographic order, i.e., if $i > k$ then a_{ij} $(i < j)$ is before $a_{k\ell}\ (k < \ell)$ and $a_{ti}\ (t < i)$ is before $a_{tk}\ (t < k)$. Note that the basis (2.13) icludes also the unit element $1_{\mathcal{A}_g}$ of $\mathcal{A}_g$ when all $\{\ell_i\}$, $\{p_{ij}\}$, $\{n_{ij}\}$ are equal to zero, i.e.:

$$f_{\bar{0},\bar{0},\bar{0}} \;=\; 1_{\mathcal{A}_g} \;. \tag{2.14}$$

We need the dual algebra of $GL_q(n)$. This is the algebra $\mathcal{U}_g = U_q(sl(n)) \otimes U_q(\mathcal{Z})$, where $U_q(\mathcal{Z})$ is central in $\mathcal{U}_g$ [21], [22]. Let us denote the Chevalley generators of $sl(n)$ by H_i, $X_i^{\pm}$, $i = 1,\ldots,n-1$. Then we take for the 'Chevalley' generators of $\mathcal{U} = U_q(sl(n))$: $k_i = q^{H_i/2}$, $k_i^{-1} = q^{-H_i/2}$, $X_i^{\pm}$, $i = 1,\ldots,n-1$, with the following *algebra* relations:

$$k_i k_j \;=\; k_j k_i \;, \quad k_i k_i^{-1} \;=\; k_i^{-1} k_i \;=\; 1_{\mathcal{U}_g} \;, \quad k_i X_j^{\pm} \;=\; q^{\pm c_{ij}} X_j^{\pm} k_i \tag{2.15a}$$

$$[X_i^+, X_j^-] \;=\; \delta_{ij}\left(k_i^2 - k_i^{-2}\right)/\lambda \;, \tag{2.15b}$$

$$\left(X_i^{\pm}\right)^2 X_j^{\pm} - [2]_q X_i^{\pm} X_j^{\pm} X_i^{\pm} + X_j^{\pm}\left(X_i^{\pm}\right)^2 \;=\; 0 \;, \quad |i-j| = 1 \;, \tag{2.15c}$$

$$[X_i^{\pm}, X_j^{\pm}] \;=\; 0 \;, \quad |i-j| \neq 1 \;, \tag{2.15d}$$

where c_{ij} is the Cartan matrix of $sl(n)$, and *coalgebra* relations :

$$\delta_{\mathcal{U}}(k_i^{\pm}) \;=\; k_i^{\pm} \otimes k_i^{\pm} \;, \tag{2.16a}$$

$$\delta_{\mathcal{U}}(X_i^{\pm}) \;=\; X_i^{\pm} \otimes k_i + k_i^{-1} \otimes X_i^{\pm} \;, \tag{2.16b}$$

$$\varepsilon_{\mathcal{U}}(k_i^{\pm}) \;=\; 1 \;, \quad \varepsilon_{\mathcal{U}}(X_i^{\pm}) \;=\; 0 \;, \tag{2.16c}$$

$$\gamma_{\mathcal{U}}(k_i) \;=\; k_i^{-1} \;, \quad \gamma_{\mathcal{U}}(X_i^{\pm}) \;=\; -q^{\pm 1} X_i^{\pm} \;, \tag{2.16d}$$

where $k_i^+ = k_i$, $k_i^- = k_i^{-1}$. Further, we denote the generator of $\mathcal{Z}$ by H and the generators of $U_q(\mathcal{Z})$ by $k = q^{H/2}$, $k^{-1} = q^{-H/2}$, $kk^{-1} = k^{-1}k = 1_{\mathcal{U}_g}$. The generators k, k^{-1} commute with the generators of $\mathcal{U}$, and their coalgebra relations are as those of any k_i. From now on we shall give most formulae only for the generators

k_i, $X_i^{\pm}$, k, since the analogous formulae for k_i^{-1}, k^{-1} follow trivially from those for k_i, k, resp.

The bilinear form giving the duality between $\mathcal{U}_g$ and $\mathcal{A}_g$ is given by [18]:

$$\langle\, k_i\,,\,a_{j\ell}\,\rangle \;=\; \delta_{j\ell}\, q^{(\delta_{ij}-\delta_{i,j+1})/2}\,, \tag{2.17a}$$

$$\langle\, X_i^+\,,\,a_{j\ell}\,\rangle \;=\; \delta_{j+1,\ell}\delta_{ij}\,, \tag{2.17b}$$

$$\langle\, X_i^-\,,\,a_{j\ell}\,\rangle \;=\; \delta_{j-1,\ell}\delta_{i\ell}\,, \tag{2.17c}$$

$$\langle\, k\,,\,a_{j\ell}\,\rangle \;=\; \delta_{j\ell}\, q^{1/2}\,. \tag{2.17d}$$

The pairing between arbitrary elements of $\mathcal{U}_g$ and f follows then from the properties of the duality pairing. The pairing (2.17) is standardly supplemented with

$$\langle\, y\,,\,1_{\mathcal{A}_g}\,\rangle \;=\; \varepsilon_{\mathcal{U}_g}(y)\,. \tag{2.18}$$

It is well know that the pairing provides the fundamental representation of $\mathcal{U}_g$:

$$F(y)_{j\ell} \;=\; \langle\, y\,,\,a_{j\ell}\,\rangle\,, \quad y \;=\; k_i, X_i^{\pm}, k\,. \tag{2.19}$$

Of course, $F(k) = q^{1/2}I_n$, where I_n is the unit $n \times n$ matrix.

3. REPRESENTATIONS OF $\mathcal{U}_g$ AND $\mathcal{U}$

We begin by defining *two actions* of the dual algebra $\mathcal{U}_g$ on the basis (2.13) of $\mathcal{A}_g$.

First we introduce the *left regular representation* of $\mathcal{U}_g$ which in the $q = 1$ case is the infinitesimal version of :

$$\pi(Y)\, M \;=\; Y^{-1}\, M\,, \quad Y, M \in GL(n)\,. \tag{3.1}$$

Explicitly, we define the action of $\mathcal{U}_g$ as follows (cf. also (1.5)):

$$\pi(y)\, a_{i\ell} \;\doteq\; \left(F\left(y^{-1}\right) M\right)_{i\ell} \;=\; \sum_j F\left(y^{-1}\right)_{ij} a_{j\ell} \;=\; \sum_j \langle\, y^{-1}\,,\,a_{ij}\,\rangle\, a_{j\ell}\,, \tag{3.2}$$

where y denotes the generators of $\mathcal{U}_g$ and y^{-1} is symbolic notation, the possible pairs being given explicitly by:

$$(y, y^{-1}) \;=\; (k_i, k_i^{-1}),\; (X_i^{\pm}, -X_i^{\pm}),\; (k, k^{-1})\,. \tag{3.3}$$

From (3.2) we find the explicit action of the generators of $\mathcal{U}_g$:

$$\pi(k_i)\, a_{j\ell} \;=\; q^{(\delta_{i+1,j}-\delta_{ij})/2}\, a_{j\ell}\,, \tag{3.4a}$$

$$\pi(X_i^+)\, a_{j\ell} \;=\; -\delta_{ij}\, a_{j+1\ell}\,, \tag{3.4b}$$

$$\pi(X_i^-)\, a_{j\ell} \;=\; -\delta_{i+1,j}\, a_{j-1\ell}\,, \tag{3.4c}$$

$$\pi(k)\, a_{j\ell} \;=\; q^{-1/2}\, a_{j\ell}\,. \tag{3.4d}$$

The above is supplemented with the following action on the unit element of $\mathcal{A}_g$:

$$\pi(k_i)\, 1_{\mathcal{A}_g} \;=\; 1_{\mathcal{A}_g}\,, \quad \pi(X_i^{\pm})\, 1_{\mathcal{A}_g} \;=\; 0\,, \quad \pi(k)\, 1_{\mathcal{A}_g} \;=\; 1_{\mathcal{A}_g}\,. \tag{3.5}$$

In order to derive the action of $\pi(y)$ on arbitrary elements of the basis (2.13), we use the twisted derivation rule consistent with the coproduct and the representation structure, namely, we take: $\pi(y)\varphi\psi \;=\; \pi(\delta'_{\mathcal{U}_g}(y))(\varphi \otimes \psi)$, where $\delta'_{\mathcal{U}_g} \;=\; \sigma \circ \delta_{\mathcal{U}_g}$ is the opposite coproduct, (σ is the permutation operator). Thus, we have:

$$\pi(k_i)\varphi\psi \;=\; \pi(k_i)\varphi \cdot \pi(k_i)\psi\,, \tag{3.6a}$$

$$\pi(X_i^{\pm})\varphi\psi \;=\; \pi(X_i^{\pm})\varphi \cdot \pi(k_i^{-1})\psi \;+\; \pi(k_i)\varphi \cdot \pi(X_i^{\pm})\psi\,, \tag{3.6b}$$

$$\pi(k)\varphi\psi \;=\; \pi(k)\varphi \cdot \pi(k)\psi\,. \tag{3.6c}$$

From now on we suppose that q is not a nontrivial root of unity. Applying the above rules one obtains:

$$\pi(k_i)\, (a_{j\ell})^n \;=\; q^{n(\delta_{i+1,j}-\delta_{ij})/2}\, (a_{j\ell})^n\,, \tag{3.7a}$$

$$\pi(X_i^{+})\, (a_{j\ell})^n \;=\; -\delta_{ij}\, c_n\, (a_{j\ell})^{n-1}\, a_{j+1\ell}\,, \tag{3.7b}$$

$$\pi(X_i^{-})\, (a_{j\ell})^n \;=\; -\delta_{i+1,j}\, c_n\, a_{j-1\ell}\, (a_{j\ell})^{n-1}\,, \tag{3.7c}$$

$$\pi(k)\, (a_{j\ell})^n \;=\; q^{-n/2}\, (a_{j\ell})^n\,, \tag{3.7d}$$

where

$$c_n \;=\; q^{(n-1)/2}\, [n]_q\,, \quad [n]_q = (q^n - q^{-n})/\lambda\,. \tag{3.8}$$

Note that (3.5) and (3.4) are partial cases of (3.7) for $n = 0$ and $n = 1$ resp. (cf. (2.14)).

Analogously, we introduce the *right action* (see also [15]) which in the classical case is the infinitesimal counterpart of :

$$\pi_R(Y)\, M \;=\; M\, Y\,, \quad Y, M \in GL(n)\,. \tag{3.9}$$

Thus, we define the right action of $\mathcal{U}_g$ as follows (cf. (1.7)):

$$\pi_R(y)\, a_{i\ell} \;=\; (MF(y))_{i\ell} \;=\; \sum_j a_{ij}\, F(y)_{j\ell} \;=\; \sum_j a_{ij}\, \langle\, y\,,\, a_{j\ell}\,\rangle\,, \tag{3.10}$$

where y denotes the generators of $\mathcal{U}_g$.

From (3.10) we find the explicit right action of the generators of $\mathcal{U}_g$:

$$\pi_R(k_i)\, a_{j\ell} \;=\; q^{(\delta_{i\ell}-\delta_{i+1,\ell})/2}\, a_{j\ell}\,, \tag{3.11a}$$

$$\pi_R(X_i^{+})\, a_{j\ell} \;=\; \delta_{i+1,\ell}\, a_{j,\ell-1}\,, \tag{3.11b}$$

$$\pi_R(X_i^{-})\, a_{j\ell} \;=\; \delta_{i\ell}\, a_{j,\ell+1}\,, \tag{3.11c}$$

$$\pi_R(k)\, a_{j\ell} \;=\; q^{1/2}\, a_{j\ell}\,, \tag{3.11d}$$

supplemented by the right action on the unit element:

$$\pi_R(k_i)\, 1_{\mathcal{A}_g} \;=\; 1_{\mathcal{A}_g}\,, \quad \pi_R(X_i^{\pm})\, 1_{\mathcal{A}_g} \;=\; 0\,, \quad \pi_R(k)\, 1_{\mathcal{A}_g} \;=\; 1_{\mathcal{A}_g}\,. \tag{3.12}$$

The twisted derivation rule is now given by $\pi_R(y)\varphi\psi = \pi_R(\delta u_g(y))(\varphi \otimes \psi)$, i.e.,

$$\pi_R(k_i)\varphi\psi = \pi_R(k_i)\varphi \cdot \pi_R(k_i)\psi \ , \qquad (3.13a)$$

$$\pi_R(X_i^{\pm})\varphi\psi = \pi_R(X_i^{\pm})\varphi \cdot \pi_R(k_i)\psi + \pi_R(k_i^{-1})\varphi \cdot \pi_R(X_i^{\pm})\psi \ , \qquad (3.13b)$$

$$\pi_R(k)\varphi\psi = \pi_R(k)\varphi \cdot \pi_R(k)\psi \ , \qquad (3.13c)$$

Using this, we find:

$$\pi_R(k_i)\,(a_{j\ell})^n = q^{n(\delta_{i\ell}-\delta_{i+1,\ell})/2}\,(a_{j\ell})^n \ , \qquad (3.14a)$$

$$\pi_R(X_i^{+})\,(a_{j\ell})^n = \delta_{i+1,\ell}\,c_n\,a_{j,\ell-1}\,(a_{j\ell})^{n-1} \ , \qquad (3.14b)$$

$$\pi_R(X_i^{-})\,(a_{j\ell})^n = \delta_{i\ell}\,c_n\,(a_{j\ell})^{n-1}\,a_{j,\ell+1} \ , \qquad (3.14c)$$

$$\pi_R(k)\,(a_{j\ell})^n = q^{n/2}\,(a_{j\ell})^n \ . \qquad (3.14d)$$

Let us now introduce the elements φ as formal power series of the basis (2.13):

$$\varphi = \sum_{\bar{\ell},\bar{m},\bar{n}\in\mathbb{Z}_+} \mu_{\bar{\ell},\bar{m},\bar{n}}\,(a_{21})^{m_{21}}\ldots(a_{n,n-1})^{m_{n,n-1}}(a_{11})^{\ell_1}\ldots(a_{nn})^{\ell_n}\times$$

$$\times\,(a_{n-1,n})^{n_{n-1,n}}\ldots(a_{12})^{n_{12}} \ . \qquad (3.15)$$

By (3.7) and (3.14) we have defined left and right action of $\mathcal{U}_g$ on φ. As in the classical case the left and right actions commute, and as in [5] we shall use the right covariance to reduce the left regular representation. In particular, we would like the right action to mimic some properties of a highest weight module, i.e., annihilation by the raising generators X_i^+ and scalar action by the (exponents of the) Cartan operators k_i, k. However, first we have to make a change of basis using the q-analogue of the classical Gauss decomposition. For this we have to suppose that the principal minor determinants of M :

$$D_m = \sum_{\rho\in S_m} \epsilon(\rho)\,a_{1,\rho(1)}\ldots a_{m,\rho(m)} =$$

$$= \sum_{\rho\in S_m} \epsilon(\rho)\,a_{\rho(1),1}\ldots a_{\rho(m),m} \ , \qquad m \le n \ , \qquad (3.16)$$

are invertible; note that $D_n = D$, $D_{n-1} = A_{nn}$.

Further, for the ordered sets $I = \{i_1 < \cdots < i_r\}$ and $J = \{j_1 < \cdots < j_r\}$, let ξ_J^I be the r-minor determinant with respect to rows I and columns J such that

$$\xi_J^I = \sum_{\rho\in S_r} \epsilon(\rho)\,a_{i_{\rho(1)}j_1}\cdots a_{i_{\rho(r)}j_r} \ . \qquad (3.17)$$

Note that $\xi_{1\,\ldots\,i}^{1\,\ldots\,i} = D_i$. Then one has [23] $(i,j,\ell = 1,\ldots,n)$:

$$a_{i\ell} = \sum_j B_{ij}Z_{j\ell} \ , \qquad B_{i\ell} = \xi_{1\,\ldots\,\ell}^{1\,\ldots\,\ell-1\,i}\,D_{\ell-1}^{-1} \ , \qquad Z_{i\ell} = D_i^{-1}\xi_{1\,\ldots\,i-1\,\ell}^{1\,\ldots\,i} \ , \qquad (3.18)$$

$B_{i\ell} = 0$ for $i < \ell$, $Z_{i\ell} = 0$ for $i > \ell$, (which follows from the obvious extension of (3.17) to the case when I, resp. J, is not ordered). Then Z_{ij}, $i < j$, may be regarded as a q-analogue of local coordinates of the flag manifold $B\backslash GL(n)$.

For our purposes we need a refinement of this decomposition :

$$B_{i\ell} \;=\; \tilde{Y}_{i\ell} D_{\ell\ell}\,, \quad \tilde{Y}_{i\ell} \;=\; \xi_{1\cdots\ell}^{1\cdots\ell-1\,i}\, D_\ell^{-1}\,, \quad D_{\ell\ell} \;=\; D_\ell D_{\ell-1}^{-1}\,, \quad (D_0 \equiv 1_{\mathcal{A}_g})\,, \quad (3.19)$$

where $\tilde{Y}_{j\ell},\, j > \ell$, may be regarded as a q-analogue of local coordinates of the flag manifold $GL(n)/DZ$.

Clearly, we can replace the basis (2.13) of $\mathcal{A}_g$ with a basis in terms of $\tilde{Y}_{i\ell},\, i > \ell$, D_ℓ, $Z_{i\ell},\, i < \ell$. (Note that $\tilde{Y}_{ii} = Z_{ii} = 1_{\mathcal{A}_g}$.) Thus, we consider formal power series:

$$\varphi \;=\; \sum_{\substack{\bar m, \bar n \in \mathbb{Z}_+ \\ \bar\ell \in \mathbb{Z}}} \mu'_{\bar\ell,\bar m,\bar n}\, (\tilde{Y}_{21})^{m_{21}} \dots (\tilde{Y}_{n,n-1})^{m_{n,n-1}} (D_1)^{\ell_1} \dots (D_n)^{\ell_n} \times$$
$$\times\, (Z_{n-1,n})^{n_{n-1,n}} \dots (Z_{12})^{n_{12}}\,. \tag{3.20}$$

Now, let us impose right covariance (cf. [5] and (1.10b)) with respect to X_i^+, [9] i.e., we require:

$$\pi_R(X_i^+)\, \varphi \;=\; 0\,. \tag{3.21}$$

First we notice that:

$$\pi_R(X_i^+)\, \xi_J^I \;=\; 0\,, \quad \text{for } J = \{1,\dots,j\}\,, \; \forall\, I\,, \tag{3.22}$$

from which follow:

$$\pi_R(X_i^+)\, D_j \;=\; 0\,, \quad \pi_R(X_i^+)\, \tilde{Y}_{j\ell} \;=\; 0\,. \tag{3.23}$$

On the other hand $\pi_R(X_i^+)$ acts nontrivially on $Z_{j\ell}$:

$$\pi_R(X_i^+)\, Z_{j\ell} \;=\; \delta_{i+1,\ell}\, q^{\delta_{ij}/2}\, Z_{j,\ell-1}\,. \tag{3.24}$$

Thus, (3.21) simply means that our functions φ do not depend on $Z_{j\ell}$. Thus, the functions obeying (3.21) are:

$$\varphi \;=\; \sum_{\bar\ell \in \mathbb{Z}\,,\; \bar m \in \mathbb{Z}_+} \mu_{\bar\ell,\bar m}\, (\tilde{Y}_{21})^{m_{21}} \dots (\tilde{Y}_{n,n-1})^{m_{n,n-1}} (D_1)^{\ell_1} \dots (D_n)^{\ell_n}\,. \tag{3.25}$$

Next, we impose right covariance with respect to k_i, k :

$$\pi_R(k_i)\, \varphi \;=\; q^{r_i/2}\, \varphi\,, \tag{3.26a}$$
$$\pi_R(k)\, \varphi \;=\; q^{\hat r/2}\, \varphi\,, \tag{3.26b}$$

where $r_i, \hat r$ are parameters to be specified below. On the other hand using (3.13a, c), (3.14a, c) we have:

$$\pi_R(k_i)\, \xi_J^I \;=\; q^{\delta_{ij}/2}\, \xi_J^I\,, \quad \pi_R(k)\, \xi_J^I \;=\; q^{j/2}\, \xi_J^I\,, \quad \text{for } J = \{1,\dots,j\}\,, \; \forall\, I\,, \tag{3.27}$$

from which follows:

$$\pi_R(k_i)\, D_j \;=\; q^{\delta_{ij}/2}\, D_j\,, \quad \pi_R(k)\, D_j \;=\; q^{j/2}\, D_j\,, \tag{3.28a}$$
$$\pi_R(k_i)\, \tilde{Y}_{j\ell} \;=\; \tilde{Y}_{j\ell}\,, \quad \pi_R(k)\, \tilde{Y}_{j\ell} \;=\; \tilde{Y}_{j\ell}\,, \tag{3.28b}$$

---

[9] Note that here we shall work with *highest* weight modules instead of lowest weight modules used in [5] and here in the Introduction. This is just a matter of choice.

and thus we have:

$$\pi_R(k_i)\,\varphi \;=\; q^{\ell_i/2}\,\varphi\;, \tag{3.29a}$$

$$\pi_R(k)\,\varphi \;=\; q^{\sum_{j=1}^{n} j\ell_j/2}\,\varphi\;. \tag{3.29b}$$

Remark: For $q = 1$ the elementary representations (in particular, the right covariance conditions) for a complex semisimple Lie group G_c are given by (instead of (1.5)) [24] :

$$\mathcal{C}_{\Lambda,\Lambda'} \;=\; \{\mathcal{F} \in C^{\infty}(G_c) \mid \mathcal{F}(gxn) = e^{\Lambda(X)+\Lambda'(\bar{X})}\cdot\mathcal{F}(g)\}\;,\quad \Lambda(X)-\Lambda'(X)\in \mathbb{Z}\;, \tag{3.30}$$

where $x = \exp(X)$, $X \in \mathcal{H}_c$, $n \in G_c^{+} = \exp(\mathcal{G}_c^{+})$, using the triangular (Gauss) decomposition $\mathcal{G}_c = \mathcal{G}_c^{+}\mathcal{H}_c\mathcal{G}_c^{-}$ of the Lie algebra $\mathcal{G}_c$ of G_c, and the last condition in (3.30) is necessary to ensure uniqueness on the Cartan subgroup $H_c = \exp(\mathcal{H}_c)$ of G_c. In the quantum group setting above, for simplicity, we are using holomorphic representations for which $\Lambda' = 0$. For $U_q(sl(2))$ with $\Lambda' \neq 0$ we refer to [17] where this construction was carried out for a q-deformed Lorentz algebra.

Comparing right covariance conditions (3.26) with the direct calculations (3.29) we obtain $\ell_i = r_i$, for $i < n$, $\sum_{j=1}^{n} j\ell_j = \hat{r}$. This means that $r_i, \hat{r} \in \mathbb{Z}$ and that there is no summation in ℓ_i, also $\ell_n = (\hat{r} - \sum_{i=1}^{n-1} ir_i)/n$.

Thus, the reduced functions obeying (3.21) and (3.26) are:

$$\varphi \;=\; \sum_{\bar{m}\in\mathbb{Z}_{+}} \mu_{\bar{m}}\,(\tilde{Y}_{21})^{m_{21}}\ldots(\tilde{Y}_{n,n-1})^{m_{n,n-1}}(D_1)^{r_1}\ldots(D_{n-1})^{r_{n-1}}(D_n)^{\hat{\ell}}\;, \tag{3.31}$$

where $\hat{\ell} = (\hat{r} - \sum_{i=1}^{n-1} ir_i)/n$.

Next we would like to derive the $\mathcal{U}_g$ - action π on φ. First, we notice that $\mathcal{U}$ acts trivially on $D_n = D$:

$$\pi(X_i^{\pm})\,D \;=\; 0\;,\quad \pi(k_i)\,D \;=\; D\;. \tag{3.32}$$

Then we note:

$$\pi(k)\,D_j \;=\; q^{-j/2}\,D_j\;,\quad \pi(k)\,\tilde{Y}_{j\ell} \;=\; \tilde{Y}_{j\ell}\;, \tag{3.33}$$

from which follows:

$$\pi(k)\,\varphi \;=\; q^{-\hat{r}/2}\,\varphi\;. \tag{3.34}$$

Thus, the action of $\mathcal{U}$ involves only the parameters r_i, $i < n$, while the action of $U_q(\mathcal{Z})$ involves only the parameter $\hat{r}$. Thus we can consistently also from the representation theory point of view restrict to the matrix quantum group $SL_q(n)$, i.e., we set:

$$D \;=\; D^{-1} \;=\; 1_{\mathcal{A}_g}\;. \tag{3.35}$$

Then the dual algebra is $\mathcal{U} = U_q(sl(n))$. This is justified as in the $q = 1$ case [5] since for our considerations only the semisimple part of the algebra is important. (This would not be possible for the multiparameter deformation of $GL(n)$ [25], [26], since there D is not central. Nevertheless, we expect most of the essential features of our approach to be preserved since the dual algebra can be transformed as a commutation algebra to the one-parameter $\mathcal{U}_g$, with the extra parameters entering only the co-algebra structure [21], [22].)

Thus, the reduced functions for the $\mathcal{U}$ action are:

$$\hat{\varphi}(\bar{Y},\bar{D}) \;=\; \sum_{\bar{m}\in\mathbb{Z}_+} \mu_{\bar{m}}\,(\tilde{Y}_{21})^{m_{21}}\ldots(\tilde{Y}_{n,n-1})^{m_{n,n-1}}\;\times$$

$$\times\;(D_1)^{r_1}\ldots(D_{n-1})^{r_{n-1}} \;= \qquad (3.36a)$$

$$=\;\hat{\varphi}(\bar{Y})\,(D_1)^{r_1}\ldots(D_{n-1})^{r_{n-1}}\;, \qquad (3.36b)$$

where $\bar{Y},\bar{D}$ denote the variables $\tilde{Y}_{il}$, $i>\ell$, D_i, $i<n$.

Further we note the commutation relations of the $\tilde{Y}_{ij}$ and D_i variables:

$$\tilde{Y}_{i\ell}\tilde{Y}_{ij} \;=\; q\tilde{Y}_{ij}\tilde{Y}_{i\ell}\;,\quad i>\ell>j\;, \qquad (3.37a)$$

$$\tilde{Y}_{kj}\tilde{Y}_{ij} \;=\; q\tilde{Y}_{ij}\tilde{Y}_{kj}\;,\quad k>i>j\;, \qquad (3.37b)$$

$$\tilde{Y}_{kj}\tilde{Y}_{i\ell} \;=\; \tilde{Y}_{i\ell}\tilde{Y}_{kj}\;,\quad k>i>\ell>j\;, \qquad (3.37c)$$

$$\tilde{Y}_{k\ell}\tilde{Y}_{ij} \;=\; \tilde{Y}_{ij}\tilde{Y}_{k\ell}+\lambda\tilde{Y}_{i\ell}\tilde{Y}_{kj}\;,\quad k>i,\ell>j\;,\quad i\neq\ell\;, \qquad (3.37d)$$

$$\tilde{Y}_{ki}\tilde{Y}_{ij} \;=\; q^{-1}\tilde{Y}_{ij}\tilde{Y}_{ki}+q^{-1}\lambda\tilde{Y}_{kj}\;,\quad k>i>j\;, \qquad (3.37e)$$

$$Y_{j\ell}D_i \;=\; D_iY_{j\ell}\;,\quad j>\ell>i\;, \qquad (3.38a)$$

$$Y_{j\ell}D_i \;=\; qD_iY_{j\ell}\;,\quad j>i\geq\ell\;, \qquad (3.38b)$$

$$Y_{j\ell}D_i \;=\; D_iY_{j\ell}\;,\quad i\geq j>\ell\;, \qquad (3.38c)$$

where in $(3.37d)$ we use $\tilde{Y}_{i\ell}=0$ when $i<\ell$. Note that $(3.37a-d)$ may be obtained by replacing $a_{i\ell}$ with $\tilde{Y}_{i\ell}$ in $(2.1a-d)$. Note that the structure of the q - flag manifold for general n is exibited already for $n=4$, while for $n=3$ relations $(3.37c,d)$ are not present. The commutation relations between the Z and D variables are obtained from (3.37), (3.38), by just replacing Y_{st} by Z_{ts} in all formuale.

Note that for real q the q - flag manifold is invariant under the anti-linear anti-involution $\tilde{\omega}$ acting as:

$$\tilde{\omega}(\tilde{Y}_{j\ell}) \;=\; \tilde{Y}_{n+1-\ell,n+1-j}\;\cdot \qquad (3.39)$$

Thus it can be considered as a q - flag manifold of the quantum group $SU_q([(n+1)/2]_{\text{int}},[n/2]_{\text{int}})$, where $[x/2]_{\text{int}}$ is the biggest integer number not greater than x. The same invariance holds for the Z coordinate q - flag manifold.

Next we calculate:

$$\pi(k_i)\,D_j \;=\; q^{-\delta_{ij}/2}\,D_j\;, \qquad (3.40a)$$

$$\pi(X_i^+)\,D_j \;=\; -\delta_{ij}\,\tilde{Y}_{j+1,j}\,D_j\;, \qquad (3.40b)$$

$$\pi(X_i^-)\,D_j \;=\; 0\;, \qquad (3.40c)$$

$$\pi(k_i)\,\tilde{Y}_{j\ell} \;=\; q^{\frac{1}{2}(\delta_{i+1,j}-\delta_{ij}-\delta_{i+1,\ell}+\delta_{i\ell})}\,\tilde{Y}_{j\ell} \qquad (3.41a)$$

$$\pi(X_i^+)\,\tilde{Y}_{j\ell} \;=\; -\delta_{ij}\,\tilde{Y}_{j+1,\ell}+\delta_{i\ell}\,q^{1-\delta_{j,\ell+1}/2}\,\tilde{Y}_{\ell+1,\ell}\,\tilde{Y}_{j\ell}\;+$$

$$+\;\delta_{i+1,\ell}\left(q^{-1}\,\tilde{Y}_{j,\ell-1}-\tilde{Y}_{\ell,\ell-1}\,\tilde{Y}_{j\ell}\right)\;, \qquad (3.41b)$$

$$\pi(X_i^-)\,\tilde{Y}_{j\ell} \;=\; -\delta_{i+1,j}\,q^{-\delta_{i\ell}/2}\,\tilde{Y}_{j-1,\ell}\;\cdot \qquad (3.41c)$$

These results have the important consequence that the degrees of the variables D_j are not changed by the action of $\mathcal{U}$. Thus, the parameters r_i indeed characterize the action of $\mathcal{U}$, i.e., we have obtained representations of $\mathcal{U}$. To obtain this representation more explicitly one just applies (3.40), (3.41) to the basis in (3.36) using (3.6). In particular, we have:

$$\pi(k_i)\,(D_j)^n \;=\; q^{-n\delta_{ij}/2}\,(D_j)^n\;,\quad n\in\mathbb{Z}\;,\tag{3.42a}$$

$$\pi(X_i^+)\,(D_j)^n \;=\; -\,\delta_{ij}\,\bar{c}_n\,\tilde{Y}_{j+1,j}\,(D_j)^n\;,\quad n\in\mathbb{Z}\;,\tag{3.42b}$$

$$\pi(X_i^-)\,(D_j)^n \;=\; 0\;,\quad n\in\mathbb{Z}\;,\tag{3.42c}$$

$$\pi(k_i)\,(\tilde{Y}_{j\ell})^n \;=\; q^{\frac{n}{2}(\delta_{i+1,j}-\delta_{ij}-\delta_{i+1,\ell}+\delta_{i\ell})}\,(\tilde{Y}_{j\ell})^n\;,\quad n\in\mathbb{Z}_+\;,\tag{3.43a}$$

$$
\begin{aligned}
\pi(X_i^+)\,(\tilde{Y}_{j\ell})^n \;=\;& -\,\delta_{ij}\,c_n\,(\tilde{Y}_{j\ell})^{n-1}\,\tilde{Y}_{j+1,\ell} \;+\\[2mm]
& +\,\delta_{i+1,\ell}\,\bar{c}_n\,\Big(q^{-1}\,\tilde{Y}_{j,\ell-1}\,(\tilde{Y}_{j\ell})^{n-1} \;-\; \tilde{Y}_{\ell,\ell-1}\,(\tilde{Y}_{j\ell})^{n}\Big) \;+\\[2mm]
& +\,\delta_{i\ell}\,q^{1-n\delta_{j,\ell+1}/2}\,c_n\,\tilde{Y}_{\ell+1,\ell}\,(\tilde{Y}_{j\ell})^{n}\;,\quad n\in\mathbb{Z}_+
\end{aligned}
\tag{3.43b}
$$

$$\pi(X_i^-)\,(\tilde{Y}_{j\ell})^n \;=\; -\,\delta_{i+1,j}\,q^{-\delta_{j,\ell+1}n/2}\,c_n\,\tilde{Y}_{j-1,\ell}\,(\tilde{Y}_{j\ell})^{n-1}\;,\quad n\in\mathbb{Z}_+\;,\tag{3.43c}$$

where

$$\bar{c}_n \;=\; q^{(1-n)/2}\,[n]_q\;.\tag{3.44}$$

It is easy to check that $\pi_{\bar{r}}$ satisfy (2.15).

We shall denote by $\mathcal{C}_{\bar{r}}$ the representation space of functions in (3.36) which have covariance properties (3.21), (3.26a). The representation acting in $\mathcal{C}_{\bar{r}}$ we denote by $\tilde{\pi}_{\bar{r}}$ doing also a renormalization to simplify things later, namely, we set:

$$\tilde{\pi}_{\bar{r}}(k_i) \;=\; \pi(k_i)\;,\qquad \tilde{\pi}_{\bar{r}}(X_i^{\pm}) \;=\; q^{\pm(r_i-1)/2}\,\pi(X_i^{\pm})\;.\tag{3.45}$$

Then $\tilde{\pi}_{\bar{r}}$ also satisfy (2.15).

Further, since the action of $\mathcal{U}$ is not affecting the degrees of D_i, we introduce (as in [5]) the restricted functions $\hat{\varphi}(\bar{Y})$ by the formula which is prompted in (3.36b) :

$$\hat{\varphi}(\bar{Y}) \;\equiv\; \big(\mathcal{A}\tilde{\varphi}\big)(\bar{Y}) \;\doteq\; \tilde{\varphi}(\bar{Y}, D_1 = \cdots = D_{n-1} = 1_{\mathcal{A}_g})\;,\tag{3.46a}$$

$$\hat{\varphi}(\bar{Y}) \;=\; \sum_{\bar{m}\in\mathbb{Z}_+}\mu_{\bar{m}}\,(\tilde{Y}_{21})^{m_{21}}\ldots(\tilde{Y}_{n,n-1})^{m_{n,n-1}}\;.\tag{3.46b}$$

We denote the representation space of $\hat{\varphi}(\bar{Y})$ by $\hat{\mathcal{C}}_{\bar{r}}$ and the representation acting in $\hat{\mathcal{C}}_{\bar{r}}$ by $\hat{\pi}_{\bar{r}}$. Thus, the operator $\mathcal{A}$ acts from $\mathcal{C}_{\bar{r}}$ to $\hat{\mathcal{C}}_{\bar{r}}$. The properties of $\hat{\mathcal{C}}_{\bar{r}}$ follow from the intertwining requirement for $\mathcal{A}$ [5]:

$$\hat{\pi}_{\bar{r}}\circ\mathcal{A} \;=\; \mathcal{A}\circ\tilde{\pi}_{\bar{r}}\;.\tag{3.47}$$

We have defined the representations $\hat{\pi}_{\bar{r}}$ for $r_i\in\mathbb{Z}$. However, notice that we can consider the restricted functions $\hat{\varphi}(\bar{Y})$ for arbitrary complex r_i. We shall make these extension from now on, since this gives the same set of representations for $U_q(sl(n))$ as in the case $q=1$.

For the more compact exposition of the representation formulae we shall need below also the following operators (corresponding to each of the variables $\tilde{Y}_{j\ell}$) :

$$\hat{M}_{j\ell}\,\hat{\varphi}(\bar{Y}) \;=\; \sum_{\bar{m}\in\mathbb{Z}_+}\mu_{\bar{m}}\,\hat{M}_{j\ell}\,(\tilde{Y}_{21})^{m_{21}}\ldots(\tilde{Y}_{n,n-1})^{m_{n,n-1}}\tag{3.48a}$$

$$T_{j\ell}\,\hat{\varphi}(\bar{Y}) \;=\; \sum_{\bar{m}\in\mathbb{Z}_+}\mu_{\bar{m}}\,T_{j\ell}\,(\tilde{Y}_{21})^{m_{21}}\ldots(\tilde{Y}_{n,n-1})^{m_{n,n-1}}\tag{3.48b}$$

$$\hat{M}_{j\ell}\,\tilde{f}_{\bar{m}} \;=\; (\tilde{Y}_{21})^{m_{21}}\ldots(\tilde{Y}_{j\ell})^{m_{j\ell}+1}\ldots(\tilde{Y}_{n,n-1})^{m_{n,n-1}}\tag{3.49a}$$

$$T_{j\ell}\,\tilde{f}_{\bar{m}} \;=\; q^{m_{j\ell}}\,\tilde{f}_{\bar{m}}\tag{3.49b}$$

$$\tilde{f}_{\bar{m}} \;=\; (\tilde{Y}_{21})^{m_{21}}\ldots(\tilde{Y}_{n,n-1})^{m_{n,n-1}}\tag{3.49c}$$

Using this we define the q-difference operators by:

$$\hat{\mathcal{D}}_{j\ell}\,\hat{\varphi}(\bar{Y}) \;=\; \frac{1}{\lambda}\,\hat{M}_{j\ell}^{-1}\left(T_{j\ell} - T_{j\ell}^{-1}\right)\hat{\varphi}(\bar{Y})\,, \tag{3.50}$$

from which follows:

$$\hat{\mathcal{D}}_{j\ell}\,(\tilde{Y}_{21})^{m_{21}}\ldots(\tilde{Y}_{n,n-1})^{m_{n,n-1}} \;=\; [m_{j\ell}]_q\,(\tilde{Y}_{21})^{m_{21}}\ldots(\tilde{Y}_{j\ell})^{m_{j\ell}-1}\ldots(\tilde{Y}_{n,n-1})^{m_{n,n-1}}\,. \tag{3.51}$$

Note that although $\hat{M}_{j\ell}^{-1}$ is not defined on $(\tilde{Y}_{21})^{m_{21}}\ldots(\tilde{Y}_{n,n-1})^{m_{n,n-1}}$ for $m_{j\ell}=0$, the operator $\hat{\mathcal{D}}_{j\ell}$ is well defined on such terms, and the result is zero (given by the action of $(T_{j\ell} - T_{j\ell}^{-1})$). Of course, for $q\to 1$ we have $\hat{\mathcal{D}}_{j\ell}\to\partial_{Y_{j\ell}}\equiv\partial/\partial Y_{j\ell}$. (Note that the above operators for different variables commute, i.e., with these we have actually passed to commuting variables.)

For the intertwining operators between partially equivalent representations we need the action of $\pi_R(X_i^-)$ on $\tilde{Y}_{j\ell}$ and D_ℓ. Using (3.11) and (3.13) we obtain:

$$\pi_R(X_i^-)\,(D_\ell)^n \;=\; \delta_{i\ell}\,c_n\,(D_\ell)^n\,Z_{\ell,\ell+1}\,, \tag{3.52a}$$

$$\pi_R(X_i^-)\,(\tilde{Y}_{j\ell})^n \;=\; \delta_{i\ell}\,q^{n-3/2}\,[n]_q\,(\tilde{Y}_{j\ell})^{n-1}\,\tilde{Y}_{j,\ell+1}\,D_{\ell+1}\,D_\ell^{-2}\,D_{\ell-1}\,, \tag{3.52b}$$

where, as usual, we use $\tilde{Y}_{jj}=1_{\mathcal{A}}=D_0$. We shall use also the repeated action of $\pi_R(X_i^-)$ so in addition we need:

$$\pi_R(X_i^-)\,Z_{j\ell} \;=\; \delta_{i\ell}\,Z_{j,\ell+1} - \delta_{ij}\,q^{-\delta_{j+1,\ell}/2}\,Z_{j,j+1}\,Z_{j\ell} + \delta_{i,j-1}\,D_j^{-1}\,\xi_{1\ldots j-2,j,\ell}^{1\ldots j}\,, \tag{3.53}$$

$$\pi_R(k_i)\,Z_{j\ell} \;=\; q^{(\delta_{i+1,j}-\delta_{ij}+\delta_{i\ell}-\delta_{i+1,\ell})/2}\,Z_{j\ell}\,. \tag{3.54}$$

4. THE CASE OF $U_q(\mathrm{sl}(4))$

In this Section we consider in more detail the case $n=4$. (For $n=2,3$, we refer to [16], [18], resp., though by a different method the case $n=2$ was given already in [27].)

It is convenient (also for the comparison with the $q=1$ case) to make the following change of variables:

$$\begin{aligned}
Y_{31} &= \tilde{Y}_{31} - q\tilde{Y}_{21}\tilde{Y}_{32}\,, \qquad Y_{41} = \tilde{Y}_{41} - q\tilde{Y}_{21}\tilde{Y}_{42}\,,\\
Y_{21} &= -q\tilde{Y}_{21}\,, \qquad Y_{43} = q\tilde{Y}_{43}\,, \qquad Y_{ij} = \tilde{Y}_{ij}\,, \quad \text{for } (ij)=(32),(42)\,.
\end{aligned} \tag{4.1}$$

Using (3.37) we have:

$$Y_{i\ell}Y_{ij} \;=\; q^{1-\delta_{\ell 2}}Y_{ij}Y_{i\ell}\,, \qquad 4\geq i>\ell>j\geq 1\,, \tag{4.2a}$$

$$Y_{kj}Y_{ij} \;=\; q^{1-\delta_{i2}}Y_{ij}Y_{kj}\,, \qquad 4\geq k>i>j\geq 1\,, \tag{4.2b}$$

$$Y_{41}Y_{32} \;=\; Y_{32}Y_{41} + \lambda Y_{31}Y_{42}\,, \tag{4.2c}$$

$$Y_{4i}Y_{j1} \;=\; Y_{j1}Y_{4i}\,, \qquad (ij)=(23),(32)\,, \tag{4.2d}$$

$$Y_{ki}Y_{ij} \;=\; q^{1-2\delta_{i3}}Y_{ij}Y_{ki} - (-1)^{\delta_{i3}}\lambda Y_{kj}\,, \qquad 4\geq k>i>j\geq 1\,, \tag{4.2e}$$

(each of $(4.2a,b,e)$ has four cases). Note that (3.38) holds also for $Y_{j\ell}$ replacing $\tilde{Y}_{j\ell}$.

Note that for q a phase ($|q| = 1$) the q - flag manifold in the Y coordinates is invariant under the anti-linear anti-involution ω acting as $\tilde{\omega}$ (cf. (3.39)) with $n = 4$:

$$\omega(Y_{j\ell}) = Y_{5-\ell,5-j} . \tag{4.3}$$

Thus it can be considered as a q - flag manifold of the quantum group $SU_q(2,2)$.

The reduced functions for the $\mathcal{U}$ action are (cf. (3.36)):

$$\tilde{\varphi}(\bar{Y}, \bar{D}) = \sum_{i,j,k,\ell,m,n \in \mathbb{Z}_+} \mu_{ijk\ell mn} \, \tilde{\varphi}_{ijk\ell mn} \tag{4.4a}$$

$$\tilde{\varphi}_{ijk\ell mn} = (Y_{21})^i \, (Y_{31})^j \, (Y_{32})^k \, (Y_{41})^\ell \, (Y_{42})^m \, (Y_{43})^n \times$$
$$\times \, (D_1)^{r_1} \, (D_2)^{r_2} \, (D_3)^{r_3} \tag{4.4b}$$

Now the action of $U_q(sl(4))$ on (4.4) is given explicitly by:

$$\hat{\pi}_{\bar{r}}(k_1) \, \tilde{\varphi}_{ijk\ell mn} = q^{i+(j-k+\ell-m-r_1)/2} \, \tilde{\varphi}_{ijk\ell mn} , \tag{4.5a}$$

$$\hat{\pi}_{\bar{r}}(k_2) \, \tilde{\varphi}_{ijk\ell mn} = q^{k+(-i+j+m-n-r_2)/2} \, \tilde{\varphi}_{ijk\ell mn} , \tag{4.5b}$$

$$\hat{\pi}_{\bar{r}}(k_3) \, \tilde{\varphi}_{ijk\ell mn} = q^{n+(-j-k+\ell+m-r_3)/2} \, \tilde{\varphi}_{ijk\ell mn} , \tag{4.5c}$$

$$\hat{\pi}_{\bar{r}}(X_1^+) \, \tilde{\varphi}_{ijk\ell mn} = q^{-1+(-j+k-\ell+m)/2} \, [r_1 - i]_q \, \tilde{\varphi}_{i+1,jk\ell mn} +$$
$$+ \, q^{i-r_1-1+(j-k-\ell+m)/2} \, [k]_q \, \tilde{\varphi}_{i,j+1,k-1,\ell mn} + \tag{4.6a}$$
$$+ \, q^{i-r_1-1+(j-k+\ell-m)/2} \, [m]_q \, \tilde{\varphi}_{ijk,\ell+1,m-1,n} \ ,$$

$$\hat{\pi}_{\bar{r}}(X_2^+) \, \tilde{\varphi}_{ijk\ell mn} = q^{r_2-k+(i-j-m+n)/2} \, [i]_q \, \tilde{\varphi}_{i-1,j+1,k\ell mn} +$$
$$+ \, q^{(i+j+m-n)/2} \, [j-i+k+m-n-r_2]_q \, \tilde{\varphi}_{ij,k+1,\ell mn} +$$
$$+ \, q^{-r_2+(-i+j+k+3m-3n)/2} \, [\ell]_q \, \tilde{\varphi}_{i,j+1,k,\ell-1,m+1,n} + \tag{4.6b}$$
$$+ \, q^{k-r_2+(-i+j+m-n)/2} \, [n]_q \, \tilde{\varphi}_{ijk\ell,m+1,n-1} \ ,$$

$$\hat{\pi}_{\bar{r}}(X_3^+) \, \tilde{\varphi}_{ijk\ell mn} = - \, q^{r_3-1-n+(j+k-\ell-m)/2} \, [j]_q \, \tilde{\varphi}_{i,j-1,k,\ell+1,mn} -$$
$$- \, q^{r_3-1-n+(3j+k-3\ell-m)/2} \, [k]_q \, \tilde{\varphi}_{ij,k-1,\ell,m+1,n} + \tag{4.6c}$$
$$+ \, q^{-1+(-j-k+\ell+m)/2} \, [n - r_3]_q \, \tilde{\varphi}_{ijk\ell m,n+1} \ ,$$

$$\hat{\pi}_{\bar{r}}(X_1^-) \, \tilde{\varphi}_{ijk\ell mn} = q^{1+(-j+k-\ell+m)/2} \, [i]_q \, \tilde{\varphi}_{i-1,jk\ell mn} +$$
$$+ \, q^{i+2+(-j+k-\ell+m)/2} \, [j]_q \, \tilde{\varphi}_{i,j-1,k+1,\ell mn} + \tag{4.7a}$$
$$+ \, q^{i+2+(j-k-\ell+m)/2} \, [\ell]_q \, \tilde{\varphi}_{ijk,\ell-1,m+1,n} \ ,$$

$$\hat{\pi}_{\bar{r}}(X_2^-) \, \tilde{\varphi}_{ijk\ell mn} = - \, q^{(-i+j-m+n)/2} \, [k]_q \, \tilde{\varphi}_{ij,k-1,\ell mn} , \tag{4.7b}$$

$$\hat{\pi}_{\bar{r}}(X_3^-) \, \tilde{\varphi}_{ijk\ell mn} = - \, q^{-n+(-j-3k+\ell+3m)/2} \, [\ell]_q \, \tilde{\varphi}_{i,j+1,k,\ell-1,mn} -$$
$$- \, q^{-n+(-j-k+\ell+m)/2} \, [m]_q \, \tilde{\varphi}_{ij,k+1,\ell,m-1,n} - \tag{4.7c}$$
$$- \, q^{1+(-j-k+\ell+m)/2} \, [n]_q \, \tilde{\varphi}_{ijk\ell m,n-1} \ .$$

It is easy to check that $\hat{\pi}_{\bar{r}}(k_i)$, $\hat{\pi}_{\bar{r}}(X_i^{\pm})$ satisfy (2.15).

From (4.6), (4.7) one can easily write down the explicit action of the non-simple root generators. These are defined as follows [28], [29] :

$$X_{ab}^{\pm} \;=\; \pm q^{\mp 1/2}\big(q^{1/2}X_a^{\pm}X_b^{\pm} - q^{-1/2}X_b^{\pm}X_a^{\pm}\big)\,, \quad (ab)\;=\;(12),(23)\,, \qquad (4.8a)$$

$$\begin{aligned}
X_{13}^{\pm} &= \pm q^{\mp 1/2}\big(q^{1/2}X_1^{\pm}X_{23}^{\pm} - q^{-1/2}X_{23}^{\pm}X_1^{\pm}\big) = \\
&= \pm q^{\mp 1/2}\big(q^{1/2}X_{12}^{\pm}X_3^{\pm} - q^{-1/2}X_3^{\pm}X_{12}^{\pm}\big)\,.
\end{aligned} \qquad (4.8b)$$

We give only the negative roots action, since these formulae will be used below:

$$\begin{aligned}
\hat{\pi}_{\bar{r}}(X_{12}^{-})\,\tilde{\varphi}_{ijk\ell mn} &= -\,q^{(i-k-\ell+n+3)/2}\,[j]_q\,\tilde{\varphi}_{i,j-1,k\ell mn} + \\
&\quad +\,q^{j+(i-k-\ell+n+5)/2}\,\lambda\,[k]_q\,[\ell]_q\,\tilde{\varphi}_{ij,k-1,\ell-1,m+1,n}\,,
\end{aligned} \qquad (4.9a)$$

$$\hat{\pi}_{\bar{r}}(X_{23}^{-})\,\tilde{\varphi}_{ijk\ell mn} = -\,q^{(-i+k+\ell-n+3)/2}\,[m]_q\,\tilde{\varphi}_{ijk\ell,m-1,n}\,, \qquad (4.9b)$$

$$\hat{\pi}_{\bar{r}}(X_{13}^{-})\,\tilde{\varphi}_{ijk\ell mn} = -\,q^{3+(i+j-m-n)/2}\,[\ell]_q\,\tilde{\varphi}_{ijk,\ell-1,mn}\,. \qquad (4.9c)$$

Then we consider the restricted functions (cf. (3.46)):

$$\hat{\varphi}(\bar{Y}) \;=\; \sum_{i,j,k,\ell,m,n\in\mathbb{Z}_+} \mu_{ijk\ell mn}\,\hat{\varphi}_{ijk\ell mn}\,, \qquad (4.10a)$$

$$\hat{\varphi}_{ijk\ell mn} \;=\; (Y_{21})^i\,(Y_{31})^j\,(Y_{32})^k\,(Y_{41})^{\ell}\,(Y_{42})^m\,(Y_{43})^n\,. \qquad (4.10b)$$

As a consequence of the intertwining property (3.47) we obtain that $\hat{\varphi}_{ijk\ell mn}$ obey the same transformation rules (4.5), (4.6), (4.7), (4.9), as $\tilde{\varphi}_{ijk\ell mn}$.

Recall that we consider the representations $\hat{\pi}_{\bar{r}}$ for arbitrary complex r_i and we know from the general analysis of [18] that whenever some $m_i = r_i + 1$ or $m_{ij} = m_i + \cdots + m_j$, $(i < j)$ is a positive integer the representations are reducible and there exist invariant subspaces. We give now two simple examples.

Let $m_1 = r_1 + 1 \in \mathbb{N}$. Then it is clear that functions $\tilde{\varphi}$ with $\mu_{ijk\ell mn} = 0$ if $i \geq m_1$ form an invariant subspace since:

$$\begin{aligned}
\hat{\pi}_{\bar{r}}(X_1^{+})\,\tilde{\varphi}_{r_1,jk\ell mn} &= q^{(j+m-\ell-2-k)/2}\,[k]_q\,\tilde{\varphi}_{r_1,j+1,k-1,\ell mn} + \\
&\quad +\,q^{(j+\ell-k-2-m)/2}\,[m]_q\,\tilde{\varphi}_{r_1,jk,\ell+1,m-1,n}\,,
\end{aligned} \qquad (4.11)$$

and all other operators in (4.5), (4.6), (4.7) either preserve or lower the index i. The same is true for the functions $\hat{\varphi}$. In particular, for $r_1 = 0$ the functions in the invariant subspace do not depend on the variable Y_{21}.

Analogously if $m_3 = r_3 + 1 \in \mathbb{N}$ the functions $\tilde{\varphi}$ with $\mu_{ijk\ell mn} = 0$ if $n \geq m_3$ form an invariant subspace since:

$$\begin{aligned}
\hat{\pi}_{\bar{r}}(X_3^{+})\,\tilde{\varphi}_{ijk\ell m,r_3} &= -\,q^{(k+j+m-\ell-2)/2}\,[j]_q\,\tilde{\varphi}_{i,j-1,k,\ell+1,m,r_3} - \\
&\quad -\,q^{(k+3j+m-3\ell-2)/2}\,[m]_q\,\tilde{\varphi}_{ij,k-1,\ell,m+1,r_3}\,,
\end{aligned} \qquad (4.12)$$

and all other operators in (4.5), (4.6), (4.7) either preserve or lower the index n, the same holding for the functions $\hat{\varphi}$. In particular, for $r_3 = 0$ the functions in the invariant subspace do not depend on the variable Y_{43}.

It will be convenient to use also the following notation for the coordinates of the flag manifold:

$$\xi = Y_{21}, \quad x = Y_{31}, \quad u = Y_{32}, \quad w = Y_{41}, \quad y = Y_{42}, \quad \eta = Y_{43} \,. \quad (4.13)$$

The above notation we shall employ also for the operators (3.48), (3.50). In terms of the latter operators we rewrite the transformation rules (4.5), (4.6), (4.7), (4.9) for the functions $\hat{\varphi}$ as follows :

$$\hat{\pi}_{\bar{r}}(k_1)\,\hat{\varphi}(\bar{Y}) \;=\; q^{-r_1/2}\,T_\xi\,(T_x T_w)^{1/2}\,(T_u T_y)^{-1/2}\,\hat{\varphi}(\bar{Y}) \,, \qquad (4.14a)$$

$$\hat{\pi}_{\bar{r}}(k_2)\,\hat{\varphi}(\bar{Y}) \;=\; q^{-r_2/2}\,T_u\,(T_x T_y)^{1/2}\,(T_\xi T_\eta)^{-1/2}\,\hat{\varphi}(\bar{Y}) \,, \qquad (4.14b)$$

$$\hat{\pi}_{\bar{r}}(k_3)\,\hat{\varphi}(\bar{Y}) \;=\; q^{-r_3/2}\,T_\eta\,(T_w T_y)^{1/2}\,(T_x T_u)^{-1/2}\,\hat{\varphi}(\bar{Y}) \,, \qquad (4.14c)$$

$$\hat{\pi}_{\bar{r}}(X_1^+)\,\hat{\varphi}(\bar{Y}) \;=\; (1/\lambda)\,q^{-1}\,\hat{M}_\xi\,(T_u T_y)^{1/2}\,(T_x T_w)^{-1/2}\,\left(q^{r_1}T_\xi^{-1} - q^{-r_1}T_\xi\right)\,\hat{\varphi}(\bar{Y}) \; +$$
$$+\; q^{-r_1-1}\,\hat{M}_x\,\hat{D}_u\,T_\xi\,(T_x T_y)^{1/2}\,(T_u T_w)^{-1/2}\,\hat{\varphi}(\bar{Y}) \; + \qquad (4.15a)$$
$$+\; q^{-r_1-1}\,\hat{M}_w\,\hat{D}_y\,T_\xi\,(T_x T_w)^{1/2}\,(T_u T_y)^{-1/2}\,\hat{\varphi}(\bar{Y}) \;\; ,$$

$$\hat{\pi}_{\bar{r}}(X_2^+)\,\hat{\varphi}(\bar{Y}) \;=\; q^{r_2}\,\hat{M}_x\,\hat{D}_\xi\,T_u^{-1}\,(T_\xi T_\eta)^{1/2}\,(T_x T_y)^{-1/2}\,\hat{\varphi}(\bar{Y}) \; +$$
$$+\; (1/\lambda)\,\hat{M}_u\,(T_\xi T_y)^{1/2}\,(T_x T_\eta)^{-1/2}\;\times$$
$$\times\;\left(q^{-r_2}T_x T_u T_y(T_\xi T_\eta)^{-1} - q^{r_2}T_\xi T_\eta(T_x T_u T_y)^{-1}\right)\,\hat{\varphi}(\bar{Y}) \; + \qquad (4.15b)$$
$$+\; q^{-r_2}\,\hat{M}_x\,\hat{M}_y\,\hat{D}_w\,(T_x\,T_u T_y^3)^{1/2}\,(T_\xi T_\eta^3)^{-1/2}\,\hat{\varphi}(\bar{Y}) \;\; +$$
$$+\; q^{-r_2}\,\hat{M}_y\,\hat{D}_\eta\,T_u\,(T_x T_y)^{1/2}\,(T_\xi T_\eta)^{-1/2}\,\hat{\varphi}(\bar{Y}) \;\; ,$$

$$\hat{\pi}_{\bar{r}}(X_3^+)\,\hat{\varphi}(\bar{Y}) \;=\; -\,q^{r_3-1}\,\hat{M}_w\,\hat{D}_x\,T_\eta^{-1}\,(T_x T_u)^{1/2}\,(T_w T_y)^{-1/2}\,\hat{\varphi}(\bar{Y}) \; -$$
$$-\; q^{r_3-1}\,\hat{M}_y\,\hat{D}_u\,T_\eta^{-1}\,(T_x^3 T_u)^{1/2}\,(T_w^3 T_y)^{-1/2}\,\hat{\varphi}(\bar{Y}) \; + \qquad (4.15c)$$
$$+\; (1/\lambda)\,q^{-1}\,\hat{M}_\eta\,(T_w T_y)^{1/2}\,(T_x T_u)^{-1/2}\,\left(q^{-r_3}T_\eta - q^{r_3}T_\eta^{-1}\right)\,\hat{\varphi}(\bar{Y}) \;\; ,$$

$$\hat{\pi}_{\bar{r}}(X_1^-)\,\hat{\varphi}(\bar{Y}) \;=\; q\,\hat{D}_\xi\,(T_u T_y)^{1/2}\,(T_x T_w)^{-1/2}\,\hat{\varphi}(\bar{Y}) \; +$$
$$+\; q^2\,\hat{M}_u\,\hat{D}_x\,T_\xi\,(T_u T_y)^{1/2}\,(T_x T_w)^{-1/2}\,\hat{\varphi}(\bar{Y}) \; + \qquad (4.16a)$$
$$+\; q^2\,\hat{M}_y\,\hat{D}_w\,T_\xi\,(T_x T_y)^{1/2}\,(T_u T_w)^{-1/2}\,\hat{\varphi}(\bar{Y})$$

$$\hat{\pi}_{\bar{r}}(X_2^-)\,\tilde{\varphi}_{ijk\ell mn} \;=\; -\,\hat{D}_u\,(T_x T_\eta)^{1/2}\,(T_\xi T_y)^{-1/2}\,\hat{\varphi}(\bar{Y}) \,, \qquad (4.16b)$$

$$\hat{\pi}_{\bar{r}}(X_3^-)\,\hat{\varphi}(\bar{Y}) \;=\; -\,\hat{M}_x\,\hat{D}_w\,T_\eta^{-1}\,(T_w T_y^3)^{1/2}\,(T_x T_u^3)^{-1/2}\,\hat{\varphi}(\bar{Y}) \; -$$
$$-\; \hat{M}_u\,\hat{D}_y\,T_\eta^{-1}\,(T_w T_y)^{1/2}\,(T_x T_u)^{-1/2}\,\hat{\varphi}(\bar{Y}) \; - \qquad (4.16c)$$
$$-\; q\,\hat{D}_\eta\,(T_w T_y)^{1/2}\,(T_x T_u)^{-1/2}\,\hat{\varphi}(\bar{Y}) \,,$$

$$\hat{\pi}_{\bar{r}}(X_{12}^{-}) \; \hat{\varphi}(\bar{Y}) \;\; = \;\; - \; q^{3/2} \; \hat{D}_x \; (T_\xi T_\eta)^{1/2} \; (T_u T_w)^{-1/2} \; \hat{\varphi}(\bar{Y}) \; +$$
$$+ \; \lambda q^{5/2} \; \hat{M}_y \; \hat{D}_u \; \hat{D}_w \; T_x \; (T_\xi T_\eta)^{1/2} \; (T_u T_w)^{-1/2} \; \hat{\varphi}(\bar{Y}) \; , \tag{4.17a}$$

$$\hat{\pi}_{\bar{r}}(X_{23}^{-}) \; \hat{\varphi}(\bar{Y}) \;\; = \;\; - \; q^{3/2} \; \hat{D}_y \; (T_u T_w)^{1/2} \; (T_\xi T_\eta)^{-1/2} \; \hat{\varphi}(\bar{Y}) \; , \tag{4.17b}$$

$$\hat{\pi}_{\bar{r}}(X_{13}^{-}) \; \hat{\varphi}(\bar{Y}) \;\; = \;\; - \; q^{3} \; \hat{D}_w \; (T_\xi T_x)^{1/2} \; (T_y T_\eta)^{-1/2} \; \hat{\varphi}(\bar{Y}) \; . \tag{4.17c}$$

5. INTERTWINING OPERATORS

The general prescription for finding the intertwining operators was presented in [18]. In order to apply this procedure we need the explicit action of $\pi_R(X_i^-)$ on our functions. First we have to calculate the action on the new basis $Y_{j\ell}$. We have instead of (3.52b):

$$\pi_R(X_i^-) \, (Y_{j\ell})^n \;=\; (-1)^{\delta_{i1}} \; \delta_{i\ell} \; \delta_{i+1,j} \; q^{n-1/2} \; [n]_q \; (Y_{i+1,i})^{n-1} \; D_{i+1} \; D_i^{-2} \; D_{i-1}, \quad i=1,3$$
$$\pi_R(X_2^-) \, (Y_{j\ell})^n \;=\; (-1)^{\ell} \; q^{(n-2)(\ell-1)+1/2} \; [n]_q \; Y_{2\ell} \; (Y_{j\ell})^{n-1} \; Y_{j3} \; D_2 \; D_1^{-2} \; , \tag{5.1}$$

where we again use $D_4 = D_0 = Y_{jj} = 1_A$, $Y_{j\ell} = 0$ for $j < \ell$.

Using (5.1) and (3.52a) we obtain:

$$\pi_R(X_1^-) \, \tilde{\varphi}_{ijk\ell mn}^{r_1,r_2,r_3} \;=\; - \; q^{i-j-k-\ell-m+(r_1-1)/2} \; [i]_q \; \tilde{\varphi}_{i-1,jk\ell mn}^{r_1-2,r_2+1,r_3} \; +$$
$$+ \; q^{(r_1-1)/2} \; [r_1]_q \; \tilde{\varphi}_{ijk\ell mn}^{r_1,r_2,r_3} \; Z_{12} \; , \tag{5.2a}$$

$$\pi_R(X_2^-) \, \tilde{\varphi}_{ijk\ell mn}^{r_1,r_2,r_3} \;=\; q^{2k+\ell+m-n+(r_2-1)/2} \; [j]_q \; \tilde{\varphi}_{i+1,j-1,k\ell mn}^{r_1+1,r_2-2,r_3+1} \; +$$
$$+ \; q^{k+\ell+m-n+(r_2-3)/2} \; [k]_q \; \tilde{\varphi}_{ij,k-1,\ell mn}^{r_1+1,r_2-2,r_3+1} \; +$$
$$+ \; q^{k-j+2m-n+(r_2-3)/2} \; [\ell]_q \; \tilde{\varphi}_{i+1,jk,\ell-1,m,n+1}^{r_1+1,r_2-2,r_3+1} \; +$$
$$+ \; q^{m-n+(r_2-5)/2} \; [m]_q \; \tilde{\varphi}_{ijk\ell,m-1,n+1}^{r_1+1,r_2-2,r_3+1} \; -$$
$$- \; q^{2m-n+(r_2-3)/2} \; \lambda \; [k]_q \; [\ell]_q \; \tilde{\varphi}_{i,j+1,k-1,\ell-1,m,n+1}^{r_1+1,r_2-2,r_3+1} \; +$$
$$+ \; q^{(r_2-1)/2} \; [r_2]_q \; \tilde{\varphi}_{ijk\ell mn}^{r_1,r_2,r_3} \; Z_{23} \; , \tag{5.2b}$$

$$\pi_R(X_3^-) \, \tilde{\varphi}_{ijk\ell mn}^{r_1,r_2,r_3} \;=\; q^{n+(r_3-1)/2} \; [n]_q \; \tilde{\varphi}_{ijk\ell m,n-1}^{r_1,r_2+1,r_3-2} \; +$$
$$+ \; q^{(r_3-1)/2} \; [r_3]_q \; \tilde{\varphi}_{ijk\ell mn}^{r_1,r_2,r_3} \; Z_{34} \; , \tag{5.2c}$$

where we have labelled the functions also with the representation parameters r_s. As in the classical case [5] the right action is taking out from the representation space $\mathcal{C}_{\bar{r}}$, and while some of the terms are functions from other representation spaces (depending on which X_s^- is acting), there are terms involving the $Z_{j\ell}$ variables which do not belong to any of our representation spaces. These terms vanish only when the respective r_s is equal to zero, and in these cases (5.2) describe three different intertwining operators corresponding to the simple roots of the root system of $sl(4)$. If $r_s \in I\!N$ then the terms with $Z_{j\ell}$ vanish exactly when we take $(\pi_R(X_s^-))^{m_s}$ [5], [18], $m_s = r_s + 1$.

Indeed, we know from the general prescription (cf. (1.15), [5], [18]) that if $m_s \in I\!N$ then there exist an intertwining operator $I_s^{m_s} = (\pi_R(X_s^-))^{m_s}$. We have the following intertwining properties:

$$I_1^{m_1} \circ \pi_{m_1,m_2,m_3} = \pi_{-m_1,m_{12},m_3} \circ I_1^{m_1} , \quad m_1 \in I\!N , \qquad (5.3a)$$

$$I_2^{m_2} \circ \pi_{m_1,m_2,m_3} = \pi_{m_{12},-m_2,m_{23}} \circ I_2^{m_2} , \quad m_2 \in I\!N , \qquad (5.3b)$$

$$I_3^{m_3} \circ \pi_{m_1,m_2,m_3} = \pi_{m_1,m_{23},-m_3} \circ I_3^{m_3} , \quad m_3 \in I\!N , \qquad (5.3c)$$

where we label the representations with the numbers m_s instead of $r_s = m_s - 1$ to simplify the notation. The explicit expressions for two of these operators are:

$$(\pi_R(X_1^-))^{m_1} \, \tilde{\varphi}_{ijk\ell mn}^{m_1,m_2,m_3} = (-1)^{m_1} \, q^{m_1(i-m_1/2)} \, \frac{[i]_q!}{[i-m_1]_q!} \, \tilde{\varphi}_{i-m_1,jk\ell mn}^{-m_1,m_{12},m_3} , \qquad (5.4)$$

$$(\pi_R(X_3^-))^{m_3} \, \tilde{\varphi}_{ijk\ell mn}^{m_1,m_2,m_3} = q^{m_3(i-m_3/2)} \, \frac{[n]_q!}{[n-m_3]_q!} \, \tilde{\varphi}_{ijk\ell m,n-m_3}^{m_1,m_{23},-m_3} . \qquad (5.5)$$

Having in mind the preceding discussion let us introduce the following q-difference operators (using notation (3.48), (3.50), (4.13)):

$$\hat{I}_1 \equiv - q^{(r_1-1)/2} \, \hat{\mathcal{D}}_\xi \, T_\xi \, (T_x T_u T_w T_y)^{-1} \qquad (5.6a)$$

$$\hat{I}_2 \equiv q^{(r_2-3)/2} \left(q \, \hat{M}_\xi \, \hat{\mathcal{D}}_x \, T_u + \hat{\mathcal{D}}_u + \right.$$

$$+ \hat{M}_\xi \, \hat{M}_\eta \, \hat{\mathcal{D}}_w \, (T_x T_w)^{-1} \, T_y + q^{-1} \, \hat{M}_\eta \, \hat{\mathcal{D}}_y \, (T_u T_w)^{-1} - \qquad (5.6b)$$

$$\left. - \lambda \, \hat{M}_x \, \hat{M}_\eta \, \hat{\mathcal{D}}_u \, \hat{\mathcal{D}}_w \, (T_u T_w)^{-1} \, T_y \right) T_u \, T_w \, T_y \, T_\eta^{-1}$$

$$\hat{I}_3 \equiv q^{(r_3-1)/2} \, \hat{\mathcal{D}}_\eta \, T_\eta \qquad (5.6c)$$

It is not difficult to see that if $m_s \in I\!N$ we have:

$$\hat{I}_s^{m_s} = I_s^{m_s} = (\pi_R(X_s^-))^{m_s} . \qquad (5.7)$$

Let us consider now the intertwinig operators corresponding to the two non-simple non-highest roots α_{12}, α_{23} which are realized when $m_{12} \in I\!N$, $m_{23} \in I\!N$, resp. In these cases the intertwining operators (up to an overall multiplicative constant) are given by :

$$I_{ij}^m = \sum_{k=0}^{m} a_k \, (\pi_R(X_i^-))^{m-k} \, (\pi_R(X_j^-))^m \, (\pi_R(X_i^-))^k , \qquad (5.8a)$$

$$m = m_{ij} , \quad (ij) = (12),(23),$$

$$a_k = (-1)^k \, a \, \frac{[m_i]_q}{[m_i - k]_q} \binom{m}{k}_q , \quad k = 0,\ldots,m, \quad a \neq 0, \qquad (5.8b)$$

or equivalently, by :

$$I_{ij}^m = \sum_{k=0}^{m} a_k' \, (\pi_R(X_j^-))^{m-k} \, (\pi_R(X_i^-))^m \, (\pi_R(X_j^-))^k , \qquad (5.8c)$$

$$m = m_{ij} , \quad (ij) = (12),(23),$$

$$a_k' = (-1)^k \, a' \, \frac{[m_j]_q}{[m_j - k]_q} \binom{m}{k}_q , \quad k = 0,\ldots,m, \quad a' \neq 0, \qquad (5.8d)$$

where we are using the singular vector given in formula (27) of [29].

Let us illustrate the resulting intertwining operators in the cases $m_{12} = 1$, $m_{23} = 1$. We have (after a suitable renormalization) :

$$I^1_{12}|_{r_1+r_2=-1} = -[r_1]_q\, \pi_R(X_1^-)\, \pi_R(X_2^-) + [r_1+1]_q\, \pi_R(X_2^-)\, \pi_R(X_1^-)\,, \qquad (5.9a)$$

$$\begin{aligned}
I^1_{12}\, \tilde{\varphi}^{r_1,r_2,r_3}_{ijk\ell mn}|_{r_1+r_2=-1} = {}& q^{2i-j-n}\, [r_1]_q\, \left(q^{k+1}\, [j]_q\, \tilde{\varphi}^{r_1-1,r_2-1,r_3+1}_{i,j-1,k\ell mn} \right. \\
& + q^{-j-\ell+m}\, [\ell]_q\, \tilde{\varphi}^{r_1-1,r_2-1,r_3+1}_{ijk,\ell-1,m,n+1} \Bigg) - \\
& - q^{i-j-n-r_1-2}\, [i]_q\, \left(q^{k+1}\, [j]_q\, \tilde{\varphi}^{r_1-1,r_2-1,r_3+1}_{i,j-1,k\ell mn} \right. + \\
& + [k]_q\, \tilde{\varphi}^{r_1-1,r_2-1,r_3+1}_{i-1,j,k-1,\ell mn} + \\
& + q^{-j-\ell+m}\, [\ell]_q\, \tilde{\varphi}^{r_1-1,r_2-1,r_3+1}_{ijk,\ell-1,m,n+1} + \\
& + q^{-k-\ell-1}\, [m]_q\, \tilde{\varphi}^{r_1-1,r_2-1,r_3+1}_{i-1,jk\ell,m-1,n+1} - \\
& - q^{-k-\ell+m}\, \lambda\, [k]_q\, [\ell]_q\, \tilde{\varphi}^{r_1-1,r_2-1,r_3+1}_{i-1,j+1,k-1,\ell-1,m,n+1} \Bigg)
\end{aligned} \qquad (5.9b)$$

$$I^1_{23}|_{r_2+r_3=-1} = -[r_3]_q\, \pi_R(X_3^-)\, \pi_R(X_2^-) + [r_3+1]_q\, \pi_R(X_2^-)\, \pi_R(X_3^-)\,, \qquad (5.10a)$$

$$\begin{aligned}
I^1_{12}\, \tilde{\varphi}^{r_1,r_2,r_3}_{ijk\ell mn}|_{r_2+r_3=-1} = {}& - q^{k+\ell+m+n-1}\, [r_3]_q\, \left(q^{-k-\ell-1}\, [m]_q\, \tilde{\varphi}^{r_1+1,r_2-1,r_3-1}_{ijk\ell,m-1,n} \right. + \\
& + q^{-j-\ell+m}\, [\ell]_q\, \tilde{\varphi}^{r_1+1,r_2-1,r_3-1}_{i+1,jk,\ell-1,mn} - \\
& - q^{-k-\ell+m}\, \lambda\, [k]_q\, [\ell]_q\, \tilde{\varphi}^{r_1+1,r_2-1,r_3-1}_{i,j+1,k-1,\ell-1,mn} \Bigg) - \\
& + q^{k+\ell+m+r_3-1}\, [n]_q\, \left(q^{k+1}\, [j]_q\, \tilde{\varphi}^{r_1+1,r_2-1,r_3-1}_{i+1,j-1,k\ell m,n-1} \right. + \\
& + [k]_q\, \tilde{\varphi}^{r_1+1,r_2-1,r_3-1}_{ij,k-1,\ell m,n-1} + \\
& + q^{-j-\ell+m}\, [\ell]_q\, \tilde{\varphi}^{r_1+1,r_2-1,r_3-1}_{i+1,jk,\ell-1,mn} + \\
& + q^{-k-\ell-1}\, [m]_q\, \tilde{\varphi}^{r_1+1,r_2-1,r_3-1}_{ijk\ell,m-1,n} - \\
& - q^{-k-\ell+m}\, \lambda\, [k]_q\, [\ell]_q\, \tilde{\varphi}^{r_1+1,r_2-1,r_3-1}_{i,j+1,k-1,\ell-1,mn} \Bigg) \,.
\end{aligned} \qquad (5.10b)$$

Using the operators $\hat{I}_s$ the above formulae can be rewritten as:

$$I^1_{12}|_{r_1+r_2=-1} = -[r_1]_q\, \hat{I}_1\, \hat{I}_2 + [r_1+1]_q\, \hat{I}_2\, \hat{I}_1\,, \qquad (5.11a)$$

$$\begin{aligned}
I^1_{12}|_{r_1+r_2=-1} = {}& [r_1]_q\, \left(q\, \hat{D}_x\, T_u + \hat{M}_\eta\, \hat{D}_w\, (T_x T_w)^{-1}\, T_y \right)\, T_\xi^2\, (T_x T_\eta)^{-1} - \\
& - q^{-r_1-2}\, \left(q\, \hat{M}_\xi\, \hat{D}_x\, T_u + \hat{D}_u + \right. \\
& + \hat{M}_\xi\, \hat{M}_\eta\, \hat{D}_w\, (T_x T_w)^{-1}\, T_y + q^{-1}\, \hat{M}_\eta\, \hat{D}_y\, (T_u T_w)^{-1} - \\
& - \lambda\, \hat{M}_x\, \hat{M}_\eta\, \hat{D}_u\, \hat{D}_w\, (T_u T_w)^{-1}\, T_y \Bigg)\, \hat{D}_\xi\, T_\xi\, (T_x T_\eta)^{-1}
\end{aligned} \qquad (5.11b)$$

$$I^1_{23}|_{r_2+r_3=-1} = -[r_3]_q\, \hat{I}_3\, \hat{I}_2 + [r_3+1]_q\, \hat{I}_2\, \hat{I}_3\,, \qquad (5.12a)$$

$$I^1_{23}|_{r_2+r_3=-1} = - q^{-1} [r_3]_q \left(q^{-1} \hat{\mathcal{D}}_y (T_u T_w)^{-1} + \hat{M}_\xi \hat{\mathcal{D}}_w (T_x T_w)^{-1} T_y - \right.$$

$$- \lambda \hat{M}_x \hat{\mathcal{D}}_u \hat{\mathcal{D}}_w (T_u T_w)^{-1} T_y \left. \right) T_u T_w T_y T_\eta +$$

$$+ q^{r_3-1} \left(q \hat{M}_\xi \hat{\mathcal{D}}_x T_u + \hat{\mathcal{D}}_u + \right. \tag{5.12b}$$

$$+ \hat{M}_\xi \hat{M}_\eta \hat{\mathcal{D}}_w (T_x T_w)^{-1} T_y + q^{-1} \hat{M}_\eta \hat{\mathcal{D}}_y (T_u T_w)^{-1} -$$

$$- \lambda \hat{M}_x \hat{M}_\eta \hat{\mathcal{D}}_u \hat{\mathcal{D}}_w (T_u T_w)^{-1} T_y \left. \right) \hat{\mathcal{D}}_\eta T_u T_w T_y .$$

6. NEW q - MINKOWSKI SPACE-TIME AND q - MAXWELL EQUATIONS HIERARCHY FROM q - CONFORMAL INVARIANCE

6.1. The present Section reviews mostly [19]. We consider the construction of q - deformed analogs of some conformally invariant equations, in particular, the Maxwell equations, following our approach. We start with the $q = 1$ situation and we first write the Maxwell equations in an indexless formulation, trading the indices for two conjugate variables $z, \bar{z}$. This formulation has two advantages. First, it is very simple, and in fact, just with the introduction of an additional parameter, we can describe a whole infinite hierarchy of equations, which we call the *Maxwell hierarchy* . Second, we can easily identify the variables $z, \bar{z}$ and the four Minkowski coordinates with the six local coordinates of a flag manifold of $SU(2,2)$, or of $SL(4)$ with the appropriate conjugation. Thus, one may look at this as a nice example of unifying internal and external degrees of freedom.

Next we give the q - analogs of the above constructions. We recall that the specifics of our approach is that one needs also the complexification of the algebra in consideration. Thus for the q - conformal algebra we can use the $U_q(sl(4))$ apparatus of Sections 4 and 5. Thus, we can propose new q - *Minkowski coordinates* as part of the appropriate q - flag manifold. Using the corresponding representations and intertwiners of $U_q(sl(4))$ we can finally write down the infinite hierarchy of q - Maxwell equations. At the end we make some comments on the existing literature and on further developments.

6.2. It is well known that Maxwell equations

$$\partial^\mu F_{\mu\nu} = J_\nu , \quad \partial^{\mu *}F_{\mu\nu} = 0 \tag{6.1}$$

or, equivalently

$$\partial_k E_k = J_0 \; (= 4\pi\rho), \quad \partial_0 E_k - \varepsilon_{k\ell m}\partial_\ell H_m = J_k \; (= -4\pi j_k),$$
$$\partial_k H_k = 0 , \quad \partial_0 H_k + \varepsilon_{k\ell m}\partial_\ell E_m = 0 , \tag{6.2}$$

where $E_k \equiv F_{k0}$, $H_k \equiv (1/2)\varepsilon_{k\ell m}F_{\ell m}$, can be rewritten in the following manner:

$$\partial_k F^\pm_k = J_0 , \quad \partial_0 F^\pm_k \pm i\varepsilon_{k\ell m}\partial_\ell F^\pm_m = J_k , \tag{6.3}$$

where

$$F^\pm_k \equiv E_k \pm iH_k . \tag{6.4}$$

Not so well known is the fact that the eight equations in (6.3) can be rewritten as two conjugate scalar equations in the following way:

$$I^+ \, F^+(z) \; = \; J(z, \bar{z}) \, , \tag{6.5a}$$

$$I^- \, F^-(\bar{z}) \; = \; J(z, \bar{z}) \, , \tag{6.5b}$$

where

$$I^+ \; = \; \bar{z}\partial_+ + \partial_v - \frac{1}{2}\Big(\bar{z}z\partial_+ + z\partial_v + \bar{z}\partial_{\bar{v}} + \partial_- \Big)\partial_z \, , \tag{6.6a}$$

$$I^- \; = \; z\partial_+ + \partial_{\bar{v}} - \frac{1}{2}\Big(\bar{z}z\partial_+ + z\partial_v + \bar{z}\partial_{\bar{v}} + \partial_- \Big)\partial_{\bar{z}} \, , \tag{6.6b}$$

$$x_\pm \equiv x_0 \pm x_3, \quad v \equiv x_1 - ix_2, \quad \bar{v} \equiv x_1 + ix_2, \tag{6.7a}$$

$$\partial_\pm \equiv \partial/\partial x_\pm, \quad \partial_v \equiv \partial/\partial v, \quad \partial_{\bar{v}} \equiv \partial/\partial \bar{v}, \tag{6.7b}$$

$$F^+(z) \; \equiv \; z^2(F_1^+ + iF_2^+) - 2zF_3^+ - (F_1^+ - iF_2^+) \, , \tag{6.8a}$$

$$F^-(\bar{z}) \; \equiv \; \bar{z}^2(F_1^- - iF_2^-) - 2\bar{z}F_3^- - (F_1^- + iF_2^-) \, , \tag{6.8b}$$

$$J(z, \bar{z}) \; \equiv \; \bar{z}z(J_0 + J_3) + \bar{z}(J_1 - iJ_2) + z(J_1 + iJ_2) + (J_0 - J_3) \, , \tag{6.8c}$$

where we continue to suppress the x_μ, resp., $x_\pm, v, \bar{v}$, dependence in F and J. (The conjugation mentioned above is standard and in our terms it is : $I^+ \longleftrightarrow I^-$, $F^+(z) \longleftrightarrow F^-(\bar{z})$.)

It is easy to recover (6.3) from (6.5) - just note that both sides of each equation are first order polynomials in each of the two variables z and $\bar{z}$, then comparing the independent terms in (6.5) one gets at once (6.3).

Writing the Maxwell equations in the simple form (6.5) has also important conceptual meaning. The point is that each of the two scalar operators I^+, I^- is indeed a single object, namely it is an intertwiner of the conformal group, while the individual components in (6.1) - (6.3) do not have this interpretation. This is also the simplest way to see that the Maxwell equations are conformally invariant, since this is equivalent to the intertwining property.

Let us be more explicit. The physically relevant representations T^χ of the 4-dimensional conformal algebra $su(2,2)$ may be labelled by $\chi = [n_1, n_2; d]$, where n_1, n_2 are non-negative integers fixing finite-dimensional irreducible representations of the Lorentz subalgebra, (the dimension being $(n_1 + 1)(n_2 + 1)$), and d is the conformal dimension (or energy). (In the literature these Lorentz representations are labelled also by $(j_1, j_2) = (n_1/2, n_2/2)$.) Then the intertwining properties of the operators in (6.6) are given by:

$$I^+ \; : \; C^+ \longrightarrow C^0 \, , \quad I^+ \circ T^+ \; = \; T^0 \circ I^+ \, , \tag{6.9a}$$

$$I^- \; : \; C^- \longrightarrow C^0 \, , \quad I^- \circ T^- \; = \; T^0 \circ I^- \, , \tag{6.9b}$$

where $T^a = T^{\chi^a}$, $a = 0, +, -$, $C^a = C^{\chi^a}$ are the representation spaces, and the signatures are given explicitly by:

$$\chi^+ = [2, 0; 2] \, , \quad \chi^- = [0, 2; 2] \, , \quad \chi^0 = [1, 1; 3] \, , \tag{6.10}$$

as anticipated. Indeed, $(n_1, n_2) = (1, 1)$ is the four-dimensional Lorentz representation, (carried by J_μ above), and $(n_1, n_2) = (2, 0), (0, 2)$ are the two conjugate three-dimensional Lorentz representations, (carried by $F_k^\pm$ above), while the conformal dimensions are the canonical dimensions of a current ($d = 3$), and of the Maxwell field ($d = 2$). We see that the variables $z, \bar{z}$ are related to the spin properties and we shall call them 'spin variables'. More explicitly, a Lorentz spin-tensor $G(z, \bar{z})$ with signature (n_1, n_2) is a polynomial in $z, \bar{z}$ of order n_1, n_2, resp.

Formulae (6.9), (6.10) are part of an infinite hierarchy of couples of first order intertwiners given already in [30] for the Euclidean conformal group $SU^*(4)$, and then for the conformal group $SU(2,2)$ in [31], [32]. (Note that [30], [31] use a different approach, while [32] already uses the essential features of [5] in the context of the conformal group.) Explicitly, instead of (6.9), (6.10) we have [32] :

$$I_n^+ \; : \; C_n^+ \longrightarrow C_n^0 \, , \quad I_n^+ \circ T_n^+ \; = \; T_n^0 \circ I_n^+ \, , \tag{6.11a}$$
$$I_n^- \; : \; C_n^- \longrightarrow C_n^0 \, , \quad I_n^- \circ T_n^- \; = \; T_n^0 \circ I_n^- \, , \tag{6.11b}$$

where $T_n^a = T^{\chi_n^a}$, $C_n^a = C^{\chi_n^a}$, and the signatures are:

$$\chi_n^+ = [n + 2, n; 2] \, , \quad \chi_n^- = [n, n + 2; 2] \, , \quad \chi_n^0 = [n + 1, n + 1; 3] \, , \quad n \in \mathbb{Z}_+ \, , \tag{6.12}$$

while instead of (6.5) we have:

$$I_n^+ \, F_n^+(z, \bar{z}) \; = \; J_n(z, \bar{z}) \, , \tag{6.13a}$$
$$I_n^- \, F_n^-(z, \bar{z}) \; = \; J_n(z, \bar{z}) \, , \tag{6.13b}$$

where

$$I_n^+ \; = \; \frac{n + 2}{2} \Big(\bar{z} \partial_+ + \partial_v \Big) - \frac{1}{2} \Big(\bar{z} z \partial_+ + z \partial_v + \bar{z} \partial_{\bar{v}} + \partial_- \Big) \partial_z \, , \quad n \in \mathbb{Z}_+ \tag{6.14a}$$

$$I_n^- \; = \; \frac{n + 2}{2} \Big(z \partial_+ + \partial_{\bar{v}} \Big) - \frac{1}{2} \Big(\bar{z} z \partial_+ + z \partial_v + \bar{z} \partial_{\bar{v}} + \partial_- \Big) \partial_{\bar{z}} \, , \quad n \in \mathbb{Z}_+ \tag{6.14b}$$

while $F_n^+(z, \bar{z})$, $F_n^-(z, \bar{z})$, $J_n(z, \bar{z})$, are polynomials in $z, \bar{z}$ of degrees $(n + 2, n)$, $(n, n + 2)$, $(n + 1, n + 1)$, resp., as explained above. If we want to use the notation with indices as in (6.1), then $F_n^+(z, \bar{z})$ and $F_n^-(z, \bar{z})$ correspond to $F_{\mu\nu, \alpha_1, \ldots, \alpha_n}$ which is antisymmetric in the indices μ, ν, symmetric in $\alpha_1, \ldots, \alpha_n$, and traceless in every pair of indices,[10] while $J_n(z, \bar{z})$ corresponds to $J_{\mu, \alpha_1, \ldots, \alpha_n}$ which is symmetric and traceless in every pair of indices. Note, however, that the analogs of (6.1) would be much more complicated if one wants to write explicitly all components. The crucial advantage of (6.13) is that the operators $I_n^\pm$ are given just by a slight generalization of $I^\pm = I_0^\pm$. (In another form these operators may be obtained [31] from those for the Euclidean conformal group in [30] using the Weyl unitary trick.)

---

[10] In 4D conformal field theory the families of mixed tensors $F_{\mu\nu, \alpha_1, \ldots, \alpha_n}$ appear, e.g., in the operator product expansion of two spin $1/2$ fields [33].

We shall call the hierarchy of equations (6.13) the **Maxwell hierarchy** . The Maxwell equations are the zero member of this hierarchy.

To proceed further we rewrite (6.14) in the following form:

$$I_n^+ \;=\; \frac{1}{2}\Big((n+2)I_1 I_2 - (n+3)I_2 I_1\Big) \;, \tag{6.15a}$$

$$I_n^- \;=\; \frac{1}{2}\Big((n+2)I_3 I_2 - (n+3)I_2 I_3\Big) \;, \tag{6.15b}$$

where

$$I_1 \;\equiv\; \partial_z \;, \quad I_2 \;\equiv\; \bar z z \partial_+ + z \partial_v + \bar z \partial_{\bar v} + \partial_- \;, \quad I_3 \;\equiv\; \partial_{\bar z} \;. \tag{6.16}$$

We note in passing that group-theoretically the operators I_a correspond to the three simple roots of the root system of $sl(4)$, while the operators $I_n^\pm$ correspond to the two non-simple non-highest roots [32], [5].

This is the form that we generalize for the q - deformed case. In fact, we can write at once the general form, which follows from (5.11a), (5.12a) :

$$_q I_n^+ \;=\; \frac{1}{2}\Big([n+2]_q I_1^q I_2^q - [n+3]_q I_2^q I_1^q\Big) \;, \tag{6.17a}$$

$$_q I_n^- \;=\; \frac{1}{2}\Big([n+2]_q I_3^q I_2^q - [n+3]_q I_2^q I_3^q\Big) \;. \tag{6.17b}$$

It is our task (using the previous Sections) to make this form explicit by first generalizing the variables, then the functions and the operators.

6.3. The variables $x_\pm, v, \bar v, z, \bar z$ have definite group-theoretical meaning, namely, they are six local coordinates on the flag manifold $\mathcal{Y} = SL(4)/B$, where B is the Borel subgroup of $SL(4)$ consisting of all upper diagonal matrices. (Equally well one may take the flag manifold $SL(4)/B^-$, where B^- is the Borel subgroup of lower diagonal matrices.) Under the natural conjugation (cf. also below) this is also a flag manifold of the conformal group $SU(2,2)$.

We know from Sections 3. and 4. what are the properties of the non-commutative coordinates on the $SL_q(4)$ flag manifold. We make the following identification (compare with (4.13)) :

$$x_+ = w = Y_{41} \;, \quad x_- = u = Y_{32} \;, \quad v = x = Y_{31} \;, \quad \bar v = y = Y_{42} \tag{6.18a}$$

$$z = \xi = Y_{21} \;, \quad \bar z = \eta = Y_{43} \;, \tag{6.18b}$$

for the q-Minkowski space-time coordinates and for the spin coordinates, which we denote as their classical counterparts. Thus, we obtain for the commutation rules of the q-Minkowski space-time coordinates (cf. (4.2)) :

$$x_\pm v \;=\; q^{\pm 1} v x_\pm \;, \quad x_\pm \bar v \;=\; q^{\pm 1} \bar v x_\pm \;,$$
$$x_+ x_- - x_- x_+ \;=\; \lambda v \bar v \;, \quad \bar v v \;=\; v \bar v \;. \tag{6.19}$$

As expected, relations (6.19) coincide with the commutation relations between the translation generators P_μ of the q-conformal algebra [7]. It is also easy to notice that these relations are as the $GL_q(2)$ commutation relations [20], if we identify our coordinates with the standard a, b, c, d generators of $GL_q(2)$ as follows:

$$M \;=\; \begin{pmatrix} a & b \\ c & d \end{pmatrix} \;=\; \begin{pmatrix} x_+ & v \\ \bar v & x_- \end{pmatrix} \;. \tag{6.20}$$

The q-Minkowski length is defined as the $GL_q(2)$ q-determinant :

$$\ell_q \doteq \det_q M = ad - qbc = x_+ x_- - q\bar{v}v ,\qquad (6.21)$$

and hence it commutes with the q-Minkowski coordinates. It has the correct classical limit $\ell_{q=1} = x_0^2 - \vec{x}^2$.

We know from (4.3) that for q phase ($|q| = 1$) the commutation relations (6.19) are preserved by an anti-linear anti-involution ω acting as :

$$\omega(x_\pm) = x_\pm , \quad \omega(v) = \bar{v} ,\qquad (6.22)$$

from which follows also that $\omega(\ell_q) = \ell_q$.

Remarks :
1. Note that relations (6.19) are different from the commutation relations of q-Minkowski space-time (with q real) in [34], [35], [36], (cf. also [37]). Recently, [38], it was shown that the q-Minkowski space of [34], [35], [36] can be obtained by a quantum Wick rotation (twisting) from a q-Euclidean space. The latter is also related to $GL_q(2)$, as our q-Minkowski space, however, for q real and under a different anti-linear anti-involution: $\tilde{\omega}_E(a) = d$, $\tilde{\omega}_E(b) = -q^{-1}c$, i.e., for the matrix M (cf. (6.20)) this is the unitary $*$, [38], while with our conjugation (6.22) M is hermitean.
2. Another proposal for deformed space-time may be obtained by extension of a new operator realization of $SU(2)$ quantum group representation matrices over non-commuting coordinates [39].
3. In the framework of algebraic field theory different proposals for quantum space-times were put forward in [40], [41].

The commutation rules of the spin variables $\bar{z}, z$ between themselves, with the q-Minkowski coordinates and with the q-Minkowski length are (cf. (4.2)) :

$$\begin{aligned}
\bar{z}z &= z\bar{z} , \\
x_+ z &= q^{-1}zx_+ , \quad x_- z = qzx_- - \lambda v , \\
vz &= q^{-1}zv , \quad \bar{v}z = qz\bar{v} - \lambda x_+ , \\
\bar{z}x_+ &= qx_+\bar{z} , \quad \bar{z}x_- = q^{-1}x_-\bar{z} + \lambda\bar{v} , \\
\bar{z}v &= q^{-1}v\bar{z} + \lambda x_+ , \quad \bar{z}\bar{v} = q\bar{v}\bar{z} , \\
z\ell_q &= \ell_q z , \quad \bar{z}\ell_q = \ell_q\bar{z} .
\end{aligned}\qquad (6.23)$$

Certainly, the commutation relations (6.23) are also preserved (for q phase) by the conjugation ω which acts (cf. (4.3)) by : $\omega(z) = \bar{z}$. Thus, with this conjugation $\mathcal{Y}_q$ becomes a flag manifold of $SU_q(2,2)$.

From (4.4) we know the normally ordered basis of the q - flag manifold $\mathcal{Y}_q$ considered as an associative algebra :

$$\hat{\varphi}_{ijk\ell mn} = z^i\,v^j\,x_-^k\,x_+^\ell\,\bar{v}^m\,\bar{z}^n , \quad i,j,k,\ell,m,n \in \mathbb{Z}_+ .\qquad (6.24)$$

Let us denote by $\mathcal{Z}$, $\bar{\mathcal{Z}}$, and $\mathcal{M}_q$ the associative algebras with unity generated by z, $\bar{z}$, and $x_\pm, v, \bar{v}$, resp. These three algebras are subalgebras of $\mathcal{Y}_q$, and we notice the following structure of $\mathcal{Y}_q$:

$$\mathcal{Y}_q \cong \mathcal{Z} \ltimes \mathcal{M}_q \rtimes \bar{\mathcal{Z}} ,\qquad (6.25)$$

where $A \mathbin{\hat{\otimes}} B$ denotes the tensor product of A and B with A acting on B.

We introduce now the representation spaces C^χ, $\chi = [n_1, n_2; d]$. The elements of C^χ, which we shall call (abusing the notion) functions, are polynomials in $z, \bar{z}$ of degrees n_1, n_2, resp., and formal power series in the q - Minkowski variables. (In the general $U_q(sl(n))$ situation the signatures n_1, n_2 are complex numbers and the functions are formal power series in $z, \bar{z}$ too, cf. (3.46b).) Namely, these functions are given by:

$$\hat{\varphi}_{n_1, n_2}(\bar{Y}) \;=\; \sum_{\substack{i,j,k,\ell,m,n \in \mathbb{Z}_+ \\ i \le n_1, \; n \le n_2}} \mu^{n_1, n_2}_{ijk\ell mn}\, \hat{\varphi}_{ijk\ell mn} \;, \tag{6.26}$$

where $\bar{Y}$ denotes the set of the six coordinates on $\mathcal{Y}_q$. Thus the analogs of $F_n^{\pm}$, J_n, cf. (6.13), are :

$$ {}_qF_n^{+} \;=\; \hat{\varphi}_{n+2,n}(\bar{Y}) \;, \quad {}_qF_n^{-} \;=\; \hat{\varphi}_{n,n+2}(\bar{Y}) \;, \quad {}_qJ_n \;=\; \hat{\varphi}_{n+1,n+1}(\bar{Y}) \;. \tag{6.27}$$

Next we introduce the following operators acting on our functions (cf. (3.48), (3.49)) :

$$\hat{M}_\kappa\, \hat{\varphi}(\bar{Y}) \;=\; \sum_{i,j,k,\ell,m,n \in \mathbb{Z}_+} \mu^{n_1, n_2}_{ijk\ell mn}\, \hat{M}_\kappa\, \hat{\varphi}_{ijk\ell mn} \;, \tag{6.28a}$$

$$T_\kappa\, \hat{\varphi}(\bar{Y}) \;=\; \sum_{i,j,k,\ell,m,n \in \mathbb{Z}_+} \mu^{n_1, n_2}_{ijk\ell mn}\, T_\kappa\, \hat{\varphi}_{ijk\ell mn} \;, \tag{6.28b}$$

where $\kappa = z, \pm, v, \bar{v}, \bar{z}$, and the explicit action on $\hat{\varphi}_{ijk\ell mn}$ is defined by:

$$\hat{M}_z\, \hat{\varphi}_{ijk\ell mn} \;=\; \hat{\varphi}_{i\pm1, jk\ell mn} \;, \tag{6.29a}$$

$$\hat{M}_v\, \hat{\varphi}_{ijk\ell mn} \;=\; \hat{\varphi}_{i, j\pm1, k\ell mn} \;, \tag{6.29b}$$

$$\hat{M}_-\, \hat{\varphi}_{ijk\ell mn} \;=\; \hat{\varphi}_{ij, k\pm1, \ell mn} \;, \tag{6.29c}$$

$$\hat{M}_+\, \hat{\varphi}_{ijk\ell mn} \;=\; \hat{\varphi}_{ijk, \ell\pm1, mn} \;, \tag{6.29d}$$

$$\hat{M}_{\bar{v}}\, \hat{\varphi}_{ijk\ell mn} \;=\; \hat{\varphi}_{ijk\ell, m\pm1, n} \;, \tag{6.29e}$$

$$\hat{M}_{\bar{z}}\, \hat{\varphi}_{ijk\ell mn} \;=\; \hat{\varphi}_{ijk\ell m, n\pm1} \;, \tag{6.29f}$$

$$T_z\, \hat{\varphi}_{ijk\ell mn} \;=\; q^i\, \hat{\varphi}_{ijk\ell mn} \;, \tag{6.30a}$$

$$T_v\, \hat{\varphi}_{ijk\ell mn} \;=\; q^j\, \hat{\varphi}_{ijk\ell mn} \;, \tag{6.30b}$$

$$T_-\, \hat{\varphi}_{ijk\ell mn} \;=\; q^k\, \hat{\varphi}_{ijk\ell mn} \;, \tag{6.30c}$$

$$T_+\, \hat{\varphi}_{ijk\ell mn} \;=\; q^\ell\, \hat{\varphi}_{ijk\ell mn} \;, \tag{6.30d}$$

$$T_{\bar{v}}\, \hat{\varphi}_{ijk\ell mn} \;=\; q^m\, \hat{\varphi}_{ijk\ell mn} \;, \tag{6.30e}$$

$$T_{\bar{z}}\, \hat{\varphi}_{ijk\ell mn} \;=\; q^n\, \hat{\varphi}_{ijk\ell mn} \;. \tag{6.30f}$$

Now we define the q-difference operators by (cf. (3.50)) :

$$\hat{\mathcal{D}}_\kappa\, \hat{\varphi}(\bar{Y}) \;=\; \frac{1}{\lambda}\, \hat{M}_\kappa^{-1}\, \left(T_\kappa - T_\kappa^{-1}\right) \hat{\varphi}(\bar{Y}) \;, \quad \kappa = z, \pm, v, \bar{v}, \bar{z} \;. \tag{6.31}$$

6.4. Using the above machinery we can present explicitly a q version of the Maxwell hierarchy of equations. First, we recall that the explicit form of the operators I_a in (6.16) is obtained by the infinitesimal right action of the three simple root generators of $sl(4)$ on the flag manifold $\mathcal{Y}$ - cf. (5.7). Adapting this to our notation we have for the q-analogs of I_a (cf. (5.6)) :

$$I_1^q \;=\; \hat{\mathcal{D}}_z \, T_z \, (T_v T_- T_+ T_{\bar{v}})^{-1} \;, \tag{6.32a}$$

$$I_2^q \;=\; \Big(q \, \hat{M}_z \, \hat{\mathcal{D}}_v \, T_- \;+\; \hat{\mathcal{D}}_- \;+$$

$$+\; \hat{M}_z \, \hat{M}_{\bar{z}} \, \hat{\mathcal{D}}_+ \, (T_v T_+)^{-1} \, T_{\bar{v}} \;+\; q^{-1} \, \hat{M}_{\bar{z}} \, \hat{\mathcal{D}}_{\bar{v}} \, (T_- T_+)^{-1} \;- \tag{6.32b}$$

$$-\; \lambda \, \hat{M}_v \, \hat{M}_{\bar{z}} \, \hat{\mathcal{D}}_- \, \hat{\mathcal{D}}_+ \, (T_- T_+)^{-1} \, T_{\bar{v}} \Big) \, T_- \, T_+ \, T_{\bar{v}} \, T_{\bar{z}}^{-1} \;,$$

$$I_3^q \;=\; \hat{\mathcal{D}}_{\bar{z}} \, T_{\bar{z}} \;. \tag{6.32c}$$

With this we have now the q - Maxwell hierarchy of equations - it remains just to substitute the operators of (6.32) in (6.17). In fact, we can also rewrite these in the q - analog of (6.13). We have (cf. (5.11b), (5.12b)) :

$$_qI_n^+ \;=\; \frac{q^2}{2} \, [n+2]_q \, \Big(q \, \hat{\mathcal{D}}_v \, T_- \;+\; \hat{M}_{\bar{z}} \, \hat{\mathcal{D}}_+ \, (T_v T_+)^{-1} \, T_{\bar{v}} \Big) \, T_z^2 \, (T_v T_{\bar{z}})^{-1} \;-$$

$$-\; \frac{1}{2} q^{-n-2} \, \Big(q \, \hat{M}_z \, \hat{\mathcal{D}}_v \, T_- \;+\; \hat{\mathcal{D}}_- \;+$$

$$+\; \hat{M}_z \, \hat{M}_{\bar{z}} \, \hat{\mathcal{D}}_+ \, (T_v T_+)^{-1} \, T_{\bar{v}} \;+\; q^{-1} \, \hat{M}_{\bar{z}} \, \hat{\mathcal{D}}_{\bar{v}} \, (T_- T_+)^{-1} \;-$$

$$-\; \lambda \, \hat{M}_v \, \hat{M}_{\bar{z}} \, \hat{\mathcal{D}}_- \, \hat{\mathcal{D}}_+ \, (T_- T_+)^{-1} \, T_{\bar{v}} \Big) \, \hat{\mathcal{D}}_z \, T_z \, (T_v T_{\bar{z}})^{-1} \tag{6.33a}$$

$$_qI_n^- \;=\; \frac{q}{2} \, [n+2]_q \, \Big(q^{-1} \, \hat{\mathcal{D}}_{\bar{v}} \, (T_- T_+)^{-1} \;+\; \hat{M}_z \, \hat{\mathcal{D}}_+ \, (T_v T_+)^{-1} \, T_{\bar{v}} \;-$$

$$-\; \lambda \, \hat{M}_v \, \hat{\mathcal{D}}_- \, \hat{\mathcal{D}}_+ \, (T_- T_+)^{-1} \, T_{\bar{v}} \Big) \, T_- \, T_+ \, T_{\bar{v}} \, T_{\bar{z}} \;+$$

$$+\; \frac{1}{2} q^{n+3} \, \Big(q \, \hat{M}_z \, \hat{\mathcal{D}}_v \, T_- \;+\; \hat{\mathcal{D}}_- \;+$$

$$+\; \hat{M}_z \, \hat{M}_{\bar{z}} \, \hat{\mathcal{D}}_+ \, (T_v T_+)^{-1} \, T_{\bar{v}} \;+\; q^{-1} \, \hat{M}_{\bar{z}} \, \hat{\mathcal{D}}_{\bar{v}} \, (T_- T_+)^{-1} \;-$$

$$-\; \lambda \, \hat{M}_v \, \hat{M}_{\bar{z}} \, \hat{\mathcal{D}}_- \, \hat{\mathcal{D}}_+ \, (T_- T_+)^{-1} \, T_{\bar{v}} \Big) \, \hat{\mathcal{D}}_{\bar{z}} \, T_- \, T_+ \, T_{\bar{v}} \tag{6.33b}$$

Clearly, for $q = 1$ the operators in (6.32), (6.33) coincide with (6.16), (6.15), resp.

With this the final result for the q - Maxwell hierarchy of equations is (cf. (6.27)) :

$$_qI_n^+ \; _qF_n^+ \;=\; _qJ_n \;, \tag{6.34a}$$

$$_qI_n^- \; _qF_n^- \;=\; _qJ_n \;. \tag{6.34b}$$

Note that our free q - Maxwell equations, obtained from (6.34) for $n = 0$, and $_qJ_0 = 0$, are different from the free q - Maxwell equations of [42], [43]. (This is natural since they use different q - Minkowski space-time from [34], [35], [36].) The advantages of our equations are: 1) they have simple indexless form; 2) we have a whole hierarchy of equations; 3) we have the full equations, and not only their free counterparts; 4) our equations are q - conformal invariant, not only q - Lorentz [43], or q - Poincaré [42], invariant. (In fact, it is not clear whether the q - Lorentz algebras of [34], [35], [36], [44] or the q - Poincaré algebra of [45] are extendable to q - conformal algebras (often easy $q = 1$ things fail for $q \neq 1$).)

6.5. In this Section we presented some highlights of the application of our approach to the derivation of q - conformal invariant equations. We should stress that the indexless formulation we are employing allows easily to write whole hierarchies of equations in a compact form. We should also stress that in our approach we are getting the full equations (not only the free counterparts) automatically, since the intertwining operators always involve two representations (and more than two representations, if we take into account compositions of such operators).

Finally, we should note that some conformally invariant equations can be obtained by a more geometric procedure [46]. The adaptation of the latter procedure to the q - deformed case is an interesting problem for future study.

References

[1] A.O. Barut and R. Rączka, *Theory of Group Representations and Applications*, II edition, (Polish Sci. Publ., Warsaw, 1980).

[2] A.W. Knapp and E.M. Stein, Ann. Math. **93** (1971) 489; Inv. Math. **60** (1980) 9.

[3] B. Kostant, Lecture Notes in Math., Vol. 466 (Springer-Verlag, Berlin, 1975) p. 101.

[4] D.P. Zhelobenko, Math. USSR Izv. **10** (1976) 1003.

[5] V.K. Dobrev, Rep. Math. Phys. **25** (1988) 159.

[6] N. Bourbaki, *Groupes at algèbres de Lie, Chapitres 4,5 et 6*, (Hermann, Paris, 1968).

[7] V.K. Dobrev, J. Phys. A: Math. Gen. **26** (1993) 1317-1334; first as Göttingen University preprint, (July 1991).

[8] A.W. Knapp, *Representation Theory of Semisimple Groups (An Overview Based on Examples)*, (Princeton Univ. Press, 1986).

[9] R.P. Langlands, On the classification of irreducible representations of real algebraic groups, Mimeographed notes, Pronceton (1993).

[10] A.W. Knapp and G.J. Zuckerman, in: Lecture Notes in Math. Vol. 587 (Springer, Berlin, 1977) pp. 138-159.; Ann. math. **116** (1982) 389-501.

[11] J. Dixmier, *Enveloping Algebras*, (North Holland, New York, 1977).

[12] V.K. Dobrev, Lectures on Lie algebras and their representations : I, ICTP internal report IC/88/96 (1988), Course of 16 lectures at the ICTP (April - July 1988).

[13] I.N. Bernstein, I.M. Gel'fand and S.I. Gel'fand, Funkts. Amal. Prilozh. **5**(1)1–9(1971); English translation: Funkt. Anal. Appl. **5**, 1–8 (1971).

[14] B. Speh and D.A. Vogan, Jr., Acta Math. **145** (1980) 227-299.

[15] T. Masuda, K. Mimachi, Y. Nakagami, M. Noumi, Y. Sabuti and K. Ueno, Lett. Math. Phys. **19** (1990) 187-194; Lett. Math. Phys. **19** (1990) 195-204.

[16] V.K. Dobrev, unpublished, (May 1993).

[17] L. Dabrowski, V.K. Dobrev and R. Floreanini, J. Math. Phys. **35** (1994) 971-985.

[18] V.K. Dobrev, J. Phys. A: Math. Gen. **27** (1994) 4841-4857; first as preprint ASI-TPA/10/93, (October 1993); hepth/9405150.

[19] V.K. Dobrev, New q - Minkowski space-time and q - Maxwell equations hierarchy from q - conformal invariance, preprint ASI-TPA/15/94 (June 1994), Phys. Lett. B, to appear.

[20] Yu.I. Manin, Quantum groups and non-commutative geometry, Montreal University preprint, CRM-1561 (1988); Comm. Math. Phys. **123** (1989) 163-175.

[21] V.K. Dobrev, J. Math. Phys. **33** (1992) 3419.

[22] V.K. Dobrev and P. Parashar, J. Phys. A: Math. Gen. **26** (1993) 6991.

[23] H. Awata, M. Noumi and S. Odake, preprint YITP/K-1016 (1993).

[24] D.P. Zhelobenko, *Harmonic Analysis on Semisimple Complex Lie Groups*, (Nauka, Moscow, 1974) (in Russian).

[25] A. Sudbery, J. Phys. A : Math. Gen. **23** (1990) L697.

[26] A. Schirrmacher, Zeit. f. Physik **C50** (1991) 321.

[27] A.Ch. Ganchev and V.B. Petkova, Phys. Lett. **B233** (1989) 374-382.

[28] M. Jimbo, Lett. Math. Phys. **10** (1985) 63-69.

[29] V.K. Dobrev, in: Proceedings of the International Group Theory Conference (St. Andrews, 1989), Eds. C.M. Campbell and E.F. Robertson, Vol. 1, London Math. Soc. Lecture Note Series 159 (Cambridge University Press, 1991) pp. 87-104; first as ICTP Trieste internal report IC/89/142 (June 1989).

[30] V.K. Dobrev and V.B. Petkova, Reports Math. Phys. **13** (1978) 233-277.

[31] V.B. Petkova and G.M. Sotkov, Lett. Math. Phys. **8** (1984) 217.

[32] V.K. Dobrev, J. Math. Phys. **26** (1985) 235.

[33] V.K. Dobrev, E.H. Hristova, V.B. Petkova and D.B. Stamenov, Bulg. J. Phys. **1** (1974) 42.

[34] U. Carow-Watamura, M. Schlieker, M. Scholl and S. Watamura, Zeit. f. Physik **C48** (1990) 159.

[35] W.B. Schmidke, J Wess and B. Zumino, Zeit. f. Physik **C52** (1991) 471.

[36] S. Majid, J. Math. Phys. **32** (1991) 3246.

[37] J.A. de Azcarraga, P.P. Kulish and F. Rodenas, preprint FTUV-93-36 (1993), hepth/9309036.

[38] S. Majid, preprint DAMTP/94-03, (1994), hepth/9401112.

[39] L.C. Biedenharn and M.A. Lohe, Quantum groups as a theory of quantized space, preprint Texas Univ. at Austin, CPP-94-24 (July 1994).

[40] G. Mack and V. Schomerus, Models of quantum space-time: quantum field planes, preprint HUTMP 94-B335, hep-th/9403170 (March 1994).

[41] S. Doplicher, K. Fredenhagen and J.E. Roberts, Phys. Lett. **331B** (1994) 39-44; The quantum structure of spacetime at the Planck scale and quantum fields, preprint (1994).

[42] M. Pillin, preprint MPI-Ph/93-61 (1993), hepth/9310097.

[43] U. Meyer, preprint DAMTP/94-10 (1994), hepth/9404054.

[44] O. Ogievetsky, W.B. Schmidke, J. Wess and B. Zumino, Lett. Math. Phys. **23** (1991) 233.

[45] O. Ogievetsky, W.B. Schmidke, J. Wess and B. Zumino, Comm. Math. Phys. **150** (1992) 495.

[46] J.-D. Hennig, in: Classical and Quantum Physics, (World Sci, Singapore, 1992) 708.

A QUANTUM MECHANICAL EVOLUTION EQUATION FOR MIXED STATES FROM SYMMETRY AND KINEMATICS

H.-D. Doebner and J.D. Hennig

Arnold Sommerfeld Institute
for Mathematical Physics
Technical University of Clausthal, Germany

I. INTRODUCTION

The idea of this contribution goes back to an article published in Symmetry in Science [1]. We proposed there a quantization method for a system S localized on a smooth Riemannian manifold (M,g) and presented preliminary results for the kinematics which were developed and formulated systematically and rigorously in [2,3], with applications in [4]. This approach, the Quantum Borel Kinematics, is based geometrically on a representation of a pair $(\mathcal{B}(M), \mathrm{Vect}(M))$, or equivalently $\mathcal{S}(M) = (C^\infty(M, \mathbb{R}), \mathrm{Vect}(M))$, on some Hilbert space $\mathcal{H}$, with $\mathcal{B}(M)$ as the Borel field and Vect(M) as the Lie algebra of smooth vector fields on M, and $C^\infty(M, \mathbb{R})$ as the space of smooth functions. For $M = \mathbb{R}^3$ the results of the quantization were derived independently in connection with a representation of a certain subgroup of the diffeomorphism group Diff(M) of M in [5]. The pair is a purely *kinematical* quantity. Borel sets are *generalized positions* and vector fields are *generalized momenta*. To describe a dynamical stuation, S must be furnished with a time dependence. A conventional method to do this is to write an evolution equation or a class of evolution equations. The choice in classical mechanics for point particles is the class of second order (or Newtonian) equations. A construction of a quantum analogue of this class has to be based on the quantization of the kinematics, i.e. on the unitarily inequivalent Quantum Borel Kinematics.

In this note we collect some ideas how to furnish the representation of $\mathcal{S}(M)$ with such a time dependence (see [6] for a more detailed version). In section II we repeat the notion of second order equations with a somewhat special view on nonrelativistic classical mechanics on M with Riemannian metric; we describe the generic quantization method for systems S and the necessary representation theory for $\mathcal{S}(M)$ in section III and construct the quantum analogue of the above mentioned class in section IV. For mixed states, we get evolution equations for density matrices (generalized von Neumann equations) of Lindblad type in section V ; for pure states we find a family of nonlinear Schrödinger equations derived in another context in [7,8].

II. CLASSICAL SYSTEMS, KINEMATICAL SYMMETRIES AND THE CLASS OF NEWTONIAN EVOLUTION EQUATIONS

We consider a system S modelled as a point $m \in M$ and moving on M ; its path in M generates at each m a tangent vector v_m (formal velocity) which describes the kinematical situation. The v_m are collected in the tangent bundle TM . Each path on M can be lifted canonically to a path on TM . We choose now those systems for which all physically possible paths, parametrized through time and with $\dot{m}(t) = v_m$, can be obtained through

a) a flow Φ^Y of a vector field Y on TM such that

b) for the tangential map

$$\pi_* (Y_{\gamma(0)}) = v_m \tag{1}$$

holds ($\pi : TM \to M$ as natural projection and $\gamma(t)$ as the flow line through $v_m = \gamma(0)$) The evolution equation for $m(t)$ is under this condition of second order and of standard form $\dot{m} = v$, $\dot{v} = a(\dot{m}, m)$. The conditions a), b) carry certainly the information that the projection of $\gamma(t)$ is a solution curve on M of a Newtonian evolution equation. For general position observables $f \in C^\infty(M, \mathbb{R})$ the situation is as follows. Denote by $F_f := \pi^* f$ the canonical lift of f from M to TM . Because $\frac{d}{dt} F_f(\gamma(t))$ is a derivative along the integral curves of Y and because the canonical lift commutes with d we have on TM

$$\frac{d}{dt} F_f(\gamma(t))\big|_{t=0} = d\,F_f(Y_{v_m}) = df\,(\pi_* Y_{v_m}) = df(v_m) \quad,$$

where $df(v_m)$ is the derivative in direction v_m of f on M . With the Riemannian structure g on M and $\overline{X}$ denoting the canonical lift of $X \in \mathrm{Vect}(M)$ to TM , we get

$$\frac{d}{dt} F_f \circ \Phi_t^Y\big|_{t=0} = \overline{\mathrm{grad}_g f} \quad, \tag{2}$$

i.e., the relation is independent of Y . This reflects the fact, that the time derivative of F_f in m only depends on the initial condition $\dot{m}$, the characteristic property of *second* order equations.

The kinematics is the pair $S(M) = (C^\infty(M, \mathbb{R}), \mathrm{Vect}(M))$. The smooth vector fields X on M form an infinite dimensional Lie algebra. If one considers the function space $C^\infty(M, \mathbf{R})$ with pointwise multiplication as an abelian Lie algebra, then $C^\infty(M, \mathbb{R})$ couples semidirectly to $\mathrm{Vect}(M)$ and one gets $\mathcal{S}(M)$ as an infinite dimensional symmetry or covariance or kinematical algebra

$$\mathcal{S}(M) = C^\infty(M, \mathbb{R}) \in \mathrm{Vect}(M) \quad,$$

i.e. an algebraic object, canonical for any M . $\mathcal{S}(M)$ was used also in [8]. With (2) on TM some information on the dynamics is encoded. Because $\mathcal{S}(M)$ is an object on M and $\overline{X}$ is defined on TM we describe, how $\mathcal{S}(M)$ and objects on TM are related. Consider function spaces $\mathcal{F}_n(TM)$ over TM spanned by polynomials in momentum variables of n'th order with coefficients from $C^\infty(M, \mathbb{R})$. We use the canonical Lie algebra isomorphism

$$F : \quad \mathcal{S}(M) \quad \to \quad \mathcal{F}_0(TM) \in \mathcal{F}_1(TM)$$
$$f + X \quad \to \quad F_f + F_X \tag{3}$$

with $F_f := \pi^* f$ and the obvious definition

$$F_X(v_m) := (X_m, v_m) \quad.$$

III. QUANTIZATION OF A SYSTEM WITH KINEMATICAL SYMMETRY $\mathcal{S}(M)$

There exists a quantization procedure for general systems which is generic:
Choose a suitable realisation of a separable Hilbert space $\mathcal{H}$. Take the time fixed. Consider the set $SA(\vartheta)$ of essentially selfadjoint operators A with common dense domain ϑ, the set $\mathcal{W}$ of positive trace class operators W of trace 1 and the set $\mathcal{O}$ of observables A of S. Then the quantization of S is a map $q: \mathcal{O} \to SA(\vartheta)$. The states of S are given through $\mathcal{W}$. The expectation value, i.e. the probability to measure an observable A is

$$\mathrm{Exp}_W(A) = \mathrm{Tr}(WA) \ .$$

We utilize this quantization procedure for localized systems on M and choose $L^2(M, d\mu)$ as a realisation of $\mathcal{H}$. Furthermore we have to decide on the set $\mathcal{O}$ of observables. Because in classical mechanics each observable can be written as a function of position and momentum we choose the kinematical algebra $\mathcal{S}(M)$ as a subset of $\mathcal{O}$. Up to ordering problems, the quantization of classical observables can be calculated with a quantization of $\mathcal{S}(M)$. For $\mathcal{S}(M)$ all mappings

$$q \ : \ \mathcal{S}(M) \to SA(\vartheta)$$
$$C^\infty(M, \mathbb{R}) \ni f \to Q(f) \in SA(\vartheta)$$
$$\mathrm{Vect}(M) \ni X \to P(X) \in SA(\vartheta)$$

were classified up to unitary equivalence under the following physically justified conditions on q.

1. The map q yields a multiplication operator, $Q(f)\psi = f \cdot \psi$, $\psi \in \mathcal{H}$.

2. The map q is a Lie algebra homomorphism, i.e. the Lie structure of $\mathrm{Vect}(M)$ is respected.

3. $P(X)$ is a differential operator in respect to the differentiable structure on line bundles with base M, hermitian metric and flat connection ∇.

The result is [2] : the unitary non equivalent irreducible quantizations q are in 1:1 correspondence to the pairs (a, D), where a is an element of the dual of the first fundamental group $\pi_1^*(M)$, and D is a real quantum number. The result for $P(X)$ on the dense domain ϑ is

$$P(X) = -i \nabla_X^a + \left(D - \frac{i}{2}\right) Q(\mathrm{div}_\mu X)$$

where the index a is a reminder for the choice of the line bundle. For discussions and examples see [2,3]. For $M = \mathbb{R}^3$ we have $(X =: \vec{g} \cdot \vec{\nabla})$

$$P(X) = -i\vec{g}\vec{\nabla} + \left(D - \frac{i}{2}\right) \mathrm{div}\vec{g} \ . \tag{4}$$

IV. THE QUANTUM ANALOGUE OF THE CLASS OF SECOND ORDER EVOLUTION EQUATIONS

To furnish the system S on M with a second order dynamics through a quantum analogue of *(2)* we use an irreducible quantization of $\mathcal{S}(M)$.

Because expectation values are the objects which are connected with experiments, it is necessary to rephrase *(2)* in the language of expectation values. There are various possibilities for this. In $\mathrm{Tr}(WA)$ either the states W or operators A (possibly also both) can be time dependent; this corresponds to the Schrödinger- and Heisenberg picture (or a mixed picture, e.g. the Dirac picture). A time evolution through a unitary group $U(t)$ for $W_t = U^*(t) W_0 U(t)$ in the Schrödinger picture gives for expectation values: $Tr(W_t A_0) = Tr(W_0 A_t)$ with $A_t = U^*(t) A_0 U(t)$, i.e. the same formula in the Heisenberg picture. For a unitary time evolution with a necessarily *linear* operator $U(t)$ both pictures coincide. For a general approach we do not assume that the quantum analogue to *(2)* allows only unitary evolutions. The results for the Schrödinger- and for the Heisenberg pictures could be different, and they are. We use here the Schrödinger picture. The reason is that this picture is directly related to physics: the system, i.e. its states , is moving, and not the observables.

To rephrase *(2)*, formulated on TM, for expectation values with kinematical operators acting on functions over M we use the isomorphism *(3)* between objects living on TM and M, respectively, and define

$$Q(F_f) := Q(f)$$
$$P(F_X) := P(X) \quad .$$

Now we are prepared to formulate the quantum analogue of *(2)* in the Schrödinger picture. We insert for F_f and $\overline{\text{grad}_g f}$ the corresponding operators and arrive at

$$\frac{d}{dt} Tr(W_t Q(f)) = Tr(W_t P(\text{grad}_g f)) \quad \text{for all } f \in C^\infty(M, \mathbf{R}) \quad . \tag{5}$$

This is our key formula, which restricts the dynamics of the system, i.e. of W_t, to the second order class in the above sense, i.e. the dynamics is as near as possible to classical Newtonian dynamics.

V. APPLICATIONS IN $\mathbf{R}^3$

The physical and mathematical contents of the trace relation become clear if we insert the Q and P maps. For $M = \mathbf{R}^3$ the trace formula is (insert *(4)*)

$$\frac{d}{dt} Tr(W_t \cdot f) = Tr(W_t \cdot A)$$

$$A = -i(\vec{\nabla} f)\vec{\nabla} + \left(D - \frac{i}{2}\right)(\vec{\nabla}^2 f) \quad . \tag{6}$$

We simplify *(6)* with the operator identities

$$[\Delta, f] = \Delta f + 2(\vec{\nabla} f)\vec{\nabla} \quad , \quad [\vec{\nabla}, [\vec{\nabla}, f]] = \Delta f$$

and with the properties of the trace and the free Hamiltonian ($\hbar = 1$, m $= 1$) $H_0 = -\frac{1}{2}\Delta$,

$$Tr(f \cdot \dot{W}_t) = Tr(f \cdot (-i[H_0, W_t] + D[\vec{\nabla}, [\vec{\nabla}, W_t]])) \quad .$$

Because this relation holds for any $f \in C^\infty(\mathbf{R}^3, \mathbf{R})$, a general evolution equation for W_t emerges

$$\dot{W}_t = -i[H_0, W_t] + D \sum_{i=1}^{3} \left[\frac{\partial}{\partial x^i}, \left[\frac{\partial}{\partial x^i}, W_t\right]\right] + F \quad , \tag{7}$$

$$Tr(f \cdot F) = 0 \quad \text{for all } f \in C^\infty(\mathbf{R}^3, \mathbf{R}) \quad ,$$

where F is an operator in $L^2(\mathbf{R}^3, d^3 x)$, *arbitrary* up to the above trace condition and the requirement, that the time evolution of W_t is contained in the set of density matrices, i.e. $W_t \in \mathcal{W}$, which is in general not easy to check. For consistency F should be a functional depending (linearly) on W_t , the kinematical observables and given geometrical objects on $\mathbb{R}^3$, like potentials $V(\vec{x})$. A possible choice is $F = -i[V, W_t]$ with

$$\dot{W}_t = -i[H, W_t] + D \sum_{i=1}^{3} \left[\frac{\partial}{\partial x^i}, \left[\frac{\partial}{\partial x^i}, W_t\right]\right] \tag{8}$$

$$H = H_0 + V$$

as an evolution equation of Lindblad type [9]. The usual von Neumann equation is a special case in (8) for $D = 0$, as expected.

If we restrict the motion of W_t from $\mathcal{W}$ to $\mathcal{W}_p \subset \mathcal{W}$ defined as the subset of pure states $W = P_\psi$, $\psi \in \mathcal{H}$, with $W^2 = W$, the trace formula (6) can be rewritten as

$$\frac{d}{dt} Tr(P_{\psi(t)} \cdot f \cdot P_{\psi(t)}) = Tr(P_{\psi(t)} A P_{\psi(t)}) \quad .$$

With a complete orthonormal system $\{\psi_\nu\}$ in $L^2(\mathbb{R}^3, d^3x)$ with $\psi_1 = \psi(t)$ this gives

$$\frac{d}{dt}(\psi(t), f\psi(t)) = (\psi(t), A\psi(t))$$

and, after some calculation

$$\int (\dot{\rho} + \vec{\nabla} j_o - D\Delta\rho) f \, d^3x = 0 \quad \text{for all} \quad f \in C^\infty(\mathbb{R}^3, \mathbb{R}) \tag{9}$$

with

$$\rho(\vec{x}, t) = \overline{\psi}(\vec{x}, t)\psi(\vec{x}, t) \quad , \quad \vec{j}_o(\vec{x}, t) = \frac{1}{2}(\overline{\psi}(\vec{x}, t)\vec{\nabla}\psi(\vec{x}, t) - \psi(\vec{x}, t)\vec{\nabla}\overline{\psi}(\vec{x}, t)) \quad .$$

From (9) we get a generalized continuity equation of Fokker Planck type with a diffusion current proportional to D in addition to the usual quantum mechanical current $\vec{j}_o$,

$$\dot{\rho} = -\vec{\nabla}\vec{j}_o + \vec{\nabla}(D\vec{\nabla}\rho) \quad . \tag{10}$$

This continuity equation carries a class of evolution equations for ψ , i.e. generalized Schrödinger equations. The class is specified through the ansatz

$$i\partial_t\psi = H\psi + G[\overline{\psi}, \psi]\psi \quad ,$$

where $G[\overline{\psi}, \psi]$ is a non linear functional of $\overline{\psi}, \psi$. Then (10) gives

$$\text{Im} G = \frac{1}{2}D\frac{\Delta\rho}{\rho} \quad ,$$

$$\text{Re} G \text{ not restricted through (10)} \quad ,$$

i.e. an evolution equation with a *non linear* term. This class was first derived together with *(10)* in [7] with a somewhat different method together with a suitable choice for Re G in [8], which yields a 5 parameter $(c_1, ..., c_5)$ family of non linear Schrödinger equations,

$$i\partial_t \psi = H\psi + i\frac{1}{2}D\frac{\Delta\rho}{\rho}\psi + R[\psi]$$

$$R[\psi] := c_1\frac{\vec{\nabla}\vec{j}_o}{\rho} + c_2\frac{\vec{\nabla}^2\rho}{\rho} + c_3\frac{\vec{j}^2}{\rho^2} + c_4\frac{\vec{j}\cdot\vec{\nabla}\rho}{\rho^2} + c_5\frac{(\vec{\nabla}\rho)^2}{\rho}\quad. \tag{11}$$

The family *(11)* was analysed as well from a mathematical as from a physical point of view
[10]. The evolution equation *(7)* for W_t on $\mathcal{W}$ was first presented at the XX. International Colloquium on Group Theoretical Methods in Physics 1994 [11].

REFERENCES

[1]   H.-D. DOEBNER, J. TOLAR in *Symmetries in Sciences*, 475-486,
       Eds.: B. Gruber, R.S. Millman, Plenum Press (1980)

[2]   B. ANGERMANN, H.-D. DOEBNER, J. TOLAR Lecture Notes in Math., Vol. 1037,
       171-208, Springer (1983)

[3]   H.-D. DOEBNER, U.A. MÜLLER J. Phys. A **26**, 719-730 (1993)

[4]   see e.g.:
       H.-D. DOEBNER, J.TOLAR, P. STOVICEK Czech J. Phys. B **31**,
       110-119 (1981)
       H.-D. DOEBNER, J. TOLAR *Symmetry in Science II*, 115-126,
       Plenum Press (1986)
       H.-D. DOEBNER, H. ELMERS, W. HEIDENREICH J. Math. Phys. **30**,
       1053-1057 (1089)

[5]   R.F. DASHEN, D.H. SHARP PHYS. REV. **165**, 1857 (1968)
       G.A. GOLDIN *Current Algebras as Unitary Representations of Groups*,
       Princeton University, PhD thesis (1969)
       G.A. GOLDIN J. Math. Phys. **12**, 462-487 (1971)

[6]   H.-D. DOEBNER, J.D. HENNIG, P. NATTERMANN to be published

[7]   H.-D. DOEBNER, G.A. GOLDIN Phys. Lett. A **162**, 397-401 (1992)

[8]   H.-D. DOEBNER, G.A. GOLDIN J. Phys. A **27**, 1771-1780 (1994)

[9]   G. LINDBLAD Comm. Math. Phys. **48**, 119-130 (1976)

[10]   G. AUBERTSON, P.C. SABATIER J. Math. Phys. **35**, 4028-4040 (1994)
       V.V. DODONOV, S.S. MIZRAHI Phys. Lett. A **181**, 129-134 (1993)
       P. NATTERMANN in *Proceedings of the XXVI Symposium on Mathematical Physics*,
       47-54, Nicolas Copernicus University Press, Torún (1994)
       P. NATTERMANN, W. SCHERER, A.G. USHVERIDZE Phys.Lett. A **184**, 234-240
       (1994)
       A.G. USHVERIDZE Phys. Lett. A **185**, 123-127 (1994)

[11]   H.-D. DOEBNER, J.D. HENNIG, P. NATTERMANN Lecture at the XX. International
       Colloquium on Group Theoretical Physics, Osaka (1994)

QUANTUM MECHANICAL MOTIONS OVER THE GROUP MANIFOLDS AND RELATED POTENTIALS

I. H. Duru

Trakya University
Department of Mathematics
P.K. 126, Edirne, Turkey
and
TUBITAK, Marmara Research Center
P.O. Box 21, 41470 Gebze, Turkey

INTRODUCTION

Schrödinger equations for many potentials are solved in terms of the special functions. Almost all of these special functions are the matrix elements of the representations of the Lie groups, with their arguments being the group parameters [1,2]. Even the simplest "special functions" namely the elementary transcendentals are related to the Lie groups, i.e., the one parameter Abelian Lie groups. The connection between the group representations and the special functions explains the mystrious properties of them, such as the recurance relations and addition theorems.

Schrödinger equations for many potentials are the same as the Schrödinger equations written for the point particle motion over suitable group manifolds. These Schrödinger equations in their time independent forms are equivalent to the eigen value equations written for the Casimir operators of the Lie algebra of some Lie groups.

Although it is possible to solve several Schrödinger equations with the classical analytical methods, i.e., without considering the symmetry aspects of them, discovering the underlying symmetries always provides deeper understandings. Suppose a quantum mechanical problem is solved in terms of the hypergeometric functions. If on the other hand we learn that the equation can be brought into a Jacobi equation we know that it is related to the motion over the *SU(2)* manifold, thus we gain a better insight.

Studying the path integrals over the group manifolds is especially important. In fact very few problems can be solved directly exacuting their path integrals. They are the free motion over the open space or in the box and the harmonic oscillator problem. Some other problems can be brought into the oscillator form by suitable coordinate and time transformations. H-atom [3], Morse potential [4] and

$V = 1/r^2 + r^2$ potential [5] are of these types. A large class of path integrals on the other hand, are solved by transforming them into the path integrations over the group manifolds.

All the above considerations are related to the non-relativistic particle motions. However the knowledge we obtain in these studies is very useful for some field theoretical problems too. For example,

(i) pair production calculations in the external electromagnetic fields or in the cosmological backgrounds, in practise are reduced to the studies of Schrödinger equations in some formal "space-times"; and,

(ii) the Green functions employed in the vacuum structure studies for different geometries (i.e., the Casimir interaction studies) are formally related to the Green functions corresponding to some non-relativistic particle motions.

In the coming sections we present examples related to the *SU(2)* group. We then mention some other group manifolds and the related potentials. At the end we outline some field theoretical examples.

II. PARTICLE MOTION OVER S^3

We parametrize the *SU(2)* manifold S^3 in terms of the Euler angles:

$$u_1 = \cos\frac{\theta}{2}\cos\frac{\varphi+\psi}{2} \quad , \quad u_2 = \cos\frac{\theta}{2}\sin\frac{\varphi+\psi}{2}$$
$$u_3 = \sin\frac{\theta}{2}\cos\frac{\varphi-\psi}{2} \quad , \quad u_4 = \sin\frac{\theta}{2}\sin\frac{\varphi-\psi}{2} \tag{1}$$

The ranges of the angles are $0 \le \theta \le \pi$, $0 \le \varphi \le 2\pi$, $-2\pi \le \psi \le 2\pi$.

Schrödinger equation for the particle moving on S^3 is (with $\hbar=1$)

$$-\frac{1}{2I}\left[\frac{1}{\sin\theta}\partial_\theta(\sin\theta\,\partial_\theta) + \frac{\partial_\varphi^2 - 2\cos\theta\,\partial_\psi\partial_\varphi + \partial_\psi^2}{\sin^2\theta}\right]\psi = E\psi \tag{2}$$

where I is the moment of inertia. The wave functions and the energy spectrum are

$$\psi_{nm}^l(\theta,\varphi,\psi) = \frac{\sqrt{2l+1}}{4\pi}\,e^{-im\varphi}\,e^{-in\psi}\,P_{mn}^l(\cos\theta) \tag{3}$$

$$E = \frac{1}{2I}\,l(l+1) \quad ; \quad l = 0,1,2,..., \quad ; \quad m,n = -l\,,\,-l+1,...,l-1\,,\,l\,. \tag{4}$$

Here P_{mn}^l are the matrix elements of the *SU(2)* representation. Note that the Schrödinger equation (2) is same as the e-value equation written for the Casimir operator of the *SU(2)* Lie algebra.

The classical Hamiltonian corresponding to (2) is

$$H_{SU(2)} = \frac{P_\theta^2}{2I} + \frac{P_\varphi^2 + P_\psi^2 - 2P_\varphi P_\psi\cos\theta}{2I\sin^2\theta} \tag{5}$$

The path integration for the motion over the *SU(2)* manifold is also exactly calculable [6]: One first writes the short time interval path integration in the Lagrangian formulation, then expands it in terms of the *SU(2)* matrix elements. By making use of the orthonormality relations of the *SU(2)* matrix elements one obtains the finite time interval Green function. In the phase space formulation the path integral is written as

$$K(u_a , u_b ; T) = \int D (\theta , \varphi , \psi) \, D (p_\theta , p_\varphi , p_\psi)$$

$$\exp \left[i \int_0^T dt \, (p_\theta \dot\theta + p_\varphi \dot\varphi + p_\psi \dot\psi \; - \; \frac{p_\theta^2}{2I} \; - \; \frac{p_\varphi^2 + p_\psi^2 - 2 p_\psi p_\psi \cos\theta - 1/4}{2I \sin^2\theta}) \right] \tag{6}$$

where T is the time interval. $(8I \sin^2\theta)^{-1}$ term in the action is the usual ordering term which appear in the polar coordinate representations of the path integrals. The solutions of (6) is [2]:

$$K(u_a , u_b ; T) = - \frac{e^{iT/8I}}{4\sin\dfrac{\odot}{2}} (\frac{I}{2\pi iT})^{3/2} \sum_{l=-\infty}^{\infty} (-1)^l \, (\odot + 2\pi l) \, e^{i \frac{I}{2T} (\odot + 2\pi il)} \tag{7}$$

$$= - \frac{e^{iT/8I}}{16\pi^2 \sin\dfrac{\odot}{2}} \frac{\partial}{\partial \odot} \, \theta_2 \, (\frac{\odot}{2\pi} , -\frac{T}{2\pi I}) \tag{8}$$

θ_2 is the Jacobi Theta function. The angle $\odot$ is defined by the relation:

$$\cos \frac{\odot}{2} = \cos \frac{\theta_{ab}}{2} \cos \frac{\varphi_{ab} + \psi_{ab}}{2} \tag{9}$$

The angles in the above expression are given in terms of the initial and final points with

$$\cos \theta_{ab} = \cos \theta_a \cos\theta_b + \sin \theta_a \sin\theta_b \cos(\psi_a - \psi_b) \tag{10}$$

$$e^{i\varphi_{ab}} = \frac{e^{i\varphi_a}}{\sin\theta_{ab}} [\sin\theta_a \cos\theta_b - \cos\theta_a \sin\theta_b \cos(\psi_a - \psi_b) - i \sin\theta_b \sin(\psi_a - \psi_b)] \tag{11}$$

$$e^{i(\varphi_{ab} + \psi_{ab})/2} = \frac{1}{\cos\dfrac{\theta_{ab}}{2}} [\cos\frac{\theta_a}{2} \cos \frac{\theta_b}{2} e^{\frac{i}{2}(\psi_a - \psi_b + \varphi_a - \varphi_b)}$$

$$+ \sin \frac{\theta_a}{2} \sin\frac{\theta_b}{2} e^{\frac{i}{2}(-\psi_a + \psi_b + \varphi_a - \varphi_b)}] \tag{12}$$

The Green function (7) can also be decomposed into the wave functions of (3):

$$K(u_a, u_b; T) = \frac{1}{16\pi^2} \sum_{l=0}^{\infty} (2l+1)\, e^{-i\frac{T}{2I}l(l+1)}$$
$$\sum_{n,m=-l}^{l} \psi^l_{mn}(\theta_a\varphi_a\psi_a)\, \psi^{l*}_{nm}(\theta_b\varphi_b\psi_b) \tag{13}$$

III. POTENTIALS RELATED TO THE MOTION OVER THE SU(2) MANIFOLD

(i) We first consider the Poschl-Teller *(P-T)* potential given by

$$V(x) = \frac{1}{2\mu}\left(\frac{K(K-1)}{\sin^2 x} + \frac{\lambda(\lambda-1)}{\cos^2 x}\right) \tag{14}$$

with

$$K, \lambda \prec 1 \quad , \quad 0 \le x \le \pi/2 \ .$$

By the variable change $x=\theta/2$ the Schrödinger equation for the potential (14)

$$\left(-\frac{1}{2\mu}\frac{d^2}{dx^2} + V(x)\right)\psi(x) = E\psi(x) \tag{15}$$

is transformed into

$$\left(-\frac{1}{2I}\frac{d^2}{d\theta^2} + \frac{\alpha^2+\beta^2-2\alpha\beta\cos\theta-1/4}{2I\sin^2\theta}\right)\psi\left(\frac{\theta}{2}\right) = E\psi\left(\frac{\theta}{2}\right) \tag{16}$$

with

$$\alpha = \frac{\lambda+K-1}{2} \quad , \quad \beta = \frac{\lambda-K}{2} \quad , \quad I = \mu/4 \tag{17}$$

Writing the wave function as

$$\psi(\theta/2) = (\sin\theta)^{1/2}\, \phi(\theta) \tag{18}$$

Eq. (16) becomes

$$-\frac{1}{2I}\left[\frac{1}{\sin\theta}\frac{d}{d\theta}\left(\sin\theta\frac{d}{d\theta}\right) + \frac{\alpha^2+\beta^2-2\alpha\beta\cos\theta}{\sin^2\theta}\right]\phi(\theta) = \left(E-\frac{1}{8I}\right)\phi(\theta) \tag{19}$$

which is same as the Schrödinger equation (2) with the momenta p_φ, p_ψ having the fixed values:

$$p_\varphi^2 = \alpha^2 \quad , \quad p_\psi^2 = \beta^2 \tag{20}$$

Dynamics described by the *P-T* potential is equivalent to the motion restricted to the section of *SU(2)* phase space corresponding to the above fixed momenta values. The wave functions and the energy spectrum is then written by using (3), (4) and by (17), (18) and (19) as:

$$\psi_n(x) = \sqrt{2(K+\lambda+2\mu)}\ P_{\alpha\beta}^{\alpha+n}\ (1-2\sin^2 x) \tag{21}$$

and

$$E_n = \frac{1}{2\mu}\ (K+\lambda+2\mu)^2 \tag{22}$$

Path integral treatment of the *P-T* potential is also available [6]. In general, since it employs the classical dynamical variables the path integration method is very convenient in converting a given quantum mechanical problem into another one: Canonical transformations of classical mechanics can easily be adopted to quantum mechanics by means of path integrals.

A brief presentations of *P-T* path integration as follows:

The Green function

$$K(x_a\ ,\ x_b\ ;\ T) = \int DxDp\ \exp\ [i\int_0^T dt(p\dot{x}-\frac{p^2}{2\mu}-V(x))] \tag{23}$$

is represented by the usual time graded formula:

$$K(x_a\ ,\ x_b\ ;\ T) = \lim_{n\to\infty}\ \prod_{j=1}^{n+1} dx_j\ \prod_{j=1}^{n+1}\frac{dp_j}{2\pi}$$

$$\prod_{j=1}^{n+1}\{\ \exp\ [\ i\sum_{j=1}^{n+1}(p_j(x_j-x_{j-1})-\frac{\epsilon p_j^2}{2\mu}\ -\ \epsilon V(x_j))]\}$$

with $x_a=x_0$, $x_{n+1} = x_b$; $(n+1)\epsilon = T$.   After the introduction of variable $\theta=2x$, the potential term in the action of (23) can be reexpressed as [6]:

$$\exp\ [\ i\int_0^T dt\ (-\ \frac{\alpha^2+\beta^2-2\alpha\beta\cos\theta}{2I\sin^2\theta}\ )\ ]$$

$$=\int_0^{2\pi} d\varphi_b\ e^{-i\alpha(\varphi_b-\varphi_a)}\int_{-\infty}^{\infty} d\psi_b e^{-i\beta(\psi_b-\psi_a)}$$

$$\int D\ (\varphi,\psi)\ D\ (p_\varphi,p_\psi)\ \exp\ [\ i\int_0^T dt\ (\ p_\varphi\dot{\varphi}+p_\psi\dot{\psi}\ -\ \frac{p_\varphi^2+p_\psi^2-2p_\varphi p_\psi\cos\theta}{2I\sin^2\theta}\ )\ ]$$

$$\tag{25}$$

Inserting the above identity into the path integral of (23) one obtains

$$K(x_a\ ,\ x_b\ ;\ T) = 2\int_0^{2\pi} d\varphi_b\ e^{-i\alpha(\varphi_b-\varphi_a)}\int_{-2\pi}^{2\pi} d\psi_b\ e^{-i\beta(\psi_b-\psi_a)}\ K_{SU(2)}\ (a\ ,\ b\ ;\ T) \tag{26}$$

where $K_{SU(2)}$ is the *SU(2)* Green function given by (6). The factor 2 in (26) is due to the transformation $dp_x = 2 \, dp_\theta$ at $(n+1)^{th}$ point in (24).

(ii) Second example we like to discuss is the Wood-Saxon *(W-S)* potential:

$$V(x) = -\frac{V_0}{1+e^{x/a}} \quad ; \quad -\infty < x < \infty \; . \tag{27}$$

To convert the above potential into the *SU(2)* form we have to employ canonical transformation which is not trivial as it was in *P-T* case. In the light of the considerations of the previous subsection, we restrict our presentations to the path integral formalism [7]. Parallel Schrödinger treatment can always be obtained. Following the canonical transformation

$$\frac{1}{1+e^{x/a}} = \cos^2\theta \quad , \quad p_x = \frac{p_\theta}{2a} \sin\theta \, \cos\theta \quad ; \quad 0 \le \theta \le \pi/2 \tag{28}$$

the path integral for *W-S* potential becomes

$$K(x_a \, , \, x_b \, ; \, T) = \frac{1}{2a} \sin\theta_b \, \cos\theta_b \int D\theta Dp_\theta$$
$$\exp\left[\, i\int_0^T dt \, (\, p_\theta\dot\theta \, - \, \frac{\sin^2\theta\cos^2\theta}{4a^2} \, \frac{p_\theta^2}{2\mu} \, + \, V_0\cos^2\theta \,) \, \right] \tag{29}$$

The factor depending on the index b, is due to $(n+1)^{th}$ momentum transformation in the measure of the path integral in time graded formula. To get rid of the coordinate dependence of the kinetic energy one has to parametrize the dynamical variables in terms of the new "time" *s* given by

$$dt = \frac{4a^2}{\sin^2\theta a^2\theta} \, ds \quad , \quad t = \int^s ds \, \frac{4a^2}{\sin^2\theta\cos^2\theta} \tag{30}$$

Introduction of *s* together with the identity

$$1 = \int_0^\infty dS \; \delta(T - \int_0^s ds \, \frac{4a^2}{\sin^2\theta\cos^2\theta}) \, \frac{4a^2}{\sin^2\theta_b\cos^2\theta_b}$$
$$= \int_0^\infty dS \, \frac{4a^2}{\sin^2\theta_b\cos^2\theta_b} \int_{-\infty}^\infty \frac{dE}{2\pi} \, e^{iET} \, \exp\left[\, i\int_0^s ds \, (-\frac{4a^2E}{\sin^2\theta\cos^2\theta}) \right] \tag{31}$$

Eq. (29) can be written as

$$K(x_a \, , \, x_b \, ; \, T) = \int_0^\infty dS \int_{-\infty}^\infty \frac{dE}{2\pi} \, e^{iET} \, \frac{2a}{\sin\theta_b\cos\theta_b}$$
$$\int D\theta Dp_\theta \, \exp\left[\, i\int_0^T dt \, (\, p_\theta\dot\theta - \frac{p_\theta^2}{2\mu} \, - \, \frac{4a^2(E-V_0)}{\sin^2\theta} \, - \, \frac{4a^2E}{\cos^2\theta} \,) \, \right] \tag{32}$$

Here the prime denotes the derivative with respect to *s*. After the symmetrization of the above expression in terms of points **a** and **b** [*] one arrives at

$$K(x_a , x_b ; T) = \frac{4a}{\sqrt{\sin^2\theta_a \cos^2\theta_b}} \int_0^\infty ds\, e^{iS/2\mu} \int_{-\infty}^\infty \frac{dE}{2\pi} e^{iET} K(\theta_a , \theta_b ; S) \tag{33}$$

where

$$\overline{K}(\theta_a , \theta_b ; S) = \int D\theta Dp_\theta \exp\left[i\int_0^S ds\, \left(p_\theta \dot\theta - \frac{p_\theta^2}{2\mu} - \frac{1}{2\mu} \left(\frac{K(K-1)}{\sin^2\theta} + \frac{\lambda(\lambda-1)}{\cos^2\theta} \right) \right)\right] \tag{34}$$

with

$$K = \frac{1}{2}\left[1+\sqrt{32\mu a^2(E-V_0)}\right] \quad , \quad \lambda = \frac{1}{2}(1+\sqrt{32\mu a^2 E}) \tag{35}$$

Inserting the solution of (34) which is of **P-T** type we obtain the solution of **W-S** Green function as [7]

$$K(x_a , x_b ; T) = \sum_{n=0}^\infty e^{-iE_n T} \varphi_n(x_a)\varphi_n^*(x_b) \tag{36}$$

Here the energy spectrum and the wave functions are given by

$$E_n = -\frac{1}{3\mu a^2(n+1)^2} [(n+1)^2+2\mu a^2 V_0]^2 \tag{37}$$

and

$$\varphi(x) = \frac{i\sqrt{4(n+1)^2-(\lambda_n-K_n)^2}}{2\sqrt{2}\,\sqrt{n+1}} \sqrt{\frac{\Gamma(n+1)\,\Gamma(-n-1)}{\Gamma(K_n+n+1/2)\,\Gamma(\lambda_n+n+1/2)}}$$
$$\frac{\exp\left[(K_n-\frac{1}{2})\,x/2a\right]}{(1+e^{x/a})^{K_n+\lambda_n-1/2}} P_n^{(K_n-1/2,\lambda_n-1/2)}((1-e^{x/a})/(1+e^{x/a})) \tag{38}$$

where

$$K_n = \frac{1}{2} + \frac{1}{n+1}\left[(n+1)^2-2\mu a^2 K_0 \right] \quad , \quad \lambda_n = \frac{1}{2} + \frac{1}{n+1}\left[(n+1)^2+2\mu a^2 V_0 \right] \tag{39}$$

(iii) Another potential which is of **SU(2)** type is the Rosen-Morse **(R-M)** potential:

$$V(x) = B\, th\, ax - U\, ch^{-2} a\, x \quad ; \qquad -\infty < x < \infty \tag{40}$$

Following the coordinate and time parameter transformations given by [7]

---

*There are several recepies to obtain the symmetric path integral measure following the point canonical transformations. See for example [7,8] .

$$th\ ax = \cos 2y \quad , \quad p_x = -\frac{a}{2}\, p_y \sin 2y \quad ; \quad 0 \le y \le \pi/2 \tag{41}$$

$$dt = \frac{4a}{a^2 \sin 2y}\, ds \tag{42}$$

RM problem takes the form of **PT** with

$$K = \frac{1}{2} + \sqrt{(2\mu/a^2)\,(E+B_0)} \quad , \quad \lambda = \frac{1}{2} + \sqrt{(2\mu/a^2)\,(E-B_0)} \tag{43}$$

Note that the symmetric form of the **R-S** potential which is obtained by **B=0** or **U=0**, or with **B=U** corresponds to the particle motion over the **SO(3)** manifold, and thus solved in terms of the Legendre polynomials [9].

(iv) Magnetic top is another example to the **SU(2)** type problems [10]. Classical Hamiltonian is given by

$$H = \frac{1}{2I}\, (\,\vec{S} - g\, I\, \vec{B}\,)^2 \tag{44}$$

Here **I** is the moment of inertia and **B** is the external magnetic field. Components of the "canonical" angular momentum are given in terms of the Euler angles as

$$\begin{aligned}
S_x &= p_\theta \cos\varphi + p_\psi \frac{\sin\varphi}{\sin\theta} - p_\varphi \frac{\sin\varphi\cos\theta}{\sin\theta} \\
S_y &= p_\theta \sin\varphi - p_\psi \frac{\cos\varphi}{\sin\theta} + p_\varphi \frac{\cos\varphi\cos\theta}{\sin\theta} \\
S_z &= p_\varphi
\end{aligned} \tag{45}$$

Total angular momentum operator is equivalent to the **SU(2)** Hamiltonian

$$S^2 = p_\theta^2 + \frac{1}{\sin^2\theta}\,(p_\varphi^2 + p_\psi^2 - 2p_\varphi p_\psi \cos\theta) \tag{46}$$

The path integral for the Hamiltonian (44) is exactly solvable [10]. The time dependent wave function is

$$\Psi_{enm}(\theta,\varphi,\psi;t) = \frac{\sqrt{l+1/2}}{2\sqrt{2\pi}}\, e^{-i\frac{t}{2I}\,(\,l\,(\,l+1)-g^2 B^2 I^2 + 2\,I\,n\,gB\,)}\, e^{im\psi}\, e^{in\varphi}\, p_{mn}^{l}(\cos\theta) \tag{47}$$

The ranges of the quantum numbers **l,m,n** are same as of the free motion over the **SU(2)** manifold.

IV. POTENTIALS RELATED TO THE NON-COMPACT GROUPS

If the sign of the **P-T** potential is reversed, or the parameters **K**, λ are given as complex numbers, the corresponding group is not *SU(2)* but *SU(1,1)* [11]. As an example we can mention the Hulthen potential:

$$V(r) = -V_0 \frac{e^{-r/a}}{1-e^{r/a}} \quad ; \quad r = 0 \rightarrow \infty \tag{48}$$

Coordinate and time parameter transformations given by [7]

$$r = -2a \ln(\sin\theta) \quad , \quad p_r = \frac{\sin\theta}{2\cos\theta} \, p_\theta \quad ; \quad 0 \leq \theta \leq \pi/2 \tag{49}$$

$$dt = -4a^2 \frac{\cos^2\theta}{\sin^2\theta} \, ds \tag{50}$$

bring the problem into the *P-T* form with

$$K = \frac{1}{2} + 2\sqrt{-\mu a^2 E} \quad , \quad \lambda = -\frac{1}{4} \tag{51}$$

V. FIELD THEORETICAL EXAMPLES

Studying the non-relativistic particle motions with Schrödinger equation or in path integral approach is often very useful in solving the problems of field theory too. Here we briefly present two examples:

To investigate particle pair production in external electromagnetic fields or in the cosmological backgrounds one needs to calculate the Green functions.

(i) To discuss the scalar particle creation in an external e-m field A_μ, the recuired Green function for the Klein-Gordon (KG) particle is given by the phase-space path integral (with x standing for the four vector (x_μ)) [9]

$$G(x, x') = \int_0^\infty dW \, e^{-i\mu^2 W} F(x, x'; W)$$

Here

$$F(x, x'; W) = \int D^4x \, D^4p \, \exp \left[i\int_0^W dw \, (p \cdot \dot{x} + p_0^2 - p^2 + 2eA \cdot p + e^2 A^2) \right] \tag{52}$$

is formally the same as the Green function for the non-relativistic particle motion in the 5-dimensional "space-time" represented by *(x,w)*.

(ii) For studying the scalar pair production in a given cosmological background, for example in a Robertson-Walker spacetime with the metric

$$ds^2 = -dt^2 + a^2(t) \, dx^2 \tag{53}$$

the Green function is given by [10]

$$G(x, x') = \int_0^\infty dW\, e^{-i\mu^2 W}$$

$$\int D^4x\, D^4p\, \exp\left[\, i\int_0^W dw\, (\, p\cdot \dot{x} + p_0^2 - \frac{p^2}{a^2(t)}\,)\, \right] \tag{54}$$

After integrating the trivial spatial part of the above path integral one obtains

$$G(x, x') = \int_0^\infty dW\, e^{-i\mu^2 W} \int \frac{d^3p}{(2\pi)^3}\, e^{i\vec{p}\cdot(\vec{x}-\vec{x}')}$$

$$\int Dt\, Dp_0\, \exp\left[\, i\int_0^W dw\, (\, p_0\dot{t} + p_0^2 - \frac{p^2}{a^2(t)}\,)\, \right] \tag{55}$$

which is equivalent to the non-relativistic motion in "space-time" *(t,s)* under the influance of the potential $p^2/a^2(t)$ with $p^2 = constant$.

REFERENCES

[1] N. Ja Vilenkin and A.U. Klimyk, "Representation of Lie Groups and Special Functions" (Kluwer Academic Publishers, Dordrecht, 1992).

[2] J.D. Talman, "Special Functions; A Group Theoretical Approach" (Benjamin, New York, 1968).

[3] I.H. Duru and H. Kleinert, Phys.Lett. 84B, 185 (1979); and Forschr.Phys. 30, 401 (1982).

[4] I.H. Duru, Phys.Rev. D28, 2689 (1983).

[5] I.H. Duru, Phys.Lett. 112A, 421 (1985).

[6] I.H. Duru, Phys.Rev. D30, 2121 (1984).

[7] I.H. Duru, Phys.Lett. A119, 163 (1986).

[8] S.N. Storchak, Phys.Lett. A135, 77 (1989).

[9] I.H. Duru, ICTP, Trieste, Report No: IC/83/178 (unpublished).

[10] A.O. Barut and I.H. Duru, Phys.Lett. 158A, 441 (1991).

[11] G. Junker and M. Bohm, Phys.Lett. 117A, 375 (1986).

QUANTUM VIOLATION OF WEAK EQUIVALENCE PRINCIPLE IN THE BRANS-DICKE THEORY

Yasunori Fujii*

Nihon Fukushi University
Okuda, Chita-gun, Aichi, 470-32 Japan, and
Institute for Cosmic Ray Research, University of Tokyo
Tanashi, Tokyo, 188 Japan

A quantum correction to the Brans-Dicke theory due to interactions among matter fields is calculated, resulting in violation of WEP, hence giving a constraint on the parameter ω far more stringent than accepted so far.

Many "alternative" theories of gravitation have been proposed. Among them, however, the Brans-Dicke theory [1] is certainly one of the best known. This is because of its theoretical simplicity and also because it has been subject to the experimental tests since Dicke himself made an effort by measuring the solar oblateness. Recently the theory seems to be revisited because it offers a sensible approximate model to study the consequences of unification, an ambitious program to unify particle physics and gravitation. In fact in many of the theoretical models of unification scalar fields participate as important ingredients, in addition to other many fields most of which will remain undetected.

In spite of this new development, the BD theory has been discussed mostly in the classical level, in both theoretical and experimental fronts. The purely classical theory has been the basis of the analysis, for example, of the solar-system experiment which yields the widely accepted constraint $\omega \gtrsim 500$ [2]. Notice that the theory reduces to the standard theory in the limit of $\omega \to \infty$. I also point out that the result depends crucially on the classical assumption that the scalar field is massless, or at least the force-range is larger than the size of the solar system.

In this talk I am going to show that there is an important quantum effect which seems to have been ignored but resulting in a much stronger constraint when the result is compared with the null tests of Weak Equivalence Principle (WEP), or sometimes called the tests of composition-independence of the gravitational force.

Let me start with giving the fundamental Lagrangian

$$\mathcal{L} = \sqrt{-g}\left[\frac{1}{8\omega}\phi^2 R - \frac{1}{2}g^{\mu\nu}\partial_\mu\phi\partial_\nu\phi + L_{\text{matter}}\right].\tag{1}$$

---

*E-mail address: ysfujii@tansei.cc.u-tokyo.ac.jp

My scalar field ϕ is related to BD's original notation by $\varphi_{BD} = \phi^2/8\omega$; I choose this definition because it is in accordance with the standard notation in the relativistic quantum field theory. ω is the only parameter that characterizes the theory. It seems reasonable to expect that ω is not far away from unity.

The basic assumptions in (1) may be summarized as the three fundamental premises:

(A) The presence of a "nonminimal coupling", the first term instead of the standard Einstein-Hilbert action $\sqrt{-g}(1/16\pi G)R$. This implies that the gravitational "constant" G is not a true constant but spacetime-dependent, because $G \sim \phi^{-2}$. This is precisely what BD wanted to have in order to realize the Machian principle. However, from a more modern point of view, it is remarkable to notice that the same type of coupling is rather common in many of the theoretical models of unification [3].

In realistic applications we expect that ϕ has a VEV, $\phi_0 \sim G^{-1/2}$, and the rest $\phi' = \phi - \phi_0$ behaves as a fluctuating field. Then after some analysis we come to find the property:

(A') The fluctuating part of the scalar field couples to the matter with the strength essentially of the gravitational strength.

(B) The scalar field is massless. BD imposed this probably just for simplicity. This would be acceptable, however, only classically. From a quantum theoretical point of view, we know no principle to prevent the scalar field from acquiring a nonzero mass. It is also rather likely that the acquired mass is very small, such that the range of the force derived from the scalar field is of a macroscopic distances. Let me call this modified property (B').

The properties (A') and (B') gave me part of the motivations to propose non-Newtonian gravity, later called a fifth force [4].

(C) BD assumed that ϕ has no *direct* coupling to the matter. The nonminimal coupling mentioned in (A) is the only coupling that ϕ may have. BD imposed this because otherwise Equivalence Principle will be violated. Without this direct coupling, one derives the geodesic equation for any matter particle, a manifestation of the geometrical nature of the theory, thus assuring composition-independence even with the effect of the scalar force included.

The main purpose of my today's talk is to show that this last assumption (C) is also subject to a quantum modification. If my argument is correct, the BD field is qualified to be a candidate of a fifth force field with respect to yet another point, composition-dependence. I emphasize that I consider the quantum effects due to the interaction among matter fields, keeping away from quantum gravity which is still not quite tractable.

The essential point is that, just as with the mass of the scalar field, there is no principle according to which composition-independence imposed in the classical level is kept preserved in the quantum level. Both of the premises (B) and (C) are fragile against their violation as quantum effects. There is, however, a subtle difference between the ways these premises are broken. In fact to compute the mass belongs to the hardest part of relativistic quantum field theory; the result is "divergent," or depends crucially on the cutoff in the theory.

On the other hand, it is quite fortunate that the calculated violation of composition-independence turns out to be *finite*. This shares the same nature as in the various kinds of "quantum anomalies," which represent violation of symmetries valid in the classical level. The breakdown of WEP is, in this sense, much less ambiguous than the mass, and hence would result in a new type of constraint on BD's parameter ω, as will be shown later.

As a simple illustration, let me consider a matter system of nucleons and the electromagnetic field; I essentially compute the QED effect to the protons and the electromagnetic field. The fact that the scalar field coupling to the matter occurs only in the field equations not in the Lagrangian according to (C) poses some difficulty when I try to apply the conventional technique of quantum field theory. For this reason I apply a conformal transformation, as discussed by Dicke himself some months later than the original BD paper[†] [5]:

$$g_{\mu\nu} \rightarrow g_{*\mu\nu} = \frac{2\pi G}{\omega}\phi^2 g_{\mu\nu}.$$ (2)

In the new starred conformal frame the Lagrangian (1) is cast into the form

$$\mathcal{L} = \sqrt{-g_*}\left[\frac{1}{16\pi G}R_* - \frac{1}{2}g_*^{\mu\nu}\partial_\mu\sigma\partial_\nu\sigma + L_{*\mathrm{matter}}\right],$$ (3)

where σ is a redefined canonical scalar field related to ϕ by

$$\phi = \sqrt{\frac{\omega}{2\pi G}}e^{\beta\sigma}, \qquad \mathrm{with} \qquad \beta = \sqrt{\frac{4\pi G}{3 + 2\omega}}.$$ (4)

For the matter Lagrangian with the proton field ψ and the electromagnetic field A_μ, the transformed Lagrangian differs in form only in the mass term:

$$L_{*\mathrm{mass}} = -me^{-\beta\sigma}\overline{\psi}_*\psi_*,$$ (5)

where $\psi_* = (2\pi G/\omega)^{-3/4}\phi^{-3/2}\psi$, while A_μ remains unchanged. The exponential factor in (5) shows the presence of the direct matter coupling in the new conformal frame. Consequently, the geodesic equation is modified but in such a universal manner that WEP remains valid. In the following we suppress the symbol $*$ to simplify the notation.

We focus upon the linear term

$$L_\sigma = \beta m\overline{\psi}\psi\sigma.$$ (6)

This is the coupling to the trace of the matter energy-momentum tensor $T = -m\overline{\psi}\psi$, with no contribution from A_μ. This interaction, corresponding to a vertex in Feynman diagrams, may be represented by a "mass insertion" to a proton line in the limit of a vanishing momentum transferred to σ.

To compute the correction to this vertex to the lowest non-trivial order with respect to e, the electromagnetic coupling constant, I calculate the following diagrams: (a) a photon line emitted on one side of the vertex and re-absorbed on the other side; (b) emission and re-absorption occurring only on one side.

The first diagram is shown to be equivalent to applying $m(\partial/\partial m)$ to the proton self-energy diagram; differentiating with respect to m generates another proton line while multiplying with m provides the mass in the vertex. The proton self-energy part gives the self-mass δm only on the mass shell. However, the contribution from the diagrams (b), of the nature of wavefunction renormalization, cancels the effect off the mass shell, hence the correction term is represented by replacing m in (6) by

$$\mathcal{M} = \left(m\frac{d}{dm}\right)\delta m.$$ (7)

[†]The final results remain the same in both conformal frames before and after the transformation in our calculation to the lowest order with respect to G, though this is not the case in general.

The self-mass is known to be

$$\delta m = (3\alpha/2\pi)\left(\ln\frac{\Lambda}{m} + \frac{1}{4}\right),$$

where Λ is a cutoff. By using this in (7) I find

$$\mathcal{M} = \delta m - \frac{3\alpha}{2\pi}m. \tag{8}$$

The first term δm is going to be absorbed into the zeroth order term m in (6), naturally replacing the "bare" mass by the "observed renormalized" mass $m + \delta m$, while the second term represents an extra contribution.

I can arrive at the same result without using an expression depending on the cut-off; all the calculations can be made on the formal integrals. Nevertheless, the above derivation is instructive because it reveals an origin of the extra term. The operation $m(\partial/\partial m)$ implies a dimensional analysis. The result would have been simply δm if it were expressed only in terms of m. It depends, however, on Λ as well, hence invalidating a naive dimensional analysis. I can say that the extra term arises simply because the underlying theory is divergent, although the result itself is obviously finite. This is one of the examples of anomalies in the relativistic quantum field theory.[‡]

It may appear that the extra term $-(3\alpha/2\pi)m$ is also included in the trace T. A crucial observation is that the energy-momentum tensor is renormalization-free due to the Ward-Takahashi identity as long as effects of higher order with respect to G are ignored. This implies that there is a difference between what the scalar feels and what the tensor metric field does, hence violating WEP. It is also worth noticing that the extra term depends on one of the "internal parameters" α. In fact neutrons do not have this difference, hence differentiating proton and neutron beyond the mass difference.

Now I will move on to phenomenology. Suppose σ acquires a finite mass $\mu = 1/\lambda$. An exchange of σ will give a non-Newtonian potential

$$V_{ij} = -G\frac{M_i M_j}{r}\left[1 + \alpha_5(i,j)e^{-r/\lambda}\right], \tag{9}$$

where

$$\alpha_5(i,j) = \xi q_i q_j, \tag{10}$$

with

$$\xi = \frac{1}{3 + 2\omega}, \tag{11}$$

as will be read off from (4) and (6).

The "charge" for the nucleon can also be read off as

$$\left\{\begin{array}{ll} q_p = 1 - \frac{3\alpha}{2\pi}, & \text{for proton,} \\ q_n = 1, & \text{for neutron.} \end{array}\right. \tag{12}$$

Consider a nucleus i with the mass number A_i and the atomic number Z_i. I assume that the nuclear binding energy respects composition-independence, expecting its breaking due only to the second term of q_p. Then the charge for this nucleus is given by

---

[‡]The closest analogue is the trace anomaly, which is, however, by itself consistent with Equivalence Principle at the fundamental level. It can be relevant to the experimental WEP violation only if it comes together with the occurrence of a sufficiently long-ranged force.

$$q_i = 1 - \frac{3\alpha}{2\pi} \frac{Z_i m}{M_i}, \tag{13}$$

where M_i is the mass of the nucleus. It would be sufficient to approximate the proton mass m by $m \approx M_i/A_i$, thus giving

$$q_i \approx 1 - \frac{3\alpha}{2\pi} \frac{Z_i}{A_i}. \tag{14}$$

The difference of the acceleration between the two nuclei i and j toward a common source mass M_S is given by

$$\delta a_{ij} = \xi \left(q_i - q_j \right) q_S \frac{GM_S}{r^2} \mathcal{F}, \tag{15}$$

where the charge of the source q_S may be replaced by the composition-independent component 1, while $\mathcal{F}(r, \lambda)$ takes care of the finite force-range; $\mathcal{F} \to 1$ for $\lambda \gg r$.

Suppose first that $\lambda \gtrsim 1$AU. Then a crucial constraint comes from the null experiments by Roll, Krotkov and Dicke et al and by Braginski and Panov [6]. These results with the Sun as the source mass impose the condition $|\delta a_{ij}/(GM_S/r^2)| \lesssim 10^{-12}$ for Al and Au. This translates into

$$\left| \xi \frac{3\alpha}{2\pi} \Delta \left(\frac{Z}{A} \right) \right| \lesssim 10^{-12}. \tag{16}$$

For Al and Au we find $\Delta(Z/A) \approx 0.08$, hence giving

$$|\xi| \lesssim 3.5 \times 10^{-9}, \qquad \text{or} \qquad \omega \gtrsim 1.5 \times 10^{8}, \tag{17}$$

far stronger than $\omega \gtrsim 500$, as obtained from the solar-system experiments.

If $\lambda \ll 1$AU, the free-fall experiments using the Earth as the source mass are available [7]. Kuroda and Mio gave upper limits $0.43 \pm 1.23, -0.13 \pm 0.78, -0.18 \pm 1.38$ in units of 10^{-8}ms^{-2} for the pairs Al-Be, Al-Cu and Al-C, respectively. The differences $q_i - q_j$ are relatively large; 0.033236, 0.025738, and -0.01414, respectively. I then derive the lower bound on ω (due to K. Kuroda):

λ(km)	100	10^3	10^4
$\mathcal{F}$	0.01677	0.19051	0.89641
ξ	1.08×10^{-3}	9.55×10^{-5}	2.03×10^{-5}
ω	461	5.23×10^3	2.46×10^4

The force-range can be even shorter. I then resort to the composition-dependent experiments of the laboratory type using torsion-balance with the nearby sources. Adelberger et al [8] reached the accuracy of $\sim 10^{-11}$ for the upper bound of the fractional acceleration difference between Cu and Be, giving $\omega \gtrsim 2.2 \times 10^6$. This is true for $\lambda \gtrsim 1$m, thus surpassing the limits set by the free-fall experiments.

In this way I saw that the constraints from the WEP experiments impose much severer limits to ω if the BD theory is corrected by the quantum effects due to the QED interaction of the matter. The value of ω too large to be "natural" may be avoided either if (i) λ is sufficiently short; $\lambda \ll 1$m, or (ii) a suppression mechanism is at work, or (iii) there is a yet-to-be-discovered symmetry ensuring a cancellation among terms from different fields.

It might be criticized that my analysis is not fully realistic because I ignored fields that mediate the nuclear force. It should be better if I repeat the same analysis at the level of QCD. I argue, however, that the effect differentiating proton and neutron should remain with nearly the same order of magnitude unless a new symmetry is discovered. From this point of view, the simplified analysis presented here would be justified as a start for further elaboration.

I should include another extra term

$$L_{\sigma(c)} = -\frac{2\alpha}{3\pi}\beta\frac{1}{4}F_{\mu\nu}F^{\mu\nu}\sigma. \tag{18}$$

This, coming from the mass insertion to the photon self-energy diagram, is again finite. Notice that this represents a coupling of the scalar field to the electromagnetic field, obviously not of the type of the trace coupling. Adding this to the kinetic term of the electromagnetic field, and applying a "wavefunction renormalization," I find a correction to the fine-structure constant due to the presence of the scalar field, as given by

$$\frac{\delta\alpha}{\alpha} = -\frac{2\alpha}{3\pi}\beta\sigma. \tag{19}$$

The Coulomb energy of a relatively heavy nucleus is known to be

$$E_c = C\alpha m Z_i^2 A_i^{-1/3}, \tag{20}$$

where $C\alpha m = 0.712\text{MeV}$, giving $C\alpha = 0.75 \times 10^{-3}$. The composition-dependent part of the mass of the nucleus due to the Coulomb energy is then given by

$$\delta M_{i(c)} = -C\alpha\frac{2\alpha}{3\pi}m\beta\sigma Z_i^2 A_i^{-1/3} \approx -C\alpha\frac{2\alpha}{3\pi}M_i Z_i^2 A_i^{-4/3}\beta\sigma, \tag{21}$$

which may also be included in the charge q_i. As it turns out, however, this is smaller than the contribution from q_p by the factor $(4/9)\alpha C Z_i A_i^{-1/3}$, which is almost negligible.

I emphasize that the BD theory as it stands is probably only the effective or approximate model of what one should expect from unified theories. Potentially crucial is that the "dilaton" field in these theories likely couples explicitly to most of the matter fields, serving another source of WEP violation [3]. As one of the consequences, the fine-structure constant could be time-dependent [9]. The scalar field might also be relevant to the cosmological constant problem [10]§. Many aspects of the scalar field are still left for better understanding. The issue is related to the phenomena of the fifth-force type in some way or the other. This will justify further experimental studies which will open a way to explore the physics characterized by the Planck mass by means of experiments at extremely low-energies.

REFERENCES

[1] C. Brans and R.H. Dicke, Phys. Rev. **124**, 925 (1961).

[2] See, for example, C. Will, Phys. Rep. **113**, 346 (1984).

[3] See, for example, C.G. Callan, D. Friedan, E.J. Martinec and M.J. Perry, Nucl. Phys. **B262**, 593 (1985); C.G. Callan, I.R. Klebanov and M.J. Perry, Nucl. Phys. **B278**, 78 (1986).

---

§According to the model of this reference, the strength of the σ coupling is suppressed by the age of the Universe.

[4] Y. Fujii, Nature Phys. Sci, **234**, 5 (1971); Phys. Rev. **D9**, 874 (1974). See, for the latest review, E. Fischbach and C. Talmadge, Nature, **356**, 207 (1992).

[5] R.H. Dicke, Phys. Rev. **125**, 2163 (1962).

[6] P.G. Roll, R. Krotkov and R.H. Dicke, Ann. Phys. (N.Y.) **26**, (1964) 442: V.B. Braginski and V.I. Panov, Sov. Phys. JETP **34**, (1972) 463.

[7] T.M. Niebauer, M.P. McHugh and J.E. Faller, Phys. Rev. Lett. **59**, 609 (1987): K. Kuroda and N. Mio, Phys. Rev. Lett. **62**, 1941 (1989).

[8] E.G. Adelberger et al., Phys. Rev. **D42**, 3267 (1990).

[9] Y. Fujii, M. Omote and T. Nishioka, Time-dependent coupling constants in unified theories? to be published in Prog. Theor. Phys: T. Damour and A.M. Polyakov, The string dilaton and a least coupling principle, preprint: Y. Fujii, K. Kuroda and N. Kanda, New proposed experiment on time-variability of the fine-structure constant, to be published in Proceedings of the XXIX Rencontre de Moriond, January 22-29, 1994, Villars sur Ollon, Switzerland.

[10] Y. Fujii and T. Nishioka, Phys. Rev. **D42**, 361 (1990).

QUANTUM UNITARY AND PSEUDOUNITARY GROUPS AND GENERALIZED HADRON MASS RELATIONS

A. M. Gavrilik

Institute for Theoretical Physics
Ukrainian National Academy of Sciences
Kiev 252143, Ukraine

1. INTRODUCTION

Applications of quantum algebras (of $su_q(2)$ as most widespread example) to phenomenological description of rotational spectra of deformed heavy nuclei and diatomic molecules, have appeared a couple of years ago and seem to be encouraging [1-3] (concerning physical applications of quantum groups/algebras in a wider context see ref.[4] and references therein).

Use of the quantum algebra $su_q(2)$ in nuclear spectroscopy is based on such ingredients as Casimir operator, Clebsch-Gordan coefficients (CGC's) [1,2]. In attempts to find application of higher rank quantum algebras, one encounters some new features absent in $su_q(2)$ case. Among such features are, e.g., the necessity to deal with nonsimple-root elements (equivalently, with q-Serre relations) of those algebras, nontriviality of the concepts of (and formulas for) Casimirs, CGC's.

Recently, the use of higher rank quantum algebras $su_q(n)$ (or $U_q(u_n)$) in order to replace conventional unitary groups $SU(n)$ and their irreducible representations (irreps) in describing flavor symmetries of hadrons (namely, vector mesons) has been proposed [5]. With the help of the corresponding algebras $U_q(u_{n+1})$ of 'dynamical' symmetry, one can realize necessary breaking of flavor symmetries up to exact (for strong interactions alone) isospin symmetry $su_q(2)_I$ and obtain some q-analogs of mass relations. Such an application of the q-algebras $U_q(u_n)$ to obtaining q-analogs of hadron mass relations (MR's) uses generators corresponding both to simple-root elements and nonsimple-root ones (thus, the q-Serre relations play definite role [6]); from another side, it eploits rather simple model which allows one to circumvent difficulties related with q-CGC's and q-Casimirs.

There exist different approaches to (non-deformed) $SU(n)$-symmetry breaking necessary for obtaining mass sum rules (MSR's) for hadrons with n quark flavors. The approach based on *dynamical unitary groups* [7] allows one to obtain the following series of MSR's [8] for vector mesons 1^- $(2 < n \leq 6)$:

$$\frac{k(k-1)}{2}m_{\omega_{k^2-1}} + \frac{k(k-1)-4}{2}m_\rho = (k-1)^2 m_{D_k^*} + \sum_{i=3}^{k-1} m_{D_i^*}, \quad k = 3, ..., n, \quad (1)$$

where D_k^* denote isodoublets: $D_3^* \equiv K^*$, $D_4^* \equiv D^*$, $D_5^* \equiv D_b^*$, $D_6^* \equiv D_t^*$. This series begins, when $n = 3$, with the famous Gell-Mann–Okubo (GMO) mass relation [9,10] $3m_{\omega_8} + m_\rho = 4m_{K^*}$. A comparison of this octet MSR with existing data *requires mixing* between isosinglet ω_8 and $SU(3)$ singlet, that is, ω_8 is considered as a superposition of ϕ and ω with some mixing angle, *determined from fit*. Likewise, in cases of more flavors $n > 3$ one needs $n - 2$ mixing angles.

Radically different situation, concerning the problem of mixing, arises if one considers some modifications ('deformed analogs') of mass relations. Let us stress that for some set $\{V_i\}$ of hadrons (more specifically, vector mesons) the MSR which is generically of the form

$$\sum_i a_i\, m(V_i) = 0 \qquad (2)$$

with $\sum_i a_i = 0$, may be deformed in two possible ways. The *first possibility* consists in modifying the r.h.s. of eq.(2). Such *'inhomogeneous' (i-)deformation* can be performed for usual group of flavor symmetry using Riemannian geometry of the group manifold as it was demonstrated for the $SU(3)$ case in [11] where the relation

$$m_\rho - 4m_{K^*} + 3m_{\omega_8} = -4(\alpha - 4\beta + 3\gamma) \qquad (3)$$

was obtained. Masses of octet mesons, in the purely geometric approach of [11], follow from eigenvalues of the Laplacian on the $SU(3)$ group manifold equipped with the $SU(3)_{\text{right}}$ times $\{SU(2)_I \times U(1)_Y\}_{\text{left}}$ invariant metric. The i-deformation *parameters* α, β, γ (*with dimension of mass*) describe in what manner full bi-invariance of the metric is reduced to $\{SU(2) \times U(1)\}_L \times \{SU(3)\}_R$; just the combination in curly brackets in (3) multiplies the 27-plet contribution to the metric (using decomposition $(8 \otimes 8)_{sym} = 1 + 8 + 27$). Obviously, standard mesonic GMO mass sum rule is contained in the deformed MR (3) as particular case for which $\alpha - 4\beta + 3\gamma = 0$ holds. An important feature here is the *ability to avoid* manifest introducing of mixing between ϕ and ω. Evidently, two different appropriate (nonzero) values of the combination in the r.h.s. of MSR (3) provide its validity either with (the mass of) ω or ϕ put directly in place of ω_8.

An *alternative possibility* is to deform the coefficients a_i in MSR (2) by making them dependent on some *dimensionless parameter* q ('*homogeneous*' or q-*deformation*). Quantum groups/quantum algebras appear to be a natural framework for doing this, and such q-deformed MSR's are the subject of our present consideration. Extending the approach of [7,8] to quantum algebras $U_q(su_n)$, one can derive for vector mesons the q-deformed MSR's which contain eqns.(1) as $q = 1$ limit, but which also admit (if $|q| = 1$, $q \neq \pm 1$) some principally different (from the $q = 1$ case) treatment [5,6] which allows, like in the case of i-deformation, to avoid manifest introducing of singlet mixing.

As a point of interest, it turns out that all the q-dependence in vector meson masses and in coefficients of their MSR's is expressible in terms of some Lorant-type polynomials of q which were noticed to be related with some knot invariants [6]. A comparison with empirical data certainly requires appropriate fixation of deformation parameter, and it appears that to every number of flavors n, $n \geq 3$, there corresponds a prime root of unity $q = q(n) = e^{i\pi/(2n-1)}$. This root turns into zero the polynomial $P_n(q) \equiv [n]_q - [n-1]_q$ that coincides with respective Alexander polynomial of the torus $(2n-1)$-knot. In a sense, the polynomial $P_n(q)$ through its root $q(n)$ determines the strength of deformation at every fixed n, and due this property may be called a *defining polynomial* for the corresponding vector meson MSR. This way is principally different from the choice of q by fitting [1-3]. Further, utilizing the *quantum* groups instead of conventional unitary groups of flavor symmetries, together with 'dynamical' *quantum* algebras, we get as a result that the collection of torus knqts 5_1, 7_1, 9_1, 11_1 is put into correspondence [6] with vector quarkonia $s\bar{s}$, $c\bar{c}$, $b\bar{b}$, and $t\bar{t}$ respectively. Thus, application of the embedding $U_q(u_n) \subset U_q(u_{n+1})$ to (vector) meson masses provides an appealing possibility of certain *topological characterization of heavy flavors*, since the number n just corresponds to $2n-1$ overcrossings of 2-strand braids whose closures give these $(2n-1)$-torus knots. Equivalently, using (a,b)-presentation of these same knots with $a = 2n-1$, $b = 2$, we are led to the correspondence: $n \longleftrightarrow w \equiv 2n-1$, where w (or a) is nothing but the winding number around the body (tube) of torus (winding number around the hole of torus being equal to 2 for all $n \geq 3$).

The approach of [5,6] was recently extended in order to treat the case of baryons $\frac{1}{2}^+$ (including charmed ones), by adopting again the algebra $U_q(u_4)$ for the 4-flavor symmetry. However, unlike the case of vector mesons were we used 'compact' q-algebras $U_q(u_{n+1})$ for dynamical symmetry, and in analogy to the case of baryon MR's obtained with (non-deformed) dynamical $u(4,1)$-symmetry [7], it is convenient to exploit now representations of the 'noncompact' dynamical symmetry, realized by the quantum algebra $U_q(u_{4,1})$, in order to effect necessary symmetry breakings.

On the base of evaluations within concrete irrep $D_{12}^+(p-1, p-3, p-4; p, p-2)$ (here p is some fixed integer, unessential in the sense that it will not enter final expressions for masses), it is demonstrated [12] that the resulting q-analog of baryon octet MR yields either the usual Gell-Mann–Okubo (GMO) mass sum rule [9,10]

$$m_N + m_\Xi = \frac{3}{2}m_\Lambda + \frac{1}{2}m_\Sigma \tag{4}$$

or a (very successful, see section 6 below) new MSR if one fixes respectively $q = 1$ or $q = e^{\frac{i\pi}{6}}$. These values are nothing but two distinct roots of one and the same *defining polynomial* A_q appearing in the q-analog. With the two alternative choices of deformation parameter, consequences for masses of charmed baryons may be obtained and compared to each other.

Another q-analog (with different defining q-polynomial) of octet MSR is obtained by calculations within the other specific representation $\tilde{D}_{32}^+(p-1, p-3, p-4; p-4, p-2)$ of $U_q(u_{4,1})$. One can show, however, that from the q-dependent expressions for octet baryon masses obtained in $\tilde{D}_{32}^+(\ldots)$, the same q-analog with the A_q mentioned in previous paragraph (modulo change $B_q \rightarrow$ certain $\tilde{B}_q$, see sect.7) can also be derived.

The author would like to acknowledge Dr. I.I.Kachurik and A.V.Tertychnyj who participated in this project.

2. QUANTUM ALGEBRAS $U_q(u_n)$, $U_q(u_{n,1})$, AND THEIR REPRESEN- TATIONS

We will use the denotion $[B]_q \equiv [B] \equiv (q^B - q^{-B})/(q - q^{-1})$ where B is either a number or an operator. The elements $\mathbf{1}$, A_{jj+1}, A_{j+1j}, A_{jj}, $j = 1, 2, \ldots, n-1$,

A_{nn} that generate the quantum (universal enveloping) algebra $U_q(gl_n)$, satisfy the relations [13]

$$[A_{ii}, A_{jj}] = 0, \qquad [A_{ii}, A_{jj+1}] = \delta_{ij} A_{ij+1} - \delta_{ij+1} A_{ji},$$
$$[A_{ii}, A_{j+1j}] = \delta_{ij+1} A_{ij} - \delta_{ij} A_{j+1i},$$
$$[A_{ii+1}, A_{j+1j}] = \delta_{ij}[A_{ii} - A_{i+1i+1}]_q,$$
$$[A_{ii+1}, A_{jj+1}] = [A_{i+1i}, A_{j+1j}] = 0 \qquad \text{for} \qquad |i-j| \geq 2, \tag{5}$$

and the trilinear (q-Serre) relations

$$(A_{i\mp 1i})^2 A_{ii\pm 1} - [2]_q A_{i\mp 1i} A_{ii\pm 1} A_{i\mp 1i} + A_{ii\pm 1}(A_{i\mp 1i})^2 = 0,$$
$$(A_{ii\pm 1})^2 A_{i\mp 1i} - [2]_q A_{ii\pm 1} A_{i\mp 1i} A_{ii\pm 1} + A_{i\mp 1i}(A_{ii\pm 1})^2 = 0. \tag{6}$$

In what follows, we'll need both compact and non-compact real forms of $U_q(gl_n)$.

The *'compact'* quantum algebra $U_q(u_n)$ is singled out by means of the *-operation

$$(A_{jj})^* = A_{jj}, \qquad (A_{j+1j})^* = A_{jj+1}, \qquad (A_{jj+1})^* = A_{j+1j}. \tag{7}$$

The *'noncompact'* quantum algebra $U_q(u_{n,1})$ is singled out from $U_q(gl_{n+1})$ by introducing another *-operation which includes relations (7) of 'maximal compact' subalgebra $U_q(u_n)$ and, in addition, the relations

$$(A_{n+1n})^* = -A_{nn+1}, \qquad (A_{nn+1})^* = -A_{n+1n}, \qquad (A_{n+1\,n+1})^* = A_{n+1\,n+1}. \tag{8}$$

Finite-dimensional representations of $U_q(u_n)$, similarly to those of the non-deformed algebra u_n, are given by sets of ordered integers $\mathbf{m}_n = (m_{1n}, m_{2n}, ..., m_{nn})$ and, since standard branching rules survive through q-deformation, realized by means of (q-analog of) Gel'fand-Tsetlin basis and formulas. Representation formulas for A_{ii} remain unchanged, and A_{kk+1}, A_{k+1k}, $k = 1, ..., n-1$, act according to formulas given in [13]. Action formulas for the operators which represent nonsimple-root elements must be consistent with q-Serre relations (6). We use A_{ij} for $|i-j| = 2$ in the form (see e.g. [5,6])

$$A_{kk+2} = A_{kk+2}(q) \equiv q^{1/2} A_{kk+1} A_{k+1k+2} - q^{-1/2} A_{k+1k+2} A_{kk+1}, \tag{9a}$$
$$A_{k+2,k} = A_{k+2,k}(q) \equiv q^{1/2} A_{k+1k} A_{k+2,k+1} - q^{-1/2} A_{k+2,k+1} A_{k+1k}, \tag{9b}$$

such that the q-Serre relations corresponding to upper signs in (6) follow from (9a) combined with the commutation rules (CR's)

$$q^{1/2} A_{k+1k+2} A_{kk+2} - q^{-1/2} A_{kk+2} A_{k+1k+2} = 0,$$
$$q^{1/2} A_{kk+2} A_{kk+1} - q^{-1/2} A_{kk+1} A_{kk+2} = 0; \tag{10a}$$

whereas those corresponding to lower signs in (6) follow from (9b) and the CR's

$$q^{1/2} A_{k+2,k+1} A_{k+2,k} - q^{-1/2} A_{k+2,k} A_{k+2,k+1} = 0,$$
$$q^{1/2} A_{k+2,k} A_{k+1k} - q^{-1/2} A_{k+1k} A_{k+2,k} = 0. \tag{10b}$$

Dual definition $\tilde{A}_{kk+2} \equiv -A_{kk+2}(q^{-1})$, $\tilde{A}_{k+2,k} \equiv -A_{k+2,k}(q^{-1})$ is paired with respective dual CR's. Operators A_{ij} for $(|i-j| > 2)$ are treated analogously.

Now let us consider some details concerning irreducible representations (irreps) of the 'noncompact' quantum algebra $U_q(u_{n,1})$, $2 \leq n < \infty$ (see e.g. [14,15]). First of all, we have to stress that the construction of the 'principal nonunitary series' of representations, analysis of their (ir)reducibility as well as the classification of irreps and infinitesimally 'unitary' irreps of $U_q(u_{n,1})$ runs in much analogous way to that of the non-deformed (that is, $u(n,1)$) case. We refer to [8,16] and references given therein for the non-deformed case.

Throughout this section, q is considered to be generic (not equal to a root of unity). The representations of the algebra $U_q(u_{n,1})$ are characterized by their signatures χ, that is, by the sets of $n+1$ numbers: $\chi \equiv (l_1, l_2, ..., l_{n-1}; c_1, c_2)$. Here c_1, c_2 are complex numbers such that $c_1 + c_2 \in \mathbf{Z}$, and all the l_i, $i = 1, ... n - 1$, are integers related with the components $m_1, m_2, .., m_{n-1} \equiv \mathbf{m}$ of the highest weight $\mathbf{m}$ of irrep of the subalgebra $U_q(u_{n-1})$, namely, $l_i = m_i - i - 1$. The condition on the components of highest weight in terms of l_i reads: $l_1 > l_2 > ... > l_{n-1}$. Under restriction to the 'compact' subalgebra $U_q(u_n)$, the representation T_χ decomposes into direct sum of all those irreps $T_{\mathbf{l}_n}$ ($\mathbf{l}_n \equiv (l_{1n}, l_{2n}, ..., l_{nn})$, $l_{jn} = m_{jn} - j$, $j = 1, ..., n$, where $m_{1n}, m_{2n}, ..., m_{nn}$ form the highest weight $\mathbf{m}_n$ of irrep of $U_q(u_n)$) for which the condition

$$l_{1n} > l_1 \geq l_{2n} > l_2 \geq ... \geq l_{n-1n} > l_{n-1} \geq l_{nn} \tag{11}$$

is satisfied. All $T_{\mathbf{l}_n}$ which satisfy eq.(11) are contained in T_χ with unit multiplicity.

Action of irrep T_χ is determined in the carrier Hilbert space taken as a direct sum of finite-dimensional carrier spaces of irreps of $U_q(u_n)$. In the carrier space of T_χ, we choose a canonical orthonormal basis formed by the union of canonical (Gel'fand – Tsetlin) bases of $T_{\mathbf{l}_n}$, in accordance with the reduction chain $U_q(u_{n,1}) \supset U_q(u_n) \supset U_q(u_{n-1}) \supset ... \supset U_q(u_2)$. An (orthonormalized) basis vector is completely characterized by the set $\chi, \mathbf{l}_n, \mathbf{l}_{n-1}, ..., \mathbf{l}_2, \mathbf{l}_1$ (here $\mathbf{l}_k \equiv (l_{1k}, l_{2k}, ..., l_{kk})$; $l_{ik} \equiv m_{ik} - i$, $i = 1, ..., k$) and will be denoted as

$$|\chi; \mathbf{l}_n, \mathbf{l}_{n-1}, ..., \mathbf{l}_1\rangle. \tag{12}$$

When restricted to subalgebra $U_q(u_n)$, representation operators act according to formulas of ref. [13]. Operators $T_\chi(A_{nn+1})$ and $T_\chi(A_{n+1n})$ that represent 'noncompact' generators of $U_q(u_{n,1})$ act according to formulas

$$T_\chi(A_{nn+1})|\chi; \mathbf{l}_n, \mathbf{l}_{n-1}, ..., \mathbf{l}_1\rangle =$$

$$\sum_{r=1}^{n} \left| \frac{[c_1 - l_{rn}][l_{rn} - c_2]\Pi_{j=1}^{n-1}[l_{jn-1} - l_{rn} - 1][l_{rn} - l_j]}{\Pi_{s=1, s\neq r}^{n}[l_{rn} - l_{sn} + 1][l_{rn} - l_{sn}]} \right|^{1/2} |\chi, \mathbf{l}_n^{+r}, \mathbf{l}_{n-1}, ..., \mathbf{l}_1\rangle \tag{13}$$

and

$$T_\chi(A_{n+1n})|\chi; \mathbf{l}_n, \mathbf{l}_{n-1}, ..., \mathbf{l}_1\rangle =$$

$$\sum_{r=1}^{n} \left| \frac{[c_1 - l_{rn} + 1][l_{rn} - c_2 - 1]\Pi_{j=1}^{n-1}[l_{jn-1} - l_{rn}][l_{rn} - l_j - 1]}{\Pi_{s=1, s\neq r}^{n}[l_{rn} - l_{sn}][l_{rn} - l_{sn} - 1]} \right|^{1/2} |\chi, \mathbf{l}_n^{-r}, \mathbf{l}_{n-1}, ..., \mathbf{l}_1\rangle \tag{14}$$

where $\mathbf{l}_n^{\pm r}$ means that the l_{rn} in $\mathbf{l}_n$ is to be replaced respectively by $l_{rn} \pm 1$.

To have representation formulas for other 'noncompact' operators $T_\chi(A_{jn+1})$ and $T_\chi(A_{n+1j})$, $1 \leq j \leq n - 1$, one has to utilize relations analogous to eqs. (9).

By means of eqns. (13)-(14) it is proved [14] that the representation T_χ is irreducible if and only if c_1 and c_2 are not integers or c_1 and c_2 coincide with some of the numbers $l_1, l_2, .., l_{n-1}$.

When both c_1 and c_2 are integers not coinciding simultaneously with any two of the integers $l_1, l_2, .., l_{n-1}$, representations from the 'principal nonunitary series' are no longer irreducible, and the corresponding irreps are extracted from these reducible representations (we call them *irreps of integer type*).

Two irreps T_χ and $T_{\chi'}$ from the Theorem with χ and χ' differing only by interchange $(c_1, c_2) \longleftrightarrow (c_2, c_1)$, are equivalent. Periodicity of the function $f(w) = [w]_q$ implies the following: the representations T_χ and $T_{\chi'}$ with $\chi' = (l_1, ..., l_{n-1}; c_1 + \frac{i\pi k}{h}, c_2 - \frac{i\pi k}{h})$, $k \in \mathbf{Z}$, are equivalent for $q = \exp h$, $h \in \mathbf{R}$; the representations T_χ and $T_{\chi'}$ with $\chi' = (l_1, ..., l_{n-1}; c_1 + \frac{\pi k}{h}, c_2 - \frac{\pi k}{h})$, $k \in \mathbf{Z}$, are equivalent for $q = \exp ih$, $h \in \mathbf{R}$. For this reason, we impose the restriction $0 \leq \mathrm{Im}\, c_1 < \frac{\pi}{h}$ (respectively the restriction $0 \leq \mathrm{Re}\, c_1 < \frac{\pi}{h}$) in the case of $q = \exp h$ (resp. of $q = \exp ih$) ($h \in \mathbf{R}$ in both cases).

The classification of all irreducible representations of the algebra $U_q(u_{n,1})$ is completely analogous to that of the non-deformed algebra $u(n, 1)$ (see e.g. [8,16]).

For our purposes it will be useful to reproduce here the list of different classes of irreps of $U_q(u_{n,1})$ which are 'unitary' for $q = e^h$, $h \in \mathbf{R}$ (the sequence of numbers $a_1, a_2, ..., a_k$ will be called *contracted* if $a_{i-1} - a_i = 1$ for $i = 2, 3, ..., k$).

I. Principal continuous series of irreps T_χ: c_1 and c_2 are such that $c_1 = \bar{c}_2$.

II. Supplementary series of irreps T_χ: $c_1, c_2 \in \mathbf{R}$ and, moreover, there exist such l_k and l_s $(k, s = 1, 2, ..., n - 1)$ that $|c_1 - l_k| < 1$, $|l_s - c_2| < 1$, and the sequence $l_k, l_{k+1}, ..., l_s$, if $c_1 > c_2$, or the sequence $l_s, l_{s+1}, ..., l_k$, if $c_1 < c_2$, is contracted.

III. Strange series of irreps T_χ: $\mathrm{Im}\, c_1 = \mathrm{Im}\, c_2 = \frac{\pi}{h}$.

For these three continuous 'unitary' series, the irreps T_χ when restricted to subalgebra $U_q(u_n)$ contain those irreps $T_{\mathbf{l}_n}$ for which the condition (11) is satisfied.

Classes of *irreps of integer type* (c_1, c_2 not both coincide with some of $l_1, ..., l_{n-1}$; we use the denotion $l_0 = \infty, l_n = -\infty$).

IV. Irreps $D_+^{ij}(l_1, ..., l_{n-1}; c_1, c_2)$ and $D_-^{ij}(l_1, ..., l_{n-1}; c_1, c_2)$ where $l_{i-1} > c_1 > l_i$, $l_{j-1} > c_2 > l_j$, $1 \leq i \leq j \leq n$. Moreover, either $i = j$ holds, or the sequence $c_1, l_i, l_{i+1}, ..., l_{j-1}$ for D_+^{ij} (the sequence $l_i, l_{i+1}, ..., l_{j-1}, c_2$ for D_-^{ij}) is contracted. The irrep $D_+^{ij}(l_1, ..., l_{n-1}; c_1, c_2)$ (resp. irrep $D_-^{ij}(l_1, ..., l_{n-1}; c_1, c_2)$) contains with unit multiplicity those and only those irreps of $U_q(u_n)$ for which the condition (11) and the conditions $l_{in} > c_1$, $l_{jn} > c_2$ (resp. $l_{in} \leq c_1$, $l_{jn} \leq c_2$) are satisfied.

V. Irreps $\tilde{D}_+^{ij}(l_1, ..., l_{n-1}; c_1, c_2)$ and $\tilde{D}_-^{ij}(l_1, ..., l_{n-1}; c_1, c_2)$ where $c_1 = l_i$, $1 \leq i \leq n - 1$, and c_2 is an integer such that $l_{j-1} > c_2 > l_j$, $1 \leq j \leq n$. For $\tilde{D}_+^{ij}$, moreover, either $i < j$ and the sequence $l_i, l_{i+1}, ..., l_{j-1}, c_2$ is contracted, or $i \geq j$ and the sequence $l_j, l_{j+1}, ..., l_i$ is contracted. For $\tilde{D}_-^{ij}$, either $i < j$ and the sequence $l_i, l_{i+1}, ..., l_{j-1}$ is contracted, or $i \geq j$ and the sequence $c_2, l_j, l_{j+1}, ..., l_i$ is contracted. The $\tilde{D}_+^{ij}(l_1, ..., l_{n-1}; c_1, c_2)$ (resp. $\tilde{D}_-^{ij}(l_1, ..., l_{n-1}; c_1, c_2)$) contains with unit multiplicity those and only those irreps of $U_q(u_n)$ for which the condition (11) and the condition $l_{jn} > c_2$ (resp. $l_{jn} \leq c_2$) are satisfied.

VI. Irreps $D_+^i(l_1, ..., l_{n-1}; c_1, c_2)$ and $D_-^i(l_1, ..., l_{n-1}; c_1, c_2)$ where $c_1 = c_2 = c$ is an integer such that $l_{i-1} > c > l_i, 1 \leq i \leq n$. The D_+^i (resp. D_-^i) contains with unit multiplicity those and only those irreps of $U_q(u_n)$ for which the condition (11) and the condition $l_{in} > c$ (resp. $l_{in} \leq c$) is satisfied.

There exist additional equivalence relations between irreps from different classes IV–VI completely analogous to the equivalence relations of the non-deformed case (we do not give them here, see e.g. [8]). Remark that two reducible representations T_χ and $T_{\chi'}$ with $\chi = (l_1, ..., l_{n-1}; c_1, c_2)$ and $\chi = (l_1, ..., l_{n-1}; c_2, c_1)$ contain equivalent irreps (from classes IV-VI) of $U_q(u_{n,1})$.

At $q = e^{ih}$, $h \in \mathbf{R}$, the classes I and III (with modification: Re c_1 = Re c_2 = $\frac{\pi}{h}$ instead of Im c_i) are the only classes that survive in classification of 'unitary' irreps.

It is worth to mention the following: the only class from the above presented list of irreps which is absent in the classical limit (disappears at $q \to 1$ or, equivalently, at $h \to 0$) is the class III (strange series) of 'unitary' irreps.

3. q-ANALOGS OF VECTOR MESON MSR's

Since finite dimensional representations of $U_q(u_n)$, $q^N \neq 1$, are parallel to those of their classical prototypes, we may apply direct extension of the dynamical unitary group-based approach of [7,8], main points of which (say, in the case of three flavors) are the following. (i) Assign to vector meson states from the octet (isotriplet, two different isodoublets, and isosinglet) their corresponding (ortho)normalized vectors of the Gelfand-Tsetlin basis. For example, $|\rho\rangle = |\{8\}_3; \{3\}_2; \alpha_\rho\rangle$, $|\omega_8\rangle = |\{8\}_3; \{1\}_2; \alpha_8\rangle$, with α_ρ characterizing states of different charges within isotriplet, $\{8\}_3 \longleftrightarrow (m+2, m+1, m)$, $\{3\}_2 \longleftrightarrow (m+2, m)$, $\{1\}_2 \longleftrightarrow (m+1, m+1)$, and $\alpha_8 = m+1$); (ii) embed octet of $U_q(u_3)$ into the adjoint 15-plet representation $(m+2, m+1, m+1, m)$ of dynamical $U_q(u_4)$; (iii) take mass operator for $U_q(u_3)$ symmetry breaking in terms of appropriate generators of $U_q(u_4)$, namely $\hat{M}_3 = M_0 + \alpha_3 A_{34} A_{43} + \beta_3 A_{43} A_{34}$; (iv) calculate matrix elements $\langle\rho|\hat{M}_3|\rho\rangle$, $\langle\omega_8|\hat{M}_3|\omega_8\rangle$, etc.

Mass operator, commuting with the 'isospin and hypercharge' q-algebra $U_q(u_2)$, for $3 \leq n \leq 6$ is constructed in terms of appropriate generators of the 'dynamical' algebra $U_q(u_{n+1})$ and has the form [5,6]

$$\hat{M}_n = M_o^{(n)} + \gamma_n A_{nn+1} A_{n+1n} + \delta_n A_{n+1n} A_{nn+1}$$
$$+ \sum_{i=3}^{n-1} (\gamma_i A_{in+1} \tilde{A}_{n+1i} + \delta_i \tilde{A}_{n+1i} A_{in+1} + \tilde{\gamma}_i \tilde{A}_{in+1} A_{n+1i} + \tilde{\delta}_i A_{n+1i} \tilde{A}_{in+1}). \quad (15)$$

It is hermitean, term by term, if q is real. For $q = e^{ih}$, $h \in \mathbf{R}$, hermiticity of mass operator requires that $\gamma_i = \tilde{\gamma}_i$, $\delta_i = \tilde{\delta}_i$. The latter choice is preferable for us.

Using (Gelfand-Tsetlin basis) state vectors for mesons from $(n^2 - 1)$-plet of 'flavor' $U_q(u_n)$ embedded into $\{(n+1)^2 - 1\}$-plet of 'dynamical' $U_q(u_{n+1})$, one performs on the base of (15) necessary calculations and obtains

$$m_\rho = M_o \qquad m_{K^*} = M_o - \gamma_3 \qquad m_{K^*} = M_o - \delta_3$$

$$m_{\omega_8} = M_o - \frac{[2]_q}{[3]_q}(\gamma_3 + \delta_3)$$

$$m_{D^*} = M_o + \gamma_4 \qquad m_{D^*} = M_o + \delta_4$$

$$m_{F^*} = M_o - \delta_3 + \gamma_4 \qquad m_{F^*} = M_o - \gamma_3 + \delta_4$$

$$m_{\omega_{15}} = M_o + \left(\frac{4}{[2]_q} - \frac{[3]_q}{[4]_q} - \frac{[4]_q}{[3]_q}\right)(\gamma_3 + \delta_3) + \frac{[3]_q}{[4]_q}(\gamma_4 + \delta_4) \quad (16)$$

in the 4-flavor case and analogous expressions for $n = 5$ and $n = 6$ (the first four relations in (16) reproduce also the 3-flavor case). q-Dependence appears only in the masses of ω_8, ω_{15}, ω_{24}, ω_{35}. Since (isodoublet) particles and their anti's must have equal masses, $\gamma_3 = \delta_3$, $\gamma_4 = \delta_4$ in (16), and likewise for $n = 5, 6$. The resulting q-MSR's [5,6] are

$$[n]_{(q)}\, m_{\omega_{n^2-1}} + (b_{n;q} + 2n - 4)\, m_\rho = 2\, m_{D_n^*} + (c_{n;q} + 2) \sum_{r=3}^{n-1} m_{D_r^*} \quad (17)$$

where the denotion $[n]_q/[n-1]_q \equiv [n]_{(q)}$ is used and

$$b_{n;q} \equiv n\, c_{n;q} - 6\,[n]_{(q)}^2 + \left(\frac{24}{[2]_q} - 1\right)[n]_{(q)}$$

$$c_{n;q} \equiv 2\,[n]_{(q)}^2 - \frac{8}{[2]_q}[n]_{(q)}.$$

This set of q-deformed MSR's contains at $q \to 1$ the relations (1), as it should. The q-analogs show that coefficients at masses are obtained from their "classical" prototypes in a more complex way than simply by replacing $a \to [a]_q$.

At $n = 3$ the eqn.(17) yields the q-analog of GMO relation:

$$m_{\omega_8} + \left(2\frac{[2]_q}{[3]_q} - 1\right)m_\rho = 2\frac{[2]_q}{[3]_q}m_{K^*}. \tag{18}$$

An essential difference is between the case $n = 3$ and MSR's (17) at more flavors: the q-GMO relation depends on q through the *ratio* $[3]_{(q)}$ *only*, while higher MSR's ($n \geq 4$) contain both the *ratio* $[n]_{(q)}$ *and the quantity* $[2]_q$. This difference is caused by presence of nonsimple-root elements in the mass operator (15) in all cases except for $n = 3$. As mentioned above, definitions of nonsimple-root elements and rules of their commutation with simple-root ones are controlled by the q-Serre relations (6).

If $[3]_q = [2]_q$, the q-deformed mass formula (18) simplifies and yields

$$m_{\omega_8} + m_\rho = 2m_{K^*}. \tag{19}$$

Putting $m_{\omega_8} \equiv m_\phi$, one recognizes in eqn.(19) the nonet mass formula of Okubo [17.10]. This relation perfectly agrees with data (up to errors of experiment and of averaging over isoplets). What are *higher analogs of Okubo's relation*? We put $[n]_q = [n-1]_q$, $n = 4, 5, 6$, in eqn.(17) and obtain them:

$$m_{\omega_{15}} + (5 - 8/[2]_{q_4})m_\rho = 2\,m_{D^*} + (4 - 8/[2]_{q_4})m_{K^*} \tag{20}$$

$$m_{\omega_{24}} + (9 - 16/[2]_{q_5})m_\rho = 2\,m_{D_b^*} + (4 - 8/[2]_{q_5})(m_{D^*} + m_{K^*}) \tag{21}$$

$$m_{\omega_{35}} + (13 - 24/[2]_{q_6})m_\rho = 2\,m_{D_t^*} + (4 - 8/[2]_{q_6})(m_{D_b^*} + m_{D^*} + m_{K^*}). \tag{22}$$

Here q_n denote the values that solve eqns. $[n]_q - [n-1]_q = 0$, namely,

$$q_n = e^{i\pi k/(2n-1)} \qquad k = \pm 1, \pm 2, \dots. \tag{23}$$

The difference between (19) ($n = 3$) and (20)-(22) is still manifest. At fixed $n \geq 4$, values of additional q-number $[2]_{q_n} = 2\cos\frac{k\pi}{2n-1}$, which is present in (20)-(22), obviously differ for different $|k|$'s from (23). However, specific data for masses [18] putted into the MSR (20) or (21) (with J/ψ, Υ respectively in place of ω_{15}, ω_{24}) can satisfy the MSR only with one value of q_n. For example, the relation (21) holds nicely just *with the primitive* 18-th *root of unity taken for* q_5, that is, with $k = 1$. Thus, the q-deuce in MSR's (17) which originates from q-Serre relations (6), serves to "select" unique appropriate value of deformation parameter from the set (23). In the case of $U_q(u_3)$, q-Serre relations are out of play, so the (extra) q-deuce is absent in eqn.(18) and all the values $q_3 = e^{i\pi k/5}$, $k = \pm 1, \pm 2, \dots$, are appropriate.

4. KNOT STRUCTURES ASSOCIATED WITH HEAVY VECTOR QUARKONIA

Let us observe that the quantities $[n]_q - [n-1]_q$ (which, through their roots, extract from the q-analogs (17) those realistic mass relations (19), (21) and others) being such polynomials $P_n(q)$ that satisfy the conditions [19]

$$(i)\quad P_n(q) = P_n(q^{-1}),$$
$$(ii)\quad P_n(1) = 1,$$

coincide (we hope, not only formally) with Alexander polynomials $\Delta(q)\{(2n-1)_1\}$ of toroidal $(2n-1)_1$-knots. Namely, we have that

$$[3]_q - [2]_q = q^2 + q^{-2} - q - q^{-1} + 1 \equiv \Delta(q)\{5_1\},$$
$$[4]_q - [3]_q = q^3 + q^{-3} - q^2 - q^{-2} + q + q^{-1} - 1 \equiv \Delta(q)\{7_1\},$$
$$[5]_q - [4]_q = q^4 + q^{-4} - q^3 - q^{-3} + q^2 + q^{-2} - q - q^{-1} + 1 \equiv \Delta(q)\{9_1\},$$
$$[6]_q - [5]_q = q^5 + q^{-5} - q^4 - q^{-4} + q^3 + q^{-3} - q^2 - q^{-2} + q + q^{-1} - 1 \equiv \Delta(q)\{11_1\}$$

do correspond to the 5_1-, 7_1-, 9_1-, and 11_1- knots, respectively. The 'extra' q-deuce in formulas (17) may be formally related with the trefoil (or 3_1-) knot, since $[2]_q - 1 = q + q^{-1} - 1 \equiv \Delta(q)\{3_1\}$. As a consequence, *all the q-dependence* in masses of ω_{n^2-1} and in coefficients of MSR's (17) can be expressed in terms of various Alexander polynomials:

$$\frac{[3]_q}{[2]_q} = 1 + \frac{\Delta\{5_1\}}{[2]_q} = 1 + \frac{\Delta\{5_1\}}{\Delta\{3_1\}+1},$$

and

$$\frac{[n]_q}{[n-1]_q} = 1 + \frac{\Delta\{(2n-1)_1\}}{[n-1]_q} = 1 + \frac{\Delta\{(2n-1)_1\}}{1 + \sum_{r=2}^{n-1} \Delta\{(2r-1)_1\}}, \qquad (24)$$

where $n = 4, 5, 6$.

Thus, the values (23) may be viewed as roots of respective Alexander polynomials (for every fixed n, just the 'senior' polynomial from those appearing in eq.(24) serves to determine, by means of its root, the corresponding MSR) and are such that reduce the q-GMO formula (18) to (simple and successful) Okubo's relation (19), as well as general formula (17) with $n = 4, 5, 6$ reduce to the higher analogs (20)-(22) of Okubo's relation.

5. (q-DEPENDENT) EXPRESSIONS FOR BARYON MASSES AND A q-ANALOG OF BARYON OCTET MSR

In order to form state vectors for baryons $\frac{1}{2}^+$ that constitute *20*-plet of $U_q(u_4)$ whose decomposition with respect to $U_q(u_3)$ is $20 = 8 + 3 + 3^* + 6$, we use the (orthonormalized) Gel'fand-Tsetlin basis elements (12) constructed in accordance with the aforementioned canonical chain, by fixing $n = 4$. For instance, for the isodoublet of nucleons (contained in octet) we have:

$$|N\rangle \longleftrightarrow |\chi; l_4, l_3, l_2, l_Q\rangle$$

where $\chi \equiv (l_1, l_2, l_3; c_1, c_2)$ labels some appropriate (that is, such that contains the 20-plet of $U_q(u_4)$) irrep of the 'dynamical' $U_q(u_{4,1})$; $l_4 \equiv (p+1, p-1, p-3, p-4)$ and $l_3 \equiv (p+1, p-1, p-3)$ label 20-plet of $U_q(u_4)$ and 8-plet of its subalgebra $U_q(u_3)$ respectively; l_2 means isodoublet $(p+1, p-1)$ of nucleons, and l_Q labels charge states.

Mass operator, according to the concept of *pseudounitary* dynamical group [7,8] adapted to present (q-deformed) case, is constructed in terms of those 'noncompact' generators of $U_q(u_{4,1})$ which break higher (flavour) symmetries, but which preserve as unbroken the isospin-hypercharge symmetry $U_q(u_2)$. Namely, we take it in the form

$$\hat{M}_4 = M_o^{(4)} + \gamma A_{45} A_{54} + \delta A_{54} A_{45}$$
$$+ \alpha A_{35} \tilde{A}_{53} + \beta \tilde{A}_{53} A_{35} + \tilde{\alpha} \tilde{A}_{35} A_{53} + \tilde{\beta} A_{53} \tilde{A}_{53}) \tag{25}$$

and put $\alpha = \tilde{\alpha}$, $\beta = \tilde{\beta}$, to reduce the number of independent parameters. As a result of calculation of the quantities $\langle N| \hat{M}_4 |N\rangle \equiv m_N$, etc., within the representation $D_+^{12}(p-1, p-3, p-4; p, p-2)$ (p being an arbitrary fixed integer), we obtain the following expressions for masses of baryons $\frac{1}{2}^+$ belonging to 20-plet of $U_q(u_4)$.

(i) Octet baryons:

$$m_N = M_8 + \frac{[2][3]}{[6]}([5]\alpha + \beta),$$

$$m_\Lambda = M_8 + \frac{1}{[6]}\left(\frac{[5][4]^2}{[2]^2} + [2]^2\right)\alpha + \frac{[2]^2[3]}{[6]}\beta,$$

$$m_\Xi = M_8 + \frac{1}{[6]}\left([2]([5] + [3]) + \frac{[5][3]}{[2]}([5] - [3] - 2)\right)\alpha + \frac{[2]^2[4]}{[6]}\beta,$$

$$m_\Sigma = M_8 + \frac{[4]^2[5]}{[2]^2[6]}\alpha + \frac{[2][4]}{[6]}\beta; \tag{26}$$

(ii) triplet baryons:

$$m_{\Xi_{cc}} = M_3 + \frac{[2][3][5]}{[6]}\alpha + \frac{[2]}{[6]}\left(([4] - [2])^2 - 1\right)\beta,$$

$$m_{\Omega_{cc}} = M_3 + \frac{[2]^2[5]}{[6]}\alpha + \frac{[2][4]}{[6]}([3] - 2)\beta; \tag{27}$$

(iii) antitriplet baryons:

$$m_{\Xi_c'} = M_{3^*} + \frac{[2][3][5]}{[6]}\alpha + \frac{[2][3]}{[6]}([3] - 2)\beta,$$

$$m_{\Lambda_c} = M_{3^*} + \frac{[3]}{[6]}([3] - 2)([5] + 2)\alpha + \frac{[3]^2}{[6]}([3] - 2)\beta; \tag{28}$$

(iv) sextet baryons:

$$m_{\Sigma_c} = M_6 + \frac{[2][3][5]}{[6]}\alpha + \frac{[2]}{[6]}\left(1 + ([3] - 1)([3] - 2)\right)\beta,$$

$$m_{\Xi_c} = M_6 + \frac{[3] + 2([3] - 1)[5]}{[6]}\alpha + \frac{2[5] - [3] + 4}{[6]}\beta,$$

$$m_{\Omega_c} = M_6 + \frac{[2]}{[6]}\left([3] + ([3] - 2)[5]\right)\alpha + \frac{[2]^2[4]}{[6]}\beta. \tag{29}$$

In the above expressions for masses, we have used the notations

$$M_8 = M_0 + \frac{[3]}{[6]}([5]\gamma + \delta), \qquad M_{3^*} = M_0 + \frac{[3]}{[6]}([4]\gamma + [2]\delta),$$

$$M_3 = M_0 + \frac{[2][4]}{[6]}(\gamma + \delta), \qquad M_6 = M_0 + \frac{1}{[6]}([2][5]\gamma + [4]\delta).$$

Let us check that in the 'classical' limit $q \rightarrow 1$ octet masses satisfy the GMO-relation. Indeed, for $q = 1$ we have that $m_N = \tilde{M}_8 + 5\alpha + \beta$, $m_\Lambda = \tilde{M}_8 + 4\alpha + 2\beta$, $m_\Xi = \tilde{M}_8 + \frac{8}{3}(\alpha + \beta)$, $m_\Sigma = \tilde{M}_8 + \frac{10}{3}\alpha + \frac{4}{3}\beta$ (here $\tilde{M}_8 = M_0 + \frac{15}{6}\gamma + \frac{1}{2}\delta$), and the relation (4) is obviously satisfied.

We are predominantly interested in an octet MSR with q-dependent coefficients. From eqns. (26), by excluding the parameters M_8, α and β, the desired q-analog MSR is obtained in the form

$$[2]m_N + \frac{[2]}{[2]-1}m_\Xi = [3]m_\Lambda + \left(\frac{[2]^2}{[2]-1} - [3]\right)m_\Sigma$$

$$+ \frac{A_q}{B_q}\left([2]m_N + m_\Xi - m_\Lambda - [2]m_\Sigma\right) \qquad (30)$$

where

$$A_q = [2]^4 + [2]^3([5] - [4]) + [2]^2([6] - [5]) - [2]([6] + [4]^2) + [4]^2, \qquad (31)$$

$$B_q = \left([2]^3 - [2]^2[4] + 3[5] - [3]\right)([2] - 1). \qquad (32)$$

The relation (30) constitutes our main result. Here we observe nontriviality of the coefficients at m_Ξ and m_Σ and, as most unexpected thing, the appearance of that additional structure (second line in eq.(30)) with A_q/B_q as its coefficient.

Strictly speaking, the relation just obtained is not a mass relation. However, at any fixed value of q it yields some 'candidate MSR'. For this reason the q-analog relation (30) may be viewed as a continuum of candidate MSR's for baryon masses, *only few of which may be considered as realistic mass relations*.

6. DEFINING q-POLYNOMIAL AND A NOVEL MASS RELATION

Now the problem consists in finding the value(s) of deformation parameter at which the relation (30) yields most realistic MSR(s). Clearly, a straightforward way to proceed would be to insert the empirical data for the octet baryon masses into the eq.(30) and then solve the equation with respect to q. However, we think of this way as not the best one for two reasons: (i) the equation for q this way appears to be rather complicated; (ii) so obtained values of deformation parameter would be neccessarily 'non-rigid' ones reflecting approximate procedure of solving the equation as well as errors of experimental data and averaging over isomultiplets. Fortunately, there exists another approach which is somewhat analogous to reasonings used in [5,6] for the case of vector mesons (see sect.3 above). To this end, let us return again to the 'classical' case. As already mentioned, the value $q = 1$ must result in the standard GMO-relation (a kind of the 'correspondence principle'). Indeed, $A_{q=1} = 0$, $(B_{q=1} \neq 0)$, and $m_N + m_\Xi = \frac{3}{2}m_\Lambda + \frac{1}{2}m_\Sigma$ results. Here we observe the point of oversimplification (due to vanishing of A_q at $q = 1$) when reducing the q-analog to GMO-relation. Adopting this as a hint of *how to search other candidate values of* q, we rewright A_q (which is 7-th order polynomial in q-deuce $[2]_q$) in its factorized form:

$$A_q = ([2] - 2)[2]^3([4] - [2])$$
$$= ([2] - 2)[2]^4([2]^2 - 3) \qquad (31')$$

(the recursion $[n]_q = [2]_q[n-1]_q - [n-2]_q$ for q-numbers is useful in doing this). Since the 'classical' GMO-relation corresponds to vanishing of A_q because of

(i) $([2] - 2) = 0$,

it is natural to examine the remaining cases when A_q turns into zero:

(ii) $[2] = 0$;

(iii) $[4] - [2] \equiv [2]([3] - 2) \equiv [2]([2]^2 - 3) = 0$, $\quad [2] \neq 0$.

The case (ii) leads to the MR $\quad m_\Lambda = m_\Sigma \quad$ which is not very accurate since, experimentally, the accuracy $\approx 6.5\%$). It is interesting, nevertheless, to compare this case with the second nonet Okubo's formula $m_\rho = m_\omega$ (which was shown to follow from the q-analog of vector meson MR, see [5], just if the same restriction $[2]_q = 0$ has been applied). Note also that in both meson and baryon cases, these MR's relate isosinglet and isotriplet masses.

Now let us consider the most interesting case (iii), that is, the values of q that solve $[2]_q = \pm\sqrt{3}$. Those are respectively $q_+ = e^{\frac{i\pi}{6}}$ and $q_- = e^{\frac{i5\pi}{6}}$. At these values, $[4]_{q_\pm} = \pm\sqrt{3}$ (that is, both q_+ and q_- solve the equation $[4]_q - [2]_q = 0$) and also $[3]_{q_\pm} = 2$ (q_+ and q_- both solve the equation $[3]_q - 2 = 0$).

Two candidate MSR's follow from (30) at $q_\pm$: the relation

$$m_N + \frac{1+\sqrt{3}}{2} m_\Xi = \frac{2}{\sqrt{3}} m_\Lambda + \frac{9-\sqrt{3}}{6} m_\Sigma \qquad (33)$$

and the relation obtained from this one by replacing $\sqrt{3} \to (-\sqrt{3})$. The second candidate MR (which corresponds to q_-) shows bad agreement with data. However, fixing q_+, *we get a surprizingly good mass relation*: with empirical values [18] for octet $\frac{1}{2}^+$ baryon masses ($m_N = 938.9\ Mev$, $m_\Xi = 1318.1\ Mev$, $m_\Lambda = 1115.6\ Mev$, and $m_\Sigma = 1193.1\ Mev$) we have $2739.5\ Mev \approx 2733.4\ Mev$. That is, eq.(33) holds with 0.22% accuracy! For comparison recall that usual GMO-relation (4) is satisfied within 0.57%.

7. DOES q-ALGEBRA $U_q(u_{4,1})$ 'REMOVE DEGENERACY' ?

Calculations analogous to those of section 5 were performed also for another specific representation of dynamical algebra. Namely, with the *20*-plet $(p+1, p-1, p-3, p-4)$ of $U_q(u_4)$ embedded into the (integer type) irrep $\tilde{D}_+^{32}(p-1, p-3, p-4; p-4, p-2)$ of $U_q(u_{4,1})$, we have obtained the expressions for octet baryon masses,

$$m_N = M_8{}' + \frac{[2][3]}{[6]}(\alpha + [5]\beta),$$

$$m_\Lambda = M_8{}' + \frac{[3]}{[6]}([3]^2 + [3] - 4)\alpha + \frac{[2]^2[3][5]}{[6]}\beta,$$

$$m_\Xi = M_8{}' + \frac{[2]}{[6]}([3]^3 - 4[3] + 1)\alpha + \frac{[2]^2[4][5]}{[6]}\beta,$$

$$m_\Sigma = M_8{}' + \frac{([3] - 1)^2}{[6]}\alpha + \frac{[2][4][5]}{[6]}\beta, \qquad (34)$$

(here $M_8' \equiv M_0 + \frac{[3]}{[6]}a + \frac{[3][5]}{[6]}b$), as well as the following q-analog of octet MR:

$$m_N - m_\Lambda - \frac{[2]-1}{[2]^2}M_C + \frac{[3]([2]-1)}{[2]^2}M_D = \frac{\tilde{A}_q}{\tilde{B}_q}M_C \qquad (35)$$

where

$$M_C = [2](m_N - m_\Sigma) + m_\Xi - m_\Lambda,$$
$$M_D = ([2]+1)m_\Lambda - [2]m_N - m_\Xi,$$
$$\tilde{B}_q = [2][3]^2 - [3]^2 - [2][3] - 2[2] + 5,$$

and the defining q-polynomial now is of the form

$$\tilde{A}_q = ([2]-2)([3]^2 - 5). \qquad (36)$$

Besides the 'classical' root $[2]_q = 2$ (equivalent to $q = 1$) which determines the standard GMO-relation (4), the polynomial $\tilde{A}_q$ has four roots more, corresponding to $[3]_q = \pm\sqrt{5}$. With $[2]_q = \sqrt{1 + \sqrt{5}}$ (this gives $[3]_q = \sqrt{5}$) we have the MSR

$$2\left(\sqrt{5} - \sqrt{\sqrt{5}+1}\right)m_N + 2\left(\sqrt{\sqrt{5}+1} - 1\right)m_\Xi =$$
$$\left(2\sqrt{5} - 4 + \sqrt{\sqrt{5}-1}\right)m_\Lambda + \left(2 - \sqrt{\sqrt{5}-1}\right)m_\Sigma. \qquad (37)$$

Examination of this MSR with inserted empirical masses shows that it is not very satisfactory (the accuracy is $\approx 2.7\%$). Nevertheless, one could conclude from this second example that distinct dynamical representations may yield essentially different q-analogs of octet MSR, with different defining q-polynomials. While the presence of factor $[2]_q - 2$ is common feature of all defining polynomials, the difference would lie in sets of extra roots (compare $[2]_q = 0$, $[2]_q = \pm\sqrt{3}$ of the polynomial $(31')$ and the roots $[2]_q^2 = 1 \pm \sqrt{5}$ of $\tilde{A}_q$ in this section). Since when applying classical Lie algebra $u(4,1)$ *all the representations* yield [7] the GMO-relation and nothing else (a kind of 'degeneracy'), we could say that application of dynamical q-algebra $U_q(u_{4,1})$, for octet baryon mass relations, removes this degeneracy.

It turns out, however, that by processing the expressions (34) in some another way we arrive at the q-analog of octet MR which coincides with our first q-analog, eq.(30), in all the points (coefficients at masses, the combination of masses in curly brackets and, most inportant, the defining polynomial A_q in 2nd line of (30)) except for change $B_q \to \tilde{B}_q$ of polynomial in denominator (it plays no role in obtaining MSR's (4) and (33)). We conclude that application of the q-algebra $U_q(u_{4,1})$ as dynamical one 'removes degeneracy' at least in the sense that here we get, besides eq.(4), mass sum rules of novel type (including such successful one as eq.(33)). Those novel MSR's seem to reflect some unusual hadronic dynamics, and this certainly deserves further study.

8. CONCLUDING REMARKS

To summarize, we have considered some implications of applying, instead of unitary groups $SU(n)$ (conventional in description of flavor symmetries and classifying hadrons into multiplets), the corresponding quantum algebras $U_q(u_n)$.

For mesons 1^-, necessary symmetry breaking was performed by the 'dynamical' q-algebra $U_q(u_{n+1})$. A series of q-analogs (17) of MSR's (1) were obtained from which the 'ideal' MSR (19) and its higher analogs (20)-(22) deduced. Our method of *fixation of q is 'rigid one'* (it uses roots of q-polynomials) and radically differs from fitting procedure in other phenomenological applications [1-3] of q-algebras.

It is shown that all singlet masses $m_{\omega_{n^2-1}}$, $n = 3, ..., 6$, and all the coefficients in q-analog MSR's depend on q just through the topological invariants of torus knots $(2n-1)_1$ realized by the Alexander polynomials $\Delta(q)\{(2n-1)_1\}$. The fact that the number of flavors $n \geq 3$ is correlated with the order of specific torus knot opens the possibility of topological characterization of heavy flavors ($n \geq 3$) through the (masses of) corresponding vector quarkonia.

Let us remark that, although we use (the $(4n-2)$-th) roots of unity (23) for q, the specific representations exploited within this approach remain irreducible.

Extending the approach of dynamical *pseudounitary* groups to the q-algebras $U_q(u_4)$ (4-flavor symmetry) and $U_q(u_{4,1})$ ('dynamical' symmetry) we have obtained q-dependent expressions for masses of baryons $\frac{1}{2}^+$ from which q-MR's follow.

The q-deformed relation (30) is of interest as a 'continuum of candidate' MR's and also due to unusual mass-dependent term (second line in (30)) with A_q/B_q. It is just this point where the concept of *defining q-polynomial* arises: the polynomial A_q, by means of its roots, determines concrete octet baryon MR's (the classical GMO-relation (4) and our new, surprisingly accurate, MSR (33) with $\sqrt{3}$ in some of its coefficients, as most successful ones).

Let us stress once more that, *due to irrational coefficients*, the MSR (33) is of unconventional nature. This is connected with root of unity $q_+ = e^{\frac{i\pi}{6}}$ and may be reflects some 'nonperturbative' (topological) information contained in the model adopted, at such value of deformation parameter.

To make this more transparent, let us present the q-MR (30) in the form

$$(1 - \Delta_N)m_N + (1 - \Delta_\Xi)m_\Xi = \left(\frac{3}{2} - \Delta_\Lambda\right)m_\Lambda + \left(\frac{1}{2} + \Delta_\Sigma\right)m_\Sigma, \tag{38}$$

where

$$\Delta_N \equiv \frac{A_q}{B_q}, \qquad\qquad \Delta_\Xi \equiv \frac{A_q}{[2]B_q} + 1 - \frac{1}{[2]-1},$$

$$\Delta_\Lambda \equiv \frac{A_q}{[2]B_q} + \frac{3}{2} - \frac{[3]}{[2]}, \qquad \Delta_\Sigma \equiv -\frac{A_q}{B_q} + \frac{[2]}{[2]-1} - \frac{[3]+1}{2}. \tag{39}$$

Since $A_q = 0$ at $q = 1$, we have that all Δ_k (here k takes the 'values' N, Ξ, Λ and Σ) equal zero in the classical limit. Therefore, it is natural to consider the quantities Δ_k at $q \neq 1$ as 'corrections' to the classical coefficients due to q-deformation. Obviously, at values of q which are very close to unity these corrections do not deviate substantially from zero.

At $q = q_+$ again A_q vanishes. However, now all Δ_k other than Δ_N are not small ('perturbative') quantities, but become of the order of magnitude comparable with the classical coefficients (for instance, $\Delta_\Sigma = \frac{\sqrt{3}}{\sqrt{3}-1} - \frac{3}{2} \approx 0.87$ to be compared with the coefficient $\frac{1}{2}$ in classical MSR (4)).

The author would like to thank Prof. Bruno Gruber for invitation and warm hospitality at the Symposium in Bregenz. This research was supported in part by the Ukrainian State Foundation for Fundamental Research, and by the INTAS Program under the Grant INTAS-93-1038.

REFERENCES

1. Iwao S 1990 *Progr. Theor. Phys.* **83** 363.
2. Raychev P P, Roussev R P and Smirnov Yu F 1990 *J. Phys.* *G* **16** 137.
   Bonatsos D et al. 1990 *Phys. Lett.* **251B** 477.
3. Celeghini E et al. 1991 *Firenze preprint* DFF 151/11/91.
4. Biedenharn L C 1990 *An overview of quantum groups*, in: Proc. XVIIth Int.
   Coll. on Group Theor. Methods in Physics (Moscow, June 1990).
   Zachos C 1991 *Paradigms of quantum algebras*, in: Symmetry in Science V (B
   Gruber, L C Biedenharn, and H D Doebner, eds.), pp 593-609.
   Kibler M 1993 *Lyon preprint* LYCEN/9358.
5. Gavrilik A M 1993 *Physics in Ukraine. Quantum Fields and Elementary Particles*
   (Proc. Int. Conf., Kiev); Gavrilik A M, Tertychnyj A V 1993 *Kiev preprint*
   ITP-93-19E.
6. Gavrilik A M 1994 *J. Phys.* *A* **27** L91; *Kiev preprint* ITP-93-28E.
7. Gavrilik A M, Shirokov V A 1978 *Yadernaya Fizika* **28** 199.
   Yakimov G, Kalman C 1976 *Lett. Nuovo Cim.* **17** 511.
   Kalman C S 1978 *Lett. Nuovo Cimento* **21** 291.
8. Gavrilik A M, Klimyk A U 1989 *Symposia Mathematica* **31** 127.
9. Novozhilov Yu V 1975 *Introduction to elementary particle theory* (N.Y., Pergamon).
10. Gasiorowicz S 1966 *Elementary particle theory* (N.Y., J Willey).
11. Coquereaux R, Esposito-Farese G 1990 *Marseille preprint* CPT-90/P.2357.
12. Gavrilik A M, Kachurik I I and Tertychnyj A V 1994 *Kiev preprint* ITP-94-34E.
13. Jimbo M 1985 *Lett. Math. Phys.* **10** 63.
14. Klimyk A U, Groza V A *Kiev preprint* ITP-89-37R.
    Klimyk A U *Infinite dimensional representations of quantum algebras.* to be
    published.
15. Chakrabarti A 1991 *J. Math. Phys.* **32** 1227.
16. Klimyk A U, Gavrilik A M 1979 *J. Math. Phys.* **20** 1624.
17. Okubo S 1963 *Phys. Lett.* **5** 165.
18. Particle Data Group 1990 *Phys. Lett.* *B* **239** 1.
19. Birman J S 1993 *Bull. Amer. Math. Soc.* **28** 253.

LINEAR COXETER GROUPS

Jose Getino

Departamento de Fisica
Universidad de Oviedo
3007 Oviedo, Spain

1 DISCRETE REFLECTION GROUPS

Any discrete group W generated by reflections in a space of constant curvature X (discrete reflection group) can be described in terms of its fundamental region P, which is a convex polyhedron whose dihedral angles are proper submultiples of π. By immersing the space X in a linear space E (the ambient space), P extends to a convex polyhedral cone C_P and W extends to a linear group generated by reflections in the faces of C_P (linear Coxeter group).

Coxeter groups are the natural algebraic generalization of discrete reflection groups, but they form a much wider class of groups. Nevertheless, any abstract Coxeter group can be represented as a linear Coxeter group.

For the proofs of the results in this section see [1, 2].

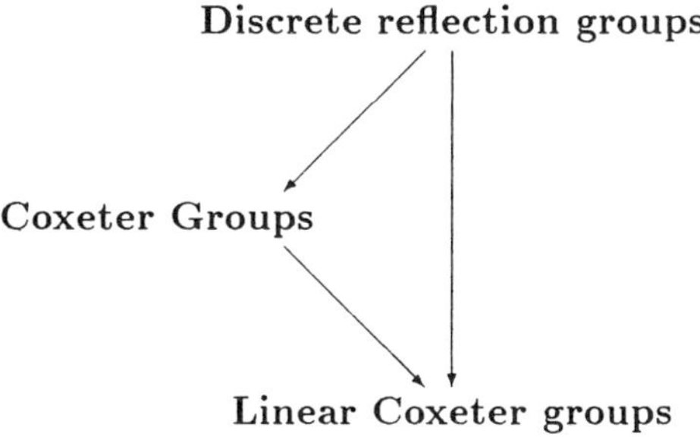

1.1   Simple examples

Let us start with some examples of discrete reflection groups in an Euclidean space E.

- The group generated by a reflection in a mirror of E. It has two elements, so it is isomorphic to $\mathbf{Z}_2$.

Symmetries in Science VII, Edited by
B. Gruber, Plenum Press, New York, 1995

- The *n-th dihedral* group D_n is generated by the reflections r_1 and r_2 in two mirrors H_1 and H_2 forming an angle π/n, where $n = 2, 3, \ldots$. Then $r_1 r_2$ is the rotation through $2\pi/n$ about the plane $H_1 \cap H_2$ and hence

$$(r_1 r_2)^n = 1.$$

  The dihedral group D_n is isomorphic to the semidirect product

$$D_n \simeq \mathbf{Z}_n \bigoplus \mathbf{Z}_2,$$

  where the cyclic group $\mathbf{Z}_n$ identifies to the group generated by the rotation $r_1 r_2$ and $\mathbf{Z}_2$ identifies to the group generated by a single reflection in any of the two hyperplanes.

- The *infinite dihedral* group D_∞ is generated by the reflections r_1 and r_2 in two different parallel mirrors. Then $r_1 r_2$ is the translation $2d\mathbf{n}$, being d the distance between the mirrors and $\mathbf{n}$ a unitary vector orthogonal to them. D_∞ is isomorphic to the semidirect product

$$D_\infty \simeq \mathbf{Z} \bigoplus \mathbf{Z}_2,$$

  where the infinite cyclic group $\mathbf{Z}$ identifies to the 1-dimensional lattice Λ formed by the translations of the Euclidean space E that are in D_∞,

$$\Lambda = 2d\mathbf{Z}\, \mathbf{n},$$

  and $\mathbf{Z}_2$ identifies to the group generated by a single reflection.

1.2 Spaces of constant curvature

Reflections and discrete reflection groups can be defined in the *spaces of constant curvature* (SCCs). An SCC is a simply connected complete Riemannian manifold of constant curvature. For a review on SCCs see [1].

The following result characterizes SCCs in terms of the Lie group $\mathrm{Aut}(X)$ of motions or isometries of X.

A simply connected complete Riemannian manifold X is a SCC iff for any $x \in X$, $f \to T_x f$ is a one to one map

$$\mathrm{Aut}(X) \overset{T_x}{\leadsto} \bigcup_{y \in X} O(T_x X, T_y X). \tag{1}$$

In the sequel X will denote a SCC. From (1) it follows that $\mathrm{Aut}(X)$ acts transitively in X and the stabilizer subgroup $\mathrm{Aut}_x(X)$ of $x \in X$ is isomorphic to the orthogonal group of the tangent space in x

$$\mathrm{Aut}_x(X) \simeq O(T_x X).$$

Any n-dimensional space of constant curvature is isomorphic to one of the three following models, consisting on hypersurfaces of a $(n+1)$-dimensional real vector space E, called the *ambient space* (Fig. 1).

Sphere. $(n \geq 2)$ formed by the points of E verifying

$$S(x, x) = 1,$$

being S a scalar product in E. It has curvature 1.

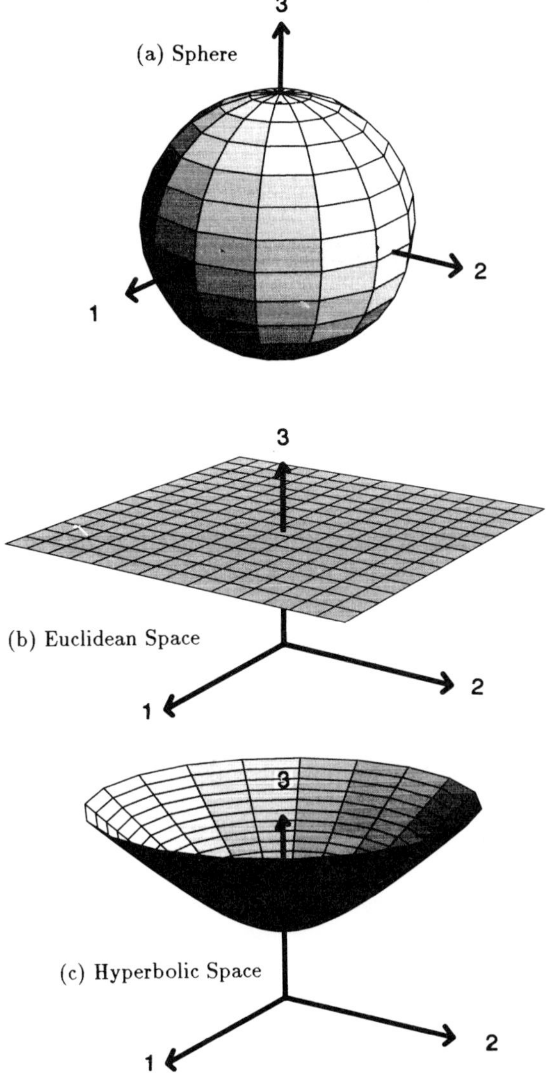

Figure 1. Spaces of constant curvature

Euclidean space. Any hyperplane X of E not containing 0, provided with a scalar product in the linear hyperplane TX of translations in X. It has curvature 0.

Hyperbolic space. One of the two connected components of the hypersurface

$$S(x,x) = -1,$$

where S denotes a Lorentzian metric in E, that is, a symmetric bilinear form with signature $(1, n)$. This space has curvature -1. It will be useful consider the space

$$\overline{X} := X \cup \partial X,$$

where ∂X denotes the points at infinite.

Any motion of X can be extended in a unique way to a linear transformation of the ambient space E, called its *linear extension*. In fact we have

Sphere. The linear extension is a one to one map

$$\mathrm{Aut}(X) \rightsquigarrow O(E).$$

Euclidean space. The linear extension is an injection

$$\mathrm{Aut}(X) \rightarrow GL(E).$$

Hyperbolic space. The linear extension is a one to one map

$$\mathrm{Aut}(X) \rightsquigarrow O_+(E),$$

where $O_+(E)$ denotes the orthocrone Lorentz group of E, formed by the Lorentz transformations preserving the napes of the light cone.

Given a group G of motions of X, call the *linear extension of G* the linear group of E formed by the linear extensions of its elements.

1.3 Planes and convex sets

Some common objects of classical geometry, whose meaning could seems to lie on the linear properties of the Euclidean space, can in fact be defined in a general SCC. Among them we have the planes, convex sets and polyhedra, that will be introduced in the sequel.

The *planes* of an SCC can be defined in several equivalent manners. A plane of X is a complete geodesic submanifold of X. We can also define a plane as the set of points of X remaining invariant under a motion of X. For our purpose, the most useful way to see a plane Π in X is as the trace in X of a (unique) linear plane F of the ambient space E, called the *linear extension* of Π

$$\Pi = X \cap F \tag{2}$$

In fact, the linear extension of a k-plane (plane of dimension k) of X is a $(k+1)$-plane of E. Call a plane F of X *admissible* if $X \cap F$ is a plane of X.

Sphere. Any plane of E is admissible.

Euclidean space. F is admissible iff it is not contained in the hyperplan TX. of translations in X.

Hyperbolic space. F is admissible iff it is an hyperbolic plane, that is, the restriction to F of the metric in E is also Lorentzian. Hyperbolic planes are the ones intersecting the interior of the light cone, i.e., containing time-like vectors.

The planes of X are also SCCs (except for the cases $X = S^2$ and $X = S^1$). As in the Euclidean case we have

- The intersection of a family of planes is also a plane. The minimal plane containing a part Y of X is called the *plane hull* of Y.

- Any set of $k+1$ points of X ($\overline{X}$. in the case of the hyperbolic space) is contained in a plane of dimension $\leq k$.

- The motions of X transform planes in planes and preserves the dimension.

Two extreme cases are particularly important

Straight lines: planes of dimension 1 (Fig. 2). The *segment* $[x,y]$ where $x,y \in X$, is the part of the straight line containing x and y, whose extremes are x and y. In the spherical case x and y must be not antipodal and we take the shortest of the two parts of the line. A segment is a geodesic of X. We can define also the non-closed segments $[x,y[$, $]x,y]$, and $]x,y[$.

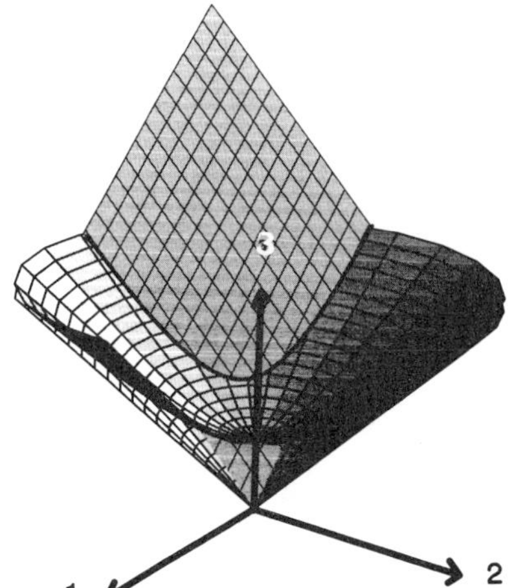

Figure 2. Straight line

Hyperplanes: planes of codimension 1. Given an hyperplane Σ, X is the union of two closed connected convex sets, denoted Σ^+ and Σ^- and called the *halfspaces* defined by Σ, whose intersection is Σ. A halfspace Σ^+ of X is the trace in X of a

linear (i.e. containing 0) halfspace H^+ of the ambient space E, called the *linear
extension* of Σ^+. For any halfspace Σ^+ define the vector $\beta \in E$

$$\left\{ \begin{array}{c} \beta \text{ is tangent to } X \text{ and orthogonal to } \Sigma \\ S(\beta, \beta) = 2 \\ \beta \in H^+. \end{array} \right\} \tag{3}$$

In the case of the sphere and the hyperbolic space, β is orthogonal to the linear
hyperplane H.

If Y is a plane and Σ^+ a halfspace in the sphere, we have

$$Y \subset \Sigma^+ \Rightarrow Y \subset \Sigma. \tag{4}$$

A subset Y of X is said to be *convex* if it contains the segment $[x, y]$ for any pair
(x, y) of (not antipodal, if X is the sphere) points of Y. The *convex hull* $\langle Y \rangle_{\text{conv}}$ of a
part Y of X is the minimal convex subset of X containing Y. The *facets* of a convex
set Y are defined as in the Euclidean, that is, a convex part F of Y is a facet of Y if
whenever $x, y \in Y$ (not antipodal if X is a sphere), if $F \cap]x, y[\neq \emptyset$, then $[x, y] \subset F$.
The facets of dimension d are called *d-facets*. A *face* is a $[\dim Y - 1]$-facet; the *walls* of
Y are the plane hulls of the faces of Y.

1.4   Polyhedra

A *convex polyhedron* P of X is the intersection of finite many halfspaces of X

$$P = \bigcap_{i \in I} \Sigma_i^+. \tag{5}$$

In the sequel we will suppose that $\text{int}(P) \neq \emptyset$ and none of the halfspaces Σ_j^+ is redundant, that is, $\cap_{I-j} \Sigma_i^+ \not\subset \Sigma_j^+$.

The walls of P are the hyperplanes $(\Sigma_i)_{i \in I}$. Two intersecting faces S_i and S_j are
said *adjacent* if $\dim(S_i \cap S_j) = \dim P - 2$. The angles (φ_{ij}) between intersecting walls
are called the *dihedral angles* of P.

The set $f_0 := \bigcap_{i \in I} \Sigma_i$ is either $\emptyset$ or a plane. In the last case f_0 is the minimal
facet of P and we say that P is a *cone* with *apex* f_0. An *strict cone* is a cone with a
0-dimensional apex.

P is the trace in X of a convex polyhedral cone C_P of E, called the *linear extension*
of P (Fig. 3).

$$P = C_P \cap X, \qquad \text{where} \qquad C_P := \bigcap_i H_i^+, \tag{6}$$

being H_i^+ the linear extension of Σ_i^+.

Define also the cone $C_0(P)$ *subtended* by P as the convex (not necessarily closed)
cone formed by the rays starting at 0 and intersecting P. We have

$$C_0(P) \subset C_P.$$

Bounded polyhedra are characterized by

$$P \text{ is bounded} \quad \Leftrightarrow \quad C_P = C_0(P) \quad \Leftrightarrow \quad P = \langle x_1, \dots, x_p \in X \rangle_{\text{conv}}. \tag{7}$$

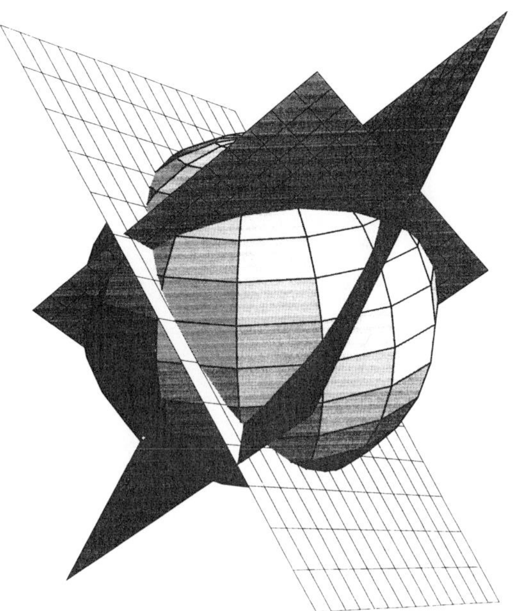

Figure 4. Unbouded polyhedron

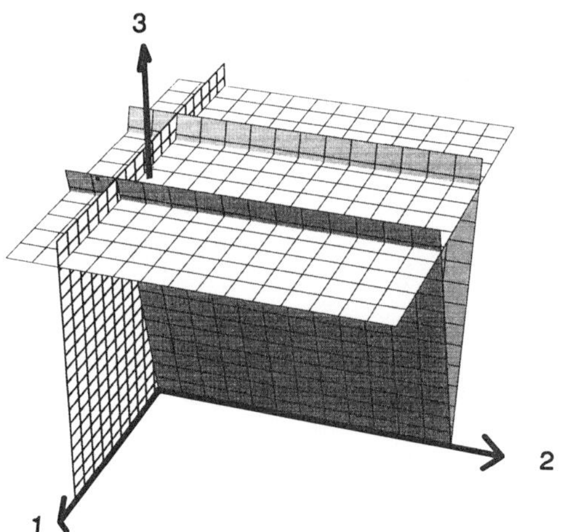

Figure 3. Spherical triangle

Define the *Gram matrix* of P, $\mathrm{Gram}(P) = (b_{ij})$, as the Gram matrix of the vectors $(\beta_i)_i$ introduced in (3)

$$b_{ij} := S(\beta_i, \beta_j). \tag{8}$$

$\mathrm{Gram}(P)$ is a symmetric matrix that stores the information of the dihedral angles (φ_{ij}) between intersecting walls and the distances (ρ_{ij}) between divergent walls (only for the hyperbolic space)

$$b_{ij} = \begin{cases} 2 & \text{if } i = j \\ -2\cos(\varphi_{ij}) & \text{if } \Sigma_i \text{ and } \Sigma_j \text{ intersect} \\ -2 & \text{if } \Sigma_i \text{ and } \Sigma_j \text{ are parallel} \\ -2\cosh(\rho_{ij}) & \text{if } \Sigma_i \text{ and } \Sigma_j \text{ diverges.} \end{cases} \tag{9}$$

Define the *rank* of P

$$\mathrm{rank}(P) := \mathrm{rank}\,[\mathrm{Gram}(P)].$$

Sphere. $\mathrm{Gram}(P)$ is semidefinite positive and $\mathrm{rank}(P) \leq \dim X + 1$. Any convex polyhedron is bounded. From (4), any plane contained in P is also contained in f_0. We say that P is *nondegenerate* if it verifies the equivalent conditions

$$\mathrm{rank}(P) = \dim X + 1 \quad \Leftrightarrow \quad P \text{ contains no antipodes} \quad \Leftrightarrow \quad P \text{ is not a cone.}$$

Euclidean space. $\mathrm{Gram}(P)$ is semidefinite positive and $\mathrm{rank}(P) \leq \dim X$. We say that P is *nondegenerate* if it verifies the equivalent conditions

$$\mathrm{rank}(P) = \dim X \quad \Leftrightarrow \quad (\beta_i) \text{ generates the hyperplane } TX,$$

in other words, there is no an hyperplane orthogonal to the (Σ_i). We have

$$P \text{ bounded} \quad \Leftrightarrow \quad P \text{ finite volume} \Rightarrow \quad P \text{ nondegenerate.}$$

Figure 4 shows an unbounded polyhedron in the Euclidean space. In this example, $C_0(P) = C_P - \{(0, x_2, 0) | x_2 > 0\}$.

Hyperbolic space. $\mathrm{Gram}(P)$ has negative index of inertia 1 and $\mathrm{rank}(P) \leq \dim X + 1$ or it is semidefinite positive and $\mathrm{rank}(P) \leq n$. We say that P is *nondegenerate* if it verifies the equivalent conditions

$$\mathrm{rank}(P) = \dim X + 1 \quad \Leftrightarrow \quad \begin{cases} P \text{ is not a cone and} \\ \text{there is no an hyerplane orthogonal to the } (\Sigma_i). \end{cases}$$

There exist finite volume unbounded polyhedra, and in fact we have

$$\mathrm{Vol}(P) < \infty \quad \Leftrightarrow \quad C_P = \overline{C_0(P)} \quad \Leftrightarrow \quad P = \left\langle x_1, \ldots, x_p \in \overline{X} \right\rangle_{\mathrm{conv}}.$$

In this case P is nondegenerate.

1.5 Coxeter Polyhedra

An *acute-angled polyhedron* is a convex polyhedron with dihedral angles $\leq \pi/2$. If P is an acute-angled polyhedron, two faces S_i and S_j are adjacent iff the walls Σ_i and Σ_j intersect. The Gram matrix of P is a nonnegative symmetric C1 matrix (see (95)). Let us see how acute-angled polyhedra looks like.

Sphere. Any nondegenerate acute-angled polyhedron is a simplex.

Euclidean space. Any acute-angled polyhedron is a direct product

$$P = X' \times P_+ \times P_0, \tag{10}$$

where X' is a Euclidean space, P_+ is a simplicial cone and P_0 is a product of simplices. The positive and zero components (see (96)) of $\mathrm{Gram}(P)$ are respectively $\mathrm{Gram}(P_+)$ and $\mathrm{Gram}(P_0)$. We have

$$P \text{ nondegenerate} \iff P = P_+ \times P_0. \tag{11}$$

In this case, if $\{Q\}$ denotes the apex of the cone P_+ and f is a facet of P_0, $\{Q\} \times f$ is a bounded facet of P. It follows that for an acute-angled polyhedron P

$$P \text{ has a bounded 1-facet} \iff \begin{cases} P \text{ is nondegenerate and} \\ \mathrm{Gram}(P) \text{ is not positive.} \end{cases} \tag{12}$$

Acute-angled bounded polyhedra are characterized by

$$P \text{ bounded} \iff \begin{cases} P \text{ is nondegenerate and} \\ \mathrm{Gram}(P) \text{ is a zero type matrix.} \end{cases} \tag{13}$$

Hyperboloc space. The situation is far more complicated, see [1].

A *Coxeter polyhedron* is a convex polyhedron with dihedral angles on the form

$$\varphi_{ij} = \frac{\pi}{m_{ij}}, \qquad m_{ij} = 2, 3, \ldots. \tag{14}$$

The classification of spherical and Euclidean Coxeter polyhedra can be found in [2]. Bounded (unbounded finite volume) hyperbolic Coxeter simplices are called *Lannèr* (*quasi-Lannèr*) polyhedra and they exist for $\dim(X) \leq 4$ ($\dim(X) \leq 9$). It is known that there are no bounded (finite volume) hyperbolic Coxeter polyhedra for $\dim(X) \geq 30$ ($\dim(X) \geq 996$), but examples are known only for $\dim(X) \leq 8$ ($\dim(X) \leq 21$).

1.6 Reflections

A *reflection* is an involutive motion r of X whose set Σ of fixed points is an hyperplane. The hyperplane Σ characterizes in fact the reflection, so we say that r is the (orthogonal) reflection in the hyperplane Σ. The linear extension of a reflection is also a reflection in a wide sense, for in general there is no metric defined in the ambient space. In this general context, an hyperplane is not enough to define a reflection, and also a straight line must be given.

A *pseudoreflection* in a linear space E is a transformation $r \in GL(E)$ whose set of fixed points is an hyperplane, i.e., $\ker(r - Id)$ is an hyperplane of E. They are on the form

$$r(x) = x - \alpha^{\vee}(x)\alpha \qquad (15)$$

where $\alpha^{\vee}$ is a nonzero linear form in E and α is nonzero vector of E. Then the hyperplane of fixed points is $H := \ker(r - Id) = \ker(\alpha^{\vee})$ and

$$D := \operatorname{Img}(r - Id) = \mathbf{R}\alpha$$

is an invariant line called the *direction* of the pseudoreflection. A *reflection* is an involutive pseudoreflection. They are characterized by the additional condition

$$\alpha^{\vee}(\alpha) = 2.$$

The invariant spaces H and D determining the reflection r, we say that r is the *reflection in H along the direction D*. The direction D of the reflection r is formed by the vectors changing its orientation under r.

Suppose a symmetric bilinear form S in E is given. A *bilinear reflection* is a reflection preserving S. Bilinear reflections along non isotropic directions are characterized by the condition

$$\alpha^{\vee} = 2S(\alpha, -)/S(\alpha, \alpha). \qquad (16)$$

When S is nondegenerate, bilinear reflections will be called *orthogonal reflections* or, for short, reflections.

Let us go back to the reflections in X. Let us consider an hyperplane Σ of X and denote by H its linear extension to E and by r the linear extension of the reflection in Σ. We have

$$r \text{ is the reflection in } \mathbf{H} \text{ along } \mathbf{R}\beta, \qquad (17)$$

where β is the vector associated to a halfspace Σ^{+} (3). Therefore, in the case of the sphere and the hyperbolic space, r is the orthogonal reflection in H.

1.7 Coxeter groups and discrete reflection groups

Firstly a short description of abstract Coxeter groups will be given. For details see [3] and [4]. Then we deal with discrete reflection groups. For a general review of discrete groups of motions, and in particular discrete reflection groups, see [2]. Discrete reflection groups in an Euclidean space are exhaustively studied in [3].

A *Coxeter group* is a pair (W, S), where W is a group and S is a part of W generating W and subject only to relations on the form

$$(st)^{m(s,t)} = 1, \qquad \text{where } m(s,s) = 1 \text{ and } m(s,t) \in 2, 3, \ldots, \infty, \text{ if } s \neq t, \qquad (18)$$

the case $m(s,t) = \infty$ meaning that there is no relation involving s and t. Call the elements in S the *principal reflections* of the Coxeter group and define the *reflections* as the conjugate wsw^{-1} of the principal reflections.

The symmetric matrix $(m(s,t))_{st}$ is called the *exponent matrix* of W and determines the group W up to isomorphism. We can represent the isomorphism class of a Coxeter group by means of a graph with a vertex for each $s \in S$ and an edge between the vertices of s and t iff $m(s,t) \geq 3$, labeled with the integer $m(s,t)$, the label 3 being usually omitted.

The *cosine matrix* of the group is the symmetric C1 and C2 matrix (see (95) and (100))

$$\mathrm{Cos}(W) := \left(-2\cos\frac{\pi}{m(s,t)} \right).$$

The following result characterizes finite Coxeter groups

$$W \text{ finite} \quad \Leftrightarrow \quad \mathrm{Cos}(W) \text{ definite positive.} \tag{19}$$

A *discrete reflection group* is a discrete group of motions generated by reflections in a SCC. Discrete reflection groups can be characterized in the following geometrical way.

"A group of motions of X is a discrete reflection group iff it is generated by the reflections in the walls of a Coxeter polyhedron in X."

Let us consider a Coxeter polyhedron P and set $W \subset \mathrm{Aut}(X)$ the discrete reflection group generated by the reflections $(r_i)_{i \in I}$ in the walls $(\Sigma_i)_i$ of P. These reflections will be called the *principal reflections* of W. We have

- P is a fundamental region of W and $\{w(P) \mid w \in W\}$ is a polyhedral complex covering X.

- W is a the group defined by the generators $(r_i)_{i \in I}$ and the relations

$$(r_i r_j)^{m_{ij}} = 1, \tag{20}$$

defined for any pair Σ_i and Σ_j of intersecting walls, and being the dihedral angle $\varphi_{ij} = \pi/m_{ij}$.

These relations summarizes the essential structural properties of discrete reflection groups and states that the pair $(W, \{r_i \mid i \in I\})$ is a Coxeter group.

The exponent matrix (m_{ij}) is given by

$$m_{ij} = \begin{cases} 1 & \text{if } i = j \\ \pi/\varphi_{ij} & \text{if } \Sigma_i \text{ and } \Sigma_j \text{ intersect} \\ \infty & \text{if } \Sigma_i \text{ and } \Sigma_j \text{ are parallel or diverges.} \end{cases} \tag{21}$$

This matrix only keeps the information on the dihedral angles of the fundamental polyhedron, saying nothing about the distances between divergent walls (compare with (9)). The cosine matrix of W is (see (102))

$$\mathrm{Cos}(W) = \mathrm{Cos}\left[\mathrm{Gram}(P)\right]. \tag{22}$$

1.8   Linear Coxeter groups

The linear extension of a discrete reflection group W with fundamental polyhedron P is generated by reflections in the walls $(H_i)_i$ of the convex polyhedral cone C_P extending P (6), along suitable directions. This group being isomorphic to W, it is also a Coxeter group.

Linear Coxeter groups, introduced by Vinberg [5], are the natural generalization of the linear extensions of discrete reflection groups. They are generated by reflections in the walls of a polyhedral convex cone which is a fundamental region for the linear group. Vinberg translated this geometrical description into an algebraic language, more suitable for computations, introducing the concept of *system*.

A *linear reflection group* in a finite dimensional real vector space E is a group generated by a set R formed by the reflections in the walls of a convex polyhedral cone C in E with nonempty interior

$$\text{int}(C) \neq \emptyset. \tag{23}$$

The elements $r \in R$ are called the *principal reflections* of the reflection group. A *linear Coxeter group* is a linear reflection group W that verifies the *fundamental region condition*

$$w[\text{int}(C)] \cap \text{int}(C) = \emptyset, \qquad \text{for any } 1 \neq w \in W. \tag{24}$$

In this case, C is a fundamental region of W and the pair

$$(W, R)$$

is a Coxeter group. Furthermore the elements of W which are reflections are the conjugates wrw^{-1} of the principal reflections.

In order to have an algebraic characterization of the linear reflection group W, let us label the faces of teh cone C with the elements of a finite set I, and set $|I|$ the cardinal of I. C can be defined by a system of inequalities

$$\alpha_i^{\vee}(x) \geq 0, \quad i \in I, \tag{25}$$

where $\alpha_i^{\vee}$ is a linear form in E vanishing in the wall H_i of C and taking nonnegative values in C. The principal reflections $(r_i)_i$ can be put in the form (15)

$$r_i(x) = x - \alpha_i^{\vee}(x)\alpha_i, \qquad i \in I, \tag{26}$$

where α_i is a vector in the direction D_i of the reflection r_i and verifying $\alpha_i^{\vee}(\alpha_i) = 2$. Therefore the triple

$$[E, (\alpha_i)_i, (\alpha_i^{\vee})_i] \tag{27}$$

determines the linear reflection group W. For characterizing linear Coxeter groups we need

- Determine the triples (27) verifying conditions (23) and (24).

- Find out when two admissible triples defines the same linear Coxeter group. For example, if $(d_i)_i$ are nonvanishing real numbers, the triple $[E, (d_i\alpha_i)_i, (d_i^{-1}\alpha_i^{\vee})_i]$ defines the same linear Coxeter group.

Then the linear Coxeter groups will correspond to admissible triples module a certain equivalence relation.

2 SYSTEMS

In this section the systems are introduced and their isomorphism classes are classified by means of the *characteristic*, as defined in [5].

2.1 Systems

Let us define a *I-system* (or simply a system) as a triple $[E, (\alpha_i)_{i \in I}, (\alpha_i^{\vee})_{i \in I}]$, where E is a finite dimensional real vector space, $(\alpha_i)_i$ is a finite family of nonvanishing vectors of E and $(\alpha_i^{\vee})_i$ is a finite family of nonvanishing linear forms in E verifying

$$\alpha_i^{\vee}(\alpha_i) = 2, \qquad i \in I.$$

Let us define the linear maps

$$\alpha \colon \mathbf{R}^I \to E, \qquad \alpha(X) = \sum_i X_i \alpha_i$$

$$\alpha^{\vee} \colon E \to \mathbf{R}^I, \qquad \alpha^{\vee}(x) = [\alpha_i^{\vee}(x)]_i.$$

For short, we will denote a system using the above maps, in the form $(E, \alpha, \alpha^{\vee})$.

The cone C associated to an I-system is the one defined by a system of inequalities as (25). The minimal facet of C is the subspace $\ker(\alpha^{\vee})$. Condition $\mathrm{int}(C) \neq \emptyset$ can be expressed in terms of the transpose map ${}^t\alpha^{\vee} \colon \mathbf{R}^I \to E^*$ as

$$\ker({}^t\alpha^{\vee}) \cap (\mathbf{R}^+)^I = \{0\}. \tag{28}$$

Let $(F, \beta, \beta^{\vee})$ be another I-system. An *isomorphism* $f \colon (E, \alpha, \alpha^{\vee}) \rightsquigarrow (F, \beta, \beta^{\vee})$ is a linear isomorphism $f \colon E \rightsquigarrow F$ such that

$$\beta = f\alpha \qquad \text{and} \qquad \beta^{\vee} = \alpha^{\vee} f^{-1},$$

i.e., for all $i \in I$, $\beta_i = f(\alpha_i)$ and $\beta_i^{\vee}(x) = \alpha_i^{\vee}(f^{-1}x)$. Two I-systems $(E, \alpha, \alpha^{\vee})$, $(F, \beta, \beta^{\vee})$ are said *isomorphic* and denoted $(E, \alpha, \alpha^{\vee}) \simeq (F, \beta, \beta^{\vee})$ if there exits an isomorphism $f \colon (E, \alpha, \alpha^{\vee}) \rightsquigarrow (F, \beta, \beta^{\vee})$. The isomorphism class of an I-system is determined by its *characteristic* χ, defined

$$\chi := [A, \ker(\alpha), \ker({}^t\alpha^{\vee}), d], \tag{29}$$

where A is the compose linear map or, equivalently, the matrix

$$A \colon \mathbf{R}^I \xrightarrow{\alpha} E \xrightarrow{\alpha^{\vee}} \mathbf{R}^I, \qquad A := (\alpha_i^{\vee}(\alpha_j))_{ij}, \tag{30}$$

called the *Cartan matrix* of the system, and d is the nonnegative integer, called the *defect*

$$d := \dim[\ker(\alpha^{\vee})/\mathrm{Img}(\alpha) \cap \ker(\alpha^{\vee})]. \tag{31}$$

The dimension of E is the sum of the nonnegative terms

$$\dim E = [|I| - \mathrm{corank}(A)] + [\mathrm{corank}(A) - \mathrm{corank}(\alpha)] + [\mathrm{corank}(A) - \mathrm{corank}({}^t\alpha^{\vee})] + d. \tag{32}$$

Reciprocally, any data (A, K_1, K_2, d), where $A \in M_I(\mathbf{R})$ verifies $a_{ii} = 2$, K_1 is a subspace of $\ker(A)$, K_2 is a subspace of $\ker({}^tA)$ and d a nonnegative integer, is the characteristic of some isomorphism class of I-systems.

2.2 The reflection group

Let us start with some concepts on G-modules. Let E be a left G-module, $\mathbf{Z}^1(G, E)$ the space of 1-cocycles (derivations from G to E) and $\partial \colon E \to \mathbf{Z}^1(G, E)$, $\partial_x(g) = g \cdot x - x$

the coboundary map. Consider the invariant subspace of the 0-cocycles, that is, the set of elements fixed by G.

$$\mathbf{Z}^0(G, E) = \ker(\partial) = \bigcap_{g \in G} \ker(g - Id). \tag{33}$$

Define the subspaces of E

$$\mathrm{Dir}(g) := \mathrm{Img}(g - Id), \qquad D(G, E) := \sum_{g \in G} \mathrm{Dir}(g) = \sum_{x \in E} \mathrm{Img}(\partial_x). \tag{34}$$

We have

a) For any $g_1, \ldots, g_n$ in G

$$\mathrm{Dir}(g_1 \ldots g_n) \subset \sum_{i=1}^{n} \mathrm{Dir}(g_i). \tag{35}$$

b) $D(G, E)$ is G-invariant and if $S \subset G$ generates G

$$D(G, E) := \sum_{g \in S} \mathrm{Dir}(g), \qquad \mathbf{Z}^0(G, E) = \bigcap_{g \in S} \ker(g - Id). \tag{36}$$

c) $\mathbf{Z}^0(G, E)$ and $D(W, E)$ verifies the duality relations,

$$\mathbf{Z}^0(G, E^*) = D(G, E)^\perp \qquad \text{and} \qquad D(G, E^*) = \mathbf{Z}^0(G, E)^\perp, \tag{37}$$

where E^* denotes the contragradient G-module.

Proof.
a) Let us proceed by induction on n and suppose $\mathrm{Dir}(g_1 \ldots g_{n-1}) \subset \sum_{i=1}^{n-1} \mathrm{Dir}(g_i)$. Then $(g_1 \ldots g_n - Id) = (g_1 \ldots g_{n-1} - Id)(g_n - Id) + (g_n - Id) + (g_1 \ldots g_{n-1} - Id)$ so $\mathrm{Dir}(g_1 \ldots g_n) \subset \mathrm{Dir}(g_1 \ldots g_{n-1}) + \mathrm{Dir}(g_n) \subset \sum_{i=1}^{n} \mathrm{Dir}(g_i)$.
b) $x \in D(G, E) \Rightarrow x = \sum_i (g_i(x_i) - x_i) \Rightarrow g(x) = \sum_i (gg_ig^{-1}(gx_i) - gx_i) \in D(G, E)$. Therefore $D(W, E)$ is G-invariant. Equations (36) follows from (35) and from $\bigcap_{i=1}^{n} \ker(g_i - Id) \subset \ker(g_1 \ldots g_n - Id)$.
c) $\varphi \in D(G, E)^\perp \Leftrightarrow \forall x \in E, g \in G, \varphi(g^{-1}x - x) = 0 \Leftrightarrow \forall g \in G, {}^t g^{-1}(\varphi) = \varphi \Leftrightarrow \varphi \in \mathbf{Z}^0(G, E^*)$. $\qquad\square$

Define the *principal reflections* of a system $(E, \alpha, \alpha^\vee)$ as the reflections (r_i) in the hyperplans (H_i) along the directions (D_i)

$$H_i := \ker(\alpha_i^\vee), \qquad D_i := \mathbf{R}\alpha_i, \qquad r_i(x) = x - \alpha_i^\vee(x)\alpha_i. \tag{38}$$

The linear group W generated by the principal reflections will be called the *reflection group* associated to the system. The following are results on the invariant subspaces $D(W, E)$ and $\mathbf{Z}^0(W, E)$ and the associated cone C

$$D(W, E) = \mathrm{Img}(\alpha) = \sum_{i} D_i, \qquad \mathbf{Z}^0(W, E) = \ker(\alpha^\vee) = \bigcap_{i} H_i. \tag{39}$$

$$D(W, E) \cap \mathbf{Z}^0(W, E) = \alpha[\ker(A)], \tag{40}$$

$$D(W, E) \cap C = \alpha[A^{-1}((\mathbf{R}^+)^I)], \tag{41}$$

$$C \cap [D(W, E) + \mathbf{Z}^0(W, E)] = C \cap D(W, E) + \mathbf{Z}^0(W, E). \tag{42}$$

Proof.

- From (36) $D(W,E) = \sum_i \mathrm{Dir}(r_i) = \sum_i D_i = \mathrm{Img}(\alpha)$ and $\mathbf{Z}^0(W,E) = \cap_{i \in I} \ker(r_i - Id) = \cap_{i \in I} H_i = \ker(\alpha^\vee)$.

- $\mathrm{Img}(\alpha) \cap \ker(\alpha^\vee) = \mathrm{Img}(\alpha) \cap \alpha^{\vee -1}(0) = \alpha[\alpha^{-1}(\alpha^{\vee -1}(0))] = \alpha[\ker(A)]$.

- $\alpha[A^{-1}((\mathbf{R}^+)^I)] = \alpha\alpha^{-1}\alpha^{\vee -1}(A) = \alpha\alpha^{-1}(C) = \mathrm{Img}(\alpha) \cap C$.

- Let $x \in C \cap [\mathrm{Img}(\alpha) + \ker(\alpha^\vee)]$, $x = x_d + x_k$ where $x_d \in \mathrm{Img}(\alpha)$ and $x_k \in \ker(\alpha^\vee)$. For all $i \in I$, $0 \le \alpha_i^\vee(x) = \alpha_i^\vee(x_d) + \alpha_i^\vee(x_k) = \alpha_i^\vee(x_d) \Rightarrow x_d \in C \cap \mathrm{Img}(\alpha)$. Then $C \cap [\mathrm{Img}(\alpha) + \ker(\alpha^\vee)] = C \cap [C \cap \mathrm{Img}(\alpha) + \ker(\alpha^\vee)]$. On the other hand, if $x \in C \cap \mathrm{Img}(\alpha) + \ker(\alpha^\vee)$, $x = x_d + x_k$, for all $i \in I$, $\alpha_i^\vee(x) = \alpha_i^\vee(x_d) + \alpha_i^\vee(x_k) = \alpha_i^\vee(x_d) \ge 0$, so $x \in C$ and hence the result. $\qquad\square$

2.3 Associated systems

The *dual* of $(E, \alpha, \alpha^\vee)$ is the system $(E^*, (\alpha_i^\vee)_i, (\alpha_i)_i)$, where E and E^{**} has been identified. In terms of the associated maps, the dual is $(E^*, {}^t\alpha^\vee, {}^t\alpha)$. Their associated cone C^* and characteristic χ^* are

$$C^* = \{\varphi \in E^* \mid \varphi(\alpha_i) \ge 0,\ \forall i \in I\}, \qquad \chi^* = [{}^tA, \ker({}^t\alpha^\vee), \ker(\alpha), d]. \qquad (43)$$

The principal reflections of $(E^*, {}^t\alpha^\vee, {}^t\alpha)$ are

$$r_i(\varphi) = \varphi - \varphi(\alpha_i)\alpha_i^\vee. \qquad (44)$$

If we identify the group generated by the reflections (44) with W, E^* is just the contragradient module of E, i.e., $w(\varphi) = {}^tw^{-1}(\varphi)$.

For any $J \subset I$ we can define the J-system $(E, \alpha_J, \alpha_J^\vee)$ in an obvious way. Their associated cone, Cartan matrix and reflection groups are

$$C_J = ({}^t\alpha^\vee)^{-1}(\mathbf{R}^J), \qquad A_J, \qquad W_J = \langle \{r_j \mid j \in J\}\rangle, \qquad (45)$$

where A_J denote the principal submatrix defined by J. From (39) it follows that for parts J and K of I,

$$J \perp K \qquad \Leftrightarrow \qquad D(W_J, E) \subset \mathbf{Z}^0(W_K, E) \quad \text{and} \quad D(W_K, E) \subset \mathbf{Z}^0(W_J, E) \qquad (46)$$

The group $GL(E)$ acts in the set $\mathrm{Sys}_I(E)$ of systems in E

$$g \cdot (E, \alpha, \alpha^\vee) := (E, g\alpha, \alpha^\vee g^{-1}). \qquad (47)$$

The orbit of the system $(E, \alpha, \alpha^\vee)$ is the isomorphism classes of I-systems in E. The principal reflections of $g \cdot (E, \alpha, \alpha^\vee)$ are the conjugates $(gr_i g^{-1})_i$ of the principal reflections of the system $(E, \alpha, \alpha^\vee)$. They are in fact the reflections in the walls $(g(H_i))_i$ of $g(C)$ along directions $(g(\alpha_i))_i$

$$gr_i g^{-1}(x) = x - \alpha_i^\vee(g^{-1}x)g(\alpha_i), \qquad i \in I.$$

Therefore the reflection group of $g \cdot (E, \alpha, \alpha^\vee)$ is the conjugate gWg^{-1} of W. So the systems $(E, \alpha, \alpha^\vee)$ and $g \cdot (E, \alpha, \alpha^\vee)$ defines the same linear group iff g is in the normalizer $\mathcal{N}(W)$ of W in $GL(E)$.

Also the abelian group $M_I PD(\mathbf{R})$ acts in $\mathrm{Sys}_I(E)$ (see Section A.2)

$$D * (E, \alpha, \alpha^\vee) := (E, \alpha D, D^{-1}\alpha^\vee). \qquad (48)$$

Two systems $(E, \alpha, \alpha^\vee)$, $(E, \beta, \beta^\vee)$ in the same orbit are said *PD-equivalent* and denoted

$$(E, \alpha, \alpha^\vee) \sim (E, \beta, \beta^\vee),$$

then $\beta := (d_i\alpha_i)_i$ and $\beta^\vee := (d_i^{-1}\alpha_i^\vee)_i$, for some positive numbers $(d_i)_i$. The orbit of $(E, \alpha, \alpha^\vee)$ will be called the *PD-class* of $(E, \alpha, \alpha^\vee)$. PD-equivalent systems defines the same principal reflections and so the same reflection group.

2.4 Invariant subspaces

The W-invariant subspaces of the E are the subspaces E' verifying

$$\sum_{j \in J} D_j \subset E' \subset \bigcap_{I-J} H_i \qquad (49)$$

for some $J \subset I$. The invariant subspace E' determines J

$$J = \{i \in I \mid \alpha_i \in E'\}. \qquad (50)$$

Proof.

If E' verifies (49), $r_i(E') = E'$ for $i \in I - J$, and $r_i(E') \subset E' + \sum_{j \in J} D_j \subset E'$ for $i \in J$. Then E' is W-invariant.

Let E' invariant and define J as in (50), so $\sum_{j \in J} D_j \subset E'$. For any $i \in I - J$ and $x \in E'$, $\alpha_i^{\vee}(x)\alpha_i = x - r_i(x) \in E'$, and as $\alpha_i \notin E'$, it follows that $\alpha_i^{\vee}(x) = 0$. Therefore $E' \subset \bigcap_{I-J} H_i$.

Let us see that E' determines J. From (49) it follows that $J \subset \{i \in I \mid \alpha_i \in E'\}$. On the other hand $i \notin J \Rightarrow \alpha_i^{\vee}(E') = 0 \Rightarrow \alpha_i \notin E'$, so $\{i \in I \mid \alpha_i \in E'\} \subset J$. $\qquad \square$

We have

a) The possible subsets J in (49) are the ones such that A stabilizes $\mathbf{R}^J$.

b) If E' is a W-invariant subspace

$$D(W, E) \subset E' \qquad \text{or} \qquad E' \subset H_i, \quad \text{for some } i \in I \qquad (51)$$

c) Any subspace E' verifying

$$D(W, E) \subset E' \qquad \text{or} \qquad E' \subset \mathbf{Z}^0(W, E) \qquad (52)$$

is W-invariant.

d) For any hyperplane H

$$D(W, E) \subset H \quad \Leftrightarrow \quad H \text{ is } W\text{-invariant and is not a wall of } C. \qquad (53)$$

Proof.

a) Using (49), $A(\mathbf{R}^J) = \alpha^{\vee}(\alpha(\mathbf{R}^J)) = \alpha^{\vee}(\sum_{j \in J} D_j) = \left(\sum_{j \in J} \alpha_i^{\vee}(D_j)\right)_i \subset \mathbf{R}^J$.

b) E' is invariant $\Rightarrow \sum_{j \in J} D_j \subset E' \subset \cap_{I-J} H_i$. If $J = I$, $D(W, E) \subset E'$. Otherwise $\cap_{I-J} H_i$ contains E', where $I - J \neq \emptyset$, so E' is contained in a wall H_i.

c) Take $J = I$ and $J = \emptyset$ in (49).

d) If $D(W, E) \subset H$, H is W-invariant by (52). Furthermore, for any $i \in I$, $\alpha_i^{\vee}(\alpha_i) = 2$, so $\alpha_i \notin \ker(\alpha_i^{\vee}) = H_i$ and then $D(W, E) \not\subset H_i$ and hence $H \neq H_i$.

If H is not a wall of C and W-invariant by (51), $D(W, E) \subset H$. $\qquad \square$

Define the *restriction* of $(E, \alpha, \alpha^{\vee})$ to the invariant subspace E' as the J-system

$$(E', \alpha_J, \alpha_J^{\vee}), \qquad \begin{cases} \alpha_J(X) & := & \sum_{j \in J} X_j \alpha_j \\ \alpha_J^{\vee}(x) & := & \left(\alpha_j^{\vee}(x)\right)_{j \in J}. \end{cases} \qquad (54)$$

The principal reflections of $(E', \alpha_J, \alpha_J^{\vee})$ are the reflections $(r_j')_{j \in J}$ in the hyperplans $(H_j') = H_j \cap E'$ along the directions (D_j). The associated cone C' is

$$C' = E' \cap C_J,$$

where C_J is given by (45). Denote by W' the reflection group, we have

$$D(W', E') = \sum_{j \in J} D_j, \qquad \mathbf{Z}^0(W', E') = E' \cap \left(\bigcap_{j \in J} H_j \right). \qquad (55)$$

In particular we have the subsystem $[D(W, E), \alpha', \alpha^{\vee\prime}]$. Its cone C' and characteristic χ' are

$$C' = D(W, E) \cap C, \qquad \chi' = \left[A, \ker(\alpha), ({}^t\alpha^\vee)^{-1} \left[D(W, E)^\perp \right], 0 \right], \qquad (56)$$

3 BILINEAR SYSTEMS

The linear extensions of discrete reflection groups admit an invariant symmetric bilinear form S. In the case of the sphere, S is the scalar product defined in E. For an Euclidean space X, S is a scalar product defined in the hyperplane TX of E formed by the translations in X. In the hyperbolic space, S is the Lorentzian metric defined in E. Note that in all the three cases, S is defined in a space containing the space of directions $D(W, E)$. *Regularity* and *coregularity* imply respectively the nondegeneracy and the extensibility of invariant bilinear forms defined in $D(W, E)$.

The usual method of representation of a Coxeter group [3] leads to *reduced* systems, or better, *simplicial* systems. Here we will use a different approach and explain how to obtain a simplicial system with a given Cartan matrix.

In this section $(E, \alpha, \alpha^\vee)$ will be an I-system and χ its characteristic. An *orthogonal form* will mean a nondegenerate symmetric bilinear form.

3.1 Regular, coregular and invertible systems

The following result defines regular systems.

The system $(E, \alpha, \alpha^\vee)$ is said *regular* if it verifies the following equivalent conditions

a) $D(W, E) \cap \mathbf{Z}^0(W, E) = \{0\}$.

b) $\mathrm{rank}(A) = \dim D(W, E)$.

c) The characteristic is $\chi = [A, \ker(A), \ker({}^t\alpha^\vee), d]$.

Proof.

a) $\Leftrightarrow$ b). A is the compose map $\mathbf{R}^I \xrightarrow{\alpha'} D(W, E) \xrightarrow{\alpha^{\vee\prime}} \mathbf{R}^I$, so we have that $D(W, E) \cap \mathbf{Z}^0(W, E) = 0 \Leftrightarrow \alpha^{\vee\prime}$ injective $\Leftrightarrow \dim D(W, E) = \mathrm{rank}(\alpha^{\vee\prime}) = \mathrm{rank}(A)$.

a) $\Leftrightarrow$ c). Using (40), $D(W, E) \cap \mathbf{Z}^0(W, E) = 0 \Leftrightarrow \ker(A) \subset \ker(\alpha)$. $\square$

The system $(E, \alpha, \alpha^\vee)$ is said *coregular* if it verifies the following equivalent conditions

a) $D(W, E) + \mathbf{Z}^0(W, E) = E$.

b) $\dim \mathbf{Z}^0(W, E) + \mathrm{rank}(A) = \dim E$.

c) The characteristic is $\chi = [A, \ker(\alpha), \ker({}^tA), d]$.

d) The dual system $(E^*, {}^t\alpha^\vee, {}^t\alpha)$ is regular.

If $(E, \alpha, \alpha^\vee)$ is coregular we have

e) $\mathrm{int}(C) \neq \emptyset \Rightarrow \mathrm{int}(C') \neq \emptyset$, where C' is the cone (56) associated to the system $[D(W, E), \alpha', \alpha^{\vee\prime}]$.

Proof.

a) $\Leftrightarrow$ b). The Cartan map A is the compose map $\mathbf{R}^I \overset{\alpha'}{\to} D(W,E) \overset{\alpha^{\vee\prime}}{\to} \mathbf{R}^I$, so $\operatorname{rank}(A) = \operatorname{rank}(\alpha^{\vee\prime}) = \dim D(W,E) - \dim[D(W,E) \cap \mathbf{Z}^0(W,E)]$. Then, $D(W,E) + \mathbf{Z}^0(W,E) = E \Leftrightarrow \dim E = \dim D(W,E) + \dim \mathbf{Z}^0(W,E) - \dim[D(W,E) \cap \mathbf{Z}^0(W,E)] = \dim \mathbf{Z}^0(W,E) + \operatorname{rank}(A)$.

b) $\Leftrightarrow$ c) $\ker({}^t\alpha^\vee) = \ker({}^tA) \Leftrightarrow \operatorname{Img}(\alpha^\vee) = \ker({}^t\alpha^\vee)^\perp = \ker({}^tA)^\perp = \operatorname{Img}(A) \Leftrightarrow \operatorname{rank}(A) = \dim[\operatorname{Img}(\alpha^\vee)] = \dim E - \dim[\ker(\alpha^\vee)]$.

c) $\Leftrightarrow$ d) follows from (43).

e) If E is coregular, $\ker({}^t\alpha^{\vee\prime}) \cap (\mathbf{R}^+)^I = \operatorname{Img}(\alpha^{\vee\prime})^\perp \cap (\mathbf{R}^+)^I = \operatorname{Img}(\alpha^\vee)^\perp \cap (\mathbf{R}^+)^I = \ker({}^t\alpha^\vee) \cap (\mathbf{R}^+)^I = \{0\}$. $\qquad\square$

The system $(E, \alpha, \alpha^\vee)$ is said *invertible* if it verifies the following equivalent conditions

a) A invertible.

b) $\operatorname{corank}(A) = 0$.

c) The system is regular, coregular and α is injective, so its characteristic is $\chi = (A, 0, 0, d)$.

d) The (α_i) and the restrictions $(\alpha_i^{\vee\prime})$ to $D(W,E)$ are linearly independent (therefore the $(\alpha_i^\vee)$ are also linearly independent).

Proof.
a) $\Leftrightarrow$ b) $\Leftrightarrow$ d) Obvious.

b) $\Leftrightarrow$ c) A invertible iff $\ker(A) = \ker(\alpha) = \ker({}^tA) = \ker({}^t\alpha^\vee) = 0$ and this is true iff the system is regular, coregular and α is injective. $\qquad\square$

3.2 Reduced and simplicial systems

Let us introduce some new classes of systems, whose cone C has certain geometric properties. The system $(E, \alpha, \alpha^\vee)$ is said *reduced* if it verifies the following equivalent conditions

a) $\{0\}$ is a facet of C, i.e., C is a strict cone.

b) $\mathbf{Z}^0(W,E) = \{0\}$.

c) The characteristic is $\chi = [A, \ker(A), \ker({}^t\alpha^\vee), 0]$, i.e., the system is regular and $d = 0$.

Proof.
a) $\Leftrightarrow$ b) It follows from the fact that $\mathbf{Z}^0(W,E)$ is the minimal facet of C.

b) $\Rightarrow$ c) Obvious. c) $\Rightarrow$ b) Being $(E, \alpha, \alpha^\vee)$ regular, $D(W,E) \cap \mathbf{Z}^0(W,E) = 0$. Therefore $0 = d = \dim(\mathbf{Z}^0(W,E) / [D(W,E) \cap \mathbf{Z}^0(W,E)]) = \dim \mathbf{Z}^0(W,E)$, so $\mathbf{Z}^0(W,E) = 0$. $\square$

Given a system $(E, \alpha, \alpha^\vee)$, set $\operatorname{Red}(E, \alpha, \alpha^\vee) := [\frac{E}{\ker(\alpha^\vee)}, \alpha^-, \alpha^{\vee-}]$ the reduced system defined in an obvious way. Its characteristic is $[A, \ker(A), \ker({}^t\alpha^\vee), 0]$.

The system $(E, \alpha, \alpha^\vee)$ is said *simplicial* if it verifies the following equivalent conditions

a) C is a simplicial cone.

b) $\alpha^\vee$ is a one to one map (so $\operatorname{int}(C) \neq \emptyset$).

c) $\chi = [A, \ker(A), 0, 0]$.

The rays of the simplicial cone C are $(\mathbf{R}^+ \alpha_i^{\vee *})_i$, where $(\alpha_i^{\vee *})_i$ is the basis of E dual of $(\alpha_i^\vee)_i$. Condition c) shows that the isomorphism class of a simplicial system is determined by its Cartan matrix. Let us see how to obtain a simplicial system with a given Cartan matrix A. Define the *simplicial representation* of A as the system

$$(\mathbf{R}^I, A, \operatorname{Id}), \tag{57}$$

where Id denotes the identity matrix. Then α_j is the $j-$th column of A and $\alpha_i^\vee$ is the canonical projector $\pi^i \colon \mathbf{R}^I \to \mathbf{R}$, that is

$$\alpha_j := (a_{1_j}, \ldots, a_{n_j}) \qquad \text{and} \qquad \alpha_i^\vee := \pi^i.$$

The associated simplicial cone is

$$C := (\mathbf{R}^+)^I.$$

The walls $(H_i)_i$ are given by the equations $(X_i = 0)_i$ and the rays are $(\mathbf{R}^+ e_i)_i$, where $(e_i)_i$ denotes the canonical basis of $\mathbf{R}^I$.

3.3 Realization of a matrix

The systems appearing in the study of Kac-Moody algebras are the ones verifying the following equivalent conditions

a) α is injective, $\alpha^\vee$ is onto (i.e., the (α_i) and the $(\alpha_i^\vee)$ are linearly independent) and $\dim E = \operatorname{corank}(A) + |I|$.

b) The characteristic is $\chi = (A, 0, 0, 0)$.

The isomorphism classes of these systems are determined by their Cartan matrices. A system $(E, \alpha, \alpha^\vee)$ of this type is called a *realization* of A [6].

3.4 Bilinear systems

A system $(E, \alpha, \alpha^\vee)$ is said *bilinear* if there exists a W-invariant symmetric bilinear form S in $D(W, E)$ verifying $S(\alpha_i, \alpha_i) > 0$, $\forall i \in I$. We have [5]

$$(E, \alpha, \alpha^\vee) \text{ bilinear} \quad \Leftrightarrow \quad A \text{ symmetrizable.} \tag{58}$$

Suppose $(E, \alpha, \alpha^\vee)$ bilinear and let S be as above. We have

a) The restriction of the $\alpha_i^\vee$ to $D(W, E)$ is

$$\alpha_i^\vee = 2S(\alpha_i, -)/S(\alpha_i, \alpha_i). \tag{59}$$

Hence, if $a_{ij} \neq 0$,

$$S(\alpha_i, \alpha_i)/S(\alpha_j, \alpha_j) = a_{ji}/a_{ij}. \tag{60}$$

b) The radical of S is

$$\operatorname{Rad}(S) = D(W, E) \cap \mathbf{Z}^0(W, E). \tag{61}$$

Proof.

a) $\alpha_i^\vee = 2S(\alpha_i, -)/S(\alpha_i, \alpha_i)$ is a consequence of (16).

b) The (α_i) generating $D(W, E)$, for any $x \in D(W, E)$ we have, $x \in \mathrm{Rad}(S) \Leftrightarrow \forall i,\ 0 = S(\alpha_i, x) = 1/2\ S(\alpha_i, \alpha_i)\alpha_i^\vee(x) \Leftrightarrow \forall i,\ \alpha_i^\vee(x) = 0 \Leftrightarrow x \in \mathbf{Z}^0(W, E)$, where (59) has been used. Therefore $\mathrm{Rad}(S) = D(W, E) \cap \mathbf{Z}^0(W, E)$.

$\square$

Let $(I_k)_k$ be the connected components of I, consider a system $(E, \beta, \beta^\vee) := D * (E, \alpha, \alpha^\vee)$ PD-equivalent (48) to $(E, \alpha, \alpha^\vee)$ and set $B = D^{-1}AD$ its Cartan matrix. From (60) we have

D symmetrizes $A \Leftrightarrow$ the vectors $(\beta_i)_{i \in I_k}$ have the same length, for any k.

From (93) we see that the matrix D symmetrizig A is not completely defined and we can choose the $(\beta_i)_i$ with the additional condition

$$S(\beta_i, \beta_i) = 2, \tag{62}$$

so $\beta_i^\vee(x) = S(\beta_i, x)$ for all $x \in D(W, E)$. We have

a) With the choice (62)

$$\beta_i := [2/S(\alpha_i, \alpha_i)]^{1/2}\alpha_i, \qquad A = D^2\mathrm{Gram}_S(\alpha), \qquad B = \mathrm{Gram}_S(\beta), \tag{63}$$

where $\mathrm{Gram}_S(\alpha) := (S(\alpha_i, \alpha_j))$ denotes the Gram matrix of the $(\alpha_i)_i$.

b) The dual system $(E^*, {}^t\alpha^\vee, {}^t\alpha)$ is also bilinear. A W-invariant bilinear form S^* in $D(W, E^*)$ is

$$S^*(\alpha_i^\vee, \alpha_j^\vee) := \frac{2}{S(\alpha_i, \alpha_i)}a_{ji} = \frac{4}{S(\alpha_i, \alpha_i)S(\alpha_j, \alpha_j)}S(\alpha_i, \alpha_j). \tag{64}$$

Therefore $S^*(\beta_i^\vee, \beta_j^\vee) = S(\beta_i, \beta_j)$.

Proof.

a) $2 = S(\beta_i, \beta_i) = S(d_i\alpha_i, d_i\alpha_i) \Rightarrow d_i = [2/S(\alpha_i, \alpha_i)]^{1/2}$. Therefore $D^2\mathrm{Gram}_S(\alpha) = (2S(\alpha_i, \alpha_j)/S(\alpha_i, \alpha_i)) = (\alpha_i^\vee(\alpha_j)) = A$.

b) A symmetrizable $\Rightarrow D^{-1}AD$ symmetric $\Rightarrow D^t A D^{-1}$ symmetric. Then ${}^tA$ is symmetrizable and D^{-1} is a symmetrization matrix. Therefore there exists a W-invariant bilinear form S^* in $D(W, E^*)$, and using (63), $\mathrm{Gram}_{S^*}({}^t\alpha^\vee) = D^{2\ t}A = (2a_{ji}/S(\alpha_i, \alpha_i))_{ij}$.

$\square$

3.5 Orthogonal systems

A system $(E, \alpha, \alpha^\vee)$ is said *orthogonal* if there exists a W-invariant orthogonal form S in $D(W, E)$ that verifies $S(\alpha_i, \alpha_i) > 0, \forall i \in I$. From (58) and (61) it follows

$$(E, \alpha, \alpha^\vee) \text{ orthogonal } \Leftrightarrow (E, \alpha, \alpha^\vee) \text{ regular and } A \text{ symmetrizable.} \tag{65}$$

Suppose $(E, \alpha, \alpha^\vee)$ orthogonal, set $(I_k)_k$ the connected components of I and $D(W_k, E) := D(W_{I_k}, E)$ (see (45)). Then for $j \neq k$, $D(W_j, E) \perp D(W_k, E)$. If S is a scalar product, $D(W, E)$ is the orthogonal sum

$$D(W, E) = \underset{k}{\perp} D(W_k, E). \tag{66}$$

For any subspace E' containing $D(W, E)$ we have

$$S \text{ admits } W\text{-invariant orthogonal extension to } E' \Leftrightarrow E' \subset D(W, E) \bigoplus \mathbf{Z}^0(W, E). \tag{67}$$

Proof.

Let S' a W-invariant orthogonal form in E' extending S. Analogously to (61), we have that the orthogonal of $D(W, E)$ in E' w.r.t. S is $D(W, E)^0 = E' \cap \mathbf{Z}^0(W, E)$. Being S' and S are nondegenerate, E' is the orthogonal sum $E' = D(W, E) \oplus D(W, E)^0 \subset D(W, E) \oplus \mathbf{Z}^0(W, E)$.

On the other hand any orthogonal form in $D(W, E) \oplus \mathbf{Z}^0(W, E)$ extending S is obviously W-invariant. $\qquad\square$

Let $P := \cap_i \Sigma_i^+$ be a convex polyhedron in a SCC X, and set (β_i) the vectors associated to the hafspaces (Σ_i^+) (3) and W the linear extension of the group generated by the reflections in the walls of P. We have

a) $\dim D(W, E) = \operatorname{rank}(P)$. Therefore

$$P \text{ nondegenerate } \Leftrightarrow D(W, E) \text{ is an hyperplan of } E. \tag{68}$$

b) The system $(E, \beta, \beta^\vee)$, where $\beta_i^\vee(x) := S(\beta_i, x)$, is orthogonal and its Cartan matrix B coincides with the Gram matrix of P

$$B = \operatorname{Gram}(P). \tag{69}$$

Proof.

a) The linear extensions (r_i) of the reflections in the walls of P generates W. As $\operatorname{Img}(r_i - Id) = \mathbf{R}\beta_i$, by (35) the vectors (β_i) generates $D(W, E)$ and hence $\dim D(W, E) = \operatorname{rank}[\operatorname{Gram}(P)]$. On the other hand, P non degenerate $\Leftrightarrow$ $\dim X = \operatorname{rank}[\operatorname{Gram}(P)]$ $\Leftrightarrow$ $\dim X = \dim D(W, E)$.

b) S is an orthogonal invariant form of the system $(E, \beta, \beta^\vee)$. The Cartan matrix $B = \beta_i^\vee(\beta_j) = S(\beta_i, \beta_j) = \operatorname{Gram}(P)$.

$\qquad\square$

A system $(E, \alpha, \alpha^\vee)$ is said *completely orthogonal* if there exists a W-invariant orthogonal form S in E whose restriction to $D(W, E)$ is also orthogonal and verifies $S(\alpha_i, \alpha_i) > 0, \forall i \in I$. From (65) and (67) it follows

$$(E, \alpha, \alpha^\vee) \text{ completely orthogonal } \Leftrightarrow \begin{cases} (E, \alpha, \alpha^\vee) \text{ regular and coregular and} \\ A \text{ symmetrizable.} \end{cases} \tag{70}$$

In this case, if S is a scalar product, E is the orthogonal sum

$$E = \left(\underset{k}{\perp} D(W_k, E) \right) \perp \mathbf{Z}^0(W, E). \tag{71}$$

4 COXETER SYSTEMS

For a system $(E, \alpha, \alpha^\vee)$ condition (28) ensures that the cone C has nonempty interior. In order to have a linear Coxeter group, the additional fundamental region condition (24) must be satisfied. This condition is equivalent to conditions C1 and C2 (see (95) and (100)) on the Cartan matrix of the system. For the case of linear extensions of discrete reflection groups, C1 reflects the fact that the fundamental polyhedron P is acute-angled. Condition C2 ensures that P is a Coxeter polyhedron.

4.1 C1 systems

A *C1 system* is a system $(E, \alpha, \alpha^\vee)$ whose Cartan matrix is C1. For the cone C we have

$$A = A_+ \bigoplus A_- \quad \Rightarrow \quad \ker({}^tA_0) \cap (\mathbf{R}^+)^I = \{0\} \quad \Rightarrow \quad \text{int}(C) \neq \emptyset. \tag{72}$$

Furthermore, if $\text{int}(C) \neq \emptyset$, the halfspaces $\alpha_i^\vee(x) \geq 0$ are not redundant, that is C is a $|I|$-sided cone. For a completely orthogonal system $(E, \alpha, \alpha^\vee)$ we have,

$$\text{int}(C) \neq \emptyset \quad \Leftrightarrow \quad A = A_+ \bigoplus A_-. \tag{73}$$

Proof.

- For a positive or negative matrix M, $\ker(M) \cap (\mathbf{R}^+)^I = \{0\}$, so we have $\ker({}^t\alpha^\vee) \cap (\mathbf{R}^+)^I \subset \ker({}^tA) \cap (\mathbf{R}^+)^I = \ker({}^tA_0) \cap (\mathbf{R}^+)^I$. Using (28), $\ker({}^tA_0) \cap (\mathbf{R}^+)^I = \{0\} \Rightarrow \ker({}^t\alpha^\vee) \cap (\mathbf{R}^+)^I \Rightarrow \text{int}(C) \neq \emptyset$.

- Suppose $(E, \alpha, \alpha^\vee)$ is completely orthogonal let S be an orthogonal form in E. Let $X \geq 0$ and denote $Y := \left(\frac{X_j}{S(\alpha_j, \alpha_j)}\right)_j \geq 0$. We have, $X \in \ker({}^t\alpha^\vee) \Leftrightarrow \sum_j X_j \alpha_j^\vee = 0 \Leftrightarrow \sum_j \frac{2X_j}{S(\alpha_j,\alpha_j)} S(\alpha_j, -) = 0 \Leftrightarrow 0 = \sum_j \frac{X_j}{S(\alpha_j,\alpha_j)} \alpha_j = \alpha(Y)$. The system being reduced, $\ker(\alpha) = \ker(A)$, so we have $\alpha(Y) = 0 \Leftrightarrow 0 = AY = A_+Y_+ \oplus A_-Y_- \oplus A_0Y_0 \Leftrightarrow Y_+ = 0$, $Y_- = 0$ and $Y_0 \in \ker(A_0)$.

In conclusion, for $X \geq 0$ we have $X \in \ker({}^t\alpha^\vee) \Leftrightarrow Y \in \ker(A_0) \cap (\mathbf{R}^+)^{I_0}$. Therefore $\text{int}(C) \neq \emptyset \Leftrightarrow \ker({}^t\alpha^\vee) \cap (\mathbf{R}^+)^I = 0 \Leftrightarrow \ker(A_0) \cap (\mathbf{R}^+)^{I_0} = 0$. But a matrix A is zero iff $\exists X > 0$ such that $AX = 0$, then $\ker(A_0) \cap (\mathbf{R}^+)^{I_0} = 0$ means that A_0 is in fact the empty matrix, i.e., $A = A_+ \bigoplus A_-$. □

By restricting the set of indices I to I_+, I_0 and I_- we define the systems $(E, \alpha_+, \alpha_+^\vee)$, $(E, \alpha_0, \alpha_0^\vee)$ and $(E, \alpha_-, \alpha_-^\vee)$ respectively. Set W_+, W_0 and W_- their associated reflection groups and C_+, C_0 and C_- their associated cones, so

$$D(W, E) = \sum_{k \in \{+,0,-\}} D(W_k, E), \quad \mathbf{Z}^0(W, E) = \bigcap_{k \in \{+,0,-\}} \mathbf{Z}^0(W_k, E), \quad C = \bigcap_{k \in \{+,0,-\}} C_k.$$

We have

- For any C1 system

$$D(W, E) \cap C = \sum_{k \in \{+,0,-\}} D(W_k, E) \cap C_k. \tag{74}$$

- If A is positive

$$D(W, E) \cap C = \left(\sum_i \mathbf{R}^+ \alpha_i\right) \cap C. \tag{75}$$

- A is zero iff

$$D(W, E) \cap C = D(W, E) \cap \mathbf{Z}^0(W, E). \tag{76}$$

- If A is negative, $[\alpha^{-1}(C) \cap (\mathbf{R}^+)^I] \subset \ker(\alpha)$.

Proof.

- Let $x \in D(W, E) \cap C$, $x = \sum_k x_k$, where $x_k \in D(W_k, E)$ for $k \in \{+, 0, -\}$. For any $i \in I_+$, $0 \leq \alpha_i^\vee(x) = \sum_k \alpha_i^\vee(x_k) = \alpha_i^\vee(x_+) \Rightarrow x_+ \in D(W_+, E) \cap C_+$. In general, $x_k \in D(W_k, E) \cap C_k$ for $k \in \{+, 0, -\}$, then $D(W, E) \cap C = [\sum_k D(W_k, E) \cap C_k] \cap C$. On the other hand, $\sum_k D(W_k, E) \cap C_k \subset D(W_+, E) \cap C_+ + D(W_0, E) + D(W_-, E) \subset$

$D(W_+, E) \cap C_+ + \mathbf{Z}^0(W, E) \subset C_+$ and, in general, $\sum_k D(W_k, E) \cap C_k \subset \cap_k C_k = C$. Therefore $D(W, E) \cap C = \sum_k D(W_k, E) \cap C_k$.

- Suppose A positive and let $x \in D(W, E) \cap C$. Then $x = \alpha(X)$ for some $X \in \mathbf{R}^I$ and $AX = \alpha^\vee(\alpha(X)) = \alpha^\vee(x) \geq 0$. Being A positive, $X \geq 0$, so $x = \alpha(X) \in \sum_i \mathbf{R}^+ \alpha_i$.

- Suppose A zero and let $x \in D(W, E) \cap C$. Then $x = \alpha(X)$ for some $X \in \mathbf{R}^I$ and $AX = \alpha^\vee(\alpha(X)) = \alpha^\vee(x) \geq 0$. Being A zero, $0 = AX = \alpha^\vee(x)$, so $x \in \mathbf{Z}^0(W, E)$. As $\mathbf{Z}^0(W, E) \subset C$, $D(W, E) \cap \mathbf{Z}^0(W, E) = D(W, E) \cap C$.

Suppose $D(W, E) \cap C \subset \mathbf{Z}^0(W, E)$. We have $A[A^{-1}((\mathbf{R}^+)^I)] = \alpha^\vee \alpha[\alpha^{-1}(C)] = \alpha^\vee[\mathrm{Img}(\alpha) \cap C] \subset \alpha^\vee[\ker(\alpha^\vee)] = 0$. Therefore $A^{-1}[(\mathbf{R}^+)^I] = \ker(A)$, that is, $AX \geq 0 \Rightarrow AX = 0$.

- If A is negative, $A^{-1}[(\mathbf{R}^+)^I] \cap (\mathbf{R}^+)^I = 0$ and then $0 = \alpha[A^{-1}((\mathbf{R}^+)^I) \cap (\mathbf{R}^+)^I] = \alpha[\alpha^{-1}(C) \cap (\mathbf{R}^+)^I]$. $\qquad\square$

4.2 C1 and C2 systems

A *C1&C2 system* is a system whose Cartan matrix is C1&C2. Let $(E, \alpha, \alpha^\vee)$ be a C1&C2 system and A its Cartan matrix. We have

$$\exists \text{ an } W\text{-invariant scalar product in } D(W, E) \Leftrightarrow \begin{cases} (E, \alpha, \alpha^\vee) \text{ is regular and} \\ A = A_+ \oplus A_0. \end{cases} \tag{77}$$

In this case define the facet f_{I_0} of the associated cone C

$$f_{I_0} := \left(\bigcap_{i \in I_0} H_i \right) \cap C. \tag{78}$$

and set N_0 the number of components of A_0. We have

$$D(W, E) \cap C \subset f_{I_0} \qquad \text{and} \qquad |I| - N_0 = \dim D(W, E) \leq \dim E. \tag{79}$$

Proof.

- Using (105) and (63), A nonnegative $\Leftrightarrow A \sim B$ where B semidefinite positive $\Leftrightarrow \exists$ invariant symmetric bilinear form S in $D(W, E)$ and (β_i) generating $D(W, E)$ such that $\mathrm{Gram}_S(\beta)$ semidefinite positive $\Leftrightarrow \exists$ invariant semidefinite positive bilinear form S in $D(W, E)$. Then S is a scalar product iff the system is regular.

Suppose that there exists a W-invariant scalar product in $D(W, E)$.

- Being A nonnegative and using (74) and (76) we have, $D(W, E) \cap C = D(W_+, E) \cap C_+ + D(W_0, E) \cap C_0 = D(W_+, E) \cap C_+ + D(W_0, E) \cap \mathbf{Z}^0(W_0, E)$. S defining a W_0-invariant scalar product in $D(W_0, E)$ and using (46), $D(W, E) \cap C = D(W_+, E) \cap C_+ \subset \mathbf{Z}^0(W_0, E) \cap C_+ = C \cap (\cap_{i \in I_0} H_i) = f_{I_0}$.

- As N_0 equals $\mathrm{corank}(A)$ and the system being regular, $\dim[D(W, E)] = \mathrm{rank}(A) = |I| - \mathrm{corank}(A) = |I| - N_0$. $\qquad\square$

Analogously to (67), for a subspace E' containing $D(W, E)$ we have

$$\exists W\text{-invariant scalar product in } E' \Leftrightarrow \begin{cases} (E, \alpha, \alpha^\vee) \text{ is regular,} \\ E' \subset D(W, E) \oplus \mathbf{Z}^0(W_0, E) \text{ and} \\ A = A_+ \oplus A_0. \end{cases} \tag{80}$$

In particular

$$\exists W\text{-invariant scalar product in } E \Leftrightarrow \begin{cases} (E, \alpha, \alpha^\vee) \text{ is regular and coregular and} \\ A = A_+ \oplus A_0. \end{cases} \tag{81}$$

4.3   Coxeter systems

A *Coxeter system* is a system whose reflection group is a linear Coxeter group. We have the following fundamental result [2]

$$(E, \alpha, \alpha^\vee) \text{ is a Coxeter system } \Leftrightarrow \begin{cases} \text{it is a C1\&C2 system and} \\ \text{int}(C) \neq \emptyset. \end{cases} \tag{82}$$

In this case the exponent matrix of W coincides with the exponent matrix of A (101) and hence

$$\mathrm{Cos}(W) = \mathrm{Cos}(A). \tag{83}$$

5   SPHERICAL COXETER GROUPS

In this section we characterize the linear Coxeter groups which are linear extensions of discrete reflection groups in the sphere.

5.1   Spherical Coxeter groups

For a linear group W in E, the following conditions, defining an *spherical Coxeter group*, are equivalent

a) W is the linear extension of the group generated by reflections in the walls of a spherical Coxeter polyhedron.

b) W is a linear Coxeter group and there exists a W-invariant scalar product S in E.

c) W is finite linear Coxeter group.

d) W is the reflection group of a regular and coregular Coxeter system such that $A = A_+ \oplus A_0$.

e) W is the reflection group of a Coxeter system such that $A = A_+$ (so it characteristic is $\chi = (A_+, 0, 0, d)$ and then $\dim E = |I| + d$).

Proof.

a) $\Rightarrow$b) Let W be the group generated by the reflections in the walls of a Coxeter polyhedron P in the sphere and set C the convex cone in the ambient space E extending P. The linear extension of W is the group generated by the orthogonal reflections in the walls of C. It preserves the scalar product of E.

b) $\Rightarrow$a) Let W be a linear Coxeter group in E and S a W-invariant scalar product. Then W is generated by orthogonal (w.r.t. S) reflections in the walls of a convex cone C. Its restriction to the sphere $X := \{x \in E \mid S(x, x) = 1\}$ is the discrete reflection group generated by the reflections in the walls of the Coxeter polyhedron $C \cap X$.

b) $\Leftrightarrow$d) See (71).

d) $\Rightarrow$e) Follows from (70) and (73).

e) $\Rightarrow$d) If $A = A_+$, A is invertible and hence E is regular and coregular.

c) $\Leftrightarrow$e) Using (19), (83) and (103), W finite $\Leftrightarrow \mathrm{Cos}(W)$ definite positive $\Leftrightarrow \mathrm{Cos}(A)$ definite positive $\Leftrightarrow A$ positive. By (79), $|I| = \dim D(W, E)$. Being regular and coregular $\dim D(W, E) = \dim E - \dim \mathbf{Z}^0(W, E) = \dim E - d.$ $\square$

5.2 Elliptic Coxeter groups

For a linear group W in E, the following conditions, defining an *elliptic Coxeter group*, are equivalent [5]

a) W is the linear extension of the group generated by reflections in the walls of a nondegenerate spherical Coxeter polyhedron.

b) W is the reflection group of a reduced Coxeter system such that $A = A_+$ (then α and $\alpha^\vee$ are isomorphisms and $\chi = (A_+, 0, 0, 0)$).

Proof. The fundamental polyhedron P has no antipodal points iff the apex of the polyhedral cone C extending P is $\{0\}$, that is, iff the system is reduced. $\square$

6 EUCLIDEAN COXETER GROUPS

In this section we deal with the linear extensions of discrete reflection groups in the Euclidean space, called Euclidean Coxeter groups. The number of facets of the cone C will be related with other invariants of the linear group, namely, the dimension of the space E and the number of components of the zero-component of the Cartan matrix. We will deal also with the relative position of the cone C and the invariant subspace $D(W, E)$. In particular, for infinite Euclidean groups $D(W, E)$ do not intersect the interior of C.

An important feature of reflection groups in a Euclidean space X is that they can be expressed as a semidirect product [3]

$$W = \Lambda \bigoplus W_0,$$

where Λ is a discrete abelian group of translations in X, and W_0 is the stabilizer subgroup of an *special* point of X. Here we will not deal with this decomposition, but is easy to see that the linear extension of the elements in Λ belong to an special kind of pseudoreflections, characterized by the fact that the reflection hyperplan contains the direction of the pseudoreflection.

6.1 Euclidean Coxeter groups

The following result introduce Euclidean Coxeter groups.

For a linear group W in E, the following conditions, defining an *Euclidean Coxeter group*, are equivalent

a) W is the linear extension of the group generated by reflections in the walls of an Euclidean Coxeter polyhedron.

b) W is a linear Coxeter group, there exists a W-invariant hyperplane H of E that is not a wall of the fundamental cone C and a W-invariant scalar product S in H.

c) W is the reflection group of a Coxeter system $(E, \alpha, \alpha^\vee)$ such that

$$\left\{ \begin{array}{ll} (E, \alpha, \alpha^\vee) & \text{is regular} \\ A & \text{is nonnegative} \\ D(W, E) \subset H \subset D(W, E) \oplus \mathbf{Z}^0(W, E) & \text{for some hyperplane } H. \end{array} \right\} \quad (84)$$

Systems verifying condition c) are called *Euclidean systems*.

Proof.

a) $\Rightarrow$ b) Let P be a Coxeter polyhedron in X and set W linear extension of the group generated by the relections in the walls of P. It is clear that the hyperplan TX of translations in X and the scalar product of TX remains invariant under the action of W.

b) $\Rightarrow$ a) Choose a system $(E, \alpha, \alpha^\vee)$ defining W, so W is generated by reflections in the walls $(H_i)_i$ of the associated convex cone $C := \cap_i H_i^+$ along the directions $(\alpha_i)_i$. Choose an affine hyperplane

$$X := x + H \qquad \text{where} \ \ x \in \mathrm{int}(C), \ x \notin H.$$

For any $x + h \in X$ and $i \in I$, $r_i(x+h) = x - \alpha_i^\vee(x)\alpha_i + r_i(h) \in x + D(W, E) + H = X$, so W stabilizes X. Define in X the hyperplanes $(\Sigma_i)_i := (H_i \cap X)_i$, the halfspaces $(\Sigma_i^+)_i := (H_i^+ \cap X)_i$ and the polyhedron

$$P := C \cap X = \cap_i \Sigma_i^+.$$

Let us see that the cone C is the linear extension of P. Given $j \in I$, by (98) the system

$$(AY)_j = -\alpha_j^\vee(x), \qquad (AY)_{i \in I - \{j\}} > 0$$

has a solution. For such a solution Y, define the element $x_j \in X$

$$x_j := x + \sum_i Y_i \alpha_i.$$

Then $\alpha_j^\vee(x_j) = \alpha_j^\vee(x) + \sum_i Y_i \alpha_j^\vee(\alpha_i) = -(AY)_j + (AY)_j = 0$. For $i \neq j$, $\alpha_i^\vee(x_j) = \alpha_i^\vee(x) + \sum_k Y_k \alpha_i^\vee(\alpha_k) = \alpha_i^\vee(x) + (AY)_i > 0$. We have proved that for all $j \in I$,

$$\left[\bigcap_{i \in I - \{j\}} (\Sigma_i^+ - \Sigma_i) \right] \bigcap \Sigma_j \neq \emptyset.$$

Therefore the $(\Sigma_j)_j$ are the walls of P, and hence the cone C is the linear extension of the polyhedron P.

As W preserves a scalar product S of X, the restriction to X of principal reflections $(r_i)_i$ of W are orthogonal reflections in the walls $(\Sigma_i)_i$ of P. Then the vectors $(\alpha_i)_i$ are orthogonal to the faces $(\Sigma_i)_i$ of P and points to the inside of P. For the dihedral angles (φ_{ij}) between adjacent faces of P we have, $\cos \varphi_{ij} = -S(\alpha_i, \alpha_j)/[S(\alpha_i, \alpha_i)S(\alpha_j, \alpha_j)]^{1/2} = -1/2 \, \alpha_i^\vee(\alpha_j)[S(\alpha_i, \alpha_i)/S(\alpha_j, \alpha_j)]^{1/2}$. From Proposition 58, $\cos \varphi_{ij} = -1/2 \, a_{ij}(a_{ji}/a_{ij})^{1/2} = 1/2(a_{ij} \, a_{ji})^{1/2}$. Being $A = A_+ \oplus A_0$ a C1&C2 matrix, A is PD-similar to the cosine matrix B, so $1/2(a_{ij} \, a_{ji})^{1/2} = \cos(\pi/m_{ij})$ and hence P is a Coxeter polyhedron. In conclusion, W is the linear extension of the group generated by the (orthogonal) reflections in the walls $(\Sigma_i)_i$ of the Coxeter polyhedron P.

b) $\Leftrightarrow$ c). Follows from (53) and (80). $\qquad\qquad\qquad\qquad\qquad\qquad\qquad\qquad\qquad$ $\square$

Let W be a Euclidean Coxeter group in E. We have

$$|I|/2 + 1 \leq |I| - N_0 + 1 \leq \dim E. \tag{85}$$

where N_0 denotes the number of components of A_0. Furthermore

$$W \ \text{infinite} \quad \Leftrightarrow \quad D(W, E) \cap \mathrm{int}(C) = \emptyset \tag{86}$$

Proof.

By (79), $|I| - N_0 = \dim[D(W, E)] \leq \dim E - 1$. As the minimal order of a zero type matrix is 2, $N_0 \leq |I|/2$ and then $|I|/2 + 1 \leq |I| - N_0 + 1$.

We have seen (spherical Coxeter groups) that a linear Coxeter group W is finite iff A is positive. For Euclidean Coxeter groups A is nonnegative, so A is positive iff there exists $X > 0$ such that $AX > 0$, that is, iff $D(W, E) \cap \mathrm{int}(C) = \emptyset$. $\qquad\qquad$ $\square$

6.2 Essential Coxeter groups

Now we deal with discrete reflection groups in the Euclidean space whose fundamental polyhedron is nondegenerate. Finiteness conditions will be given.

For a linear group W in E, the following conditions, defining an *essential Coxeter group*, are equivalent

a) W is the linear extension of the group generated by reflections in the walls of a nondegenerate Euclidean Coxeter polyhedron.

b) W is the reflection group of a Coxeter system $(E, \alpha, \alpha^\vee)$ such that

$$\left\{ \begin{array}{ll} (E, \alpha, \alpha^\vee) & \text{is regular} \\ A & \text{is nonnegative} \\ D(W, E) & \text{is an hyperplane.} \end{array} \right\} \qquad (87)$$

System verifying these conditions are called *essential systems*.

c) W is the reflection group of a Euclidean system $(E, \alpha, \alpha^\vee)$ and the subsystem induced (54) by the W-invariant hyperplane H is reduced.

d) W is the reflection group of a Euclidean system $(E, \alpha, \alpha^\vee)$ and

$$\dim E = |I| - N_0 + 1. \qquad (88)$$

Proof.
a) $\Leftrightarrow$ b) Use (84) and (68).
b) $\Leftrightarrow$ c) As there exists an invariant scalar product in H, using (81), the subsystem $(H, \alpha', \alpha^{\vee\prime})$ induced by H is regular and coregular, so $H = D(W', H) \oplus \mathbf{Z}^0(W', H) = D(W, E) \oplus \mathbf{Z}^0(W', H)$. Then $D(W, E)$ is an hyperplane iff $\mathbf{Z}^0(W', H) = 0$.
b) $\Leftrightarrow$ d) follows from (79). $\qquad\square$

For an Euclidean Coxeter group W, the following conditions are equivalent

a) W is essential and finite.

b) The characteristic is $\chi = (A, 0, 0, 1)$, where A is positive (so these groups are determined but isomorphism by its positive Cartan matrix A).

Proof.
If W essential and finite, $A = A_+$ so $\chi = (A_+, 0, 0, d)$. From (32) $\dim E = |I| + d = \dim D(W, E) - \mathrm{corank}(\alpha) + d = \dim D(W, E) + d \Rightarrow d = 1$.
Reciprocally, if $\chi = (A_+, 0, 0, 1)$, W is finite, regular, and $\dim E = |I| + 1 = \dim D(W, E) + 1$, then $D(W, E)$ is an hyperplane of E. Therefore W is essential and finite. $\qquad\square$

Let us now characterize essential infinite Coxeter groups. For an linear Coxeter group W, the following conditions are equivalent

a) W is the linear extension of the group generated by reflections in the walls of a Euclidean Coxeter polyhedron P having a bounded 1-facet.

b) W is an infinite essential Coxeter group.

c) W is the reflection group of a reduced Euclidean system.

d) $\chi = [A, \ker(A), K_2, 0]$, where A nonnegative and K_2 is an hyperplane of $\ker({}^tA)$.

Proof.

a) $\Leftrightarrow$ b) From (12), P has a bounded 1-facet $\Leftrightarrow$ P nondegenerate and $\mathrm{Gram}(P)$ is not definite positive. Using (69) and (103), this is equivalent to W essential and A non positive, that is, W essential and infinite.

b) $\Rightarrow$ c) $D(W, E)$ is an hyperplane, so using (73), $\mathbf{Z}^0(W, E) \neq 0$ implies $E = D(W, E) \oplus \mathbf{Z}^0(W, E)$ and then $N_0 \neq 0$, so W infinite. Therefore $\mathbf{Z}^0(W, E) = 0$.

c) $\Rightarrow$ b) If W is reduced, as $D(W, E) \subset H \subset D(W, E) \oplus \mathbf{Z}^0(W, E)$, $H = D(W, E)$ so W is essential. Therefore by (88) and being W reduced, $N_0 = |I| + 1 - \dim E = |I| + 1 - \mathrm{rank}(\alpha^\vee) \geq |I| + 1 - |I| = 1$. Therefore A has components of zero type, so W is infinite.

b), c) $\Rightarrow$ d) The system being Euclidean and reduced, $\chi = [A, \ker(A), K_2, 0]$ with A nonnegative, so $E \simeq D(W, E) \oplus \ker[({}^tA)/K_2]^*$. The system being essential we have, $\dim \ker[({}^tA)/K_2]^* = 1$, so K_2 is an hyperplane of $\ker({}^tA)$.

d) $\Rightarrow$ c) In this case W is regular, reduced and $E \simeq D(W, E) \oplus \ker[({}^tA)/K_2]^*$, so $D(W, E)$ is an hyperplane of E. Then it is Euclidean and reduced. $\qquad\square$

6.3 Parabolic Coxeter groups

The following result introduce parabolic Coxeter groups.

For a linear Coxeter group W in E, the following conditions, defining an *parabolic Coxeter group*, are equivalent [5].

a) W is the linear extension of the group generated by reflections in the walls of a bounded Euclidean Coxeter polyhedron.

b) W is the reflection group of a essential system such that $A = A_0$.

c) W is the reflection group of a reduced Euclidean system such that $D(W, E) \cap C = \{0\}$.

Proof.
a) $\Leftrightarrow$ b) It follows from (13).
b) $\Leftrightarrow$ c) It follows from (76). $\qquad\square$

APPENDIX: Real Matrices

In this appendix we collect the results on real matrices that are used along this work. For the non proved results see [5].

Matrices satisfying conditions C1 and C2 can be viewed as a short of generalized Cartan matrices, whose elements need not be longer integer. The geometrical properties of the Cartan matrices related with the reflection group and the chamber geometry deepens only on conditions C1 and C2. But these conditions do not ensures the existence of invariant lattices.

Condition C1 implies that, for an irreducible matrix A, the polyhedral cone $AX \geq 0$ intersects the boundary of the polyhedral cone $X \geq 0$ only at the origin. This fact limits the possible relative positions of these two cones to just three possibilities, and generalizes the usual classification of irreducible Cartan matrices into positive, zero and negative types.

I will denote a finite set. If $X, Y \in \mathbf{R}^I$, the notation $X > Y$ $(X \geq Y)$ will mean that $X_i > Y_i$ $(X_i \geq Y_i)$ for all $i \in I$. If $J \subset I$, we will consider $\mathbf{R}^J$ as a subspace of $\mathbf{R}^I$, with the natural embedding $\mathbf{R}^J \subset \mathbf{R}^I$. Denote the set of nonnegative real numbers by $\mathbf{R}^+$.

Real matrices

The linear space $M_I(\mathbf{R})$ formed by the real square matrices with indices in I will be identified to the space of linear maps $\mathbf{R}^I \to \mathbf{R}^I$. Let $A := (a_{ij}) \in M_I(\mathbf{R})$ be a real matrix. Write

$$J \perp K \quad \text{if} \quad a_{jk} = a_{kj} = 0\,, \ \forall (j,k) \in J \times K$$

where $J, K \subset I$. Denote $J^\perp = \{i \in I \mid i \perp J\}$.

It will be convenient to associate a graph to the matrix A. Each vertex of the graph represents a element of I and there is an edge between the vertices of i and j iff $i \neq j$ and not $i \perp j$.

A matrix is said *acyclic* if its graph has no circuits. Then for any cyclic sequence $i_1, i_2, \ldots, i_p, i_1$ with $p \geq 3$

$$a_{i_1 i_2} a_{i_2 i_3} \ldots a_{i_p i_1} = a_{i_2 i_1} a_{i_3 i_2} \ldots a_{i_1 i_p} = 0. \tag{89}$$

A *principal submatrix* of A is a matrix of the form $A_J := (a_{ij})_{i,j \in J}$, where $J \subset I$. We say that A_J is a *direct submatrix* of A if it verifies the equivalent conditions

$$J \perp (I - J), \qquad A \text{ and } {}^t\!A \text{ stabilize } \mathbf{R}^J. \tag{90}$$

In other words, the graph of A is the disconnect union of the subgraphs of A_J and A_{I-J}. In this case we say that A is the *direct sum* of A_J and A_{I-J} and write $A = A_J \oplus A_{I-J}$.

If A has no direct submatrices, but itself and the empty matrix $A_\emptyset$, A is said *irreducible*. Every matrix A is the direct sum

$$A = \bigoplus_k A_{I_k} \tag{91}$$

of its irreducible direct submatrices, called the *components* of A, that correspond with the connected components of the graph of A. The elements of the associated partition $(I_k)_k$ of I are called the *connected components of I*.

PD-similarity

Write PD for positive diagonal. The multiplicative abelian group $M_I PD(\mathbf{R})$ formed by the PD matrices $D \in M_I(\mathbf{R})$ acts in $M_I(\mathbf{R})$ by conjugation

$$D * A := D^{-1} A D = (d_i^{-1} d_j a_{ij}),$$

where $D = \mathrm{diag}(d_i)$. The orbit of A will be called the *PD class* of A and denoted $[A]_{PD}$. Two matrices A, B in the same PD-class are said *PD-similar* and denoted $A \sim B$. The integers $\mathrm{rank}(A)$ and $\mathrm{corank}(A)$ are PD-invariants. We have

a) $A \sim B$ iff they have identical cyclic products and its elements have the same sign $(+, 0, -)$, that is

$$\mathrm{sign}(a_{ij}) = \mathrm{sign}(b_{ij})$$

$$a_{i_1 i_2} a_{i_2 i_3} \ldots a_{i_p i_1} = b_{i_1 i_2} b_{i_2 i_3} \ldots b_{i_p i_1}, \qquad \forall i_1, \ldots, i_p \in I. \tag{92}$$

In particular, A and B have identical diagonal elements.

b) The stabilizer subgroup of A is formed by the $D \in M_I PD(\mathbf{R})$ of the form

$$D = \bigoplus_k D_k, \tag{93}$$

where $I = \bigcup_k I_k$ is the decomposition of I in connected components and D_k is a positive homothety of the subspace $\mathbf{R}^{I_k} \subset \mathbf{R}^I$.

Proof.

a) Obviously $A \sim B \Rightarrow A$ and B have the same cyclic products and signs. Suppose A and B have identical cyclic products and the same signs. For $a_{ij} \neq 0 \neq b_{ij}$ define $d_{ij} := a_{ij} b_{ij}^{-1} > 0$. If if $a_{i_1 i_2} a_{i_2 i_3} \ldots a_{i_p i_1} \neq 0$, we have $1 = a_{i_1 i_2} a_{i_2 i_3} \ldots a_{i_p i_1} (b_{i_1 i_2} b_{i_2 i_3} \ldots b_{i_p i_1})^{-1} = d_{i_1 i_2} d_{i_2 i_3} \ldots d_{i_p i_1}$. Then $d_{ij} = d_i d_j^{-1}$ with the $d_i > 0$ and hence $b_{ij} = d_i^{-1} d_j a_{ij}$.

b) $D \in M_I PD(\mathbf{R})$ stabilizes A iff $A = D^{-1} AD$. Set $I = \bigcup_k I_k$ the decomposition of I in connected components. Then $A = D^{-1} AD$ iff for any k and for all $i, j \in I_k$, $d_i^{-1} d_j a_{ij} = a_{ij}$. Being I_k connected, $d_i = d_j$ and hence the result. $\square$

Symmetrizable matrices

A matrix A is said *symmetrizable* if it is PD-similar to a symmetric matrix. If $D^{-1} AD$ is a symmetric matrix for $D \in M_I PD(\mathbf{R})$, we say that D *symmetrizes* A. The following result characterizes symmetrizable matrices; condition b) is the usual one used to introduce symmetrizable Cartan matrices [6].

The following conditions are equivalent

a) A symmetrizable.

b) There exists $D' \in M_I PD(\mathbf{R})$ such that AD' is a symmetric matrix.

c) For all i, j and for any for any circuit $i_1, i_2, \ldots, i_p, i_1$ of the graph of A

$$\begin{aligned} \mathrm{sign}(a_{ij}) &= \mathrm{sign}(a_{ji}) \\ a_{i_1 i_2} a_{i_2 i_3} \ldots a_{i_p i_1} &= a_{i_2 i_1} a_{i_3 i_2} \ldots a_{i_1 i_p}. \end{aligned} \tag{94}$$

If A is symmetrizable there exist a unique symmetric matrix in the PD-class of A, namely, the matrix

$$B = \left(\mathrm{sign}(a_{ij})(a_{ij} a_{ji})^{1/2} \right).$$

Proof.

a) $\Leftrightarrow$ b) A symmetrizable iff $S := D^{-1} AD$ is symmetric, for some matrix $D \in M_I PD(\mathbf{R})$. But S is symmetric iff DSD is symmetric, i.e, iff AD^2 is symmetric.

a) $\Rightarrow$ c). Let $B := (b_{ij})$ be a symmetric matrix PD-similar to A. Using (92), for all i, j, $\mathrm{sign}(a_{ij}) = \mathrm{sign}(b_{ij}) = \mathrm{sign}(b_{ji}) = \mathrm{sign}(a_{ji})$. On the other hand, $a_{i_1 i_2} a_{i_2 i_3} \ldots a_{i_p i_1} = b_{i_1 i_2} b_{i_2 i_3} \ldots b_{i_p i_1} = b_{i_2 i_1} b_{i_3 i_2} \ldots b_{i_1 i_p} = a_{i_2 i_1} a_{i_3 i_2} \ldots a_{i_1 i p}$.

c) $\Rightarrow$ a). Define the symmetric matrix $B = (b_{ij})$, $b_{ij} := \mathrm{sign}(a_{ij})(a_{ij} a_{ji})^{1/2}$. Then for all i, j, $\mathrm{sign}(a_{ij}) = \mathrm{sign}(b_{ij})$. Let us see that A and B have the same cyclic products. It is clear that $b_{ii} = a_{ii}$ and $b_{ij} b_{ji} = a_{ij} a_{ji}$. For any circuit $i_1, i_2, \ldots, i_p, i_1$ of the graph of A we have, $b_{i_1 i_2} b_{i_2 i_3} \ldots b_{i_p i_1} = \mathrm{sign}(a_{i_1 i_2}) \, \mathrm{sign}(a_{i_2 i_3}) \ldots \mathrm{sign}(a_{i_p i_1}) \, [(a_{i_1 i_2} a_{i_2 i_3} \ldots a_{i_p i_1})(a_{i_2 i_1} a_{i_3 i_2} \ldots a_{i_1 i_p})]^{1/2}$, and by hypothesis, $\ldots = \mathrm{sign}(a_{i_1 i_2}) \, \mathrm{sign}(a_{i_2 i_3}) \ldots \mathrm{sign}(a_{i_p i_1}) \, |a_{i_1 i_2} a_{i_2 i_3} \ldots a_{i_p i_1}| = a_{i_1 i_2} a_{i_2 i_3} \ldots a_{i_p i_1}$. A and B having the same signs and cyclic products, the result follows from (92).

It is obvious that B is the unique symmetric matrix in the PD-class of A. $\square$

C1 matrices

We say that a matrix $A \in M_I(\mathbf{R})$ is a *C1 matrix* it verifies condition C1 below

$$a_{ii} = 2, \qquad a_{ij} \leq 0 \text{ if } i \neq j, \qquad \text{and} \qquad a_{ij} = 0 \Leftrightarrow a_{ji} = 0. \tag{95}$$

Let us define, in two equivalent forms, three different types of C1 matrices. These types are PD and transposition-invariant.

- *A positive*: $AX \geq 0 \Rightarrow X \geq 0$. There exists $X > 0$ such that $AX > 0$.

- *A zero*: $AX \geq 0 \Rightarrow AX = 0$. There exists $X > 0$ such that $AX = 0$.

- *A indefinite* or *negative*: $(X \geq 0 \ \& \ AX \geq 0) \Rightarrow X = 0$. There exists $X > 0$ such that $AX < 0$.

For irreducible C1 matrices, we have

$$A \text{ irreducible} \quad \Rightarrow \quad A \text{ positive, zero or negative.}$$

For a positive irreducible C1 matrix A, we have in fact

$$AX \geq 0 \quad \Rightarrow \quad X = 0 \text{ or } X > 0.$$

Any C1 matrix A decomposes

$$A = A_+ \bigoplus A_0 \bigoplus A_-, \tag{96}$$

where (A_+, A_0, A_-) is the (positive, zero, negative) submatrix formed by the sum of the irreducible submatrices of (positive, zero, negative) type. Denote by (I_+, I_0, I_-) the set of indices of the submatrix (A_+, A_0, A_-), so I is the disjoint union $I = I_+ \bigcup I_0 \bigcup I_-$.

We say that A is *nonnegative* if $A = A_+ \bigoplus A_0$. Let A be a nonnegative C1 matrix and N_0 the number of components of A_0. We have

- Denoting by N_0 the number of components of A_0

$$\operatorname{corank}(A) = N_0. \tag{97}$$

  In particular, if A is positive it is invertible.

- For any $j \in I$ and $c > 0$, there exists $Y \in \mathbf{R}^I$ such that

$$(AY)_j = -c \qquad \text{and} \qquad (AY)_{i \in I-\{j\}} > 0 \tag{98}$$

Proof. Suppose first that A is indecomposable. If $A = A_+$, it is invertible, so the result is obvious. If $A = A_0$, $\operatorname{corank}(A) = 1$, so $H := \operatorname{Img}(A)$ is an hyperplane of $\mathbf{R}^I$. As $AX \geq 0 \Rightarrow AX = 0$, $H \cap (\mathbf{R}^+)^I = \{0\}$ and then H is described by an equation on the form

$$\sum_{i \in I} v_i Y_i = 0 \qquad \qquad \text{where } v_i > 0 \text{ for all } i \in I.$$

As the system

$$\sum_{i \in I-\{j\}} v_i Y_i = v_j c, \qquad (Y_i)_{i \in I-\{j\}} > 0$$

has obviously solution, the result follows. If A is not indecomposable, we can limit to the case $A = A_0$ as above. For any connected component I_k of I consider the plane $\operatorname{Img}(A_{I_k}) := (\sum_{i \in I_k} v_i Y_i = 0)$, where $v_i > 0$ for all $i \in I_k$. Consider now the system

$$\sum_{i \in I_{k(j)}-\{j\}} v_i Y_i = v_j c - \sum_{i \in I-I_{k(j)}} v_i Y_i, \qquad (Y_i)_{i \in I-\{j\}} > 0,$$

where $I_{k(j)}$ denotes the connected component of I containing j. Making $\sum_{i \in I-I_{k(j)}} v_i Y_i < v_j c$ with a convenient choice of the $(Y_i)_{i \in I-I_{k(j)}} > 0$, the system above has a solution, and the result follows. $\qquad \square$

Symmetrizable C1 matrices

From (92) it follows that two C1 matrices $A, B \in M_I(\mathbf{R})$ are PD similar iff they have the same cyclic products. It is also clear that a C1 matrix A is symmetrizable iff it verifies second condition of (94). In particular, any acyclic C1 matrix is symmetrizable.

Let A be a symmetrizable C1 matrix. The symmetric matrix $B = (b_{ij})$ in PD-class of A is given by

$$b_{ij} = -(a_{ij}a_{ji})^{1/2} \quad \text{if } i \neq j, \qquad \text{and} \qquad b_{ii} = a_{ii}.$$

For a symmetric C1 matrix A we have

$$\begin{aligned} A \text{ nonnegative} \quad &\Leftrightarrow \quad A \text{ semidefinite positive} \\ A \text{ positive} \quad &\Leftrightarrow \quad A \text{ definite positive.} \end{aligned} \tag{99}$$

C2 matrices

We say that a matrix $A \in M_I(\mathbf{R})$ is a C2 matrix if it verifies condition C2, namely,

$$a_{ij}a_{ji} = 4\cos^2\left(\frac{\pi}{m_{ij}}\right), \quad m_{ij} = 2, 3, \ldots \qquad \text{if } a_{ij}a_{ji} < 4. \tag{100}$$

A C1&C2 matrix is a matrix satisfying conditions C1 and C2. For such a matrix A define the *exponent matrix* as the matrix $M = (m_{ij})$, where

$$m_{ij} = \begin{cases} 1 & \text{if } i = j \\ \infty & \text{if } a_{ij}a_{ji} \geq 4 \\ \text{given by (100)} & \text{otherwise.} \end{cases} \tag{101}$$

Define also the *cosine matrix* of A, depending only in the PD-class of A, as the C1&C2 symmetric matrix

$$\mathrm{Cos}(A) = (c_{ij}) \qquad c_{ij} := -2\cos(\pi/m_{ij}) = \begin{cases} \sup\left\{-(a_{ij}a_{ji})^{1/2}, -2\right\} & \text{if } i \neq j \\ 2 & \text{if } i = j \end{cases} \tag{102}$$

Write $A \approx B$, where $A \in M_I(\mathbf{R})$ and $B \in M_J(\mathbf{R})$, if for some identification of the sets of indices I and J, we have $A \sim B$. Let A be a C1&C2 matrix. We have

$$A \text{ positive} \Leftrightarrow \mathrm{Cos}(A) \text{ definite positive} \Rightarrow A \sim \mathrm{Cos}(A) \text{ and } A \text{ is acyclic.} \tag{103}$$

For zero matrices we have

$$\begin{aligned} A \text{ zero} \quad \Leftrightarrow \quad &\mathrm{Cos}(A) \text{ zero and } A_{I_k} \sim \mathrm{Cos}(A)_{I_k} \text{ if } \mathrm{Cos}(A)_{I_k} \approx \tilde{A}_l \ (l \geq 1) \Rightarrow \\ &\Rightarrow A \sim \mathrm{Cos}(A). \end{aligned} \tag{104}$$

where I_k denotes an arbitrary connected component of I. From (103) and (104) it follows

$$A \text{ nonnegative} \quad \Leftrightarrow \quad \mathrm{Cos}(A) \text{ semidefinite positive and } A \sim \mathrm{Cos}(A). \tag{105}$$

Also, if $\mathrm{Cos}(A)$ is nonnegative and has no direct submatrices of type $\tilde{A}_l \ (l \geq 1)$, A is nonnegative.

$$A \text{ zero, irreducible and cyclic} \Leftrightarrow A \approx \tilde{A}_l, \ (l \geq 2). \tag{106}$$

REFERENCES

[1] Alekseevskij D.V., Vinberg E.B. and Solodovnikov A.S., Geometry of spaces of constant curvature, in Geometry II, (Vinberg ed.), Springer-Verlag 1993.

[2] Vinberg E.B. and Shvartsman O.V., Discrete groups of motions of spaces of constant curvature, in Geometry II (Vinberg ed.), Springer-Verlag, 1993.

[3] Bourbaki N., Groupes et Algèbres de Lie, Chap. 4-6, Herman, Paris, 1968.

[4] Humphreys J.E., Reflection groups and Coxeter groups, Cambridge University Press 1992.

[5] Vinberg E.B., Discrete linear groups generated by reflections, Math USSR-Izvestija **5** (1971), 1083-1119.

[6] Kac V.G., Infinite dimensional Lie algebras, Cambridge University Press, 1990.

DIFFEOMORPHISM GROUPS, QUASI-INVARIANT MEASURES, AND INFINITE QUANTUM SYSTEMS

Gerald A. Goldin[1] and Ugo Moschella[2]

[1] Departments of Mathematics and Physics, Rutgers University
New Brunswick, New Jersey 08903 USA

[2] Service de Physique Théorique, CEA-Saclay
F-91191 Gif-sur-Yvette Cedex, Fraence

INTRODUCTION

This paper provides a brief introduction to how unitary representations of diffeomorphism groups can describe certain quantum systems having infinitely many degrees of freedom. It is a partial report of our joint work [1], based on the August 1994 talk by the first author at the Symmetries in Science VIII conference in Bregenz, Austria. We would like to express appreciation to the conference organizers, especially Professor Bruno Gruber, for the opportunity to present our results.

We begin in the next section by presenting the background for our construction, describing how unitary representations of diffeomorphism groups enter into quantum physics and why the problem of quasi-invariant measures is physically fundamental. In the following section we introduce our method for constructing measures on certain infinite-dimensional configuration spaces in one space dimension. We show through some elementary examples why the resulting measures are quasi-invariant for $Diff(\mathbf{R})$, the group of diffeomorphisms of the real line. Next we sketch the extension of our method to obtain the major physical result–that a parameterized family of these models, regarded as describing a quantum gas of point particles in one-dimensional space, actually shows a phase transition at a critical value of the correlation parameter κ. In the final section we discuss our findings and mention some directions of our ongoing research.

DIFFEOMORPHISM GROUPS IN QUANTUM MECHANICS

The work we are doing advances the concept that quantum theories are described by means of unitary representations of diffeomorphism groups. This idea, developed in the late 1960's by Goldin and Sharp, was originally motivated by the study of local current algebras [2]. It is an approach that has already produced quite a few interesting and unexpected results [3].

Symmetries in Science VII, Edited by
B. Gruber, Plenum Press, New York, 1995

One way to see how diffeomorphism group representations occur in quantum mechanics is the following. Let ψ and ψ^* be second-quantized, nonrelativistic fields satisfying the canonical commutation relations (-) or anticommutation relations (+) at fixed time,

$$[\psi(\mathbf{x}), \psi^*(\mathbf{y})]_{\pm} = \delta(\mathbf{x} - \mathbf{y}),\tag{1}$$

for $\mathbf{x} \in \mathbf{R}^3$. Dashen and Sharp [4] suggested that quantum mechanics be described by a certain formal, singular, fixed-time Lie algebra, which is satisfied by the mass density operator $\rho(\mathbf{x})$ and the momentum density operator $\mathbf{J}(\mathbf{x})$ defined in terms of the fields:

$$\rho(\mathbf{x}) = m\psi^*(\mathbf{x})\psi(\mathbf{x}),\tag{2}$$

$$\mathbf{J}(\mathbf{x}) = (\hbar/2i)\,[\psi^*(\mathbf{x})\,\nabla\psi(\mathbf{x}) - (\nabla\psi^*(\mathbf{x}))\psi(\mathbf{x})]\,.$$

Goldin and Sharp then proposed to interpret $\rho(\mathbf{x})$ and $\mathbf{J}(\mathbf{x})$ rigorously as operator-valued distributions in Hilbert space, modeled on Schwartz-space functions–i.e., on spaces of C^∞ scalar functions and vector fields on $\mathbf{R}^3$ that, together with all derivatives, vanish faster than the inverse of any power of $|\mathbf{x}|$ at infinity. Defining

$$\rho(f) = \int \rho(\mathbf{x})f(\mathbf{x})d\mathbf{x}, \qquad J(\mathbf{g}) = \int \mathbf{J}(\mathbf{x}) \cdot \mathbf{g}(\mathbf{x})d\mathbf{x},\tag{3}$$

where f is a Schwartz-space function and $\mathbf{g}$ is a vector field with components in Schwartz space, the resulting Lie algebra of gauge-invariant quantum observables is

$$[\rho(f_1),\, \rho(f_2)] = 0, \qquad [\rho(f),\, J(\mathbf{g})] = i\hbar\rho(\mathbf{g} \cdot \nabla f),\tag{4}$$

$$[J(\mathbf{g}_1),\, J(\mathbf{g}_2)] = -i\hbar J([\mathbf{g}_1,\, \mathbf{g}_2])\,.$$

Equation (4) is understood as a representation by self-adjoint operators of the natural semidirect sum of the Abelian algebra of functions $\mathcal{S}(\mathbf{R}^3)$, with the algebra of vector fields $Vect(\mathbf{R}^3)$. Here $\mathbf{g} \cdot \nabla f$ is the Lie derivative of f in the direction of $\mathbf{g}$, and $[\mathbf{g}_1,\, \mathbf{g}_2] = \mathbf{g}_1 \cdot \nabla\mathbf{g}_2 - \mathbf{g}_2 \cdot \nabla\mathbf{g}_1$ is the Lie bracket of the vector fields.

Note that (4) is a *commutator* algebra, regardless of whether ψ and ψ^* are assumed to satisfy commutation or anticommutation relations. If one begins with either the Bose or the Fermi Fock representations of ψ and ψ^*, satisfying commutation or anticommutation relations respectively, $\rho(f)$ and $J(\mathbf{g})$ are defined as self-adjoint operators and obey (4) on an appropriate common, dense, invariant domain of the Hilbert space. The N-particle Fock subspaces of totally symmetric (or totally antisymmetric) wave functions are reducing subspaces for these representations.

Thus the infinite-dimensional Lie algebra of rapidly decreasing scalar functions and vector fields on physical space is already represented in nonrelativistic quantum mechanics as it is usually understood, with the interpretation of spatially-averaged mass and momentum density operators. Nothing new has been put in "by hand". The N-particle spaces describing different numbers of Bose and Fermi particles in $\mathbf{R}^3$ correspond to distinct, unitarily inequivalent irreducible representations of (4). Let us write these explicitly. Taking $\Psi_N^{(s)}$ (respectively, $\Psi_N^{(a)}$) to be a totally symmetric (respectively, antisymmetric) square-integrable wave function of the N particle positions $(\mathbf{x}_1, \ldots \mathbf{x}_N)$, we have

$$\rho_N(f)\Psi^{(s,a)} = m\sum_{j=1}^{N} f(\mathbf{x}_j)\Psi^{(s,a)},$$

$$J_N(\mathbf{g})\Psi^{(s,a)} = \frac{\hbar}{2i}\sum_{j=1}^{N}[\mathbf{g}(\mathbf{x}_j) \cdot \nabla_j + \nabla_j \cdot \mathbf{g}(\mathbf{x}_j)]\Psi^{(s,a)},\tag{5}$$

where ∇_j acts on $\mathbf{x}_j$. Note that the operators ρ and J preserve the particle number and the wave function symmetry. In particular when $\mathbf{g}$ approximates a constant vector field, $J(\mathbf{g})$ approximates the usual expression for total momentum in the direction of $\mathbf{g}$. If one begins with a given self-adjoint representation of (4), the kinematical physical content is contained in the interpretation of ρ and J as mass and momentum densities respectively.

Exponentiating this Lie algebra, Goldin considered quantum theories arising from continuous unitary representations of the resulting infinite-dimensional group. The group is the natural semidirect product of the Schwartz-space $\mathcal{S}$ of scalar functions on $\mathbf{R}^3$ under pointwise addition, with $Diff(\mathbf{R}^3)$, a group of diffeomorphisms of $\mathbf{R}^3$ under composition. Let us explain this in a self-contained way.

A *diffeomorphism* of a C^∞ manifold M is, very simply, an invertible, C^∞ mapping, $\phi : M \to M$, whose inverse is also C^∞. It is thus a general coordinate transformation. The composition $\phi_1 \circ \phi_2$ of two diffeomorphisms is also a diffeomorphism, defining a group operation for which the identity element is the diffeomorphism $e(\mathbf{x}) \equiv \mathbf{x}$. The relation between vector fields on M and diffeomorphisms of M is the usual one: given a C^∞ vector field $\mathbf{g}$, one asks about the existence of a *flow* associated with the ordinary differential equation $\dot{\mathbf{x}}(t) = \mathbf{g}(\mathbf{x}(t))$, with $\mathbf{x}(0) = \mathbf{x}_0$ (for all $\mathbf{x}_0 \in M$). When M is compact, such a flow always exists. In the case $M = \mathbf{R}^s$ (for $s = 1, 2, 3$, or more space dimensions), we use not only the fact that $\mathbf{g}$ is C^∞, but that its behavior is such that together with all derivatives it vanishes rapidly as $|\mathbf{x}| \to \infty$. Then the solution $\mathbf{x}(t)$ exists for all $\mathbf{x}_0$ and for all t, and it is C^∞ in both variables. We thus obtain, for each vector field $\mathbf{g}$, a one-parameter group of C^∞ diffeomorphisms (the flow), which we write $\phi_t^{\mathbf{g}} : \mathbf{R}^s \to \mathbf{R}^s$ ($\forall t \in \mathbf{R}$). We have $\phi_t^{\mathbf{g}}$ satisfying the differential equation $(\partial/\partial t)\phi_t^{\mathbf{g}}(\mathbf{x}) = \mathbf{g}(\phi_t(\mathbf{x}))$, together with the boundary condition $\phi_{t=0}^{\mathbf{g}}(\mathbf{x}) = \mathbf{x}$, and the group law $\phi_{t_1}^{\mathbf{g}} \circ \phi_{t_2}^{\mathbf{g}} = \phi_{t_1+t_2}^{\mathbf{g}}$. The conditions on $\mathbf{g}$ at infinity are important to the existence of this flow, and naturally they constrain the diffeomorphisms that are obtained to approach the identity mapping (rapidly, in all derivatives) as $|\mathbf{x}| \to \infty$. When we refer to the group $Diff(\mathbf{R}^s)$, we shall always understand this condition at infinity to hold. $Diff(\mathbf{R}^s)$ is a topological group, and the flows are continuous one-parameter subgroups.

Returning to the self-adjoint operators obeying (4), set $U(f) = \exp[(i/m)\rho(f)]$ and $V(\phi_t^{\mathbf{g}}) = \exp[(it/\hbar)J(\mathbf{g})]$. Then U and V obey a semidirect product law,

$$U(f_1)V(\phi_1)U(f_2)V(\phi_2) = U(f_1 + f_2 \circ \phi_1)V(\phi_1\phi_2), \tag{6}$$

where $\phi_1\phi_2 = \phi_2 \circ \phi_1$ denotes the composition of diffeomorphisms. The corresponding N-particle representations of U and V are:

$$U_N(f)\Psi^{(s,a)}(\mathbf{x}_1, \ldots \mathbf{x}_N) = \prod_{j=1}^{N} \exp[if(\mathbf{x}_j)]\,\Psi^{(s,a)}(\mathbf{x}_1, \ldots \mathbf{x}_N)\,,$$

$$V_N(\phi)\Psi^{(s,a)}(\mathbf{x}_1, \ldots \mathbf{x}_N) = \Psi^{(s,a)}(\phi(\mathbf{x}_1), \ldots \phi(\mathbf{x}_N))\sqrt{\prod_{j=1}^{N} \mathcal{J}_\phi(\mathbf{x}_j)}\,, \tag{7}$$

where $\mathcal{J}_\phi(\mathbf{x})$ is the Jacobian of ϕ at $\mathbf{x}$. If one begins with a given continuous unitary representation of (6), the self-adjoint generators are recovered in the representation from the appropriate one-parameter unitary subgroups, permitting each such representation to be interpreted physically; thus, as an example, (5) follows from (7).

Today it is natural to take a more geometric approach to the problem of quantization. In that spirit we may consider eqs. (4) and (6) as describing quantum theory not merely in $\mathbf{R}^s$, but in a more general C^∞ manifold M. The semidirect product group is then a very general "local symmetry group" for quantum theory in M. The concept of general local symmetries provides an alternate rationale for the introduction of diffeomorphism groups into quantum mechanics.

We take M throughout to be the *physical space* for the system in which all the particles move (as do Goldin and Sharp in their development). Different possible configuration spaces then arise from the classification of the unitary representations of the group–just as the N-particle configuration spaces for bosons and fermions do. An alternative conception, subsequently proposed and developed by Doebner, Tolar, and their co-workers, is to let M actually be the configuration-space for the classical system to be quantized [5]. The theory that results, termed "quantum Borel kinematics", coincides with the Goldin-Sharp approach for the case of a single particle, and leads to many of the same predictions. However, our current interest in infinite-dimensional configuration spaces makes the latter method less attractive.

Many inequivalent continuous unitary representations of (6) have been constructed and interpreted quantum-mechanically over the past two decades. Though the group is infinite-dimensional, interesting representations exist whose configuration-spaces, like those of (7), are finite-dimensional. For $\mathbf{M} = \mathbf{R}^2$, the classification of representations led to a mathematically rigorous prediction of intermediate particle statistics by Goldin, Menikoff, and Sharp [6], as had been earlier conjectured by Leinaas and Myrheim [7]; such particles were subsequently termed "anyons" by Wilczek [8]. Important properties of anyons were discovered from the diffeomorphism group representations–the shifted angular momentum and energy spectra, the connection with configuration-space topology, the relation to the physics of a charged particle circling a region of magnetic flux, and the role of the braid group in anyon statistics. More recently, another class of representations was interpreted by Doebner and Goldin as leading to a new family of nonlinear Schrödinger equations, capable of describing dissipative processes in quantum mechanics [9].

Other interesting representations are associated with infinite-dimensional configuration spaces–for instance, infinite Bose or Fermi gases in the thermodynamic limit [10]. To understand the role played by configuration spaces, and by measures when the configuration spaces are infinite dimensional, it is necessary to present a bit of framework for studying unitary representations of the diffeomorphism group and its semidirect product. We do this in a abbreviated way.

Adapting the work of Gelfand and Vilenkin [11], a continuous unitary representation $U(f)$ of $\mathcal{S}$ can, under appropriate technical conditions, always be realized by multiplication operators in a concrete Hilbert space $\mathcal{H}$. Let $\mathcal{S}'$ be the continuous dual of $\mathcal{S}$, i.e. the space of tempered distributions, whose elements are generalized functions. Denote the value of a distribution $\gamma \in \mathcal{S}'$ on f by $\langle \gamma, f \rangle$. If Ω is a normalized cyclic vector for the representation U, then the expectation functional $(\Omega, U(f)\Omega)$ is the Fourier transform of a probability measure μ on $\mathcal{S}'$:

$$(\Omega, U(f)\Omega) \;=\; \int_{\mathcal{S}'} \exp\left[i\langle \gamma, f \rangle\right] d\mu(\gamma). \tag{8}$$

The Hilbert space $\mathcal{H}$ is realized as the space $L^2_\mu(\mathcal{S}')$, of complex valued functions on $\mathcal{S}'$ square integrable with respect to μ; and $U(f)$ is the operator of multiplication by $\exp[i\langle \gamma, f \rangle]$. Since the $U(f)$ generate a complete set of commuting physical observables, it is natural to consider the subset Δ of $\mathcal{S}'$ on which μ is concentrated to be the quantum configuration space. More generally, in the non-cyclic case, one may have $\mathcal{H} = L^2_\mu(\Delta, \mathcal{W})$, the Hilbert space of square-integrable, multi-component functions

$\Psi(\gamma)$ taking values in an inner product space $\mathcal{W}$. Here $\mathcal{W}$ may describe internal degrees of freedom.

Now the semidirect product law provides us with an action $f \to f \circ \phi$ of the group $Diff(\mathbf{R}^s)$ on $\mathcal{S}$, and hence we also have a dual action $\gamma \to \phi\gamma$ on $\mathcal{S}'$ given by $\langle \phi\gamma, f \rangle = \langle \gamma, f \circ \phi \rangle$.

We would now like to represent $Diff(\mathbf{R}^s)$ by unitary operators in $\mathcal{H}$. For a unitary representation of the diffeomorphism group to act in $\mathcal{H}$ requires that μ have the key property of *quasi-invariance* for this action of diffeomorphisms. A measure μ on a measurable space Δ is called *invariant* for a group G acting on Δ, if for any measurable set $A \subset \Delta$ and $\phi \in G$, $\mu(\phi(A)) = \mu(A)$. The weaker condition of *quasi-invariance* of μ for G, states that if A has positive measure, its image $\phi(A)$ has positive measure for any $\phi \in G$, but in general $\mu(A)$ can be unequal to $\mu(\phi(A))$. Defining the transformed measure μ_ϕ by $\mu_\phi(A) = \mu(\phi(A))$, quasi-invariance is necessary and sufficient for the existence of the Radon-Nikodym (R-N) derivative $(d\mu_\phi/d\mu)(\gamma)$ for all group elements. A quasi-invariant measure μ on the configuration space Δ then defines a class of unitary representations V of $Diff(\mathbf{R}^s)$, given by

$$[V(\phi)\Psi](\gamma) = \chi_\phi(\gamma)\Psi(\phi\gamma)\sqrt{\frac{d\mu_\phi}{d\mu}(\gamma)}, \tag{9}$$

where $\chi_\phi : \mathcal{W} \to \mathcal{W}$ is a family of unitary operators in $\mathcal{W}$ satisfying the following *cocycle equation*–for all $\phi_1, \phi_2 \in G$,

$$\chi_{\phi_1}(\gamma)\chi_{\phi_2}(\phi_1\gamma) = \chi_{\phi_1\phi_2}(\gamma) \tag{10}$$

almost everywhere in Δ. The choice $\chi_\phi(\gamma) \equiv 1$ is always permitted, while alternate choices of χ_ϕ (noncohomologous cocycles) are associated with nontrivial phase effects and particle statistics. Note that it is the quasi-invariance of μ that allows the square root of the R-N derivative to occur as a factor in (9), and this is exactly what is needed to make the representation unitary!

Let us illustrate this framework with the N-particle representations we have already written down. From the first equation of (7), it is apparent that the configuration space $\Delta^{(N)} \subset \mathcal{S}'$ can be identified with the space of sums of N distinct evaluation functionals (Dirac δ-functions):

$$\Delta^{(N)} = \{\gamma \in \mathcal{S}' \,|\, \gamma = \sum_{j=1}^{N} \delta_{\mathbf{x}_j}, \, \mathbf{x}_j \neq \mathbf{x}_k \text{ for } j \neq k\}, \tag{11}$$

where $\langle \delta_{\mathbf{x}}, f \rangle = f(\mathbf{x})$. A configuration (describing N identical particles) is thus an unordered, N-point subset of $\mathbf{R}^s$. The action of a diffeomorphism ϕ on γ is to move all the points in the configuration:

$$\phi\gamma = \sum_{j=1}^{N} \delta_{\phi(\mathbf{x}_j)}. \tag{12}$$

The measure μ, concentrated on $\Delta^{(N)}$, is (locally) a product of Lebesgue measures, which is quasi-invariant for diffeomorphisms, and the R-N derivative is the product of the Jacobians occurring in (7). The wave function symmetry is now encoded in the cocycle: for bosons, $\chi_\phi(\gamma) \equiv 1$, while for fermions, χ_ϕ is nontrivial. Anyons are possible because additional, inequivalent cocycles are possible when $s = 2$.

The quasi-invariance for diffeomorphisms of Lebesgue measure in $\mathbf{R}^s$ is just the statement that the Jacobian of the diffeomorphism is positive and finite. We see in the preceding examples how the finite-dimensionality of configuration space ensures the possibility of quasi-invariant measures that are (locally) products of Lebesgue measures. Yet the existence of quasi-invariant measures is not a minor, technical point; rather, it is essential for the unitarity of the representation. Indeed, it is the measure that contains the information about how the outcomes of physical measurements distribute in quantum mechanics.

The construction of quasi-invariant measures when the configuration spaces are infinite dimensional is an in-general unsolved problem; yet they are needed to obtain the quantum theory for spatially-extended physical systems. For example, quantized vortex configurations in ideal, incompressible fluids have been obtained from representations of groups of (area- and volume-preserving) diffeomorphisms of $\mathbf{R}^2$ and $\mathbf{R}^3$, leading to unexpected physical conclusions. For planar fluids, it turns out that pure point vortices are not permitted quantum-mechanically, but one-dimensional *filaments* of vorticity are allowed; similarly, in $\mathbf{R}^3$ pure filaments are kinematically forbidden, while two-dimensional vortex *surfaces*, e.g. ribbons and tubes, can occur [12]. However, a major gap remains–the construction of *unitary* group representations that actually describe the quantum mechanics of vortex systems hinges on finding measures, quasi-invariant for diffeomorphisms, on configuration-spaces of filaments or tubes. Naturally, a nonrelativistic quantum theory of strings also depends on quasi-invariant measures.

The construction of diffeomorphism-invariant measure is also one key point in the long-standing, major problem of finding a consistent theory for quantized gravity [13]. Recent advances in this direction have been made by Ashtekar and Lewandowski [14], who construct a faithful, diffeomorphism-invariant measure on a compactification of the space of gauge-equivalent connections. This is, however, a rather large space for our purposes. We therefore set out to explore a class of quantum-mechanical models in one space dimension, having a countable infinity of degrees of freedom.

We consider the action of $Diff(\mathbf{R})$ on a space Δ of (quantum) configurations γ, which are *infinite sequences* of points drawn from $\mathbf{R}$. In the physical models we have in mind, the points in a sequence describe a particular configuration of a gaseous system in one space dimension, by identifying the positions of the (now distinguishable) particles. For a configuration $\gamma = (x_i) \in \Delta$ the configuration $\phi\gamma = (y_i)$ is just the sequence $y_i = \phi(x_i)$, so that $(\phi_2 \circ \phi_1)\gamma = \phi_2(\phi_1\gamma)$, and the map $(\phi, \gamma) \to \phi\gamma$ is suitably smooth. For measures on Δ as they are usually constructed in the case of sequence spaces, the R-N derivative (if it exists) will assume the form of an infinite product:

$$\frac{d\mu_\phi}{d\mu}(\gamma) = \prod_{j=1}^{\infty} u_j(\gamma) \, . \tag{13}$$

Then the issue of quasi-invariance becomes that of the convergence of (13) to a nonzero, non-infinite limit.

We use a technique based on self-similar random processes to construct nontrivial families of quasi-invariant measures, thus obtaining unitary representations of the group $Diff(\mathbf{R})$ describing continuum quantum systems with infinitely many degrees of freedom. In our picture, a parameterized family of such representations describes a quantum gas of point particles undergoing a phase transition, from rarefied to condensed, at the critical value of a correlation parameter. The method offers promise for generalization to higher spatial dimensions, and we believe it is a step toward the quantum theory of more general extended objects.

The most elementary example, developed in the next section, describes the quantum mechanics of a strictly confined gas of particles. We shall also see how self-similarity enters into the description of the infinite free Bose gas at zero temperature.

MEASURES FROM SELF-SIMILAR RANDOM PROCESSES

To see concretely how the previous ideas apply, we start by reconsidering the simple but important example of the Poisson measure [10].

Suppose N identical particles are uniformly distributed on a segment of length L, which we take for instance as the interval $(-L/2, L/2) \subset \mathbf{R}$. The probability of finding exactly n particles (with $n \leq N$) in a measurable subset $A \subset (-L/2, L/2)$ whose Lebesgue measure is $m(A)$, is given by

$$p_{N,L}(n; A) = \frac{N! \, [m(A)]^n \, (L - m(A))^{N-n}}{n! \, (N - n)! \, L^N}. \tag{14}$$

The infinite-volume limit is then obtained by fixing the average density: we take the limit of the previous expression for $N, L \to \infty$ in such a way that $\lim(N/L) = \bar{\rho}$. It follows that

$$p(n; A) = \frac{\bar{\rho}^n \, [m(A)]^n}{n!} \exp[-\bar{\rho} m(A)]. \tag{15}$$

The probability of finding exactly n particles in the region A is thus given by the Poisson distribution with parameter $\bar{\rho} m(A)$. There exists a unique Borel measure μ having this property on the space $\Delta^{(u)}$ of unordered configurations in $\mathbf{R}$ (a configuration is given here by an unordered sequence of points drawn from $\mathbf{R}$). In fact, the whole construction ensures that the measure μ is concentrated on $\Delta_{\bar{\rho}}^{(u)}$, the space of (unordered) *locally finite configurations* in $\mathbf{R}$ having fixed average density $\bar{\rho}$. This measure is called the Poisson measure with parameter $\bar{\rho}$ and it is quasi-invariant under diffeomorphisms. Quasi-invariance of μ is a consequence of the behavior of ϕ at infinity, together with the finiteness of the particle density; this implies that (with probability one) $u_j = \mathcal{J}_\phi(x_j) \to 1$ rapidly as $j \to \infty$. This class of representations of $\mathit{Diff}(\mathbf{R})$ has been shown to describe the quantum theory of a nonrelativistic free Bose gas at zero temperature [10].

We would now like to consider measures that permit points in a configuration to accumulate in a bounded region. This means that we look for quasi-invariant measures on a configuration space Δ such that configurations with clusters of particles occur with nonzero probability. Existence of the R-N derivative is then a much more delicate question; the simple argument we invoked in the case of the free Bose gas will certainly not apply. In general, if the positions of the particles distribute non-identically but independently, in such a way that configurations accumulate with nonzero probability, the corresponding measure will not be quasi-invariant. Therefore we have to consider more general stochastic processes to construct quasi-invariant measures supported by configurations with cluster points, in which the positions of the particles are correlated.

The physical systems we study mimic an interacting gas of particles. However our approach differs from the usual Hamiltonian aproach to quantum statistical mechanics, in that we directly constructing the corresponding probability measure on the relevant configuration space.

From here on we consider the configuration space $\Delta = \mathbf{R}^\infty$. An element of Δ is thus an ordered set of points (i.e. a sequence) drawn from the real line $\mathbf{R}$. Elements of Δ will be denoted by the symbol $\gamma = (x_j), j = 1, 2, \ldots$, and we write $\gamma_j = p_j(\gamma) = x_j$. There is a natural action of $\mathit{Diff}(\mathbf{R})$ on $\Delta = \mathbf{R}^\infty$, namely

$$\gamma \to \phi\gamma, \qquad (\phi\gamma)_j = \phi(x_j). \tag{16}$$

We want to construct a measure μ on Δ, quasi-invariant under this action of $\mathit{Diff}(\mathbf{R})$.

A very general construction of probability measures on Δ proceeds as follows: for each natural number j write a conditional (Borel) probability measure μ_j on $\mathbf{R}$

depending measurably on $j - 1$ real parameters $x_1, \ldots, x_{j-1}$ (the positions of the first $j - 1$ particles) and absolutely continuous with respect to the Lebesgue measure on x_j:

$$d\mu_j(x_j) = f(x_j | x_1, \ldots, x_{j-1}) dx_j$$

The joint probability measure for the positions of the first j particles is then given by

$$\mu^{(j)} = \prod_{k=1}^{j} \mu_k. \tag{17}$$

$\{\mu^{(j)}, j \in \mathbf{N}\}$ is a compatible family of probability measures. Indeed, since every $\mu^{(j)}$ is absolutely continuous with respect to the j-dimensional Lebesgue measure, and $\mu^{(j)}(\mathbf{R}^j) = 1$ for all j, there is no problem in interchanging the order of integration in the iterated integrals, and this assures compatibility. Thus, by Kolmogorov's theorem, there is a unique measure μ on the infinite-dimensional configuration space Δ associated with the sequence of measures $\{\mu^{(j)}\}$. For the finite dimensional measure $\mu^{(j)}$ we obtain that the R-N derivative is given by

$$\frac{d\mu_\phi^{(j)}}{d\mu^{(j)}} = \prod_{k=1}^{j} \frac{d\mu_{k,\phi}}{d\mu_k} \tag{18}$$

with

$$\frac{d\mu_{k,\phi}}{d\mu_k} = u_{k,\phi}(\gamma) = \frac{f(\phi(x_k)|\phi(x_1), \ldots, \phi(x_{k-1}))}{f(x_k | x_1, \ldots, x_{k-1})} \mathcal{J}_\phi(x_k). \tag{19}$$

The quasi-invariance of the measure $\mu^{(j)}$ is assured if Radon-Nikodym derivative (18) is almost everywhere positive and finite. As we have anticipated, in the infinite-dimensional case quasi-invariance of the measure μ depends on the behavior of the infinite product

$$\frac{d\mu_\phi}{d\mu} = \prod_{k=1}^{\infty} u_{k,\phi}(\gamma). \tag{20}$$

Of course not every measure constructed like this will be quasi-invariant and some further input is needed to produce a measure which is. The main idea which gives a whole class of quasi-invariant measures consists in letting the probability distribution of the jth particle's position *scale* according to the *outcomes* for the particles previously chosen, establishing a self-similar random process.

Let us explore a first example where the particles accumulate with probability one, and in which the idea of scaling become more precise. We choose the positions (x_1, x_2) of the first pair of particles from a nonvanishing probability density $f_1(x) = f_2(x)$. The positions of the second pair of particle is chosen from the *uniform* density on the interval $[x_1, x_2]$; i.e.,

$$f_3(x_3 | x_1, x_2) = f_4(x_4 | x_1, x_2) = \frac{\chi_{[x_1, x_2]}(x)}{|x_2 - x_1|}, \tag{21}$$

where $\chi_{[a,b]}$ denotes the characteristic function (indicator function) of the interval $[a, b]$. Iterating this process, we choose (x_{2m+1}, x_{2m+2}) from the uniform density on $[x_{2m-1}, x_{2m}]$ (see Fig. 1); we have in this way constructed a Markov process.

It is easy to show that the measure μ is concentrated on configurations having one cluster point. Furthermore, the explicit expression for $u_{k,\phi}(\gamma)$ is in this case very simple. Since sequences in Δ are converging with probability one, we have that

$$u_{2m+1,\phi}(\gamma) = \frac{|x_{2m} - x_{2m-1}|}{|\phi(x_{2m}) - \phi(x_{2m-1})|} \mathcal{J}_\phi(x_{2m+1}) \to 1 \tag{22}$$

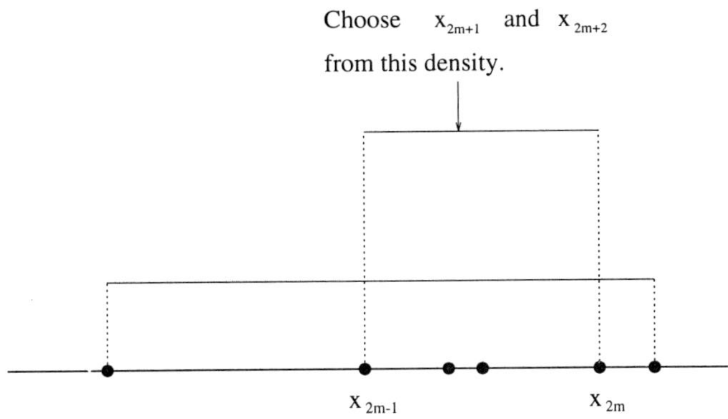

Figure 1. Self-similar random sequence based on uniform distributions.

with probability one, and similarly for $u_{2m,\phi}$. In the structure of the expression (22) for $u_{k,\phi}$ we can see clearly how the conditional probability enters: the first factor approaches the reciprocal of the Jacobian as the x_k approach their limit. The way in which the width of the distribution for each pair of points is determined directly by the outcome of choosing the previous pair builds a kind of scale invariance or self-similarity into the configurations which are selected by the measure μ, which is *just what is needed* for quasi-invariance. Condition (22) is necessary for the convergence of the infinite product (20), but it is not sufficient. The quasi-invariance of μ follows because our construction also insures that the *rate* of convergence is sufficiently rapid; indeed, we have

$$\sum_{j=1}^{\infty} |x_{j+1} - x_j| < \infty \tag{23}$$

with probability one.

The quasi-invariance of μ means that a unitary representation of the diffeomorphism group, and thus a consistent quantum mechanics, exists for these configurations! Physically, this elementary model may be interpreted as describing a strictly confined "cluster" formed from infinitely many particles.

It is interesting to note that the infinite free Bose gas in one dimension, described by Poisson measure with parameter $\bar{\rho}$, can be obtained straightforwardly in this framework as a kind of "reciprocal" of the above Markov process. This time one chooses the positions of successive particles to be *outside* the intervals established by preceding choices, with densities f_j decaying exponentially in both directions, keeping $\bar{\rho}$ independent of j in the exponential distributions (see Fig. 2).

The previous example based on uniform distributions depends critically on the fact that orientation-preserving diffeomorphisms of **R** respect "betweenness"–so that regions of positive measure (with the points x_{2m+1} and x_{2m+2} falling as they must between x_{2m-1} and x_{2m}) cannot be mapped by $\phi \in G$ into regions of zero measure (with x_{2m+1} or x_{2m+2} outside the interval). This feature is not shared by diffeomorphism groups of higher dimensional manifolds and one can see that attempts to generalize the construction based on uniform distributions to higher-dimensional manifolds fail. One way to overcome these difficulties relies one the use of successive probability densities that are *nowhere* vanishing.

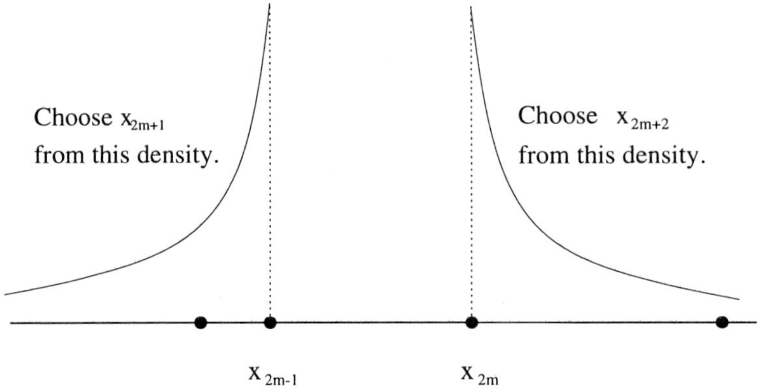

Figure 2. Self-similar random sequence based on exponential distributions.

MODELS WITH PHASE TRANSITION

It might be thought that one could proceed by choosing a point x_0 from a nonvanishing density, fixing a standard deviation σ_0, and then choosing x_j from a normal distribution with mean x_0 and standard deviation $\sigma_j = 2^{-j}\sigma_0$. Such a process describes points on a Brownian path, obtained from a Wiener measure. However, though x_j converges (with probability one) to x_0, the measure is *not* quasi-invariant for diffeomorphisms. The convergence of x_j and the condition on σ_j (independent of the outcomes $x_1, \ldots, x_{j-1}$) are not sufficient to control the behavior either of the ratio $[f(\phi(x_j)|\phi(x_1), \ldots, \phi(x_{j-1}))/f(x_j|x_1, \ldots, x_{j-1})]$ or of the Jacobian $\mathcal{J}_\phi(x_{j+1})$. Indeed, Brownian paths in $\mathbf{R}^n$ have, with probability one, fixed second variation according to the covariance matrix of the Brownian motion; while diffeomorphisms of $\mathbf{R}^n$ act so as to change the covariance matrix. Thus we cannot use Wiener measure as a means of achieving our goal.

Instead, we again use *self-similar* random process to construct measures out of nowhere vanishing densities. To illustrate the role played by self-similarity consider the following model: choose the zero-th point x_0 from a nonvanishing density f_0, and the point x_1 from a nonvanishing density f_1. Having chosen the points $x_0, \ldots, x_m$, choose x_{m+1} from a normal distribution; let the mean for this normal be x_0, and let the standard deviation be $\sigma_m = \kappa |x_m - x_0|$ (see Fig. 3). Here $\kappa > 0$ is a *correlation parameter* independent of m, and small values of κ correspond to more tightly bound systems. Thus

$$f_{m+1}^\kappa(x_{m+1}|x_m, x_0) = \frac{(2\pi)^{-\frac{1}{2}}}{\kappa|x_m - x_0|} \, \exp\left[-\frac{1}{2\kappa^2}\left(\frac{x_{m+1} - x_0}{x_m - x_0}\right)^2\right]. \tag{24}$$

We can now show that for small values of κ the sequence (x_j) converges to x_0 (with probability one), and that $u_j \to 1$ sufficiently rapidly to ensure convergence of the infinite product, where the terms u_j in (20) have been defined from (24). More precisely, we can prove there exists κ_0 such that:

(1) if $\kappa < \kappa_0$, sequences converge with probability one;

(2) if $\kappa > \kappa_0$, sequences diverge with probability one;

(3) the associated measures on Δ are quasi-invariant for diffeomorphism.

Let us prove here the first two of these facts. The situation is less intuitive than the example based on uniform densities. In that case, once the positions for the $(2m-1)$-th and the $(2m)$-th particles are chosen, the remaining particles are necessarily located in

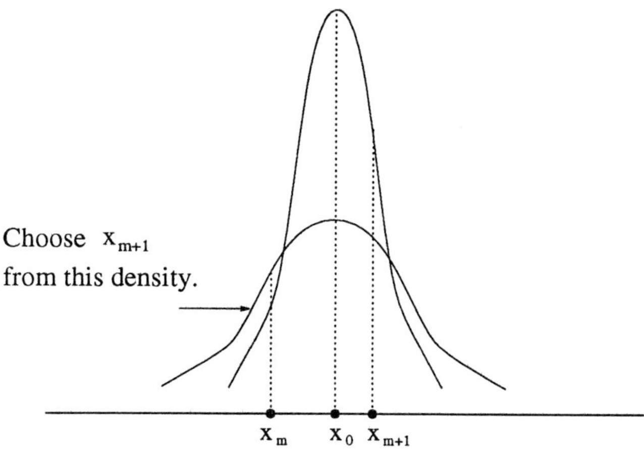

Figure 3. Self-similar random sequence based on Gaussian distributions.

the interval $[x_{2m-1}, x_{2m}]$, and therefore particles necessarily accumulate somewhere in the interval. This argument fails in the present case: after the position for the m-th particle is chosen, the $(m+1)$-th particle can be very far away with nonzero probability. Here the self-similarity of the random process comes in. The probability distribution for $y_{m+1} = (x_{m+1} - x_0)/|x_m - x_0|$, i.e. the distribution of $(x_{m+1} - x_0)$ in standard deviation units, is a *fixed normal distribution* (depending only on κ), independent of m. Consider the random variable $\log |x_m - x_0|$. Since the distributions of the y_m do not depend on m, we can identify $\log |x_m - x_0|$ with a random walk on the real line. The step of this random walk is $\xi = \log |y_{m+1}| = \log |x_{m+1} - x_0| - \log |x_m - x_0|$, and the probability density for ξ is given by

$$f(\xi) = \frac{2(2\pi)^{-\frac{1}{2}}}{\kappa} \exp \xi \exp \left(-\frac{1}{2\kappa^2} \exp 2\xi \right). \tag{25}$$

A standard application of the central limit theorem [15] then implies that if $E\xi = \int \xi f(\xi) < 0$, the random walk $(\log |x_m - x_0|)$ drifts to $-\infty$ (with probability one), which is equivalent to the result that (x_m) converges (with probability one). On the other side, if $E\xi > 0$, then $(\log |x_m - x_0|)$ drifts to $+\infty$ and $|x_m| \to \infty$ (with probability one). The critical value of κ occurs when

$$\int_{-\infty}^{\infty} \xi \exp \xi \exp[-\frac{1}{2\kappa^2} \exp(2\xi)] \, d\xi = \frac{\kappa\sqrt{2\pi}}{4} \left(\log \frac{\kappa^2}{2} - \gamma \right) = 0, \tag{26}$$

where $\gamma = 0.5772\ldots$ is the Euler-Mascheroni constant. It follows that $\kappa_0^2/2 = \exp \gamma$. The proof of the actual quasi-invariance of the measures associated with this family of Markov processes, necessary for the existence of the corresponding quantum models, is a little more complicated; it relies on showing a condition similar to (23). We note that nothing we have done requires the use of normal distributions. All that is really necessary is the scaling property, and the applicability of the central limit theorem.

In the above we have treated the particles as at the outset distinguishable, in that measures are constructed on *ordered* sequences (x_j). However, the physics does not depend on the labeling of points associated with their positions in the sequences. In one space dimension it is straightforward to "sum over permutations" by reconstructing the measures in terms of a physical labeling, in which for any configuration, points are indexed according to the positions they actually assume on the real line.

DISCUSSION AND OUTLOOK

Using a technique based on self-similar random processes, we have constructed unitary representations of the group of diffeomorphisms of R describing quantum systems with a countable infinity of degrees of freedom. This is also a step toward the quantum theory for more general extended objects. The most elementary example we have found describes the quantum mechanics of a strictly confined gas of particles. Generalizing this, we have realized parameterized families of models, where the parameter relates to correlations. The quantum systems that result from the corresponding unitary representations of $Diff(R)$ undergo a *phase change*, from a rarefied phase of locally finite configurations with zero average density of particles, to a condensed phase of configurations that have a cluster point. Classical (non-quantized) statistical-mechanics models with phase changes in one dimension include the Kac-Baker model of hard spheres on a line, interacting by means of an exponentially decaying potential, and a family of Ising models with long-range interactions described by Dyson [16]. At the quantum level, it is not necessary to have long-range interactions for phase changes to occur, as showed by Lieb, Schultz, and Mattis [17]. The latter models are set up on a lattice, as is usual in quantum statistical mechanics; in contrast, our method provides fully-quantized infinite point-particle models in the continuum, by directly constructing satisfactory probability measures on an infinite-dimensional quantum configuration space. However, it is not immediately apparent how to derive these systems from an assumed interaction, though our construction provides a good deal of intuition.

Thus for a whole class of models there is a critical value of κ. For sufficiently large κ the sequence (x_j) diverges, with probability one, and has zero average density (rarefied phase); and for sufficiently small κ it converges, with probability one, and the particles accumulate (condensed phase). We believe that other nontrivial phenomena in quantum statistical mechanics can also be modeled by continuous unitary representations of the group G.

Finally we conjecture that a procedure similar to that given by (24) will work in s space dimensions, $s > 1$, to give measures quasi-invariant for $Diff(R^s)$. Here it will be necessary to choose successive points x_{m+1} based on the outcomes for the preceding s choices $(x_{m-s}, \ldots, x_m)$. These outcomes will be used to define the covariance matrix of a multivariate normal distribution, from which x_{m+1} is selected.

ACKNOWLEDGMENTS

The authors thank Jean-Pierre Antoine, Lincoln Chayes, Carol J. Feltz, Israel M. Gel'fand, and David H. Sharp for useful discussions. We also wish to acknowledge hospitality from the Theoretical Physics Department of the Catholic University of Louvain-la-Neuve, Belgium, where we began this line of work in June 1992. The Arnold Sommerfeld Institute for Mathematical Physics at the Technical University of Clausthal, Germany, and the Laboratoire de Physique Théorique et Mathématique, Université Paris 7, have generously supported our continued collaboration.

REFERENCES

1. G. A. Goldin and U. Moschella, Saclay preprint T94/029 (1994).

2. G. A. Goldin, "Current Algebras as Unitary Representations of Groups", Ph. D. thesis, Princeton University (1969). G. A. Goldin and D. H. Sharp, in *1969 Battelle*

Rencontres: Group Representations, Lecture Notes in Physics **6**, edited by V. Bargmann (Springer, Berlin, 1970), p. 300; G. A. Goldin, J. Math. Phys. **12**, 462 (1971).

3. For more detailed reviews, see G. A, Goldin and D. H. Sharp, "Diffeomorphism Groups and Local Symmetries: Some Applications in Quantum Physics", in *Symmetries in Science III*, edited by B. Gruber and F. Iachello (New York: Plenum, 1989), p. 181; G. A. Goldin, "Predicting Anyons: The Origins of Fractional Statistics in Two-Dimensional Space", in *Symmetries in Science V*, edited by B. Gruber, L. C. Biedenharn, and H.-D. Doebner (New York: Plenum, 1991), p. 259; G. A. Goldin and D. H. Sharp, Int. J. Mod. Phys. **B5**, 2625 (1991); and G. A. Goldin, Int. J. Mod. Phys. **B6**, 1905 (1992).

4. R. Dashen and D. H. Sharp, Phys. Rev. **165** (1968), 1867.

5. H.-D. Doebner and J. Tolar, in *Symposium on Symmetries in Science, Carbondale, Illinois 1979*, edited by B. Gruber and R. S. Millman (Plenum, New York, 1980), p. 475; B. Angermann, H.-D. Doebner, and J. Tolar, in *Nonlinear Partial Differential Operators and Quantization Procedures*, Lecture Notes in Mathematics **1037**, edited by S. I. Andersson and H.-D. Doebner (Springer, Berlin, 1983), p. 171; H.-D. Doebner, H. J. Elmers, and W. Heidenreich, *J. Math. Phys.* **30** (1989), 1053.

6. G. A. Goldin, R. Menikoff, and D. H. Sharp, J. Math. Phys. **21**, 650 (1980); J. Math. Phys. **22**, 1664 (1981); Phys. Rev. Lett. **51**, 2246 (1983); G. A. Goldin and D. H. Sharp, Phys. Rev. **D28**, 830 (1983).

7. J. M. Leinaas and J. Myrheim, Nuovo Cimento **37B**, 1 (1977); J. M. Leinaas, Nuovo Cimento **4A**, 19 (1978); Fort. Phys. **28**, 579 (1980).

8. F. Wilczek, Phys. Rev. Lett. **48**, 1144 (1982); Phys. Rev. Lett. **49**, 957 (1982).

9. H.-D. Doebner and G. A. Goldin, Phys. Letts. **A162**, 397 (1992); G. A. Goldin, Int. J. Mod. Phys. **B6**, 1905 (1992); H.-D. Doebner and G. A. Goldin, J. Phys. A: Math. Gen. **27**, 1771 (1994).

10. G. A. Goldin, J. Grodnik, R. T. Powers, and D.H. Sharp, J. Math. Phys. **15**, 88 (1974); R. Menikoff, *J. Math. Phys.* **15** (1974), 1138 and 1394; A. M. Vershik, I. M. Gelfand, and M. I. Graev, Dokl. Akad. Nauk. SSSR **232**, 745 (1977).

11. I. M. Gelfand and N. Ya Vilenkin, *Generalized Functions, Vol. 4* (New York: Academic Press, 1964).

12. M. Rasetti and T. Regge, Physica **80A**, 217 (1975); J. Marsden and A. Weinstein, Physica **7D**, 305 (1983); G. A. Goldin, R. Menikoff, and D. H. Sharp, Phys. Rev. Lett. **58**, 2162 (1987); Phys. Rev. Lett. **67**, 3499 (1991). For recent reviews of related results see G. A. Goldin and D. H. Sharp, Int. J. Mod. Phys. **B5**, 2625 (1991), and G. A. Goldin, Int. J. Mod. Phys. **B6**, 1905 (1992).

13. S. W. Hawking, "The Path-Integral Approach to Quantum Gravity", in *General Relativity: An Einstein Centenary Survey*, edited by S. W. Hawking and W. Israel, Cambridge University Press (1979).

14. A. Ashtekar and J. Lewandowski, "Representation Theory of Analytic Holonomy C^*-algebras", in *Knots and Quantum Gravity*, edited by J. Baez, Oxford University Press (1994, in press); A. Ashtekar, D. Marolf, and J. Mourao, "Integration on the Space of Connections Modulo Gauge Transformations" (March 1994 preprint).

15. W. Feller *An Introduction to Probability Theory and Its Applications*, Vol. II Second Edition, Wiley, New York (1971).

16. G. Baker, Phys. Rev. **126**, 2071 (1962); M. Kac, G. Uhlenbeck, and P. Hemmer, J. Math. Phys. **4**, 216 (1963); F. Dyson, Commun. Math. Phys. **12**, 1961 (1969).

17. E. Lieb, T. Schultz, and D. Mattis, Ann. Phys. **16**, 407 (1961), Rev. Mod. Phys. **36**, 856 (1964). See also J. M. Luck, J. Stat. Phys. **72**, 417 (1993) and references therein.

Algebraic Shells and the Interacting Boson Model of the Nucleus

Bruno Gruber

Southern Illinois University at Carbondale in Niigata
Nakajo, Niigata 959-26, Japan,
and
Arnold Sommerfeld Institute for Mathematical Physics
Technical University of Clausthal, D-38678 Clausthal, Germany

Introduction

It is well known that, for the simplest possible case of a shell (spin $s=1/2$, orbital angular momentum $l=0$), the four shell states transform according to the smallest spin representation $(1/2,1/2)$ of the algebra $\mathbf{so(5)}$. In this article it is shown that, within the (space of the) Clifford algebra associated with the fermion operators, an algebraic shell can be found which is a generalization of the standard shell. The algebraic shell states go over into the standard shell states on the quotient space of the Clifford algebra with respect to a left ideal generated by the identity operator $\mathbf{1}$ (which corresponds to the standard vacuum state).

The algebraic shell transforms according to the defining representation [1000] of $\mathbf{su(4)} \sim \mathbf{so(6)}$, and the dual shell (the fermion annihilation operator states) like the complex conjugate representation [1110] of $\mathbf{su(4)}$. The algebraic vacuum state is an algebraic realization of the abstact vacuum in terms of creation-annihilation operator pairs and goes over, on the quotient space, into the standard vacuum state.

An operator valued representation theory is introduced via the algebraic representations [1000] and [1110] of $\mathbf{su(4)}$. Introducing two species of fermions (the **ordered** tensor product of two algebraic reresentations) the selfconjugate representation [1100] of $\mathbf{su(4)}$ is constructed, from two species of fermion creation operators as well as from two species of fermion annihilation operators. The 36 product operator states $b_i{}^+(F^+xF^+)$ and $b_j(FxF)$, i, j = 1,...6, of the two representations [1100] are shown to satisfy the $\mathbf{su(6)}$ commutation relations. Moreover it is shown that the $b_i{}^+$ transform according to the representation [100000] of $\mathbf{su(6)}$, while the b_j transform like the representation [111110] of $\mathbf{su(6)}$. This then implies that bosonic states have been obtained for the Interacting Boson Model of the Nucleus which have an internal fermionic structure. While the results obtained by the Interacting Boson Model of the Nucleus remain unchanged, the internal fermionic structure can be used to calculate additional properties of the model as a consequence of its microscopic structure.

In section 1 a brief review is given for the **simplest possible case** of a (atomic, nuclear) shell for spin 1/2 particles with orbital angular momentum equal to zero. This brief review serves the purpose to illustrate, and to show the contrast with, the new properties of the **algebraic shells** which will be introduced, and discussed, in sections 2 and 3.

In section 2 properties of the associated Clifford algebra C_4 of dimension $d=2^4$, and some of its ideals, are discussed. The ordered tensor product of (left and right) ideals is introduced, and their relationship to the shell S and the dual shell DS is discussed. The ordered tensor product implies two "species" of fermions, distinguished by their ordering. Identification of the two species corresponds to a map from the tensor product into the Clifford algebra C_4. The Clifford algebra C_4 is closed with respect to an associative multiplication law for the composition of two fermion operators, both with respect to the non-associative anticommutator and with respect to the non-associative commutator. With repect to the commutator the Clifford algebra C_4 becomes the Lie algebra $u(4)$. This relationship holds for arbitrary positive integer n, $C_{2n} \sim u(2^n)$, $d=2^{2n}$ (the equivalence sign indicates identical spaces on which both, the anticommutator and the commutator, are defined).

In section 3 the algebraic shells F^+ and F are introduced for fermions and antifermions. The vacuum state $|0\rangle$ of standard shell theory is replaced by an algebraic expression which represents a realization of the standard vacuum state on the space of the Clifford algebra (**CA**). It is shown that there exists a complete correspondence to standard shell theory: that is, on the quotient space $Q_L = CA / I_L$, I_L a left ideal of the **CA** generated by the relations $f_i|0\rangle=0$, $|0\rangle$ the vacuum state, the familiar results are reproduced.

However, in distinction to standard shell theory the algebraic shells F^+ and F transform like the two representations (1/4)(3-1-1-1) and (1/4)(111-3) **su(4)** ([1000] and [1110] of **u(4)** .In this article **u(4)** notation will generally be used for the **su(4)** representations). With respect to the subalgebra **so(5)** these two **su(4)**representations go over into the familiar representation (1/2,1/2) of the standard theory. In this section it is also shown that the Lie algebra **u(4)** can be defined on the space spanned by the ordered tensor product $F^+ \times F$. Finally, the map $F^+ \times F \rightarrow F^+ \cdot F$ from the ordered tensor product (two particle species) onto the **CA** (one particle species) is introduced and it is shown that this map yields again the Lie algebra **u(4)** by making use of the **su(4)** transformation properties of the fermion-antifermion operator states.

In section 4 a brief outline is given for a representation theory built upon ordered direct products of the algebraic shells F^+ and F . Operator states $b_i^+(F^+ \times F^+)$ and $b_j(F \times F)$, i, j = 1,2,...,6, are obtained, constructed from two species of fermions $F^+ \times F^+$ and two species of antifermions $F \times F$ respectively, which transform like the six dimensional representation [1100] of **su(4)**.

In section 5 it is shown that the set of operators $\left(b_i^+(F^+ \times F^+) \cdot b_j(F \times F) \right.$, i,j=1,2,...,6$\left. \right)$, with the operators b_i^+ and b_i constructed from the direct product of fermion operator creation pairs and from annihilation pairs respectively, satisfies the Lie product for the algebra **su(6)**. Moreover, it will be shown that the operator states $b_i^+(F^+ \times F^+)$ and $b_j(F \times F)$ transform like the representations [100000] and [111110] of **su(6)**. The boson operators obtained in this manner have an internal fermionic structure, in distinction to the "bare" bosons of the standard theory. They satisfy a generalized boson operator commutation relation, which goes over, on the quotient space

Q_L (i.e. upon projection onto the standard shell), into the familiar boson operator commutation relations.

Thus a picture has been obtained in which the boson operators have an internal fermionic structure, while all the standard results are maintained. This picture may provide a microscopic structure for the Interacting Boson Model of the Nucleus.

1. Standard Shell Theory

In this section a brief review is given of the simplest possible case of a shell, namely for a spin 1/2 fermion with no angular momentum. This brief review will help to illustrate the new features of the algebraic shell theory which will be introduced later on. For an excellent review of the standard (atomic) shell model, and its associated Lie algebras, see reference 1.

Consider the set

$$f^+ = \left(f_1^+, f_2^+, f_1, f_2 \right) \tag{1.1}$$

of fermion creation and annihilation operators for spin 1/2 particles, together with an associative multiplication law for the composition of the fermion operators. The (non-associative) anticommutator $\{ a , b \} \equiv a \cdot b + b \cdot a$, $a, b \in f^+$, is obtained by making use of the associative multiplpcation law $a \cdot b \equiv ab$ (the multiplication symbol $\cdot$ will be deleted whenever the context is clear). The relations

$$\{ f_i , f_j^+ \} = \delta_{ij} \mathbf{1} \quad , \qquad \{ f_i , f_j \} = 0, \qquad i,j = 1,2 \tag{1.2}$$

are the defining relations for the Clifford algebra C_4 (**CA**). The index i=1 denotes the third component of spin $m_s = 1/2$ and i=2 denotes $m_s = -1/2$. The **CA** is generated by the set f^+ and the identity $\mathbf{1}$, using the associative multiplication law and the defining relations eq(1.2),

$$\begin{aligned}
\mathbf{CA} = \quad (\,\mathbf{1}, \\
f_1^+, f_2^+, f_1, f_2, f_1^+ f_2, f_2^+ f_1, f_1^+ f_2^+, f_1 f_2, f_1^+ f_1, f_2^+ f_2, \\
f_1^+ f_1 f_2, f_2^+ f_2 f_1, f_1^+ f_1 f_2^+, f_1^+ f_2^+ f_2, f_1^+ f_1 f_2^+ f_2)
\end{aligned} \tag{1.3}$$

The **CA** is closed with respect to both, the associative multiplication law and the anticommutator.

Special elements of the **CA**, which will be used in later sections, are

$$p_i = 1 - f_i^+ f_i, \qquad p_i^2 = p_i, \qquad f_i \cdot p_i = 0, \qquad f_i^+ \cdot p_i = f_i^+ \tag{1.4}$$

$$e_i = 1 - 2 f_i^+ f_i, \qquad e_i^2 = 1, \qquad f_i^+ \cdot e_i = f_i^+, \qquad f_i \cdot e_i = -f_i, \tag{1.5}$$

In the standard shell model a vacuum state

$$|0\rangle \tag{1.6}$$

is introduced, with the property that this state is mapped to zero by the fermion annihilation operators,

$$f_i \ |0> = 0 \tag{1.7}$$

The shell S is generated by acting with the fermion creation operators upon this vacuum state, and S is spanned by the four states

$$S = (\ |0>,\ f_1^+ \ |0>,\ f_2^+ \ |0>,\ f_1^+ f_2^+ \ |0>\ ) \tag{1.8}$$

The dual shell DS is given by the dual states

$$DS = (\ <0|,\ <0|\ f_1,\ <0|\ f_2,\ <0|\ f_2\ f_1,\ ) \tag{1.9}$$

and the inner product for the four basis states satisfies the orthonormality relations

$$(\ |i>,|j>\ ) = <i||j> = \delta_{ij}\ ,\ i,j = 1,2,3,4,\qquad |j> \in S,\ \ <i| \in DS \tag{1.10}$$

The associative multiplication law of the CA can be used to to form commutators $1/2[a,b] \equiv (1/2)(ab - ba)$. With respect to the commutators the set f^+ generates the Lie algebra

$$so(5) = (\ H_1 = f_1^+ f_1 - 1/2,\ H_2 = f_2^+ f_2 - 1/2,\ f_1^+,\ f_2^+,\ f_1,\ f_2,$$
$$f_1^+ f_2^+,\ f_2 f_1,\ f_2^+ f_1,\ f_1^+ f_2\ ) \tag{1.11}$$

which contains the physically relevant spin S and quasispin Q subalgebras

$$su(2)_S = (\ S_0 = (1/2)(f_1^+ f_1 - f_2^+ f_2),\ S_+ = \tfrac{1}{\sqrt{2}} f_1^+ f_2,\ S_- = \tfrac{1}{\sqrt{2}} f_2^+ f_1\ ) \tag{1.12}$$

$$su(2)_Q = (\ Q_0 = (1/2)(f_1^+ f_1 + f_2^+ f_2 - 1),\ Q_+ = \tfrac{1}{\sqrt{2}} f_1^+ f_2^+,\ Q_- = \tfrac{1}{\sqrt{2}} f_2 f_1\ ) \tag{1.13}$$

$$su(2)_{S+Q} = (\ (S+Q)_0 = f_1^+ f_1 - 1/2,\ (S+Q)_+ = \tfrac{1}{\sqrt{2}}(f_2^+ + f_2)f_1^+,$$
$$(S+Q)_- = \tfrac{1}{\sqrt{2}}(f_2^+ + f_2)f_1\ ) \tag{1.14}$$

The operators Op of these algebras act upon the states of the shell S from the left, Op $|i>$. On the states of the dual shell DS the operators Op act from the right with a change of sign, $<i|\ (-Op)$. The action of the operators upon the states is the associative multiplication law of the CA. Both, the shell S as well as the dual shell DS carry the 4-dimensional spin representation with highest weight $(1/2, 1/2)$ of $so(5)$.

2. Algebraic Shell Model, Clifford Algebra C_4 and su(4)

The CA forms a 16-dimensional space, closed with respect to anticommutation relations. The anticommutation relations are based upon an associative multiplication law with respect to which the space **CA** is also closed as a consequence of the relations eq.(1.2). The relation $f_i \cdot \mathbf{1} = 0$ generates a left ideal I_L within the **CA** , eq(1.3). For the qotient space $\mathbf{Q}_L = \mathbf{CA} / I_L$ a basis can be chosen as

$$\mathbf{Q}_L = \left(1, f_1^+, f_2^+, f_1^+ f_2^+ \right), \qquad \text{with} \quad f_i^+ \cdot \mathbf{1} = f_i^+, \qquad (2.1)$$

The conjugate relation $\mathbf{1} \cdot f_1^+ = 0$ genetrates a right ideal I_R in **CA**. For the qotient space $\mathbf{Q}_R = \mathbf{CA} / I_R$, a basis can be chosen as

$$\mathbf{Q}_R = \left(1, f_1, f_2, f_2 f_1 \right), \qquad (2.2)$$

The quotient space $\mathbf{Q}_L$ corresponds to the shell **S**, and the identity operator **1** represents the physical vacuum state $|0\rangle$ on the quotient space $\mathbf{Q}_L$. The quotient space $\mathbf{Q}_R$ corresponds to the dual shell **DS**, and the identity operator **1** represents the physical vacuum state $\langle 0|$ in the quotient space $\mathbf{Q}_R$.

The ordered tensor product $\mathbf{Q}_L \times \mathbf{Q}_R$ spans a 16-dimensional space with basis

$$
\mathbf{Q}_L \times \mathbf{Q}_R = \begin{pmatrix}
1 \times 1, & 1 \times f_1, & 1 \times f_2, & 1 \times f_2 f_1, \\
f_1^+ \times 1, & f_1^+ \times f_1, & f_1^+ \times f_2, & f_1^+ \times f_2 f_1, \\
f_2^+ \times 1, & f_2^+ \times f_1, & f_2^+ \times f_2, & f_2^+ \times f_2 f_1, \\
f_1^+ f_2^+ \times 1, & f_1^+ f_2^+ \times f_1, & f_1^+ f_2^+ \times f_2, & f_1^+ f_2^+ \times f_2 f_1
\end{pmatrix}
$$

$$(2.3)$$

There is a one to one map from the space $\mathbf{Q}_L \times \mathbf{Q}_R$ of the tensor product onto the space **CA** of the Clifford algebra. This is seen by identifying the elements of $\mathbf{Q}_L$ and $\mathbf{Q}_R$ as elements of the Clifford algebra **CA** and using the associative multiplication law of the **CA**, i.e., by applying the map

$$\mathbf{Q}_L \times \mathbf{Q}_R \quad \rightarrow \quad \mathbf{Q}_L \cdot \mathbf{Q}_R = \mathbf{CA} \qquad (2.4)$$

Finally, since the Clifford algebra is necessarily closed with respect to its associative multiplication law it has also to form a Lie algebra with respect to commutation relations. It turns out that the space **CA** forms, with respect to commutation relations, the Lie algebra $u(4) \sim su(4) + u(1)$:

$$
\begin{aligned}
su(4) = \big(\quad & H_1 = f_1^+ f_1 f_2^+ f_2 - 1/4, \qquad H_2 = -f_1^+ f_1 f_2^+ f_2 + f_1^+ f_1 - 1/4, \\
& H_3 = -f_1^+ f_1 f_2^+ f_2 + f_2^+ f_2 - 1/4, \qquad H_4 = f_1^+ f_1 f_2^+ f_2 - f_1^+ f_1 - f_2^+ f_2 + 3/4, \\
& E(010\text{-}1) = f_1^+ p_2, \quad E(10\text{-}10) = f_1^+ f_2^+ f_2, \quad E(0\text{-}101) = f_1 p_2, \\
& E(\text{-}1010) = f_1 f_2^+ f_2, \quad E(001\text{-}1) = f_2^+ p_1, \quad E(1\text{-}100) = f_2^+ f_1^+ f_1, \\
& E(00\text{-}11) = f_2 p_1, \quad E(\text{-}1100) = f_2 f_1^+ f_1, \quad E(01\text{-}10) = f_1^+ f_2, \\
& E(100\text{-}1) = f_1^+ f_2^+, \quad E(0\text{-}110) = f_2^+ f_1, \quad E(\text{-}1001) = f_1 f_2 \quad \big)
\end{aligned}
$$

$$(2.5)$$

with $H_1+H_2+H_3+H_4=0$, and

$$\mathbf{so(6)} = (H_1{}'=f_1{}^+f_1-1/2=1/2-\mathbf{p}_1, \quad H_2{}'=f_2{}^+f_2-1/2=1/2-\mathbf{p}_2, \quad H_3{}'=(-/+)2H_1{}'H_2{}',$$
$$E(10\text{-}1)=f_1{}^+\mathbf{p}_2, \qquad E(101)=f_1{}^+f_2{}^+f_2, \qquad E(\text{-}101)=f_1\mathbf{p}_2, \qquad E(\text{-}10\text{-}1)=f_1f_2{}^+f_2,$$
$$E(01\text{-}1)=f_2{}^+\mathbf{p}_1, \qquad E(011)=f_2{}^+f_1{}^+f_1, \qquad E(0\text{-}11)=f_2\mathbf{p}_1, \qquad E(0\text{-}1\text{-}1)=f_2f_1{}^+f_1,$$
$$E(1\text{-}10)=f_1{}^+f_2, \qquad E(110)=f_1{}^+f_2{}^+, \qquad E(\text{-}110)=f_2{}^+f_1, \quad E(\text{-}1\text{-}10)=f_1f_2)$$

$$(2.6)$$

with

$$H_1{}'=H_1+H_2, \ H_2{}'=H_1+H_3, \ H_3{}'=H_1+H_4. \ \text{and} \ \mathbf{p}_i=1-f_i{}^+f_i, \quad \mathbf{p}_i{}^2=\mathbf{p}_i.= \qquad (2.7)$$

In this article the positive sign is chosen for $H_3{}'$. It is seen that for this particular realisation of $\mathbf{so(6)}$ the element $H_3{}'$ is a bilinear function of $H_1{}'$ and $H_2{}'$. The correspondence between the $\mathbf{so(6)}$ and $\mathbf{su(4)}$ roots is given by

$\mathbf{so(6)}$:	$(10\text{-}1)$	(101)	$(\text{-}101)$	$(\text{-}10\text{-}1)$
$\mathbf{su(4)}$:	$(010\text{-}1)$	$(10\text{-}10)$	$(0\text{-}101)$	$(\text{-}1010)$
	$(01\text{-}1)$	(011)	$(0\text{-}11)$	$(0\text{-}1\text{-}1)$
	$(001\text{-}1)$	$(1\text{-}100)$	$(00\text{-}11)$	$(\text{-}1100)$
	$(1\text{-}10)$	(110)	$(\text{-}110)$	$(\text{-}1\text{-}10)$
	$(01\text{-}10)$	$(100\text{-}1)$	$(0\text{-}110)$	$(\text{-}1001)$

$$(2.8)$$

3. The Algebraic Shells F^+ and F

In this section it will be shown that the states of the algebraic shell F^+ are given as, and transform like,

$F^+ =$	$(f_1{}^+f_2{}^+,$	$f_1{}^+\mathbf{p}_2,$	$f_2{}^+\mathbf{p}_1,$	$\mathbf{p}_1\mathbf{p}_2)$
	ϕ_1	ϕ_2	ϕ_3	ϕ_4
$\mathbf{u(4)}$:	$[1000]$	$[0100]$	$[0010]$	$[0001]$
$\mathbf{su(4)}$:	$(1/4)(3\text{-}1\text{-}1\text{-}1)$	$(1/4)(\text{-}13\text{-}1\text{-}1)$	$(1/4)(\text{-}1\text{-}13\text{-}1)$	$(1/4)(\text{-}1\text{-}1\text{-}13)$
$\mathbf{so(6)}$:	$(1/2)(111)$	$(1/2)(1\text{-}1\text{-}1)$	$(1/2)(\text{-}11\text{-}1)$	$(1/2)(\text{-}1\text{-}11)$
$\mathbf{so(5)}$:	$(1/2)(11)$	$(1/2)(1\text{-}1)$	$(1/2)(\text{-}11)$	$(1/2)(\text{-}1\text{-}1)$
$\mathbf{so(4)}$:	$(1/2)(11)$	$(1/2)(1\text{-}1)$	$(1/2)(\text{-}11)$	$(1/2)(\text{-}1\text{-}1)$

$$(3.1)$$

where the $[n_1,n_2,n_3,n_4]$, $n_1 \geq n_2 \geq n_3 \geq n_4$, $N=n_1+n_2+n_3+n_4$, n_i integers, denotes the partition numbers for the irreducible representations of $\mathbf{u(4)}$. (The ϕ_i represent wave functions which transform identically to the algebraic shell states. A computer code has been developed which calculates the wave functions with respect to any semisimple symmetry chain. Simple substitution yields the corresponding results for the algebraic states once their tranformation properties have been determined. See ref.2). It holds

$$f_i \cdot p_1 p_2 = 0,$$

$$(p_1 p_2)^2 = p_1 p_2, \qquad p_i^2 = p_i, \qquad f_i \cdot p_i = 0,$$

$$f_i^+ \cdot p_i = f_i^+.$$

$$(3.2)$$

The first line of eq. (3.2) shows that the algebraic shell state $p_1 p_2$ satisfies the properties which are required of a vacuum state, i. e. the operator-state $p_1 p_2$ is mapped to zero by the fermion annihilation operators. The (algebraic) shell F^+ , eq(3.1), is obtained through the action of the fermion creation operators upon the vacuum state $p_1 p_2$ in the same manner as it is the case for the standard shell states S. To the single fermion shell states $f_i^+ |0\rangle$ of the standard shell S correspond in the algebraic shell F^+ the fermion operators $f_j^+ p_i$, $j \neq i$, while the operator $f_1^+ f_2^+$ is identical in both models.

Thus, the state $p_1 p_2$ corresponds to the vacuum state $|0\rangle \sim 1$ of the standard shell model. The algebraic vacuum state $p_1 p_2$ has, however, an internal structure. It is made up of products of particle-antiparticle pairs. In fact, the second line of eq(3.2) shows that the vacuum state $p_1 p_2$ is actually the product of two separate vacua p_1 and p_2, one for each of the two fermions f_1^+ and f_2^+ (in fact, the fermion F^+ can be looked upon as the direct product $(p_1, f_1^+) \times (p_2, f_2^+)$ of two two-spinors in the indices 1 and 2 respectively). Also, the single particle states of the algebraic shell model have an additional structure. The creation operator for a single particle is multiplied by the vacuum of the other particle.

The operator $p_1 p_2$ is thus an **algebraic realization** of the vacuum state, and the other (algebraic) shell states are built from the vacuum state through the action of the fermion creation operators, in analogy with the standard shell model.

On the quotient space Q_L the four algebraic shell states of F^+ go over into the familiar S shell states. In particular, on the Q_L the state $p_1 p_2$ becomes the identity operator $1 \sim |0\rangle$.

While the shell states S tranform according to the representation $(1/2, 1/2)$ of $so(5)$, the F^+ shell states transform according to the representation $[1000]$ of $su(4)$, or equivalently, according the spin representation $(1/2, 1/2, 1/2)$ of $so(6)$. This is easily verified by an explicit calculation of the action of the algebras upon the operator states (the algebra su(4), and its semisimple subalgebras, are discussed in ref.3),

$$H_1 \cdot f_1^+ f_2^+ = (3/4)\ f_1^+ f_2^+, \quad H_2 \cdot f_1^+ f_2^+ = (-1/4)\ f_1^+ f_2^+,$$

$$H_3 \cdot f_1^+ f_2^+ = (-1/4)\ f_1^+ f_2^+, \quad H_4 \cdot f_1^+ f_2^+ = (-1/4)\ f_1^+ f_2^+,$$

$$E(1\text{-}100) \cdot f_1^+ f_2^+ = 0, \quad E(01\text{-}10) \cdot f_1^+ f_2^+ = 0, \quad E(001\text{-}1) \cdot f_1^+ f_2^+ = 0,$$

$$E(\text{-}1100) \cdot f_1^+ f_2^+ = -f_1^+ p_2, \quad E(0\text{-}110) \cdot f_1^+ p_2 = f_2^+ p_1, \quad E(00\text{-}11) \cdot f_2^+ p_1 = p_1 p_2$$

$$(3.3)$$

The third line shows that $f_1^+ f_2^+$ is an extremal vector of the representation $[1000]$ (i.e.

is mapped to zero by the raising operators), while the last line lists the nonzero matrix elements for the simple lowering operators.

Similarly one obtains for the algebraic dual shell F,

$F =$	$(p_1 p_2,$	$f_2 p_1,$	$-f_1 p_2,$	$f_1 f_2)$
	$\phi_1' = \phi_4{}^*$	$\phi_2' = \phi_3{}^*$	$\phi_3' = -\phi_2{}^*$	$\phi_4' = -\phi_1{}^*$
$u(4):$	$[1\,1\,1\,0]$	$[1\,1\,0\,1]$	$[1\,0\,1\,1]$	$[0\,1\,1\,1]$
$su(4):$	$(1/4)(1\,1\,1\,-3)$	$(1/4)(1\,1\,-3\,1)$	$(1/4)(1\,-3\,1\,1)$	$(1/4)(-3\,1\,1\,1)$
$so(6):$	$(1/2)(1\,1\,-1)$	$(1/2)(1\,-1\,1)$	$(1/2)(-1\,1\,1)$	$(1/2)(-1\,-1\,-1)$
$so(5):$	$(1/2)(1\,1)$	$(1/2)(1\,-1)$	$(1/2)(-1\,1)$	$(1/2)(-1\,-1)$
$so(4):$	$(1/2)(1\,1)$	$(1/2)(1\,-1)$	$(1/2)(-1\,1)$	$(1/2)(-1\,-1)$

$$(3.4)$$

The phases have been chosen to correspond to the phases used in physical applications.

It holds (the Lie algebra operators act to the left),

$$p_1 p_2 \cdot (-H_1') = (1/2)\ p_1 p_2,$$
$$p_1 p_2 \cdot (-H_2') = (1/2)\ p_1 p_2,$$
$$p_1 p_2 \cdot (-H_3') = (-1/2)\ p_1 p_2, \quad \text{etc,}$$

$$p_1 p_2 \cdot (-E(1\,-1\,0\,0)) = 0, \quad p_1 p_2 \cdot (-E(0\,-1\,1\,0)) = 0, \quad p_1 p_2 \cdot (-E(0\,0\,-1\,1)) = 0,$$
$$p_1 p_2 \cdot (-E(0\,0\,-1\,1)) = -f_2 p_1, \quad f_2 p_1 \cdot (-E(0\,-1\,1\,0)) = -f_1 p_2, \quad -f_1 p_2 \cdot (-E(-1\,1\,0\,0)) = f_1 f_2$$

$$(3.5)$$

The shells F^+ and F consist of algebraic states (operators) which transform like the representations $[1000]$ and $[1110]$ of $su(4)$. For the direct product of these two representations holds

$$[1000]\ \times\ [1110]\ =\ [2110]\ +\ [1111] \qquad\qquad (3.6)$$
$$4\ \ \times\ \ 4\ \ =\ \ 15\ +\ \ 1$$

Thus the basis states for the representations $[2110] + [1111]$ are given by the ordered tensor product states

$$TA \supset F^+ \times F = \big(p_1 p_2 \times p_1 p_2, \quad f_1{}^+ p_2 \times p_1 p_2, \quad f_2{}^+ p_1 \times p_1 p_2, \quad f_1{}^+ f_2{}^+ \times p_1 p_2,$$
$$p_1 p_2 \times f_1 p_2, \quad f_1{}^+ p_2 \times f_1 p_2, \quad f_2{}^+ p_1 \times f_1 p_2, \quad f_1{}^+ f_2{}^+ \times f_1 p_2,$$
$$p_1 p_2 \times f_2 p_1, \quad f_1{}^+ p_2 \times f_2 p_1, \quad f_2{}^+ p_1 \times f_2 p_1, \quad f_1{}^+ f_2{}^+ \times f_2 p_1,$$
$$p_1 p_2 \times f_1 f_2, \quad f_1{}^+ p_2 \times f_1 f_2, \quad f_2{}^+ p_1 \times f_1 f_2, \quad f_1{}^+ f_2{}^+ \times f_1 f_2 \big)$$

$$(3.7)$$

The commutator of two elements of $F^+ \times F$ is in TA, but not in $F^+ \times F$. However, the commutator of the operator states $(a \cdot x \cdot b)$ of $F^+ \times F$ yields a realization of the algebra $u(4)$ with respect to multiplication defined by

$$(a \cdot x \cdot b) \cdot (c \cdot x \cdot d) \equiv (a \cdot (b \cdot c) \cdot x \cdot d) = (a \cdot x \cdot (b \cdot c) \cdot d),$$

or, in short notation

$$(a \times b)(c \times d) \equiv (a \ (bc) \times d) = (a \times (bc) \ d),\qquad\qquad (3.8)$$

with $a, c \in F^+$, $b, d \in F$. This multiplication law reduces the four-fold tensor product back into the two-fold tensor product. It holds that either

$$(\cdot b) \cdot (c \cdot) \equiv (bc) = 0,$$

or,

$$(\cdot b) \cdot (c \cdot) = (p_1 p_2) \cdot (p_1 p_2) = (p_1 p_2)$$

with $F^+ \cdot (p_1 p_2) = F^+$ $(p_1 p_2) \cdot F = F$.

That is,

$$a \cdot (p_1 p_2) = a, \qquad (p_1 p_2) \cdot b = b$$

One obtains for the commutator, based upon the product as defined in eq(3.8)

$$(a \times b)(c \times d) - (c \times d)(a \times b) \equiv (a \times (bc) \ d) - (c \times (da) \ b)$$

Direct calculation shows that **the operators** of the set $F^+ \times F$ indeed satisfiy the commutation relations of $u(4)$ in $F^+ \times F$. Alternatively, this can also be seen by making use of the map $F^+ \times F \rightarrow F^+ \cdot F$. This map is one to one and preserves the multiplication law. But, $F^+ \cdot F \sim u(4)$ (see also eq(3.12)). This proves that the set of operators $F^+ \times F$ indeed forms the algebra $u(4)$ with respect to the commutator based upon the multiplication law eq(3.8).

The operator states of $F^+ \times F$, eq(3.7), carry, **as states**, the representations [2110] and [0000] of $u(4)$. The action of the operators (Op^X) upon the tensor product states is given by

$$(Op^X) \cdot (a \times b) \equiv (Op \cdot a) \times b + a \times (b \cdot (-Op)) \qquad a \in F^+, b \in F \qquad (3.9)$$

Using eq(3.9) the singlet state $[1111] \sim [0000]$ is extracted from $F^+ \times F$ by standard methods of representation theory. The singlet operator state is obtained as

$$-(1/2)(p_1 p_2 \times p_1 p_2 + f_2^+ p_1 \times f_2 p_1 + f_1^+ p_2 \times f_1 p_2 - f_1^+ f_2^+ \times f_1 f_2)$$
$$= -(1/2)(\phi_4 \times \phi_1{}' + \phi_3 \times \phi_2{}' - \phi_2 \times \phi_3{}' - \phi_1 \times \phi_4{}')$$
$$= -(1/2)(\phi_4 \times \phi_4{}^* + \phi_3 \times \phi_3{}^* + \phi_2 \times \phi_2{}^* + \phi_1 \times \phi_1{}^*)$$
$$= (1/2) \ \phi \ \gamma_4{}' \ \phi^* \qquad\qquad (3.10)$$

with $\qquad \phi' = \gamma_4{}' \ \phi^*,$

$$\gamma_4' = \begin{vmatrix} 0 & 0 & 0 & -1 \\ 0 & 0 & -1 & 0 \\ 0 & 1 & 0 & 0 \\ 1 & 0 & 0 & 0 \end{vmatrix}$$

$$(3.11)$$

Here the Dirac matrix $\gamma_4' = f_2^+ - f_2$ has been obtained with respect to the basis states Φ_i in the order,

$$\Phi_1 \sim \phi_2 = f_1^+ p_2, \quad \Phi_2 \sim \phi_3 = f_2^+ p_1, \quad \Phi_3 \sim \phi_1 = p_1 p_2, \quad \Phi_4 \sim \phi_4 = f_1^+ f_2^+ \quad,$$

following the conventional order in physical applications. The wave functions ϕ' are the adjoint wave functions to the Dirac spinor ϕ .

The operator states of F^+ and F are elements of the the **CA**. Thus the associative multiplication law of the the **CA** can be used and one obtains through the map $F^+ \mathrm{x} F \rightarrow F^+ \cdot F$, eq(2.4),

$$F^+ \cdot F \sim su(4) + u(1) \sim CA =$$

$$\begin{array}{llll}
(p_1 p_2, & f_1^+ p_2, & f_2^+ p_1, & f_1^+ f_2^+, \\
f_1 p_2, & f_1^+ f_1 p_2, & f_2^+ f_1, & f_1^+ f_1 f_2^+, \\
f_2 p_1, & f_1^+ f_2, & f_2^+ f_2 p_1, & f_1^+ f_2^+ f_2, \\
f_1 f_2, & f_1^+ f_1 f_2, & f_2^+ f_2 f_1, & f_1^+ f_1 f_2^+ f_2)
\end{array}$$

$$(3.12)$$

This set is closed with respect to the associative multiplication law, the anticommutator and the commutator. With respect to the commutation relations

$$[a, b] \equiv a \cdot b - a \cdot b, \qquad a,b \in F^+ \cdot F,$$

the set $F^+ \cdot F$ forms the Lie algebra $u(4)$, with the Cartan subalgebra given by

$$H_1 = f_1^+ f_1 f_2^+ f_2 - 1/4, \quad H_2 = f_1^+ f_1 p_2 - 1/4, \quad H_3 = f_2^+ f_2 p_1 - 1/4, \quad H_4 = p_1 p_2 - 1/4.$$

$$(3.13)$$

The singlet operator state becomes

$$(1/2)(p_1 p_2 + f_2^+ f_2 p_1 - f_1^+ f_1 p_2 + f_1^+ f_1 f_2^+ f_2) = (1/2)\mathbf{1} \qquad (3.14)$$

The operators eq(3.12) are seen to be equivalent to the algebra $u(4)$ as given in eq(2.5). However here the algebra $u(4)$ has been obtained in a different manner, namely via the tensor product $F^+ \mathrm{x} F$ of two $su(4)$ representations and a subsequent map of $F^+ \mathrm{x} F \rightarrow F^+ \cdot F \left(= F^+ F \sim u(4) \sim CA \right).$

4. Selflimiting Representation Theory

The F^+ and F shell states are all states built from one single fermion species f^+ and its antifermion f. Moreover, all states of F^+ and F are obtainable from each other through the operations of $su(4)$ shift operators. Thus, F^+ can be called "the fermion", and F "the antifermion".

It has been shown that the fermion F^+, and the antifermion F, transform like the representations [1000], and [1110], of $su(4)$ respectively. In fact, in the previous section the direct product of these two (operator states) representations was analysed. After forming the ordered direct product of F^+ and F, the tensor product states $a \times b$ of the resultant representations were mapped upon the **CA** states ab. In this manner $su(4)$ representations were obtained on the space of the **CA**. This map however implies an identification of particle species which in the ordered tensor product have to be considered as distinct.

In the following we will consider more then one species of fermions F^+ and F and build up larger representations of $su(4)$. The Clifford algebra properties (Pauli principle) will limit the number of representations which can be constructed once a certain number of fermion species has been fixed.

The case for one species has been covered in section 3. The following representations are obtained **on the Clifford algebra**

$$
\begin{aligned}
F^+ &: &&[1000] \\
F^+ \times F \sim u(4) &: &&[1000] \times [1110] = [2110] + [0000] \\
F &: &&[1110] \sim [000\text{-}1]
\end{aligned}
$$

$$(4.1)$$

No other representations can be obtained for a single particle species only, since identification of two F^+ and two F in the tensor product implies $F^+ \times F^+ \rightarrow F^+$, $F \times F \rightarrow F$.

For the case of two particle species the following representations are obtained:

$$
\begin{aligned}
F^+ &: &&[1000] \\
F^+ \times F^+ &: &&[1000] \times [1000] = [2000] + [1100] \\
F^+ \times F \sim u(4) &: &&[1000] \times [1110] = [2110] + [0000] \\
(F^+ \times F) \times F^+ &: &&[3110],[2210],[2111],[1000] \\
(F^+ \times F)(\times F^+ \times F) &: &&[4220],\ [4211],\ [3221],\ [2222],\ [2110],\ [0000] \\
(F^+ \times F) \times F &: &&[3220],\ [3211],\ [2221],\ [1110] \\
F \times F &: &&[1110] \times [1110] = [2220] + [2211] \\
F &: &&[1110] \sim [000\text{-}1]
\end{aligned}
$$

$$(4.2)$$

The tensor product operators (Op^X) act upon the operator tensor product states $a \times b$ of $F^+ \times F^+$ from the left as

$$(Op^X)\,(a \times b) \equiv ((Op^1 \times 1) + (1 \times Op^2))\,(a \times b) = (Op^1)a \times b + a \times (Op^1)b$$

$$(4.3)$$

The last equality sign is a consequence of the fact that the two representations of $F^+ \times F^+$

are identical, i.e. $(Op^2)=(Op^1)=(Op)$. For the representation on the **CA** this action goes over into action from the left given by

$$(Op)(ab)= ((Op)a)b + a(Op)b \qquad (4.4)$$

i.e. the action of a derivative.

On states **axb** of **FxF** the dual tensor product operator acts $(Op_D{}^X)$ as

$$(Op_D{}^X) (axb)=((Op_D{}^1)x1)+(1x(Op_D{}^2)) (axb) =$$
$$a(-Op^1) \; x \; 1+1 \; x \; b(-Op^1)$$

$$(4.5)$$

and on the **CA** this relation goes over into

$$(Op_D)(ab) = (a(-Op))b + a(b(-Op)), \qquad (4.6)$$

since $(Op_D)a = a(-Op)$.

For the states of the tensor product F^+xF , eq.(3.7), the corresponding action on the tensor states is given by

$$(Op^X) (axb)=((Op^1x1)+(1xOp^2)) (axb)=(Op^1)a \; x \; b+a \; x \; b(-Op^1)$$

$$(4.7)$$

since **F** is the complex conjugate representation to F^+. Making use of the associative multiplication law of the **CA** this product goes over into the familiar commutator,

$$((Op^1) a) b - a ((Op^1) a) b = [(Op^1), ab] \qquad (4.8)$$

$$ab \in F^+\cdot F \sim su(4)+u(1).$$

In the following the operator state bases will be constructed for some of the representations which will be of relevance later on.

The operator state $f_1{}^+f_2{}^+x f_1{}^+f_2{}^+ \equiv (f_1{}^+f_2{}^+)(f_1{}^+f_2{}^+)$ is seen to be an extremal state, i.e. to belong to highest weight [2000] of **su(4)**, and is mapped to zero by the (simple) raising operators of **su(4)**, eq(2.5). Acting upon this state with the simple lowering oprators the representation space is spanned:

$$E(-1100)(f_1{}^+f_2{}^+)(f_1{}^+f_2{}^+)= \sqrt{2} \frac{1}{\sqrt{2}} ((f_1{}^+p_2)(f_1{}^+f_2{}^+) + (f_1{}^+f_2{}^+)(f_1{}^+p_2))$$

where normalization of the state to 1 yields the matrix element $\sqrt{2}$. In this manner the states of the representation [2000] are obtained as

su(4)	F^+xF^+	$\rightarrow$ $F^+F^+ \sim F^+$
so(6)		
[2000]	$(f_1{}^+f_2{}^+)(f_1{}^+f_2{}^+)$	0
[1100]	$\frac{1}{\sqrt{2}} ((f_1{}^+p_2)(f_1{}^+f_2{}^+) + (f_1{}^+f_2{}^+)(f_1{}^+p_2))$	0

184

[0200]	$(f_1^+ p_2)(f_1^+ p_2)$	0
[1010]	$\frac{1}{\sqrt{2}}\left((f_2^+ p_1)(f_1^+ f_2^+) + (f_1^+ f_2^+)(f_2^+ p_1)\right)$	0
[0110]	$\frac{1}{\sqrt{2}}\left((f_1^+ p_2)(f_2^+ p_1) + (f_2^+ p_1)(f_1^+ p_2)\right)$	0
[1001]	$\frac{1}{\sqrt{2}}\left((p_1 p_2)(f_1^+ f_2^+) + (f_1^+ f_2^+)(p_1 p_2)\right)$	$f_1^+ f_2^+$
[0020]	$(f_2^+ p_1)(f_2^+ p_1)$	0
[0101]	$\frac{1}{\sqrt{2}}\left((p_1 p_2)(f_1^+ p_2) + (f_1^+ p_2)(p_1 p_2)\right)$	$f_1^+ p_2$
[0011]	$\frac{1}{\sqrt{2}}\left((f_2^+ p_1)(p_1 p_2) + (p_1 p_2)(f_2^+ p_1)\right)$	$f_2^+ p_1$
[0002]	$(p_1 p_2)(p_1 p_2)$	$p_1 p_2$

$$(4.9)$$

The equivalence sign in $F^+ F^+ \sim F^+$ follows from the particle species identification, i.e. from the map into the **CA** (i.e. $f^2 = 0$).

The weight subspace of the weight [1100] has dimension 2. The representation [2000] contains one state of this subspace, but not the orthogonal component. Thus the state orthogonal to state [1100] of the representation [2000] belongs to the representation [1100]. The states of the representation [1100] are again obtained obtained through the action of the lowering operators of **su(4)**. They are (the **so(6)** weights are also listed for this representation)

[1 1 0 0] (1 0 0)	$b_1^+ = d_2^+ = \frac{1}{\sqrt{2}}\left((f_1^+ p_2)(f_1^+ f_2^+) - (f_1^+ f_2^+)(f_1^+ p_2)\right)$	0
[1 0 1 0] (0 1 0)	$b_2^+ = d_1^+ = \frac{1}{\sqrt{2}}\left((f_2^+ p_1)(f_1^+ f_2^+) - (f_1^+ f_2^+)(f_2^+ p_1)\right)$	0
[0 1 1 0] (0 0 -1)	$b_3^+ = \frac{1}{\sqrt{2}}(d_0^+ - s^+) = \frac{1}{\sqrt{2}}\left((f_1^+ p_2)(f_2^+ p_1) - (f_2^+ p_1)(f_1^+ p_2)\right)$	0
[1 0 0 1] (0 0 1)	$b_4^+ = \frac{1}{\sqrt{2}}(d_0^+ + s^+) = \frac{1}{\sqrt{2}}\left((p_1 p_2)(f_1^+ f_2^+) - (f_1^+ f_2^+)(p_1 p_2)\right)$	$-f_1^+ f_2^+$
[0 1 0 1] (0 -1 0)	$b_5^+ = d_{-1}^+ = \frac{1}{\sqrt{2}}\left((p_1 p_2)(f_1^+ p_2) - (f_1^+ p_2)(p_1 p_2)\right)$	$-f_1^+ p_2$
[0 0 1 1] (-1 0 0)	$b_6^+ = d_{-2}^+ = \frac{1}{\sqrt{2}}\left((f_2^+ p_1)(p_1 p_2) - (p_1 p_2)(f_2^+ p_1)\right)$	$f_2^+ p_1$

$$(4.10)$$

The operator states of the representation [1100] of **su(4)**, or equivalently the representation (100) of **so(6)**, transform like the set of the orbital angular momentum l=2 d^+-bosons and the l=1 s^+-boson of the **su(6)** interacting boson model of the nucleus (**IBM**). It will be shown in section 5 that the set ($b_i^+ b_j$) of operator states of eq.(4.10) and eq(4.11) satisfies the commutation relations of **su(6)**. and that the set b_i^+ transforms like the defining representation [100000] of **su(6)**.

The states for the representation [2220] are obtained in an analogous manner.

s u (4)	**F x F**	$\rightarrow$ **FF ~ F**
[2220]		

Since this representation are not of relevance to the following discussion, these states will not be listed here.

The states for the representation [2211] $\sim$ [1100] are obtained as

s u (4) $\qquad\qquad$ **F x F** $\qquad\qquad\qquad\qquad$ $\rightarrow$ **FF ~ F**
so(6)

[2211]
(1 0 0) $\qquad$ $b_1 = d_2 = \frac{1}{\sqrt{2}} \left((f_1 f_2)(f_1 p_2) - (f_1 p_2)(f_1 f_2) \right)$ $\qquad$ 0

[1010]
(0 1 0) $\qquad$ $b_2 = d_1 = \frac{1}{\sqrt{2}} \left((f_1 f_2)(f_2 p_1) - (f_2 p_1)(f_1 f_2) \right)$ $\qquad$ 0

[0110]
(0 0 -1) $\qquad$ $b_3 = \frac{1}{\sqrt{2}} (d_0 - S) = \frac{1}{\sqrt{2}} \left((f_2 p_1)(-f_1 p_2) - (-f_1 p_2)(f_2 p_1) \right)$ $\qquad$ 0

[1001]
(0 0 1) $\qquad$ $b_4 = \frac{1}{\sqrt{2}} (d_0 + S) = \frac{1}{\sqrt{2}} \left((f_1 f_2)(p_1 p_2) - (p_1 p_2)(f_1 f_2) \right)$ $\qquad$ $-f_1 f_2$

[0101]
(0 -1 0) $\qquad$ $b_5 = d_{-1} = \frac{1}{\sqrt{2}} \left((-f_1 p_2)(p_1 p_2) - (p_1 p_2)(-f_1 p_2) \right)$ $\qquad$ $-f_1 p_2$

[0011]
(-1 0 0) $\qquad$ $b_6 = d_{-2} = \frac{1}{\sqrt{2}} \left((f_2 p_1)(p_1 p_2) - (p_1 p_2)(f_2 p_1) \right)$ $\qquad$ $f_2 p_1$

with $\qquad\qquad$ $(b_j^+)^+ = b_j$

$$(4.11)$$

This representation [2211] $\sim$ [1100] of **su(4)** transforms like the representation [111110] of **su(6)**. This will be shown in section 5.

In the following section it will be shown that the operators, and the operator states, discussed above provide a microscopic substructure for the Interacting Boson Model of the Nucleus in terms of spin 1/2 fermions. The bosonic properties of the standard model will be supplemented with an internal fermionic structure and with internal fermionic statistics.

5. A Microscopic Theory for the Interacting Boson Model of the Nucleus

One of the three symmetry chains used in the **su(6)** Interacting Boson Model of the Nucleus (**IBM** , for the pioneering work see ref.4) is the chain

$$su(6) \quad \rightarrow \quad so(6) \sim su(4) \quad \rightarrow \quad so(5) \quad \rightarrow \quad so(3)_L \tag{5.1}$$

This symmetry chain is one of three chains which are of significance for the **IBM** (for a classification of all semisimple symmetry chains of **su(6)** see ref.5).

In section 3 the fermion F^+ and the antifermion F were found to tranform like the representations [1000] and [1110] of **su(4)**. Then the tensor product F^+xF of the two representations was formed yielding the **su(4)** representations [2110] and [0000] in terms of operator states. Making use of the associative multiplication law upon which the anticommutator is based in the **CA** , the operators F^+F were seen to correspond to the algebra **su(4)+u(1)**.

It will now be shown that the operator state representations [1100] of su(4), in terms of the d^+, s^+ states, as well as in terms of the **d, s** states of section 4, transform according to the chain eq.(5.1) as

	su(6)	su(4)	so(5)	so(3)_L
$b_i^+(F^+xF^+):d^+, s^+:$	[100000]	[1100]	$(10)+(00)$	
$l=2,\ \ l=0$				
$b_i(FxF): \qquad d, s :$	[111110]	[1100]	$(10)+(00)$	
$l=2,\ \ l=0$				

$$\tag{5.2}$$

First it will be show that the set $\big(\ b_i^+(F^+xF^+)\cdot b_j(FxF)\ :\ i,j=1,2,3,4,5,6\ \big)$ satisfies the commutation relations of **su(6)**. Explicit calculation yields

$$b_j\,b_k^+ = \delta_{jk}\,(p_1 p_2)(p_1 p_2)\ , \tag{5.3}$$

$$b_j\,(p_1 p_2)(p_1 p_2)=0\ , \tag{5.4}$$

$$(p_1 p_2)(p_1 p_2)\,b_j = b_j\ , \tag{5.5}$$

$$(\ b_i^+ b_j\ )(\ b_k^+ b_s\ ) = \delta_{jk}\ (\ b_i^+ b_s\ ). \tag{5.6}$$

These relations hold since the boson operators $b^+(F^+xF^+)$ and $b(FxF)$ have an internal structure which is not present in the "plain" bosons of the standard theory.

Using the relations eq(5.3)-eq(5.6), and calculating the commutator, one obtains the **su(6)** commutation relations

$$[(\ b_i^+ b_j\ )\ ,(\ b_k^+ b_s\ )] = \delta_{jk}\ (\ b_i^+ b_s\ ) - \delta_{si}\ (\ b_k^+ b_j\ ) \tag{5.7}$$

The relationship between the Cartan notation for **su(6)** and the boson operator notation for **su(6)** is given by

$$H_i = b_i^+ b_i,$$
$$E(e_i - e_j\ ) = b_i^+ b_j,\quad i=1,2,3,4,5,6,$$

with e_i denoting the i-th cartesian unit vector in six dimensions.

The b_i^+ and b_j do, however, not satisfy the ordinary boson operator commutation relations. Instead they satisfy the commutation relations

$$(\mathbf{b}_i \, \mathbf{b}_j^+ - \mathbf{b}_j^+ \mathbf{b}_i) \, (\mathbf{p}_1 \mathbf{p}_2)(\mathbf{p}_1 \mathbf{p}_2) = \delta_{ij} \, (\mathbf{p}_1 \mathbf{p}_2)(\mathbf{p}_1 \mathbf{p}_2) \qquad (5.8)$$

The operator state $(\mathbf{p}_1 \mathbf{p}_2)(\mathbf{p}_1 \mathbf{p}_2)$ represents the direct product of two "physical vacua", or equivalently, two "algebraic vacua" (containing particle-antiparticle pairs). On the quotient space $\mathbf{Q}_L$ with respect to the left ideal $\mathbf{I}_L$, generated by the relation $f_i 1 = 0$, i.e. with respect to the "emty" or "bare" vacua $\mathbf{1}$ (containing no particle-antiparticle pairs) the familiar commutation relations for the boson operators are obtained,

$$(\mathbf{b}_i \, \mathbf{b}_j^+ - \mathbf{b}_j^+ \mathbf{b}_i) \, \mathbf{1} = \delta_{ij} \, \mathbf{1} \, , \qquad (5.9)$$

since all particle-antiparticle pairs of the vacuum state $(\mathbf{p}_1 \mathbf{p}_2)(\mathbf{p}_1 \mathbf{p}_2) = \mathbf{p}_1 \mathbf{p}_2 \times \mathbf{p}_1 \mathbf{p}_2$ are mapped to zero. This is not surprising since, as it was mentioned before, on the quotient space the algebraic theory goes over into the standard theory

The states of the **su(6)** representation $[n_1 n_2 n_3 n_4 n_5 n_6]$, $n_1 \geq n_2 \geq n_3 \geq n_4 \geq n_5 \geq n_6$, contain $N = n_1 + n_2 + n_3 + n_4 + n_5 + n_6$ bosons and are given by (ordered) direct products of boson operators $\mathbf{b}_j^+$. The state which belongs to highest weight is given by

$$(\mathbf{b}_1^+)^{n_1} (\mathbf{b}_2^+)^{n_2} (\mathbf{b}_3^+)^{n_3} (\mathbf{b}_4^+)^{n_4} (\mathbf{b}_5^+)^{n_5} (\mathbf{b}_6^+)^{n_6} \qquad (5.10)$$

The diagonal operators are given as

$$H_j^x = (\mathbf{b}_j^+ \mathbf{b}_j)^x =$$
$$(\mathbf{b}_j^+ \mathbf{b}_j) \times 1 \times \ldots .. \times 1 + 1 \times (\mathbf{b}_j^+ \mathbf{b}_j) \times 1 \times \ldots .. \times 1 + \ldots + 1 \times \ldots .. \times 1 \times (\mathbf{b}_j^+ \mathbf{b}_j),$$

$$(5.11)$$

consisting of a sum of $N = n_1 + n_2 + n_3 + n_4 + n_5 + n_6$ terms, each of which contains $N = n_1 + n_2 + n_3 + n_4 + n_5 + n_6$ direct product factors. The action of the operators $(\mathbf{b}_j^+ \mathbf{b}_j)^x$ on each factor $(\mathbf{b}_i^+)^{n_i}$ of eq(5.10) is given by

$$
\begin{aligned}
(\mathbf{b}_j^+ \mathbf{b}_j)^x \cdot (\mathbf{b}_i^+)^{n_i} \quad &= (\mathbf{b}_j^+ \mathbf{b}_j) \cdot \mathbf{b}_i^+ \times \mathbf{b}_i^+ \times \ldots \ldots \ldots \times \mathbf{b}_i^+ \\
&+ \mathbf{b}_i^+ \times (\mathbf{b}_j^+ \mathbf{b}_j) \cdot \mathbf{b}_i^+) \times \mathbf{b}_i^+ \times \ldots \ldots \times \mathbf{b}_i^+ \\
&+ \ldots \ldots \ldots \ldots \ldots \ldots \ldots \ldots \ldots \ldots \ldots \ldots \\
&+ \mathbf{b}_i^+ \times \ldots \ldots \ldots \ldots \times \mathbf{b}_i^+ \times (\mathbf{b}_j^+ \mathbf{b}_j) \cdot \mathbf{b}_i^+
\end{aligned}
$$

$$n_i \text{ factors}$$

$$
\begin{aligned}
&= \mathbf{b}_j^+ \, \delta_{ij} \, (\mathbf{p}_1 \mathbf{p}_2)(\mathbf{p}_1 \mathbf{p}_2) \times \mathbf{b}_i^+ \times \ldots \ldots \ldots \times \mathbf{b}_i^+ \\
&+ \mathbf{b}_i^+ \times \mathbf{b}_j^+ \, \delta_{ij} \, (\mathbf{p}_1 \mathbf{p}_2)(\mathbf{p}_1 \mathbf{p}_2) \times \mathbf{b}_i^+ \times \ldots \ldots \times \mathbf{b}_i^+ \\
&+ \ldots \ldots \ldots \ldots \ldots \ldots \ldots \ldots \ldots \ldots \ldots \ldots \ldots \\
&+ \mathbf{b}_i^+ \times \ldots \ldots \ldots \times \mathbf{b}_i^+ \times \mathbf{b}_j^+ \, \delta_{ij} \, (\mathbf{p}_1 \mathbf{p}_2)(\mathbf{p}_1 \mathbf{p}_2) = \\
\\
&= \delta_{ij} \, n_j \, (\mathbf{b}_i^+)^{n_i}
\end{aligned}
$$

$$(5.12)$$

It follows that the state eq.(5.10) belongs to weight $[n_1 n_2 n_3 n_4 n_5 n_6]$. Thus the states given by eq(4.4) indeed form a basis for the representation [100000] of **su(6)**.

One obtains

$$b_1{}^+ = d_2{}^+: \qquad [100000] \qquad (1/6)(5\text{-}1\text{-}1\text{-}1\text{-}1\text{-}1)$$

$$b_2{}^+ = d_1{}^+: \qquad [010000] \qquad (1/6)(\text{-}15\text{-}1\text{-}1\text{-}1\text{-}1)$$

$$b_3{}^+ = \tfrac{1}{\sqrt{2}}(d_0{}^+ + S^+): \qquad [001000] \qquad (1/6)(\text{-}1\text{-}15\text{-}1\text{-}1\text{-}1)$$

$$b_4{}^+ = \tfrac{1}{\sqrt{2}}(d_0{}^+ - S^+): \qquad [000100] \qquad (1/6)(\text{-}1\text{-}1\text{-}15\text{-}1\text{-}1)$$

$$b_5{}^+ = d_{-1}{}^+: \qquad [000010] \qquad (1/6)(\text{-}1\text{-}1\text{-}1\text{-}15\text{-}1)$$

$$b_6{}^+ = d_{-2}{}^+: \qquad [000001] \qquad (1/6)(\text{-}1\text{-}1\text{-}1\text{-}1\text{-}15)$$

$$(5.13)$$

For the dual states, the antibosons b_i, holds

$$(b_i)^{n_i} \bullet (-b_j{}^+ b_j)^x = -\,\delta_{ij}\, n_i\, (b_i)^{n_i} \tag{5.14}$$

and thus,

$$b_1 = d_2 \quad: \qquad [00000\text{-}1] \qquad [111110] \qquad (1/6)(11111\text{-}5)$$

$$b_2 = d_1 \qquad [0000\text{-}10] \qquad [111101] \qquad (1/6)(1111\text{-}51)$$

$$b_3 = \tfrac{1}{\sqrt{2}}(d_0 - S): \qquad [000\text{-}100] \qquad [111011] \qquad (1/6)(111\text{-}511)$$

$$b_4 = \tfrac{1}{\sqrt{2}}(d_0 + S): \qquad [00\text{-}1000] \qquad [110111] \qquad (1/6)(11\text{-}5111)$$

$$b_5 = d_{-1}: \qquad [0\text{-}10000] \qquad [101111] \qquad (1/6)(1\text{-}51111)$$

$$b_6 = d_{-2}: \qquad [\text{-}100000] \qquad [011111] \qquad (1/6)(\text{-}511111)$$

$$(5.15)$$

The relationship of the boson operator realization of **su(6)** to the Cartan notation is given by

$$H_i = b_i{}^+ b_i,$$
$$E(e_i - e_j) = b_i{}^+ b_j, \qquad i = 1,2,3,4,5,6,$$

with e_i denoting the i-th cartesian unit vector in six dimensions.

The subalgebras of the chain $su(6) \rightarrow su(4) \rightarrow so(5) \rightarrow so(3)_L$ are:

The **su(4)** subalgebra of **su(6)**:

$$H_1' = (1/4)(3H_1+3H_2+3H_3-H_4-H_5-H_6)$$
$$= (1/4)(3d_2{}^+d_2+3d_1{}^+d_1+3(1/2)(d_0{}^++S^+)(d_0+S)-(1/2)(d_0{}^+-S^+)(d_0-S)$$
$$-d_{-1}{}^+d_{-1}-d_{-2}{}^+d_{-2})$$

$$H_2' = (1/4)(3H_1-H_2-H_3+3H_4+3H_5-H_6)$$
$$= (1/4)(3d_2{}^+d_2-d_1{}^+d_1-(1/2)(d_0{}^++S^+)(d_0+S)+3(1/2)(d_0{}^+-S^+)(d_0-S)$$
$$+3d_{-1}{}^+d_{-1}-d_{-2}{}^+d_{-2})$$

$$H_3' = (1/4)(-H_1+3H_2-H_3+3H_4-H_5+3H_6)$$
$$= (1/4)(-d_2{}^+d_2+3d_1{}^+d_1-(1/2)(d_0{}^++S^+)(d_0+S)+3(1/2)(d_0{}^+-S^+)(d_0-S)$$
$$-d_{-1}{}^+d_{-1}+3d_{-2}{}^+d_{-2})$$

$$H_4' = (1/4)(-H_1-H_2+3H_3-H_4+3H_5+3H_6)$$
$$= (1/4)(-d_2{}^+d_2-d_1{}^+d_1+3(1/2)(d_0{}^++S^+)(d_0+S)-(1/2)(d_0{}^+-S^+)(d_0-S)$$
$$+3d_{-1}{}^+d_{-1}+3d_{-2}{}^+d_{-2})$$

$$E(1\text{-}100) = d_1{}^+\tfrac{1}{\sqrt{2}}(d_0\text{-}S)+\tfrac{1}{\sqrt{2}}(d_0{}^++S^+)d_{-1}$$
$$E(01\text{-}10) = d_2{}^+d_1 + d_{-1}{}^+d_{-2}$$
$$E(001\text{-}1) = d_1{}^+\tfrac{1}{\sqrt{2}}(d_0+S)+\tfrac{1}{\sqrt{2}}(d_0{}^+-S^+)d_{-1}$$

$$(5.16)$$

Only the operators corresponding to the negative simple roots have been listed.

The **so(5)** subalgebra of **su(6)**:

$$H_1'' =(1/2)\left(H_1'+H_2'-H_3'-H_4'\right)$$
$$= d_2{}^+d_2 - d_{-2}{}^+d_{-2}$$

$$H_2'' =(1/2)\left(H_1'-H_2'+H_3'-H_4'\right)$$
$$= d_1{}^+d_1 - d_{-1}{}^+d_{-1}$$

$$E(1\text{-}1) = d_2{}^+d_1 + d_{-1}{}^+d_{-2}$$
$$E(01) =\tfrac{1}{\sqrt{2}}\left(d_1{}^+\tfrac{1}{\sqrt{2}}(d_0\text{-}S) +\tfrac{1}{\sqrt{2}}(d_0{}^++S^+)d_{-1}\right.$$
$$\left.+ d_1{}^+\tfrac{1}{\sqrt{2}}(d_0+S)+\tfrac{1}{\sqrt{2}}(d_0{}^+-S^+)d_{-1}\right) \qquad (5.17)$$

The **so(3)**$_L$ orbital angular momentum subalgebra of **su(6)** is given as

$$L_0 = 2H_1''+H_2'' = 2d_2{}^+d_2 + d_1{}^+d_1 - d_{-1}{}^+d_{-1} - 2d_{-2}{}^+d_{-2}$$
$$L_+ = \sqrt{2}\left(d_2{}^+d_1 + d_{-1}{}^+d_{-2}\right) + \sqrt{3/2}\left(d_1{}^+\sqrt{1/2}(d_0\text{-}S)\right.$$
$$\left.+\sqrt{1/2}(d_0{}^++S^+)d_{-1}+d_1{}^+\sqrt{1/2}(d_0+S)+\sqrt{1/2}(d_0{}^++S^+)d_{-1}\right)$$
$$(5.18)$$

The set of six bosons $b^+(F^+xF^+)$, l=0,2 , and the set of six antibosons $b(FxF)$, l=0,2, transforming like irreducible representations, can be considered as "the boson" and "the antiboson" respectively.

There is no need to discuss the other symmetry chains of **su(6)** **IBM**. These are well known and they are not needed for the discussion in what follows.

The detailed discussion given above for the

$$\mathbf{su(6)} \quad \rightarrow \quad \mathbf{su(4)} \sim \mathbf{so(6)} \quad \rightarrow \quad \mathbf{so(5)} \quad \rightarrow \quad \mathbf{so(3)}_L$$

symmetry chain shows clearly that all results obtained by the standard **IBM** remain unchanged if the microscopic bosons $b^+(F^+xF)$ are substituted for the usual, unstructured bosons. The new features introduced by the analysis given in this article are that each set of bosons and antibosons, $b^+(F^+xF^+)$ and $b(FxF)$, is built from pairs of two species of nucleon and antinucleon states. Each fermion species F^+, (and thus F) is equipped with its own vacuum (p_1p_2) which contains fermion-antifermion pairs. These vacuum fermion-antifermion pairs dissappear as physical fermions are created. The vacuum for the two species of fermions F^+xF^+ is $p_1p_2xp_1p_2{\equiv}(p_1p_2)(p_1p_2)$. While the Lie algebras remain the same as in the standard **IBM**, the commutation relations for the microscopic boson operators are satisfied on $p_1p_2xp_1p_2$.

To summarize, the following transformation properties were found for the algebraic shell states and their direct products with respect to **su(4)** and **su(6)** :

su(4): F^+ x F^+ $b^+(F^+xF^+)$
            [1000]   x   [1000]   =[2000]+[1100]
su(6):   [100000]

$$(5.19)$$

with the **su(4)** representation [1100] tranforming according to the **su(6)** representation [100000].

su(4):   F   x   F $b(FxF)$
            [1110]   x   [1110]   =   [2220]+[1100]
su(6):   [111110]

$$(5.20)$$

with the **su(4)** representation [1100] tranforming according to the **su(6)** representation [111110].

Equations (5.19) and (5.20) illustrate the construction of the boson creation and annihilation operators.

su(4): F^+ x   F $\subset$ tensor algebra **TA**

            $F^+.F$ $=$ **su(4)** + **u(1)**
         [1000]   x   [1110]   =[2110]+[0000]

$$(5.21)$$

$$\begin{aligned}
\mathbf{su(6)}: \quad &\mathbf{b^+(F^+xF^+)\,x\ b(FxF)}=\text{tensor product basis of representation}\\
&\mathbf{b^+(F^+xF^+) \cdot b(FxF)}= \qquad \mathbf{su(6)} \qquad + \qquad \mathbf{u(1)}\\
&[100000]\ x\ [111110]= \qquad [211110] \qquad + \qquad [000000]\\
\mathbf{su(4)}: \quad &[1100] \quad x \quad [1100]= \qquad [2200]+[2110]+ \qquad [0000]\\
&\qquad\qquad\qquad \text{with } [2110]=\mathbf{su(4)}
\end{aligned}$$

$$(5.22)$$

Eq(5.21) and eq.(5.22) illustrate the construction of the algebras **su(4)** and **su(6)**, in terms of the operator boson states with internal fermionic structure which were introduced in this article. This construction is achieved by making use of the transformation properties of the operator states.

References

(1) B.R. Judd, in "Group Theory and its Applications I", ed. E.M. Loebl, (Academic Press, New York, 1968) 183

(2) B. Gruber, M. Lorente, T. Nomura and M. Ramek, "Symmetry Adaptation of Wave Functions and Matrix Elements", J. Phys. A: Math. Gen. **21** (1988) 4471

B. Gruber and M. Ramek, "Symmetry-Adapted Wave Functions and Matrix Elements: the General Case", J. Phys. A: Math. Gen. **24**, (1991) 5445

T. Nomura, M. Ramek and B. Gruber, Comput. Phys. Commun. **61** (1990) 410

M. Ramek and B. Gruber, Comput. Phys. Commun. **70** (1992) 371

(3) B. Gruber and M. Ramek, "Boson and Fermion Operator Realisations of su(4) and its Semisimple Subalgebras", in "Symmetries in Science VII", ed. B. Gruber and T. Otsuka (Plenum Press, New York, 1993)

(4) A. Arima and F. Iachello, Ann. Phys. N.Y. **99** (1976) 153

A. Arima and F. Iachello, Ann. Phys. N.Y. **111** (1978) 201

A. Arima and F. Iachello, Ann. Phys. N.Y. **115** (1978) 325

F. Iachello and S. Kuyucak, Ann. Phys. N.Y. **136** (1981) 19

(5) B. Gruber and M.T. Samuel, in "Group Theory and its Applications III", ed. E.M. Loebl (Academic Press, New York, 1975) 95

RECENT DEVELOPMENTS
IN THE APPLICATION
OF VECTOR COHERENT STATE THEORY

K. T. Hecht

Physics Department
University of Michigan
Ann Arbor, MI 48109

INTRODUCTION

In the past ten years two types of coherent state constructions have been used to great advantage to give the matrix representations of group generators and the Wigner coefficients of many higher rank symmetry groups. In both, the irreducible representations of a higher rank group are constructed by an induction process from the irreducible representations of a lower rank subgroup, the so-called core subgroup. In the more widely used first type of vector coherent state construction, [1, 2, 3], state vectors are mapped onto states of a multidimensional harmonic oscillator through a set of Bargmann variables, $\mathbf{z}$, the so-called "collective" or "orbital" variables, and a set of "intrinsic" or "spin" variables, q_i, which specify the states of the irreducible representates of the core subgroup, the full state vectors being constructed through a "vector-coupling" of the "intrinsic" and "collective" states. This VCS construction has been used for many of the mathematically natural group chains such as $U(n) \supset U(n-1)$ x $U(1) \supset U(n-2)$ x $U(1) \supset ...$ for which the subgroup chain gives a complete labelling of the state vectors. The VCS construction leads to $SU(n)$ Wigner coefficients which can be expressed in terms of recoupling (Racah or 9j) coefficients of the $U(n-1)$ sugroup, leading to a very simple buildup process for the construction of the Wigner-Racah calculus for the full group.

In the more recent second type of coherent state construction, rotor expansions are used which are particularly effective for many group chains of interest in actual physical applications for which an $SO(3)$ or $SU(2)$ subgroup, related to a physically meaningful angular momentum, is a necessary member of the group chain. This often leads to a missing label problem, since the physically meaningful subgroup chain does not give a complete labelling of the state vectors. Rotor coherent state constructions have very recently been used to advantage for three such group chains of particular interest in nuclear physics applications, the $SU(3) \supset SO(3) \supset SO(2)$ chain of the 3-dimensional harmonic oscillator [4], (with one missing label), the $SU(4) \supset SU(2)$ x $SU(2)$ Wigner supermultiplet [5], (two missing labels), and the $SO(5) \supset SO(3) \supset SO(2)$ chain [6, 7, 8], (two missing labels). Earlier solutions to the missing label problem have used the Elliott angular momentum projection technique and angular momentum projection labels, K, in all three cases. In the $SU(3) \supset SO(3)$ case, Elliott [9, 10] projects states of good angular momentum from a single intrinsic state, the highest weight state of the mathematically natural SU(3) basis. For the SU(4) $\supset$ SU(2) x SU(2) case Draayer [11] has used a spin and isospin space double angular momentum projection again from the single intrinsic SU(4) state of highest weight. For the totally symmetric irreducible representations of SO(5), of particular relevance for nuclear physics, Williams and Pursey [12] have employed the angular momentum projection technique with the use of *three* independent intrinsic states. The rotor coherent state expansion of Rowe, LeBlanc, and Repka [4] gives an elegant systematization of the Elliott angular momentum projection technique which can be taken over directly for the SU(4) $\supset$ SU(2) x SU(2) and SO(5) $\supset$ SO(3) schemes to greatly simplify calculations in these cases. The recent applications of rotor coherent state constructions [4, 5, 6] have focused on the derivation of the matrix elements of the group generators. The rotor coherent state method is, however, equally applicable for the derivation of matrix elements of arbitrary irreducible tensor operators and hence leads easily to the calculation of the higher symmetry Wigner coefficients in the physically relevant schemes. This will be illustrated by this contribution with the fundamental (4-dimensional) irreducible tensor operators of SU(4) and their conjugates. A brief synopsis will be given first of the rotor coherent state construction for the SU(3) $\supset$ SO(3) scheme and of the rotor coherent state derivation for the matrix representations of the SU(4) generators in the Wigner supermultiplet scheme.

ROTOR COHERENT STATE CONSTRUCTION FOR THE SU(3) $\supset$ SI(3) CASE

In the rotor coherent state construction of the SU(3) group an arbitrary state vector, $|\Psi>$, is transformed into its coherent state wave function, $\Psi(\Omega)$,

$$\Psi(\Omega) = \langle\phi_{(\lambda\mu)}|R(\Omega)|\Psi\rangle, \tag{1}$$

where $|\phi_{(\lambda\mu)}\rangle$ is the highest weight state in the SU(3) $\supset$ SU(2) x U(1) scheme in Elliott notation. Here, $R(\Omega)$ is the rotation operator

$$R(\Omega) = e^{i\alpha L_z}e^{i\beta L_y}e^{i\gamma L_z}, \tag{2}$$

where α, β, γ are Euler angles, L_i are space-fixed components of the orbital angular momentum, and where the scalar product is defined in terms of the angular measure

$$\int d\Omega = \int\int\int d\alpha sin\beta d\beta d\gamma. \tag{3}$$

If $|\Psi>$ is expanded in angular momentum eigenvectors $|\nu;LM>$, where ν is shorthand for all additional quantum numbers, these angular momentum base vectors are mapped into

$$\Psi_{\nu;LM}(\Omega) = \langle\phi_{(\lambda\mu)}|R(\Omega)|\nu;LM\rangle$$
$$= \sum_{K\geq0} c_K\{D^L_{KM}(\Omega) + (-1)^{\lambda+\mu+L}D^L_{-KM}(\Omega)\}\sqrt{\frac{2L+1}{16\pi^2(1+\delta_{K0})}}. \tag{4}$$

That is, c_{-K} is related to c_K by the factor $(-1)^{\lambda+\mu+L}$; and angular momentum eigenstates are mapped into a basis of symmetrized D-functions which form a simple orthonormal set (note the square-root factor) with respect to the rotational measure of eq. (3). Operators, $\mathbf{O}$, are then mapped into their coherent state realizations, $\Gamma(\mathbf{O})$, through

$$\Gamma(\mathbf{O})\Psi(\Omega) = \langle\phi_{(\lambda\mu)}|R(\Omega)\mathbf{O}|\Psi\rangle. \tag{5}$$

For spherical tensors of rank r, component m,

$$\Gamma(O_m^r)\Psi(\Omega) = \langle\phi_{(\lambda\mu)}|R(\Omega)O_m^r|\Psi\rangle$$
$$= \langle\phi_{(\lambda\mu)}|R(\Omega)O_m^r R(\Omega)^{-1}R(\Omega)|\Psi\rangle$$
$$= \sum_k D_{km}^r(\Omega)\langle\phi_{(\lambda\mu)}|O_k^r R(\Omega)|\Psi\rangle. \tag{6}$$

The SU(3) group generators are the 3 components of the orbital angular momentum operators, L_m^1, and the 5 components of the Elliott ($\lambda\mu$-preserving) quadrupole operator, Q_m^2. The rotor realizations $\Gamma(L_m^1)$ are given in terms of their usual Euler angle realizations

$$\Gamma(L_0) = \frac{1}{i}\frac{\partial}{\partial\gamma}, \qquad \Gamma(L_\pm) = e^{\pm i\gamma}\{i\cot\beta\frac{\partial}{\partial\gamma} \pm \frac{\partial}{\partial\beta}\}, \tag{7}$$

where $\Gamma(L_0)$ has eigenvalue M, while $\Gamma(L_+)$, $(\Gamma(L_-))$ are standard M-raising, (lowering) operators. Eq. (6) shows that both the standard (right-action) rotor realizations of operators O_m^r, as well as their left-action version which will be denoted by a $\bar\Gamma$,

$$\bar\Gamma(O_k^r)\Psi(\Omega) = \langle\phi_{(\lambda\mu)}|O_k^r R(\Omega)|\Psi\rangle \tag{8}$$

are useful. Now $\bar\Gamma(L_0)$ has eigenvalue K, whereas $\bar\Gamma(L_+)$, $(\bar\Gamma(L_-))$ are now K-lowering, (raising) operators, a well known property of the intrinsic (body-fixed) components of angular momentum operators.

The coherent state realizations of the quadrupole operator as given by Rowe, Le Blanc, and Repka [4] are

$$\Gamma(Q_m^2) = (2\lambda + \mu + 3)D_{0m}^2(\Omega) - \tfrac{1}{2}[\mathbf{L}^2, D_{0m}^2(\Omega)]$$
$$+\sqrt{\tfrac{3}{2}}\{D_{2m}^2(\Omega)(\mu - \bar\Gamma(L_0)) + D_{-2m}^2(\Omega)(\mu + \bar\Gamma(L_0))\}, \tag{9}$$

i.e., these are expressed in terms of the simple operators, $\mathbf{L}^2$, $\bar\Gamma(L_0)$, and simple D-functions. The well-known matrix elements of these D-functions in the orthonormal rotor basis, expressed by the symmetrized D-functions of eq. (4), at once lead to reduced matrix elements of $\Gamma(Q_m^2)$ in terms of simple angular momentum Wigner coefficients. The simplicity of this result is negated partly by the fact that the $\Gamma(Q_m^2)$ are nonunitary realizations of these operators. In order to translate the nonhermitian matrix elements of $\Gamma(Q_m^2)$ of rotor space into the hermitian matrix elements of Q_m^2 in ordinary Hilbert space, the nonunitary realizations, $\Gamma(\mathbf{O})$, of coherent state constructions must be converted to the unitary realizations, $\gamma(\mathbf{O})$ via the $\mathcal{K}$-operator transformation

$$\gamma(\mathbf{O}) = \mathcal{K}^{-1}\Gamma(\mathbf{O})\mathcal{K}. \tag{10}$$

Matrix elements of the $\mathcal{K}$ and $\mathcal{K}^{-1}$ operators can then be used to convert the nonhermitian matrix elements of $\Gamma(\mathbf{O})$ to hermitian form $\gamma(\mathbf{O})$ and hence directly to hermitian form in ordinary Hilbert space. Thus, the angular momentum reduced matrix elements of Q_m^2 are given by

$$\langle\nu'; L'||Q^2||\nu; L\rangle = \sum_{K,K'} (\mathcal{K}^{-1}(L'))_{\nu'K'}\langle K'; L'||\Gamma(Q^2)||K; L\rangle(\mathcal{K}(L))_{K\nu}, \tag{11}$$

where the new quantum numbers, ν, are defined through the eigenvalues of the hermitian matrix $(\mathcal{K}\mathcal{K}^\dagger)_{K'K}$ which can be calculated by simple recursion techniques and can be given in terms of simple analytical formulae for many of the simpler irreducible representations. The $(\mathcal{K}\mathcal{K}^\dagger)$ matrix, which (except for a normalization factor) is also related to the overlap matrix of the Elliott angular momentum projected states with K and K', can be converted into the needed matrix elements of $\mathcal{K}$ and $\mathcal{K}^{-1}$ through the unitary matrix, U, which diagonalizes the hermitian matrix $(\mathcal{K}\mathcal{K}^\dagger)$

$$(U(\mathcal{K}\mathcal{K}^\dagger)U^\dagger)_{\nu'\nu} = \Lambda_\nu \delta_{\nu'\nu} \tag{12}$$

with

$$(\mathcal{K})_{K\nu} = U^\dagger_{K\nu}\sqrt{\Lambda_\nu}; \qquad (\mathcal{K}^{-1})_{\nu K} = \frac{1}{\sqrt{\Lambda_\nu}}U_{\nu K} \tag{13}$$

defined for all states ν with non zero eigenvalue, Λ_ν. A zero eigenvalue Λ_ν signals a forbidden state, and the recursive process for the calculation of $\mathcal{K}\mathcal{K}^\dagger$ terminates at the boundary of allowed L-values of a given irreducible representation. The $\mathcal{K}$-matrix technique of coherent state theory thus effectively converts the Elliott K-projection label to the status of a good quantum number, ν.

It should, however, be stressed that the rotor coherent state construction outlined here is very closely related to the Elliott angular momentum projection technique [10] for SU(3). The matrix elements for Q^2_m are precisely those given by Elliott [10]. What then are the advantages of the coherent state rotor construction? By mapping the SU(3) angular momentum eigenstates onto the orthonormal symmetrized basis of D-functions, (see eq. (4)), of the rotor expansion, the construction of matrix elements is split into two clearly separated simple steps: In step 1, matrix elements of $\Gamma(\mathbf{O})$ are given very simply in the orthonormal rotor basis where K defines the orthonormal states. In step 2, which is the unitarization process, K is converted to the quantum number ν in ordinary Hilbert space. By relating ν to the nonzero eigenvalues of $(\mathcal{K}\mathcal{K}^\dagger)$ an author-independent choice can be made for the quantum number ν. Although some numerical work is required in the determination of the U-matrix elements which diagonalize the multi-dimensional $(\mathcal{K}\mathcal{K}^\dagger)$-matrices; no arbitrary choices are made in a Gram-Schmidt orthonormalization process.

A DOUBLE ROTOR COHERENT STATE EXPANSION FOR THE WIGNER SUPERMULTIPLET SU(4) $\supset$ SU(2) $\times$ SU(2)

Since the derivation of the generator matrix elements for this case has been published [5], only a brief synopsis will be given here. A complete labelling scheme for the Wigner supermultiplet has been achieved by Draayer [11] who used the Elliott angular momentum projection technique to augment the spin and isospin quantum numbers (SM_S), (TM_T) with the projection labels K_S and K_T. In order to calculate the generator matrix elements and SU(4) reduced Wigner coefficients in this fully labelled but nonorthogonal basis, however, Draayer first calculates the transformation coefficients to the canonical fully specified orthonormal U(4) $\supset$ U(3) $\supset$ U(2) $\supset$ U(1) basis, leading to a somewhat laborious calculational algorithm. This example therefore will fully illustrate the power of the rotor coherent state construction which leads in a very simple and direct way to the desired results.

The supermultiplet scheme is based on the four spin-charge states of a single nucleon, $|m_s m_t\rangle$, with $m_s = \pm\frac{1}{2}, m_t = \pm\frac{1}{2}$. To gain the most convenient double rotor expansion it will be useful to define the basis states $|i\rangle$, $i = 1, \ldots, 4$, by

$$|+\tfrac{1}{2}+\tfrac{1}{2}\rangle = \frac{1}{\sqrt{2}}(|1\rangle + |2\rangle), \qquad |-\tfrac{1}{2}-\tfrac{1}{2}\rangle = \frac{1}{\sqrt{2}}(-|1\rangle + |2\rangle),$$

$$|+\tfrac{1}{2}-\tfrac{1}{2}\rangle = \frac{1}{\sqrt{2}}(|3\rangle + |4\rangle), \qquad |-\tfrac{1}{2}+\tfrac{1}{2}\rangle = \frac{1}{\sqrt{2}}(-|3\rangle + |4\rangle), \tag{14}$$

196

and define the 15 supermultiplet generators [13], $\mathbf{S},\mathbf{T}$, and $\mathbf{E} = \sigma\tau$ in terms of U(4) generators, C_{ij},

$$C_{ij} = \sum_\alpha a_{i\alpha}^\dagger a_{j\alpha}, \qquad i,j = 1\ldots 4, \tag{15}$$

where the $\mathbf{a}^\dagger$, $\mathbf{a}$ could be chosen either as fermion or boson creation, annihilation operators. For later purposes it will be useful to use the boson realization so that i, j specify the spin, isospin quantum numbers in the form defined by eq. (14), and α can be interpreted as a particle index with particle numbers which also range over four possible values. In terms of these C_{ij} the generators are

$$
\begin{aligned}
S_0 &= \tfrac{1}{2}(C_{12} + C_{21} + C_{34} + C_{43})\\
S_+ &= \tfrac{1}{2}(-C_{13} - C_{23} + C_{14} + C_{24} - C_{31} + C_{32} - C_{41} + C_{42})\\
S_- &= \tfrac{1}{2}(-C_{31} - C_{32} + C_{41} + C_{42} - C_{13} + C_{23} - C_{14} + C_{24})\\
T_0 &= \tfrac{1}{2}(C_{12} + C_{21} - C_{34} - C_{43})\\
T_+ &= \tfrac{1}{2}(C_{13} + C_{23} + C_{14} + C_{24} + C_{31} - C_{32} - C_{41} + C_{42})\\
T_- &= \tfrac{1}{2}(C_{31} + C_{32} + C_{41} + C_{42} + C_{13} - C_{23} - C_{14} + C_{24})\\
E_{00} &= \tfrac{1}{2}(C_{11} + C_{22} - C_{33} - C_{44})\\
E_{10} &= \tfrac{1}{2\sqrt{2}}(C_{13} + C_{23} - C_{14} - C_{24} - C_{31} + C_{32} - C_{41} + C_{42}),\\
E_{-10} &= \tfrac{1}{2\sqrt{2}}(-C_{31} - C_{32} + C_{41} + C_{42} + C_{13} - C_{23} + C_{14} - C_{24}),\\
E_{01} &= \tfrac{1}{2\sqrt{2}}(-C_{13} - C_{23} - C_{14} - C_{24} + C_{31} - C_{32} + C_{42} - C_{41}),\\
E_{0-1} &= \tfrac{1}{2\sqrt{2}}(C_{31} + C_{32} + C_{41} + C_{42} - C_{13} + C_{23} - C_{24} + C_{14}),\\
E_{11} &= \tfrac{1}{2}(-C_{11} + C_{22} + C_{12} - C_{21}),\\
E_{-1-1} &= \tfrac{1}{2}(-C_{11} + C_{22} - C_{12} + C_{21}),\\
E_{1-1} &= \tfrac{1}{2}(C_{33} - C_{44} - C_{34} + C_{43}),\\
E_{-11} &= \tfrac{1}{2}(C_{33} - C_{44} + C_{34} - C_{43}).
\end{aligned}
\tag{16}
$$

The SU(4) irreducible representations are labelled by 4-rowed Young tableaux partition labels [f_1,f_2,f_3,f_4], by the SU(4) labels $\{\lambda_1, \lambda_2, \lambda_3\}$, or by the Wigner supermultiplet (or standard Cartan SO(6) labels (P, P', P'')), with

$$
\begin{aligned}
\lambda_1 &= f_1 - f_2, \qquad \lambda_2 = f_2 - f_3, \qquad \lambda_3 = f_3 - f_4,\\
P &= \tfrac{1}{2}(\lambda_1 + 2\lambda_2 + \lambda_3), \quad P' = \tfrac{1}{2}(\lambda_1 + \lambda_3), \quad P'' = \tfrac{1}{2}(\lambda_1 - \lambda_3).
\end{aligned}
\tag{17}
$$

These characterize the highest weight state $|\phi\rangle \equiv |\phi_{\{\lambda_1,\lambda_2,\lambda_3\}}\rangle$, with

$$
\begin{aligned}
C_{ij}|\phi\rangle &= 0 \quad \text{for} \quad i < j,\\
C_{11}|\phi\rangle &= (\lambda_1 + \lambda_2 + \lambda_3)|\phi\rangle, \qquad C_{22}|\phi\rangle = (\lambda_2 + \lambda_3)|\phi\rangle,\\
C_{33}|\phi\rangle &= \lambda_3|\phi\rangle, \qquad C_{44}|\phi\rangle = 0.
\end{aligned}
\tag{18}
$$

The double rotor expansion uses the double rotation operator $R(\Omega) \equiv R(\Omega_S)R(\Omega_T)$, with Euler angles $\alpha_S, \beta_S, \gamma_S \equiv \Omega_S$ and $\alpha_T, \beta_T, \gamma_T \equiv \Omega_T$ in the spin and isospin space. Draayer [11] has shown that the set of states, $\{R(\Omega)|\phi\rangle\}$, obtained by rotation of the highest weight state through all possible angles $\alpha_S, \ldots, \gamma_T$ span the full SU(4) space. Arbitrary state vectors $|\Psi\rangle$ in this space are now tranformed into their coherent state realizations with coherent state wave function

$$\Psi(\Omega) = \langle\phi|R(\Omega)|\Psi\rangle. \tag{19}$$

A state $|\alpha SM_STM_T\rangle$ with definite spin and isospin quantum numbers is represented by

$$\Psi_{\alpha SM_STM_T}(\Omega) = \langle\phi|R(\Omega)|\alpha SM_STM_T\rangle$$
$$= \sum_{K_S,K_T} \langle\phi|\alpha SK_STK_T\rangle D^S_{K_SM_S}(\Omega_S)D^T_{K_TM_T}(\Omega_T). \tag{20}$$

Draayer [11] has shown that the SU(4) irreducible representations $[f_1, f_2, f_3, f_4] \equiv \{\lambda_1\lambda_2\lambda_3\}$ are spanned by the double rotor wave functions with K_S, K_T-values restricted by

$$(K_S + K_T) = \pm\lambda_1, \pm(\lambda_1 - 2), \pm(\lambda_1 - 4),\ldots, 0(or \pm 1),$$
$$(K_S - K_T) = \pm\lambda_3, \pm(\lambda_3 - 2), \pm(\lambda_3 - 4),\ldots, 0(or \pm 1). \tag{21}$$

The symmetries of the $\langle\phi|\alpha SK_STK_T\rangle$ of eq. (20) again lead to symmetrized combinations of the D-functions. The double rotor coherent state wave functions are thus spanned by the symmetrized (normalized) double rotor functions

$$\frac{1}{8\pi^2}\left[\frac{(2S + 1)(2T + 1)}{2(1 + \delta_{K_S0}\delta_{K_T0})}\right]^{\frac{1}{2}}$$
$$\times\{D^S_{K_SM_S}(\Omega_S)D^T_{K_TM_T}(\Omega_T) + (-1)^{\lambda_2+\lambda_3+S+T}D^S_{-K_SM_S}(\Omega_S)D^T_{-K_TM_T}(\Omega_T)\}, \tag{22}$$

and it will therefore be sufficient to choose $K_S \geq 0$, and for $K_S = 0 : K_T \geq 0$. The requirement $S \geq |K_S|, T \geq |K_T|$ together with the structure of the $\mathcal{KK}^\dagger$-matrices will determine the multiplicity of a given S, T value. For states with low values of $S + T$, for which the eigenvalues of $\mathcal{KK}^\dagger$ are all nonzero (no redundant states), the number of occurrences of a given S, T will be determined by the number of possible K_S, K_T combinations. For larger values of $S + T$, zero eigenvalues of $\mathcal{KK}^\dagger$ will signal the appearance of forbidden states. The recursive process for the calculation of the $\mathcal{KK}^\dagger$ matrix elements terminates at the maximum possible value of S + T which is given by $S + T = \lambda_1 + \lambda_2 + \lambda_3$, with S(or T) $\geq \frac{1}{2}(\lambda_1 + \lambda_3)$; with an occurrence of 1 for these maximal S, T-values.

In the coherent state rotor expansion operators, $\mathbf{O}$, are transformed into their coherent state realizations, $\Gamma(\mathbf{O})$, through $\mathbf{O}|\Psi\rangle \to \Gamma(\mathbf{O})\Psi(\Omega)$, by the double rotor analogue of eq. (5). The SU(4) generators, $\mathbf{O} = \mathbf{S}, \mathbf{T}, \mathbf{E}$ are the first operators to be considered. Now $\Gamma(S_0)$ has the simple eigenvalue M_S whereas $\bar{\Gamma}(S_0)$ has eigenvalue K_S; while $\Gamma(S_+)$, $(\Gamma(S_-)$ are M_S-raising, (lowering) operators, whereas $\bar{\Gamma}(S_+)$, $(\bar{\Gamma}(S_-))$ are K_S-lowering, (raising) operators; with similar properties for the $\Gamma(\mathbf{T})$ and $\bar{\Gamma}(\mathbf{T})$.

The generators $\mathbf{E}$ are double spherical tensors $O^{r_Sr_T}_{m_Sm_T}$ with $r_S = r_T = 1$ and can be transformed into left-action operators via

$$\Gamma(E_{m_Sm_T})\Psi(\Omega) = \langle\phi|R(\Omega)E_{m_Sm_T}|\Psi\rangle$$
$$= \langle\phi|(R(\Omega)E_{m_Sm_T}R^{-1}(\Omega))R(\Omega)|\Psi\rangle$$
$$= \sum_{k_Sk_T} D^1_{k_Sm_S}(\Omega_S)D^1_{k_Tm_T}(\Omega_T)\langle\phi|E_{k_Sk_T}R(\Omega)|\Psi\rangle. \tag{23}$$

Using the properties of the highest weight state, eq. (18), and the specific expressions of the generators, eqs. (16), it can be seen that

$$\langle\phi|E_{00} = \tfrac{1}{2}(\lambda_1 + 2\lambda_2 + \lambda_3)\langle\phi|,$$
$$\langle\phi|E_{\pm10} = -\tfrac{1}{\sqrt{2}}\langle\phi|S_\pm, \qquad \langle\phi|E_{0\pm1} = -\tfrac{1}{\sqrt{2}}\langle\phi|T_\pm,$$
$$\langle\phi|E_{\pm1\pm1} = \tfrac{1}{2}\langle\phi|(-\lambda_1 \pm S_0 \pm T_0), \qquad \langle\phi|E_{\pm1\mp1} = \tfrac{1}{2}\langle\phi|(\lambda_3 \mp S_0 \pm T_0)). \tag{24}$$

At this stage the usefulness of the transformation (14) can be appreciated. Although it seemingly complicates the relations of the group generators in terms of the C_{ij}, it can now be seen that the transformation (14) makes it possible to express the operators $E_{k_S k_T}$ in their left actions on the single highest weight state into equivalent left actions of components of $\mathbf{S}$ or $\mathbf{T}$ or the Cartan generators C_{ii}. The relations (24) lead to

$$
\begin{aligned}
\Gamma(E_{m_S m_T})\Psi(\Omega) =\ & \{\tfrac{1}{2}(\lambda_1 + 2\lambda_2 + \lambda_3)D^1_{0m_S}(\Omega_S)D^1_{0m_T}(\Omega_T) \\
& - \tfrac{1}{\sqrt{2}}\left[D^1_{1m_S}(\Omega_S)\bar{\Gamma}(S_+) + D^1_{-1m_S}(\Omega_S)\bar{\Gamma}(S_-)\right]D^1_{0m_T}(\Omega_T) \\
& - \tfrac{1}{\sqrt{2}}D^1_{0m_S}(\Omega_S)\left[D^1_{1m_T}(\Omega_T)\bar{\Gamma}(T_+) + D^1_{-1m_T}(\Omega_T)\bar{\Gamma}(T_-)\right] \\
& + \tfrac{1}{2}D^1_{1m_S}(\Omega_S)D^1_{1m_T}(\Omega_T)(-\lambda_1 + \bar{\Gamma}(S_0) + \bar{\Gamma}(T_0)) \\
& + \tfrac{1}{2}D^1_{-1m_S}(\Omega_S)D^1_{-1m_T}(\Omega_T)(-\lambda_1 - \bar{\Gamma}(S_0) - \bar{\Gamma}(T_0)) \\
& + \tfrac{1}{2}D^1_{1m_S}(\Omega_S)D^1_{-1m_T}(\Omega_T)(\lambda_3 - \bar{\Gamma}(S_0) + \bar{\Gamma}(T_0)) \\
& + \tfrac{1}{2}D^1_{-1m_S}(\Omega_S)D^1_{1m_T}(\Omega_T)(\lambda_3 + \bar{\Gamma}(S_0) - \bar{\Gamma}(T_0))\}\langle\phi|R(\Omega)|\Psi\rangle. \quad (25)
\end{aligned}
$$

Finally, using the identity

$$
[\mathbf{S}^2, D^1_{0m_S}(\Omega_S)] = \sqrt{2}(D^1_{1m_S}(\Omega_S)\bar{\Gamma}(S_+) + D^1_{-1m_S}(\Omega_S)\bar{\Gamma}(S_-)) + 2D^1_{0m_S}(\Omega_S), \quad (26)
$$

and the similar relation for the isospin operators, we obtain

$$
\begin{aligned}
\Gamma(E_{m_S m_T}) =\ & \{\tfrac{1}{2}(\lambda_1 + 2\lambda_2 + \lambda_3) + 2\}D^1_{0m_S}(\Omega_S)D^1_{0m_T}(\Omega_T) \\
& - \tfrac{1}{2}\{[\mathbf{S}^2, D^1_{0m_S}(\Omega_S)]D^1_{0m_T}(\Omega_T) + D^1_{0m_S}(\Omega_S)[\mathbf{T}^2, D^1_{0m_T}(\Omega_T)]\} \\
& + \tfrac{1}{2}D^1_{1m_S}(\Omega_S)D^1_{1m_T}(\Omega_T)(-\lambda_1 + \bar{\Gamma}(S_0) + \bar{\Gamma}(T_0)) \\
& + \tfrac{1}{2}D^1_{-1m_S}(\Omega_S)D^1_{-1m_T}(\Omega_T)(-\lambda_1 - \bar{\Gamma}(S_0) - \bar{\Gamma}(T_0)) \\
& + \tfrac{1}{2}D^1_{1m_S}(\Omega_S)D^1_{-1m_T}(\Omega_T)(\lambda_3 - \bar{\Gamma}(S_0) + \bar{\Gamma}(T_0)) \\
& + \tfrac{1}{2}D^1_{-1m_S}(\Omega_S)D^1_{1m_T}(\Omega_T)(\lambda_3 + \bar{\Gamma}(S_0) - \bar{\Gamma}(T_0)). \quad (27)
\end{aligned}
$$

This is the analogue of eq. (9); and as in that case leads at once to the reduced matrix elements of $\Gamma(\mathbf{E})$, expressed in terms of simple spin and isospin Wigner coefficients with magnetic quantum number K_S, K_T. The detailed matrix element expressions are given in ref. [5]. This contribution will focus on the matrix elements of arbitrary irreducible tensor operators which, unlike the group generators, will in general induce changes in the irreducible representations. The most important irreducible tensors are the fundamental tensors and their conjugates, since more complicated tensors can be constructed through these by a buildup process.

THE FUNDAMENTAL TENSORS AND THEIR SU(4) ⊃ SU(2) × SU(2) MATRIX ELEMENTS

The fundamental tensors transform according to the 4-dimensional irreducible representation $[f_1 f_2 f_3 f_4]$=[1000] or $\{\lambda_1\lambda_2\lambda_3\} = \{100\}$, the conjugate tensors transform according to the irrep. [1110] or $\{\lambda_1\lambda_2\lambda_3\} = \{001\}$. The 4-components transform like double spherical tensors $O^{r_S r_T}_{m_S m_T}$ with $r_S = \tfrac{1}{2}, r_T = \tfrac{1}{2}$. Application of the analogue of eq. (23) to these double spherical tensors will thus lead to the needed matrix elements if the left actions of the $O^{r_S r_T}_{m_S m_T}$ on the highest weight states can be calculated. The fundamental tensors can add one square to one row of an arbitrary tableau characterized by $[f_1 f_2 f_3 f_4]$. It will be useful to construct very specific shift tensors of Biedenharn-Louck form, (see e.g. ref.[14]), which add one square to a specific i^{th} row of the tableau and denote these by the shift index Δ_i=1. Their conjugates have shift index Δ_i=-1.

(For economy of space we avoid the upper Gelfand pattern notation [14] of Biedenharn and Louck and use the simple shift label instead). It is now easy to give very specific constructions for both these shift tensors as well as the highest weight states $|\phi\rangle \equiv |\phi_{\{\lambda_1\lambda_2\lambda_3\}}\rangle$ in terms of the boson operators $a_{i\alpha}^\dagger, a_{j\alpha}$ of eq. (15) in order to make the needed calculation. For this purpose it is useful to classify the state vectors under U(4) x U$'$(4) symmetry; i.e. the particle index α as well as the spin, isospin index i will be allowed to range over four values with i=1,2,3,4, (see eq. (14)), and α=a,b,c,d, where the generators for U(4) are the C_{ij} given by eq. (15), while the U$'$(4) generators are given by

$$B_{pq} = \sum_{i=1}^{4} a_{ip}^\dagger a_{iq}, \qquad \text{with} \quad p,q = a,b,c,d. \tag{28}$$

For most cases it will be sufficient to consider SU(4) highest weight states given by [15, 16]

$$|\phi_{\{\lambda_1\lambda_2\lambda_3\}}\rangle = N_H (\Delta_a^1)^{\lambda_1} (\Delta_{ab}^{12})^{\lambda_2} (\Delta_{abc}^{123})^{\lambda_3} |0\rangle, \tag{29}$$

with

$$\Delta_a^1 = a_{1a}^\dagger, \qquad (\Delta_{ab}^{12}) = \begin{vmatrix} a_{1a}^\dagger & a_{1b}^\dagger \\ a_{2a}^\dagger & a_{2b}^\dagger \end{vmatrix}, \qquad (\Delta_{abc}^{123}) = \begin{vmatrix} a_{1a}^\dagger & a_{1b}^\dagger & a_{1c}^\dagger \\ a_{2a}^\dagger & a_{2b}^\dagger & a_{2c}^\dagger \\ a_{3a}^\dagger & a_{3b}^\dagger & a_{3c}^\dagger \end{vmatrix}. \tag{30}$$

For shift tensors with $\Delta_4 \neq 0$, it may also be useful to use the U(4) highest weight state

$$|\phi_{\{\lambda_1\lambda_2\lambda_3 f_4\}}\rangle = N_H (\Delta_a^1)^{\lambda_1} (\Delta_{ab}^{12})^{\lambda_2} (\Delta_{abc}^{123})^{\lambda_3} (\Delta_{abcd}^{1234})^{f_4} |0\rangle, \tag{31}$$

with normalization factors N_H [15] given in both cases by

$$N_H = \Big[\prod_{\nu>\mu=1}^{4} (f_\mu - f_\nu + \nu - \mu) / \prod_{\mu=1}^{4} (f_\mu + 4 - \mu)! \Big]^{\frac{1}{2}}. \tag{32}$$

As a very specific example we consider first the fundamental tensors which induce shifts Δ_3=1 to be denoted by $T_i^{\Delta_3=1}$ or $T_{m_S m_T}^{\Delta_3=1}$ in the double spherical tensor form. From the fact that a$_{ic}$ must annihilate an excitation in the third row of the highest weight state $|\phi_{\{\lambda_1\lambda_2\lambda_3\}}\rangle$ of eq. (30), it can be seen that right actions of a$_{ic}$ on $|\phi_{\{\lambda_1\lambda_2\lambda_3\}}\rangle$ convert this to a state $|\phi_{\{\lambda_1\lambda_2+1\lambda_3-1\}}\rangle$. In particular, straightforward calculation yields

$$\begin{aligned} a_{3c}|\phi_{\{\lambda_1\lambda_2\lambda_3\}}\rangle &= \lambda_3 \frac{N_H(\{\lambda_1\lambda_2\lambda_3\})}{N_H(\{\lambda_1\lambda_2+1\lambda_3-1\})} |\phi_{\{\lambda_1\lambda_2+1\lambda_3-1\}}\rangle \\ &= \sqrt{\frac{\lambda_3(\lambda_2+1)(\lambda_1+\lambda_2+2)}{(\lambda_2+2)(\lambda_1+\lambda_2+3)}} |\phi_{\{\lambda_1\lambda_2+1\lambda_3-1\}}\rangle, \end{aligned} \tag{33}$$

$$a_{2c}|\phi_{\{\lambda_1\lambda_2\lambda_3\}}\rangle = -\sqrt{\frac{\lambda_3(\lambda_1+\lambda_2+2)}{(\lambda_2+1)(\lambda_2+2)(\lambda_1+\lambda_2+3)}} C_{32}|\phi_{\{\lambda_1\lambda_2+1\lambda_3-1\}}\rangle, \tag{34}$$

$$\begin{aligned} a_{1c}|\phi_{\{\lambda_1\lambda_2\lambda_3\}}\rangle &= \sqrt{\frac{\lambda_3}{(\lambda_2+1)(\lambda_2+2)(\lambda_1+\lambda_2+2)(\lambda_1+\lambda_2+3)}} \\ &\quad \times \{C_{32}C_{21} - C_{31}(\lambda_2+2)\}|\phi_{\{\lambda_1\lambda_2+1\lambda_3-1\}}\rangle, \end{aligned} \tag{35}$$

$$a_{4c}|\phi_{\{\lambda_1\lambda_2\lambda_3\}}\rangle = 0. \tag{36}$$

With $T^{\Delta_3=1}_{m_S m_T} = a^{\dagger}_{m_S m_T, c}$, the analogue of eq. (23), with the use of eq. (14) to convert the $a^{\dagger}_{m_S m_T, c}$ to $a^{\dagger}_{ic}$, yields

$$\Gamma(T^{\Delta_3=1}_{m_S m_T})\Psi(\Omega) = \langle \phi_{\{\lambda_1 \lambda_2 \lambda_3\}} | R(\Omega) a^{\dagger}_{m_S m_T, c} | \Psi \rangle$$

$$= \frac{1}{\sqrt{2}} (D^{\frac{1}{2}}_{\frac{1}{2} m_S}(\Omega_S) D^{\frac{1}{2}}_{\frac{1}{2} m_T}(\Omega_T) - D^{\frac{1}{2}}_{-\frac{1}{2} m_S}(\Omega_S) D^{\frac{1}{2}}_{-\frac{1}{2} m_T}(\Omega_T)) \langle \phi_{\{\lambda_1 \lambda_2 \lambda_3\}} | a^{\dagger}_{1c} R(\Omega) | \Psi \rangle$$

$$+ \frac{1}{\sqrt{2}} (D^{\frac{1}{2}}_{\frac{1}{2} m_S}(\Omega_S) D^{\frac{1}{2}}_{\frac{1}{2} m_T}(\Omega_T) + D^{\frac{1}{2}}_{-\frac{1}{2} m_S}(\Omega_S) D^{\frac{1}{2}}_{-\frac{1}{2} m_T}(\Omega_T)) \langle \phi_{\{\lambda_1 \lambda_2 \lambda_3\}} | a^{\dagger}_{2c} R(\Omega) | \Psi \rangle$$

$$+ \frac{1}{\sqrt{2}} (D^{\frac{1}{2}}_{\frac{1}{2} m_S}(\Omega_S) D^{\frac{1}{2}}_{-\frac{1}{2} m_T}(\Omega_T) - D^{\frac{1}{2}}_{-\frac{1}{2} m_S}(\Omega_S) D^{\frac{1}{2}}_{\frac{1}{2} m_T}(\Omega_T)) \langle \phi_{\{\lambda_1 \lambda_2 \lambda_3\}} | a^{\dagger}_{3c} R(\Omega) | \Psi \rangle$$

$$+ \frac{1}{\sqrt{2}} (D^{\frac{1}{2}}_{\frac{1}{2} m_S}(\Omega_S) D^{\frac{1}{2}}_{-\frac{1}{2} m_T}(\Omega_T) + D^{\frac{1}{2}}_{\frac{1}{2} m_S}(\Omega_S) D^{\frac{1}{2}}_{-\frac{1}{2} m_T}(\Omega_T)) \langle \phi_{\{\lambda_1 \lambda_2 \lambda_3\}} | a^{\dagger}_{4c} R(\Omega) | \Psi \rangle$$

$$= \sqrt{\frac{\lambda_3}{2(\lambda_2 + 1)(\lambda_2 + 2)(\lambda_1 + \lambda_2 + 2)(\lambda_1 + \lambda_2 + 3)}}$$

$$\times \ \{(D^{\frac{1}{2}}_{\frac{1}{2} m_S}(\Omega_S) D^{\frac{1}{2}}_{\frac{1}{2} m_T}(\Omega_T) - D^{\frac{1}{2}}_{-\frac{1}{2} m_S}(\Omega_S) D^{\frac{1}{2}}_{-\frac{1}{2} m_T}(\Omega_T))$$

$$\times \ \langle \phi_{\{\lambda_1 \lambda_2 + 1 \lambda_3 - 1\}} | \{C_{12} C_{23} - C_{13}(\lambda_2 + 2)\} R(\Omega) | \Psi \rangle$$

$$- \ (\lambda_1 + \lambda_2 + 2)(D^{\frac{1}{2}}_{\frac{1}{2} m_S}(\Omega_S) D^{\frac{1}{2}}_{\frac{1}{2} m_T}(\Omega_T) + D^{\frac{1}{2}}_{-\frac{1}{2} m_S}(\Omega_S) D^{\frac{1}{2}}_{-\frac{1}{2} m_T}(\Omega_T))$$

$$\times \ \langle \phi_{\{\lambda_1 \lambda_2 + 1 \lambda_3 - 1\}} C_{23} R(\Omega) | \Psi \rangle$$

$$+ \ (\lambda_2 + 1)(\lambda_1 + \lambda_2 + 2)(D^{\frac{1}{2}}_{\frac{1}{2} m_S}(\Omega_S) D^{\frac{1}{2}}_{-\frac{1}{2} m_T}(\Omega_T) - D^{\frac{1}{2}}_{-\frac{1}{2} m_S}(\Omega_S) D^{\frac{1}{2}}_{\frac{1}{2} m_T}(\Omega_T))$$

$$\times \ \langle \phi_{\{\lambda_1 \lambda_2 + 1 \lambda_3 - 1\}} | R(\Omega) | \Psi \rangle \}, \tag{37}$$

where we have used the hermitian conjugate of eqs. (33)-(36) in the last step. Now, inverting eqs. (16) and using the highest weight properties of eq. (24), we obtain

$$\langle \phi_{\{\lambda_1 \lambda_2 + 1 \lambda_3 - 1\}} | C_{23} = \tfrac{1}{2} \langle \phi_{\{\lambda_1 \lambda_2 + 1 \lambda_3 - 1\}} | (S_- - S_+ + T_+ - T_-), \tag{38}$$

$$\langle \phi_{\{\lambda_1 \lambda_2 + 1 \lambda_3 - 1\}} | \{C_{12} C_{23} - C_{13}(\lambda_2 + 2)\} =$$
$$\tfrac{1}{2} \langle \phi_{\{\lambda_1 \lambda_2 + 1 \lambda_3 - 1\}} | \{(S_+ - T_+)(\lambda_2 + 1 - S_0 - T_0) + (S_- - T_-)(\lambda_2 + 1 + S_0 + T_0). \tag{39}$$

This leads to

$$\langle \phi_{\{\lambda_1 \lambda_2 \lambda_3\}} | R(\Omega) T^{\Delta_3=1}_{m_S m_T} | \Psi \rangle$$

$$= \frac{1}{2} \sqrt{\frac{\lambda_3}{2(\lambda_2 + 1)(\lambda_2 + 2)(\lambda_1 + \lambda_2 + 2)(\lambda_1 + \lambda_2 + 3)}}$$

$$\times \ \{(D^{\frac{1}{2}}_{\frac{1}{2} m_S}(\Omega_S) D^{\frac{1}{2}}_{-\frac{1}{2} m_T}(\Omega_T) - D^{\frac{1}{2}}_{-\frac{1}{2} m_S} D^{\frac{1}{2}}_{\frac{1}{2} m_T}(\Omega_T))[2(\lambda_2 + 1)(\lambda_1 + \lambda_2 + 2)]$$

$$+ \ D^{\frac{1}{2}}_{\frac{1}{2} m_S}(\Omega_S) D^{\frac{1}{2}}_{\frac{1}{2} m_T}(\Omega_T)[\bar{\Gamma}(S_+)(\lambda_1 + 2\lambda_2 + 3 - \bar{\Gamma}(S_0) - \bar{\Gamma}(T_0))$$

$$- \ \bar{\Gamma}(S_-)(\lambda_1 + 1 - \bar{\Gamma}(S_0) - \bar{\Gamma}(T_0)) + \bar{\Gamma}(T_-)(\lambda_1 + 1 - \bar{\Gamma}(S_0) - \bar{\Gamma}(T_0))$$

$$- \ \bar{\Gamma}(T_+)(\lambda_1 + 2\lambda_2 + 3 - \bar{\Gamma}(S_0) - \bar{\Gamma}(T_0))]$$

$$+ \ D^{\frac{1}{2}}_{-\frac{1}{2} m_S}(\Omega_S) D_{-\frac{1}{2} m_T}(\Omega_T)[\bar{\Gamma}(S_+)(\lambda_1 + 1 + \bar{\Gamma}(S_0) + \bar{\Gamma}(T_0))$$

$$
\begin{aligned}
- \quad & \bar{\Gamma}(S_-)(\lambda_1 + 2\lambda_2 + 3 + \bar{\Gamma}(S_0) + \bar{\Gamma}(T_0)) + \bar{\Gamma}(T_-)(\lambda_1 + 2\lambda_2 + 3 + \bar{\Gamma}(S_0) + \bar{\Gamma}(T_0)) \\
- \quad & \bar{\Gamma}(T_+)(\lambda_1 + 1 + \bar{\Gamma}(S_0) + \bar{\Gamma}(T_0))] \Big\} \\
\times \quad & \langle \phi_{\{\lambda_1 \lambda_2 + 1 \lambda_3 - 1\}} | R(\Omega) | \Psi \rangle .
\end{aligned} \tag{40}
$$

The matrix elements of $\bar{\Gamma}(S_+)$, $\bar{\Gamma}(S_+)\bar{\Gamma}(S_0)$, etc. follow from the expansion of $|\Psi\rangle$ in terms of basis states of good $SM_S T M_T$ as in eq. (20). E.g. the matrix element of $\bar{\Gamma}(S_+)\bar{\Gamma}(S_0)$ in the rotor basis follows from

$$
\begin{aligned}
& \sum_{K_S K_T} \langle \phi | S_+ S_0 | \gamma S K_S T K_T \rangle D^S_{K_S M_S}(\Omega_S) D^T_{K_T M_T}(\Omega_T) \\
= \quad & \sum_{K_S K_T} \langle \phi | \gamma S (K_S + 1) T K_T \rangle \sqrt{(S - K_S)(S + K_S + 1)} K_S D^S_{K_S M_S}(\Omega_S) D^T_{K_T M_T}(\Omega_T) \\
= \quad & \sum_{K_S K_T} \langle \phi | \gamma S K_S T K_T \rangle \sqrt{(S - K_S + 1)(S + K_S)} \\
\times \quad & (K_S - 1) D^S_{(K_S - 1) M_S}(\Omega_S) D^T_{K_T M_T}(\Omega_T) ,
\end{aligned} \tag{41}
$$

and shows the K_S-lowering property of $\bar{\Gamma}(S_+)$. Together with the well-known D-function matrix elements, relations such as eq. (41) lead to the reduced matrix elements of the shift operator $T^{\Delta_3 = 1}$. In particular,

$$
\begin{aligned}
& \langle \{\lambda_1 \lambda_2 \lambda_3\}(K_S \pm \tfrac{3}{2})(K_T \pm \tfrac{1}{2}); S'T' \, \| \, \Gamma(T^{\Delta_3 = 1}) \, \| \, \{\lambda_1 \lambda_2 + 1 \lambda_3 - 1\} K_S K_T; ST \rangle \\
= \quad & \mp \tfrac{1}{2} \sqrt{\frac{\lambda_3 (2S + 1)(2T + 1)(S \mp K_S)(S \pm K_S + 1)}{2(\lambda_2 + 1)(\lambda_2 + 2)(\lambda_1 + \lambda_2 + 2)((\lambda_1 + \lambda_2 + 3)}} (\lambda_1 \mp K_S \mp K_T) \\
& \times \langle S(K_S \pm 1) \tfrac{1}{2} \pm \tfrac{1}{2} | S'(K_S \pm \tfrac{3}{2}) \rangle \langle T K_T \tfrac{1}{2} \pm \tfrac{1}{2} | T'(K_T \pm \tfrac{1}{2}) \rangle ,
\end{aligned} \tag{42}
$$

$$
\begin{aligned}
& \langle \{\lambda_1 \lambda_2 \lambda_3\}(K_S \pm \tfrac{1}{2})(K_T \pm \tfrac{3}{2}); S'T' \, \| \, \Gamma(T^{\Delta_3 = 1}) \, \| \, \{\lambda_1 \lambda_2 + 1 \lambda_3 - 1\} K_S K_T; ST \rangle \\
= \quad & \pm \tfrac{1}{2} \sqrt{\frac{\lambda_3 (2S + 1)(2T + 1)(T \mp K_T)(T \pm K_T + 1)}{2(\lambda_2 + 1)(\lambda_2 + 2)(\lambda_1 + \lambda_2 + 2)((\lambda_1 + \lambda_2 + 3)}} (\lambda_1 \mp K_S \mp K_T) \\
& \times \langle S K_S \tfrac{1}{2} \pm \tfrac{1}{2} | S'(K_S \pm \tfrac{1}{2}) \rangle \langle T(K_T \pm 1) \tfrac{1}{2} \pm \tfrac{1}{2} | T'(K_T \pm \tfrac{3}{2}) \rangle ,
\end{aligned} \tag{43}
$$

$$
\begin{aligned}
& \langle \{\lambda_1 \lambda_2 \lambda_3\}(K_S \pm \tfrac{1}{2})(K_T \mp \tfrac{1}{2}); S'T' \, \| \, \Gamma(T^{\Delta_3 = 1}) \, \| \, \{\lambda_1 \lambda_2 + 1 \lambda_3 - 1\} K_S K_T; ST \rangle \\
= \quad & \pm \tfrac{1}{2} \sqrt{\frac{\lambda_3 (2S + 1)(2T + 1)}{2(\lambda_2 + 1)(\lambda_2 + 2)(\lambda_1 + \lambda_2 + 2)((\lambda_1 + \lambda_2 + 3)}} \\
\times \quad & \{ 2(\lambda_2 + 1)(\lambda_1 + \lambda_2 + 2) \langle S(K_S \tfrac{1}{2} \pm \tfrac{1}{2} | S'(K_S \pm \tfrac{1}{2}) \rangle \langle T K_T \tfrac{1}{2} \mp \tfrac{1}{2} | T'(K_T \mp \tfrac{1}{2}) \rangle \\
- \quad & (\lambda_1 + 2\lambda_2 + 4 \pm K_S \pm K_T) \sqrt{(S \mp K_S)(S \pm K_S + 1)} \\
\times \quad & \langle S(K_S \pm 1) \tfrac{1}{2} \mp \tfrac{1}{2} | S'(K_S \pm \tfrac{1}{2}) \rangle \langle T K_T \tfrac{1}{2} \mp \tfrac{1}{2} | T'(K_T \mp \tfrac{1}{2}) \rangle \\
- \quad & (\lambda_1 + 2\lambda_2 + 4 \mp K_S \mp K_T) \sqrt{(T \pm K_T)(T \mp K_T + 1)} \\
\times \quad & \langle S(K_S \tfrac{1}{2} \pm \tfrac{1}{2} | S'(K_S \pm \tfrac{1}{2}) \rangle \langle T(K_T \mp 1) \tfrac{1}{2} \pm \tfrac{1}{2} | T'(K_T \mp \tfrac{1}{2}) \rangle \} .
\end{aligned} \tag{44}
$$

As for the group generators, the rotor-space reduced matrix elements of the fundamental shift tensors are given simply in terms of spin, isospin SU(2) Wigner coefficients.

Shift tensors with $\Delta_2 = 1$ can be constructed from linear combinations of $a^\dagger_{ib}$ and $a^\dagger_{ic}$. In particular,

$$
T_i^{\Delta_2 = 1} = \lambda_2 a^\dagger_{ib} + B_{bc} a^\dagger_{ic} , \tag{45}
$$

where B_{bc} is a U$'$(4) generator as defined in eq. (28). The shift character can be seen at once from the relations

$$(\lambda_2 a_{1b} + a_{1c}B_{cb})|\phi_{\{\lambda_1\lambda_2\lambda_3\}}\rangle = -\sqrt{\frac{\lambda_2(\lambda_2+1)(\lambda_2+\lambda_3+1)}{(\lambda_1+1)(\lambda_1+2)}}C_{21}|\phi_{\{\lambda_1+1\lambda_2-1\lambda_3\}}\rangle, \qquad (46)$$

$$(\lambda_2 a_{2b} + a_{2c}B_{cb})|\phi_{\{\lambda_1\lambda_2\lambda_3\}}\rangle = \sqrt{\frac{(\lambda_1+1)\lambda_2(\lambda_2+1)(\lambda_2+\lambda_3+1)}{(\lambda_1+2)}}|\phi_{\{\lambda_1+1\lambda_2-1\lambda_3\}}\rangle, \qquad (47)$$

$$(\lambda_2 a_{ib} + a_{ic}B_{cb})|\phi_{\{\lambda_1\lambda_2\lambda_3\}}\rangle = 0; \qquad \text{for } i = 3 \text{ or } 4. \qquad (48)$$

The techniques illustrated through eqs. (37)-(41) lead to the reduced matrix elements for the $\Delta_2{=}1$ shift tensors

$$\langle\{\lambda_1\lambda_3\lambda_3\}(K_S \pm \tfrac{1}{2})(K_T \pm \tfrac{1}{2}); S'T' \parallel \Gamma(T^{\Delta_2=1}) \parallel \{\lambda_1+1\lambda_2-1\lambda_3\}K_SK_T; ST\rangle$$
$$= \sqrt{\frac{(\lambda_2+\lambda_3+1)(2S+1)(2T+1)}{2(\lambda_1+1)(\lambda_1+2)}}$$
$$\times \quad (\lambda_1+1 \mp K_S \mp K_T)\langle SK_S\tfrac{1}{2} \pm \tfrac{1}{2}|S'(K_S \pm \tfrac{1}{2})\rangle\langle TK_T\tfrac{1}{2} \pm \tfrac{1}{2}|T'(K_T \pm \tfrac{1}{2})\rangle. \qquad (49)$$

Similarly, shift tensors with $\Delta_1{=}1$ are given by

$$T_i^{\Delta_1=1} = \lambda_1(\lambda_1+\lambda_2+1)a_{ia}^\dagger + (\lambda_1+\lambda_2+1)B_{ab}a_{ib}^\dagger + B_{ab}B_{bc}a_{ic}^\dagger + \lambda_1 B_{ac}a_{ic}^\dagger, \qquad (50)$$

with the property

$$\left\{\lambda_1(\lambda_1+\lambda_2+1)a_{1a} + (\lambda_1+\lambda_2+1)a_{1b}B_{ba} + a_{1c}B_{cb}B_{ba} + \lambda_1 a_{1c}B_{ca}\right\}|\phi_{\{\lambda_1\lambda_2\lambda_3\}}\rangle$$
$$= \sqrt{\lambda_1(\lambda_1+1)(\lambda_1+\lambda_2+1)(\lambda_1+\lambda_2+2)(\lambda_1+\lambda_2+\lambda_3+2)}|\phi_{\{\lambda_1-1\lambda_2\lambda_3\}}\rangle, \qquad (51)$$

$$\left\{\lambda_1(\lambda_1+\lambda_2+1)a_{ia} + (\lambda_1+\lambda_2+1)a_{ib}B_{ba} + a_{ic}B_{cb}B_{ba} + \lambda_1 a_{ic}B_{ca}\right\}|\phi_{\{\lambda_1\lambda_2\lambda_3\}}\rangle$$
$$= 0 \qquad \text{for } i = 2, 3, 4. \qquad (52)$$

This leads to the reduced matrix elements

$$\langle\{\lambda_1\lambda_2\lambda_3\}(K_S \pm \tfrac{1}{2})(K_T \pm \tfrac{1}{2}); S'T' \parallel \Gamma(T^{\Delta_1=1}) \parallel \{\lambda_1-1\lambda_2\lambda_3\}K_SK_T; ST\rangle$$
$$= \pm\sqrt{\tfrac{1}{2}(2S+1)(2T+1)\lambda_1(\lambda_1+1)(\lambda_1+\lambda_2+1)(\lambda_1+\lambda_2+2)(\lambda_1+\lambda_2+\lambda_3+3)}$$
$$\times \quad \langle SK_S\tfrac{1}{2} \pm \tfrac{1}{2}|S'(K_S \pm \tfrac{1}{2})\rangle\langle TK_T\tfrac{1}{2} \pm \tfrac{1}{2}|T'(K_T \pm \tfrac{1}{2})\rangle. \qquad (53)$$

Shift tensors with $\Delta_4{=}1$ can be realized through $T_i^{\Delta_4=1} = a_{id}^\dagger$, if the highest weight state $|\phi\rangle$ is expressed through eq. (31) with $f_4{=}1$ by $|\phi_{\{\lambda_1\lambda_2\lambda_31\}}\rangle$. The shift property can be seen from

$$a_{1d}|\phi_{\{\lambda_1\lambda_2\lambda_3\}}\rangle =$$
$$-\frac{\{C_{43}C_{32}C_{21} - C_{42}C_{21}(\lambda_3+2) - (\lambda_2+\lambda_3+3)[C_{43}C_{31} - C_{41}(\lambda_3+2)]\}|\phi_{\{\lambda_1\lambda_2\lambda_3+1\}}\rangle}{\sqrt{(\lambda_3+1)(\lambda_3+2)(\lambda_2+\lambda_3+2)(\lambda_2+\lambda_3+3)(\lambda_1+\lambda_2+\lambda_3+3)(\lambda_1+\lambda_2+\lambda_3+4)}}, \qquad (54)$$

$$a_{2d}|\phi_{\{\lambda_1\lambda_2\lambda_3\}}\rangle = \frac{\sqrt{(\lambda_1+\lambda_2+\lambda_3+3)}\{C_{43}C_{32} - (\lambda_3+2)C_{42}\}|\phi_{\{\lambda_1\lambda_2\lambda_3+1\}}\rangle}{\sqrt{(\lambda_3+1)(\lambda_3+2)(\lambda_2+\lambda_3+2)(\lambda_2+\lambda_3+3)(\lambda_1+\lambda_2+\lambda_3+4)}}, \qquad (55)$$

$$a_{3d}|\phi_{\{\lambda_1\lambda_2\lambda_3\}}\rangle = -\frac{\sqrt{(\lambda_1+\lambda_2+\lambda_3+3)(\lambda_2+\lambda_3+2)}\,C_{43}|\phi_{\{\lambda_1\lambda_2\lambda_3+1\}}\rangle}{\sqrt{(\lambda_3+1)(\lambda_3+2)(\lambda_2+\lambda_3+3)(\lambda_1+\lambda_2+\lambda_3+4)}}, \tag{56}$$

$$a_{4d}|\phi_{\{\lambda_1\lambda_2\lambda_3\}}\rangle = \sqrt{\frac{(\lambda_1+\lambda_2+\lambda_3+3)(\lambda_2+\lambda_3+2)(\lambda_3+1)}{(\lambda_1+\lambda_2+\lambda_3+4)(\lambda_2+\lambda_3+3)(\lambda_3+2)}}|\phi_{\{\lambda_1\lambda_2\lambda_3+1\}}\rangle. \tag{57}$$

This leads to the reduced matrix elements

$$\langle\{\lambda_1\lambda_2\lambda_3\}(K_S\pm\tfrac{3}{2})(K_T\pm\tfrac{1}{2}); S'T' \,\|\, \Gamma(T^{\Delta_4=1}) \,\|\, \{\lambda_1\lambda_2\lambda_3+1\}K_SK_T; ST\rangle$$
$$= -\tfrac{1}{2}\sqrt{\frac{(2S+1)(2T+1)(S\mp K_S)(S\pm K_S+1)}{2(\lambda_3+1)(\lambda_2+\lambda_3+2)(\lambda_1+\lambda_2+\lambda_3+3)}}(\lambda_1\mp K_S\mp K_T)(\lambda_3+1\mp K_S\pm K_T)$$
$$\times\ \langle S(K_S\pm1)\tfrac{1}{2}\pm\tfrac{1}{2}|S'(K_S\pm\tfrac{3}{2})\rangle\langle TK_T\tfrac{1}{2}\pm\tfrac{1}{2}|T'(K_T\pm\tfrac{1}{2})\rangle, \tag{58}$$

$$\langle\{\lambda_1\lambda_2\lambda_3\}(K_S\pm\tfrac{1}{2})(K_T\pm\tfrac{3}{2}); S'T' \,\|\, \Gamma(T^{\Delta_4=1}) \,\|\, \{\lambda_1\lambda_2\lambda_3+1\}K_SK_T; ST\rangle$$
$$= -\tfrac{1}{2}\sqrt{\frac{(2S+1)(2T+1)(T\mp K_T)(T\pm K_T+1)}{2(\lambda_3+1)(\lambda_2+\lambda_3+2)(\lambda_1+\lambda_2+\lambda_3+3)}}(\lambda_1\mp K_S\mp K_T)(\lambda_3+1\pm K_S\mp K_T)$$
$$\times\ \langle SK_S\tfrac{1}{2}\pm\tfrac{1}{2}|S'(K_S\pm\tfrac{1}{2})\rangle\langle T(K_T\pm1)\tfrac{1}{2}\pm\tfrac{1}{2}|T'(K_T\pm\tfrac{3}{2})\rangle, \tag{59}$$

$$\langle\{\lambda_1\lambda_2\lambda_3\}(K_S\pm\tfrac{1}{2})(K_T\mp\tfrac{1}{2}); S'T' \,\|\, \Gamma(T^{\Delta_4=1}) \,\|\, \{\lambda_1\lambda_2\lambda_3+1\}K_SK_T; ST\rangle$$
$$= \tfrac{1}{2}\frac{\sqrt{(2S+1)(2T+1)}(\lambda_3+1\mp K_S\pm K_T)}{\sqrt{2(\lambda_3+1)(\lambda_2+\lambda_3+2)(\lambda_1+\lambda_2+\lambda_3+3)}}$$
$$\times\ \Big\{2(\lambda_2+\lambda_2+2)(\lambda_1+\lambda_2+\lambda_2+3)\langle SK_S\tfrac{1}{2}\pm\tfrac{1}{2}|S'(K_S\pm\tfrac{1}{2})\rangle\langle TK_T\tfrac{1}{2}\mp\tfrac{1}{2}|T'(K_T\mp\tfrac{1}{2})\rangle$$
$$-\ (\lambda_1+2\lambda_2+2\lambda_3+6\pm K_S\pm K_T)\sqrt{(S\mp K_S)(S\pm K_S+1)}$$
$$\times\ \langle S(K_S\pm1)\tfrac{1}{2}\mp\tfrac{1}{2}|S'(K_S\pm\tfrac{1}{2})\rangle\langle TK_T\tfrac{1}{2}\mp\tfrac{1}{2}|T'(K_T\mp\tfrac{1}{2})\rangle$$
$$-\ (\lambda_1+2\lambda_2+2\lambda_3+6\mp K_S\mp K_T)\sqrt{(T\pm K_T)(T\mp K_T+1)}$$
$$\times\ \langle S(K_S\tfrac{1}{2}\pm\tfrac{1}{2}|S'(K_S\pm\tfrac{1}{2})\rangle\langle T(K_T\mp1)\tfrac{1}{2}\pm\tfrac{1}{2}|T'(K_T\mp\tfrac{1}{2})\rangle\Big\}. \tag{60}$$

Shift tensors for the conjugate irreducible tensors of symmetry $\{001\}$ and shifts $\Delta_i = -1$ can be constructed in similar fashion. The results are

$$T_i^{\Delta_1=-1} = a_{ia}, \tag{61}$$

$$T_i^{\Delta_2=-1} = (\lambda_1 a_{ib} - B_{ab}a_{ia}), \tag{62}$$

$$T_i^{\Delta_3=-1} = (\lambda_1+\lambda_2+1)(\lambda_2 a_{ic} - B_{bc}a_{ib}) + (B_{ab}B_{bc} - (\lambda_2+1)B_{ac})a_{ia}, \tag{63}$$

$$\begin{aligned}
T_i^{\Delta_3=-1} &= \{B_{ab}B_{bc}B_{cd} - (\lambda_3+1)B_{ab}B_{bd} \\
&\quad + (\lambda_2+\lambda_3+2)[(\lambda_3+1)B_{ad} - B_{ac}B_{cd}]\}a_{ia} \\
&\quad + (\lambda_1+\lambda_2+\lambda_3+2)[(\lambda_3+1)B_{bd} - B_{bc}B_{cd}]a_{ib} \\
&\quad + (\lambda_2+\lambda_3+1)(\lambda_1+\lambda_2+\lambda_3+2)B_{cd}a_{ic} \\
&\quad - \lambda_3(\lambda_2+\lambda_3+1)(\lambda_1+\lambda_2+\lambda_3+2)a_{id},
\end{aligned} \tag{64}$$

where the conjugates of these tensors acting on the highest weight states $|\phi_{\{\lambda_1\lambda_2\lambda_3\}}\rangle$ can be expressed in terms of step-down generators C_{ij} acting on the shifted $|\phi_{\{\lambda'_1\lambda'_2\lambda'_3\}}\rangle$. Relations such as the eqs. (33)-(36) and the techniques illustrated for the $\{100\}$ shift tensors with $\Delta_3 = +1$ lead to the needed reduced matrix elements.

For $T^{\Delta_1=-1}$ tensors:

$$\langle\{\lambda_1\lambda_2\lambda_3\}(K_S \pm \tfrac{3}{2})(K_T \mp \tfrac{1}{2}); S'T' \| \Gamma(T^{\Delta_1=-1}) \| \{\lambda_1 + 1\lambda_2\lambda_3\}K_SK_T; ST\rangle$$

$$= \mp\tfrac{1}{2}\sqrt{\frac{(2S+1)(2T+1)(S \mp K_S)(S \pm K_S + 1)}{2(\lambda_1+1)(\lambda_1+2)(\lambda_1+\lambda_2+2)(\lambda_1+\lambda_2+3)(\lambda_1+\lambda_2+\lambda_3+3)}}$$

$$\times \ (\lambda_1 + 1 \mp K_S \mp K_T)(\lambda_3 \mp K_S \pm K_T)$$

$$\times \ \langle S(K_S \pm 1)\tfrac{1}{2} \pm \tfrac{1}{2}|S'(K_S \pm \tfrac{3}{2})\rangle\langle TK_T\tfrac{1}{2} \mp \tfrac{1}{2}|T'(K_T \mp \tfrac{1}{2})\rangle, \tag{65}$$

$$\langle\{\lambda_1\lambda_2\lambda_3\}(K_S \mp \tfrac{1}{2})(K_T \pm \tfrac{3}{2}); S'T' \| \Gamma(T^{\Delta_1=-1}) \| \{\lambda_1 + 1\lambda_2\lambda_3\}K_SK_T; ST\rangle$$

$$= \mp\tfrac{1}{2}\sqrt{\frac{(2S+1)(2T+1)(T \mp K_T)(T \pm K_T + 1)}{2(\lambda_1+1)(\lambda_1+2)(\lambda_1+\lambda_2+2)(\lambda_1+\lambda_2+3)(\lambda_1+\lambda_2+\lambda_3+3)}}$$

$$\times \ (\lambda_1 + 1 \mp K_S \mp K_T)(\lambda_3 \pm K_S \mp K_T)$$

$$\times \ \langle SK_S\tfrac{1}{2} \mp \tfrac{1}{2}|S'(K_S \mp \tfrac{1}{2})\rangle\langle T(K_T \pm 1)\tfrac{1}{2} \pm \tfrac{1}{2}|T'(K_T \pm \tfrac{3}{2})\rangle, \tag{66}$$

$$\langle\{\lambda_1\lambda_2\lambda_3\}(K_S \pm \tfrac{1}{2})(K_T \pm \tfrac{1}{2}); S'T' \| \Gamma(T^{\Delta_1=-1}) \| \{\lambda_1 + 1\lambda_2\lambda_3\}K_SK_T; ST\rangle$$

$$= \frac{\mp\tfrac{1}{2}\sqrt{(2S+1)(2T+1)}(\lambda_1 + 1 \mp K_S \mp K_T)}{\sqrt{2(\lambda_1+1)(\lambda_1+2)(\lambda_1+\lambda_2+2)(\lambda_1+\lambda_2+3)(\lambda_1+\lambda_2+\lambda_3+3)}}$$

$$\times \ \Big\{2(\lambda_1+\lambda_2+2)(\lambda_1+\lambda_2+\lambda_3+3)\langle SK_S\tfrac{1}{2} \pm \tfrac{1}{2}|S'(K_S \pm \tfrac{1}{2})\rangle\langle TK_T\tfrac{1}{2} \pm \tfrac{1}{2}|T'(K_T \pm \tfrac{1}{2})\rangle$$

$$- \ (2\lambda_1 + 2\lambda_2 + \lambda_3 + 6 \pm K_S \mp K_T)\sqrt{S \mp K_S)(S \pm K_S + 1)}$$

$$\times \ \langle S(K_S \pm 1)\tfrac{1}{2} \mp \tfrac{1}{2}|S'(K_S \pm \tfrac{1}{2})\rangle\langle TK_T\tfrac{1}{2} \pm \tfrac{1}{2}|T'(K_T \pm \tfrac{1}{2})\rangle$$

$$- \ (2\lambda_1 + 2\lambda_2 + \lambda_3 + 6 \mp K_S \pm K_T)\sqrt{T \mp K_T)(T \pm K_T + 1)}$$

$$\times \ \langle SK_S\tfrac{1}{2} \pm \tfrac{1}{2}|S'(K_S \pm \tfrac{1}{2})\rangle\langle T(K_T \pm 1)\tfrac{1}{2} \mp \tfrac{1}{2}|T'(K_T \pm \tfrac{1}{2})\rangle\Big\}. \tag{67}$$

For $T^{\Delta_2=-1}$ tensors:

$$\langle\{\lambda_1\lambda_2\lambda_3\}(K_S \pm \tfrac{3}{2})(K_T \pm \tfrac{1}{2}); S'T' \| \Gamma(T^{\Delta_2=-1}) \| \{\lambda_1 - 1\lambda_2 + 1\lambda_3\}K_SK_T; ST\rangle$$

$$= \tfrac{1}{2}\sqrt{\frac{\lambda_1(\lambda_1+1)(2S+1)(2T+1)(S \mp K_S)(S \pm K_S + 1)}{2(\lambda_2+1)(\lambda_2+2)(\lambda_2+\lambda_3+2)}}(\lambda_3 \mp K_S \pm K_T)$$

$$\times \ \langle S(K_S \pm 1)\tfrac{1}{2} \pm \tfrac{1}{2}|S'(K_S \pm \tfrac{3}{2})\rangle\langle TK_T\tfrac{1}{2} \mp \tfrac{1}{2}|T'(K_T \mp \tfrac{1}{2})\rangle, \tag{68}$$

$$\langle\{\lambda_1\lambda_2\lambda_3\}(K_S \mp \tfrac{1}{2})(K_T \pm \tfrac{3}{2}); S'T' \| \Gamma(T^{\Delta_2=-1}) \| \{\lambda_1 - 1\lambda_2 + 1\lambda_3\}K_SK_T; ST\rangle$$

$$= \tfrac{1}{2}\sqrt{\frac{\lambda_1(\lambda_1+1)(2S+1)(2T+1)(T \mp K_T)(T \pm K_T + 1)}{2(\lambda_2+1)(\lambda_2+2)(\lambda_2+\lambda_3+2)}}(\lambda_3 \pm K_S \mp K_T)$$

$$\times \ \langle SK_S\tfrac{1}{2} \mp \tfrac{1}{2}|S'(K_S \mp \tfrac{1}{2})\rangle\langle T(K_T \pm 1)\tfrac{1}{2} \pm \tfrac{1}{2}|T'(K_T \pm \tfrac{3}{2})\rangle, \tag{69}$$

$$\langle\{\lambda_1\lambda_2\lambda_3\}(K_S \pm \tfrac{1}{2})(K_T \pm \tfrac{1}{2}); S'T' \| \Gamma(T^{\Delta_2=-1}) \| \{\lambda_1 - 1\lambda_2 + 1\lambda_3\}K_SK_T; ST\rangle$$

$$
\begin{aligned}
= \ & \tfrac{1}{2}\sqrt{\frac{\lambda_1(\lambda_1+1)(2S+1)(2T+1)}{2(\lambda_2+1)(\lambda_2+2)(\lambda_2+\lambda_3+2)}} \\
\times \ & \Big\{ 2(\lambda_2+1)(\lambda_2+\lambda_3+2)\langle SK_S\tfrac{1}{2}\pm\tfrac{1}{2}|S'(K_S\pm\tfrac{1}{2})\rangle\langle TK_T\tfrac{1}{2}\pm\tfrac{1}{2}|T'(K_T\pm\tfrac{1}{2})\rangle \\
- \ & (2\lambda_2+\lambda_3+4\pm K_S\mp K_T)\sqrt{(S\mp K_S)(S\pm K_S+1)} \\
\times \ & \langle S(K_S\pm1)\tfrac{1}{2}\mp\tfrac{1}{2}|S'(K_S\pm\tfrac{1}{2})\rangle\langle TK_T\tfrac{1}{2}\pm\tfrac{1}{2}|T'(K_T\pm\tfrac{1}{2})\rangle \\
- \ & (2\lambda_2+\lambda_3+4\mp K_S\pm K_T)\sqrt{(T\mp K_T)(T\pm K_T+1)} \\
\times \ & \langle SK_S\tfrac{1}{2}\pm\tfrac{1}{2}|S'(K_S\pm\tfrac{1}{2})\rangle\langle T(K_T\pm1)\tfrac{1}{2}\mp\tfrac{1}{2}|T'(K_T\pm\tfrac{1}{2})\rangle \Big\}.
\end{aligned}
\tag{70}
$$

For $T^{\Delta_3=-1}$ tensors:

$$
\langle\{\lambda_1\lambda_2\lambda_3\}(K_S\pm\tfrac{1}{2})(K_T\mp\tfrac{1}{2});S'T'\ \|\ \Gamma(T^{\Delta_3=-1})\ \|\ \{\lambda_1\lambda_2-1\lambda_3+1\}K_SK_T;ST\rangle
$$
$$
\begin{aligned}
= \ & \pm\sqrt{\frac{\lambda_2(\lambda_2+1)(\lambda_1+\lambda_2+1)(\lambda_1+\lambda_2+2)(2S+1)(2T+1)}{2(\lambda_3+1)(\lambda_3+2)}} \\
\times \ & (\lambda_3+1\mp K_S\pm K_T)\langle SK_S\tfrac{1}{2}\pm\tfrac{1}{2}|S'(K_S\pm\tfrac{1}{2})\rangle\langle TK_T\tfrac{1}{2}\mp\tfrac{1}{2}|T'(K_T\mp\tfrac{1}{2})\rangle.
\end{aligned}
\tag{71}
$$

Finally for the $T^{\Delta_4=-1}$ tensors:

$$
\langle\{\lambda_1\lambda_2\lambda_3\}(K_S\pm\tfrac{1}{2})(K_T\mp\tfrac{1}{2});S'T'\ \|\ \Gamma(T^{\Delta_4=-1})\ \|\ \{\lambda_1\lambda_2\lambda_3-1\,1\}K_SK_T;ST\rangle
$$
$$
\begin{aligned}
= \ & -\sqrt{\frac{(\lambda_3+1)(\lambda_2+\lambda_3+2)(\lambda_1+\lambda_2+\lambda_3+3)(2S+1)(2T+1)}{2\lambda_3(\lambda_2+\lambda_3+1)(\lambda_1+\lambda_2+\lambda_3+2)}} \\
\times \ & \langle SK_S\tfrac{1}{2}\pm\tfrac{1}{2}|S'(K_S\pm\tfrac{1}{2})\rangle\langle TK_T\tfrac{1}{2}\mp\tfrac{1}{2}|T'(K_T\mp\tfrac{1}{2})\rangle.
\end{aligned}
\tag{72}
$$

The nonhermitian character of these matrix elements is immediately apparent from a comparison of the reduced matrix elements for tensors with $\Delta_i=+1$ and $\Delta_i=-1$. The hermitian matrix elements in ordinary Hilbert space can be gained through the SU(4) analogue of eq. (11) through the $\mathcal{K}$ and $\mathcal{K}^{-1}$ matrix elements of initial and final states, respectively. The operators $T^{\Delta_i=\pm1}$ of this section have no special normalization. In order to convert these to SU(4) $\supset$ SU(2) x SU(2) reduced Wigner coefficients with standard orthonormality properties the operators T^{Δ_i} of this section have to be converted to *unit* tensor operators by a renormalization factor, (see below). In terms of such *unit* tensors the SU(4) $\supset$ SU(2) $\supset$ SU(2) reduced Wigner coefficients for the coupling $\{\bar\lambda_1\bar\lambda_2\bar\lambda_3\}\times\{100\}\to\{\lambda_1\lambda_2\lambda_3\}$ are given by

$$
\langle\{\bar\lambda_1\bar\lambda_2\bar\lambda_3\}\nu;ST,\{100\}\tfrac{1}{2}\tfrac{1}{2}\ \|\ \{\lambda_1\lambda_2\lambda_3\}\nu';S'T'\rangle
$$
$$
\begin{aligned}
= \ & \sum_{K'_SK'_T}\sum_{K_SK_T}\frac{(\mathcal{K}^{-1}(S'T'))_{\nu',K'_SK'_T}\,\mathcal{K}(ST)_{K_SK_T,\nu}}{\sqrt{(2S'+1)(2T'+1)}} \\
\times \ & \Big\{ \langle\{\lambda_1\lambda_2\lambda_3\}K'_SK'_T;S'T'\ \|\ \Gamma(T^{\Delta_i=1}_{unit})\ \|\ \{\bar\lambda_1\bar\lambda_2\bar\lambda_3\}K_SK_T;ST\rangle + (-1)^{\lambda_2+\lambda_3+S'+T'} \\
\times \ & \langle\{\lambda_1\lambda_2\lambda_3\}-K'_S-K'_T;S'T'\ \|\ \Gamma(T^{\Delta_i=1}_{unit})\ \|\ \{\bar\lambda_1\bar\lambda_2\bar\lambda_3\}K_SK_T;ST\rangle \Big\},
\end{aligned}
\tag{73}
$$

where the K_SK_T values in the sums are restricted by $K_S\geq 0$, and for $K_S=0$, $K_T\geq 0$, (similarly for $K'_SK'_T$); and the expression has made full use of both pieces of the symmetrized orthonormal rotor states of eq. (22). (Note that the symmetries of the reduced matrix elements are such that, together with the symmetries of the SU(2) Wigner coefficients the pieces with negative and positive K-values combine naturally). The second term in the curly bracket, with the phase factor $(-1)^{\lambda_2+\lambda_3+S'+T'}$, is needed

206

only in rare cases, e.g. with $K_S + K_T \leq 1$. E.g. the state $K_S K_T = 01$ can connect to $-K'_S - K'_T = -\frac{1}{2} + \frac{1}{2}$.

The fundamental Wigner coefficients of eq. (73) are also related by symmetry to those for the coupling $\{\lambda_1\lambda_2\lambda_3\} \times \{001\} \to \{\bar{\lambda}_1\bar{\lambda}_2\bar{\lambda}_3\}$, given by the conjugate shift tensors with $\Delta_i = -1$.

$$\langle\{\bar{\lambda}_1\bar{\lambda}_2\bar{\lambda}_3\}\nu'; S'T', \{100\}\tfrac{1}{2}\tfrac{1}{2} \| \{\lambda_1\lambda_2\lambda_3\}\nu; ST\rangle$$

$$= \sqrt{\frac{dim\{\lambda_1\lambda_2\lambda_3\}(2S'+1)(2T'+1)}{dim\{\bar{\lambda}_1\bar{\lambda}_2\bar{\lambda}_3\}(2S+1)(2T+1)}}(-1)^{\lambda_1+\lambda_2+\lambda_3+S+T-\bar{\lambda}_1-\bar{\lambda}_2-\bar{\lambda}_3-S'-T'}$$

$$\times \quad \langle\{\lambda_1\lambda_2\lambda_3\}\nu; ST, \{001\}\tfrac{1}{2}\tfrac{1}{2} \| \{\bar{\lambda}_1\bar{\lambda}_2\bar{\lambda}_3\}\nu'; S'T'\rangle, \tag{74}$$

where $dim\{\lambda_1\lambda_2\lambda_3\}$ stands for the dimension of the irrep, and where

$$\langle\{\lambda_1\lambda_2\lambda_3\}\nu; ST, \{001\}\tfrac{1}{2}\tfrac{1}{2} \| \{\bar{\lambda}_1\bar{\lambda}_2\bar{\lambda}_3\}\nu'; S'T'\rangle$$

$$= \sum_{K'_S K'_T} \sum_{K_S K_T} \frac{(\mathcal{K}^{-1}(S'T'))_{\nu',K'_S K'_T}(\mathcal{K}(ST))_{K_S K_T,\nu}}{\sqrt{(2S'+1)(2T'+1)}}$$

$$\times \quad \left\{\langle\{\bar{\lambda}_1\bar{\lambda}_2\bar{\lambda}_3\}K'_S K'_T; S'T' \| \Gamma(T_{unit}^{\Delta_i=-1}) \| \{\lambda_1\lambda_2\lambda_3\}K_S K_T; ST\rangle + (-1)^{\bar{\lambda}_2+\bar{\lambda}_3+S'+T'}\right.$$

$$\times \quad \left.\langle\{\bar{\lambda}_1\bar{\lambda}_2\bar{\lambda}_3\} - K'_S - K'_T; S'T' \| \Gamma(T_{unit}^{\Delta_i=-1}) \| \{\lambda_1\lambda_2\lambda_3\}K_S K_T; ST\rangle\right\}, \tag{75}$$

with the same restrictions as for eqs. (73). By far the simplest expressions are obtained with the use of eq. (73) for shift tensors with $\Delta_i = +1$ and $i = 1,2$; and eq.(75) for shift tensors with $\Delta_i = -1$ and $i = 3,4$. For economy of notation the normalization factors which convert the shift tensors into *unit* tensors will be absorbed into the final expressions which are given by

$$\frac{\langle\{\lambda_1\lambda_2\lambda_3\}K'_S K'_T; S'T' \| \Gamma(T_{unit}^{\Delta_1=1}) \| \{\lambda_1 - 1\lambda_2\lambda_3\}K_S K_T; ST\rangle}{\sqrt{(2S'+1)(2T'+1)}}$$

$$= \pm N^{\Delta_1=1}\sqrt{\frac{(2S+1)(2T+1)}{2(2S'+1)(2T'+1)}}\langle SK_S\tfrac{1}{2} \pm \tfrac{1}{2}|S'K'_S\rangle\langle TK_T\tfrac{1}{2} \pm \tfrac{1}{2}|T'K'_T\rangle, \tag{76}$$

$$\frac{\langle\{\lambda_1\lambda_2\lambda_3\}K'_S K'_T; S'T' \| \Gamma(T_{unit}^{\Delta_2=1}) \| \{\lambda_1 + 1\lambda_2 - 1\lambda_3\}K_S K_T; ST\rangle}{\sqrt{(2S'+1)(2T'+1)}}$$

$$= N^{\Delta_2=1}\sqrt{\frac{(2S+1)(2T+1)}{2(2S'+1)(2T'+1)}\frac{(\lambda_1 + 1 \mp K_S \mp K_T)}{(\lambda_1 + 2)}}$$

$$\times \quad \langle SK_S\tfrac{1}{2} \pm \tfrac{1}{2}|S'K'_S\rangle\langle TK_T\tfrac{1}{2} \pm \tfrac{1}{2}|T'K'_T\rangle, \tag{77}$$

$$\frac{\langle\{\lambda_1\lambda_2 + 1\lambda_3 - 1\}K'_S K'_T; S'T' \| \Gamma(T_{unit}^{\Delta_3=-1}) \| \{\lambda_1\lambda_2\lambda_3\}K_S K_T; ST\rangle}{\sqrt{(2S'+1)(2T'+1)}}$$

$$= \pm N^{\Delta_3=-1}\sqrt{\frac{(2S+1)(2T+1)}{2(2S'+1)(2T'+1)}\frac{(\lambda_3 \mp K_S \pm K_T)}{(\lambda_3 + 1)}}$$

$$\times \quad \langle SK_S\tfrac{1}{2} \pm \tfrac{1}{2}|S'K'_S\rangle\langle TK_T\tfrac{1}{2} \mp \tfrac{1}{2}|T'K'_T\rangle, \tag{78}$$

$$\frac{\langle\{\lambda_1\lambda_2\lambda_3 + 1\}K'_S K'_T; S'T' \| \Gamma(T_{unit}^{\Delta_4=-1}) \| \{\lambda_1\lambda_2\lambda_3\}K_S K_T; ST\rangle}{\sqrt{(2S'+1)(2T'+1)}}$$

$$= -N^{\Delta_4=-1}\sqrt{\frac{(2S+1)(2T+1)}{2(2S'+1)(2T'+1)}}\langle SK_S\tfrac{1}{2} \pm \tfrac{1}{2}|S'K'_S\rangle\langle TK_T\tfrac{1}{2} \mp \tfrac{1}{2}|T'K'_T\rangle. \tag{79}$$

$$\text{Table 1. The } N^{\Delta}\text{-factors with } \Delta_i = +1$$

$\lambda_1\lambda_2\lambda_3$	$N^{\Delta_1=1}$	$N^{\Delta_2=1}$
eee	1	1
eoe	$\sqrt{\dfrac{(\lambda_1+\lambda_2+2)(\lambda_1+\lambda_2+\lambda_3+3)}{(\lambda_1+\lambda_2+1)(\lambda_1+\lambda_2+\lambda_3+2)}}$	$\sqrt{\dfrac{(\lambda_2+1)(\lambda_2+\lambda_3+2)}{\lambda_2(\lambda_2+\lambda_3+1)}}$
oeo	$\sqrt{\dfrac{(\lambda_1+1)(\lambda_1+\lambda_2+2)}{\lambda_1(\lambda_1+\lambda_2+1)}}$	$\sqrt{\dfrac{(\lambda_1+2)(\lambda_2+\lambda_3+2)}{(\lambda_1+1)(\lambda_2+\lambda_3+1)}}$
ooo	$\sqrt{\dfrac{(\lambda_1+1)(\lambda_1+\lambda_2+\lambda_3+3)}{\lambda_1(\lambda_1+\lambda_2+\lambda_3+2)}}$	$\sqrt{\dfrac{(\lambda_1+2)(\lambda_2+1)}{(\lambda_1+1)\lambda_2}}$
oee	$\sqrt{\dfrac{(\lambda_1+1)(\lambda_1+\lambda_2+2)(\lambda_1+\lambda_2+\lambda_3+3)}{\lambda_1(\lambda_1+\lambda_2+1)(\lambda_1+\lambda_2+\lambda_3+2)}}$	$\sqrt{\dfrac{(\lambda_1+2)}{(\lambda_1+1)}}$
ooe	$\sqrt{\dfrac{(\lambda_1+1)}{\lambda_1}}$	$\sqrt{\dfrac{(\lambda_1+2)(\lambda_2+1)(\lambda_2+\lambda_3+2)}{(\lambda_1+1)\lambda_2(\lambda_2+\lambda_3+1)}}$
eeo	$\sqrt{\dfrac{(\lambda_1+\lambda_2+\lambda_3+3)}{(\lambda_1+\lambda_2+\lambda_3+2)}}$	$\sqrt{\dfrac{(\lambda_2+\lambda_3+2)}{(\lambda_2+\lambda_3+1)}}$
eoo	$\sqrt{\dfrac{(\lambda_1+\lambda_2+2)}{(\lambda_1+\lambda_2+1)}}$	$\sqrt{\dfrac{(\lambda_2+1)}{\lambda_2}}$

The N^{Δ_i} factors of eqs. (76)-(79) are given in Tables 1 and 2. These combine the λ_i-dependent square root factors of eqs. (49), (53), (71), and (72) with the renormalization factors needed to convert the shift tensors to unit shift tensors. The N^{Δ_i} factors are dependent on the parity of the quantum numbers $\lambda_1\lambda_2\lambda_3$. Note, in particular, that these are tabulated for the even (e) or odd (o) character of $\lambda_1\lambda_2\lambda_3$ which sits on the left hand side of the reduced matrix elements with $\Delta_i = +1$ and on the right hand side of the matrix elements with $\Delta_i = -1$ in eqs. (76)-(79).

CONCLUDING REMARKS

The rotor coherent state constructions which have been developed recently give an elegant systematization of the Elliott angular momentum projection technique. Up to now they have been applied mainly to find the matrix representations of the group generators in bases specified by the group chains $SU(3) \supset SO(3) \supset SO(2)$, $SU(4) \supset SU(2) \times SU(2)$, and $SO(5) \supset SO(3) \supset SO(2)$, all of which contain physically relevant angular momentum subroups and hence lead to a missing label problem. It has been shown that the rotor coherent state constructions are equally useful to find the matrix representations of arbitrary irreducible tensor operators which are not group generators, provided the actions of these operators on the relevant initial highest weight states of the group can be converted to the actions of the relevant angular momentum subgroup generators on the highest weight states of the final (shifted) irreducible representation. As a very specific example matrix represenations have been given for the 4-dimensional fundamental tensors of SU(4) and their conjugates in the Wigner supermultiplet scheme of good spin, S, and isospin, T. In the first step of the calculation, extremely simple expressions (see eqs. (76)-(79)) are gained which give reduced matrix elements of the needed unit tensors in the orthonormal rotor basis in terms of ordinary spin and isospin angular momentum coefficients containing the Elliott-type labels K_S, K_T. In the second step of the calculation, the unitarization process of the coherent state technique, these K_S, K_T are converted to the quantum number ν through the $\mathcal{K}$-matrix method of coherent state theory (see eqs. (73) and (75)). Similar techniques have so far been used for nongenerators only for the SU(3)$\supset$SO(3) scheme, see ref. [17]. Effectively, the K-labels are elevated to the status of good quantum numbers in an author-independent way based on the eigenvalues of the $\mathcal{K}\mathcal{K}^\dagger$ overlap matrix.

Table 2. The N^Δ-factors with $\Delta_i = -1$

$\lambda_1\lambda_2\lambda_3$	$N^{\Delta_3=-1}$	$N^{\Delta_4=-1}$
eee	$\sqrt{\dfrac{(\lambda_2+2)(\lambda_3+1)(\lambda_1+\lambda_2+3)}{(\lambda_2+1)\lambda_3(\lambda_1+\lambda_2+2)}}$	$\sqrt{\dfrac{(\lambda_3+2)(\lambda_2+\lambda_3+3)(\lambda_1+\lambda_2+\lambda_3+4)}{(\lambda_3+1)(\lambda_2+\lambda_3+2)(\lambda_1+\lambda_2+\lambda_3+3)}}$
eoe	$\sqrt{\dfrac{(\lambda_3+1)}{\lambda_3}}$	$\sqrt{\dfrac{(\lambda_3+2)}{(\lambda_3+1)}}$
oeo	$\sqrt{\dfrac{(\lambda_2+2)}{(\lambda_2+1)}}$	$\sqrt{\dfrac{(\lambda_1+\lambda_2+\lambda_3+4)}{(\lambda_1+\lambda_2+\lambda_3+3)}}$
ooo	$\sqrt{\dfrac{(\lambda_1+\lambda_2+3)}{(\lambda_1+\lambda_2+2)}}$	$\sqrt{\dfrac{(\lambda_2+\lambda_3+3)}{(\lambda_2+\lambda_3+2)}}$
oee	$\sqrt{\dfrac{(\lambda_2+2)(\lambda_3+1)}{(\lambda_2+1)\lambda_3}}$	$\sqrt{\dfrac{(\lambda_3+2)(\lambda_2+\lambda_3+3)}{(\lambda_3+1)(\lambda_2+\lambda_3+2)}}$
ooe	$\sqrt{\dfrac{(\lambda_3+1)(\lambda_1+\lambda_2+3)}{\lambda_3(\lambda_1+\lambda_2+2)}}$	$\sqrt{\dfrac{(\lambda_3+2)(\lambda_1+\lambda_2+\lambda_3+4)}{(\lambda_3+1)(\lambda_1+\lambda_2+\lambda_3+3)}}$
eeo	$\sqrt{\dfrac{(\lambda_2+2)(\lambda_1+\lambda_2+3)}{(\lambda_2+1)(\lambda_1+\lambda_2+2)}}$	1
eoo	1	$\sqrt{\dfrac{(\lambda_2+\lambda_3+3)(\lambda_1+\lambda_2+\lambda_3+4)}{(\lambda_2+\lambda_3+2)(\lambda_1+\lambda_2+\lambda_3+3)}}$

APPENDIX

To determine the normalization factors N^Δ of section 4 it was necessary to find general ($\{\lambda_1\lambda_2\lambda_3\}$-dependent) values for some $\mathcal{K}\mathcal{K}^\dagger$-matrix elements for small ST-values. The $\mathcal{K}\mathcal{K}^\dagger$-matrix elements can be derived by a recursive process using the two relations, [5],

$$\sum_{K'_{S_2}K'_{T_2}} (\mathcal{K}\mathcal{K}^\dagger(S',T'))_{K'_{S_1}K'_{T_1};K'_{S_2}K'_{T_2}} \langle K_{S_2}K_{T_2}; ST \parallel \Gamma(E) \parallel K'_{S_2}K'_{T_2}; S'T'\rangle(-1)^{S+T-S'-T'}$$

$$= \sum_{K_{S_1}K_{T_1}} \langle K'_{S_1}K'_{T_1}; S'T' \parallel \Gamma(E) \parallel K_{S_1}K_{T_1}; ST\rangle(\mathcal{K}\mathcal{K}^\dagger(S,T))_{K_{S_1}K_{T_1};K_{S_2}K_{T_2}}, \quad (80)$$

and

$$\sum_{K'_{S_2}K'_{T_2}} \sum_{K'_{S_3}K'_{T_3}} (\mathcal{K}\mathcal{K}^\dagger(S',T'))_{K'_{S_1}K'_{T_1};K'_{S_2}K'_{T_2}} \langle K'_{S_3}K'_{T_3}; S'T' \parallel \Gamma(E) \parallel K'_{S_2}K'_{T_2}; S'T'\rangle$$

209

$$\times \ \langle K_{S_2} K_{T_2}; ST \parallel \Gamma(E) \parallel K'_{S_3} K'_{T_3}; S'T' \rangle (-1)^{S+T-S'-T'}$$

$$= \sum_{K_{S_1} K_{T_1} K'_{S_3} K'_{T_3}} \langle K'_{S_1} K'_{T_1}; S'T' \parallel \Gamma(E) \parallel K'_{S_3} K'_{T_3}; S'T' \rangle$$

$$\times \ \langle K'_{S_3} K'_{T_3}; S'T' \parallel \Gamma(E) \parallel K_{S_1} K_{T_1}; ST \rangle (\mathcal{K}\mathcal{K}^{\dagger}(S,T))_{K_{S_1} K_{T_1}; K_{S_2} K_{T_2}}. \tag{81}$$

Eq. (81) is rarely needed. Its derivation has been given in ref. [5]. It is given here because of a typographical error in eq. (42) of ref. [5]. where one set of superscript primes on the ST values associated with $K'_{S_3} K'_{T_3}$ has been omitted, and because it has been used for the results of this appendix.

The starting S, T-values for the recursive calculation process for the $\mathcal{K}\mathcal{K}^{\dagger}$-matrix elements depends on the parity of $\lambda_1 \lambda_2 \lambda_3$. For $\lambda_1 \lambda_2 \lambda_3 = eee$ the starting ST-value is 00; for $\lambda_1 \lambda_2 \lambda_3 = eoe$ the starting ST-values are 10 or 01; with $K_S K_T = 00$ for both $\lambda_2 = e$ or o. For $\lambda_1 \lambda_2 \lambda_3 = oeo$ or ooo the starting ST-values are 10 and 01 with $K_S K_T = 10$ and 01, respectively. For $\lambda_1 \lambda_2 \lambda_3 = eeo$ or eoo, the starting ST-values are $\frac{1}{2}\frac{1}{2}$ with $K_S K_T = +\frac{1}{2}, -\frac{1}{2}$. For $\lambda_1 \lambda_2 \lambda_3 = oee$ or ooe the starting ST-values are again $\frac{1}{2}\frac{1}{2}$, but now with $K_S K_T = +\frac{1}{2}, +\frac{1}{2}$. For all of these starting S,T-values the $\mathcal{K}\mathcal{K}^{\dagger}$-matrices are 1-dimensional and are normalized to $\mathcal{K}\mathcal{K}^{\dagger} = 1$.

In addition, it will be very useful to have general $\mathcal{K}\mathcal{K}^{\dagger}$-matrix elements for S,T=1,1 for the three cases $\lambda_1 \lambda_2 \lambda_3 = oeo, ooo$; and eoe; for which these matrices are all 2-dimensional. The results for the needed $(\mathcal{K}\mathcal{K}^{\dagger}(11))_{K_S K_T; K'_S K'_T}$ are:

1. For $\lambda_1 \lambda_2 \lambda_3 = oeo$:

$$(\mathcal{K}\mathcal{K}^{\dagger}(11))_{10;10} = (\mathcal{K}\mathcal{K}^{\dagger}(11))_{01;01} = \frac{[(\lambda_2 + 1)(\lambda_1 + \lambda_3 + 2) + \lambda_2(\lambda_2 + 2)]}{[(\lambda_2 + 2)(\lambda_1 + \lambda_2 + \lambda_3 + 4)]}, \tag{82}$$

$$(\mathcal{K}\mathcal{K}^{\dagger}(11))_{10;01} = (\mathcal{K}\mathcal{K}^{\dagger}(11))_{01;10} = \frac{(\lambda_1 + \lambda_3 + 2)}{[(\lambda_2 + 2)(\lambda_1 + \lambda_2 + \lambda_3 + 4)]}. \tag{83}$$

Note, in particular, that this matrix has one zero eigenvalue for the special case $\lambda_2 = 0$.

2. For $\lambda_1 \lambda_2 \lambda_3 = ooo$:

$$(\mathcal{K}\mathcal{K}^{\dagger}(11))_{10;10} = (\mathcal{K}\mathcal{K}^{\dagger}(11))_{01;01} = \frac{[(\lambda_1 + \lambda_2 + 1)(\lambda_2 + \lambda_3 + 3) + (\lambda_3 - \lambda_1)]}{(\lambda_1 + \lambda_2 + 3)(\lambda_2 + \lambda_3 + 3)}, \tag{84}$$

$$(\mathcal{K}\mathcal{K}^{\dagger}(11))_{10;01} = (\mathcal{K}\mathcal{K}^{\dagger}(11))_{01;10} = \frac{(\lambda_3 - \lambda_1)}{(\lambda_1 + \lambda_2 + 3)(\lambda_2 + \lambda_3 + 3)}. \tag{85}$$

3. For $\lambda_1 \lambda_2 \lambda_3 = eoe$:

$$(\mathcal{K}\mathcal{K}^{\dagger}(11))_{11;11} = \frac{\lambda_1}{(\lambda_1 + 2)}, \qquad (\mathcal{K}\mathcal{K}^{\dagger}(11))_{1-1;1-1} = \frac{\lambda_3}{(\lambda_3 + 2)},$$

$$(\mathcal{K}\mathcal{K}^{\dagger}(11))_{11;1-1} = (\mathcal{K}\mathcal{K}^{\dagger}(11))_{1-1;11} = 0. \tag{86}$$

Final note: In the derivation of these results an error was discovered in eqs. (26) and (27) of ref. [5]. In eq. (26) the factor $(\lambda_3 + 1)[1 + (-1)^{\lambda_2 + \lambda_3 + S' + T}]$ should be replaced by $[(\lambda_3 + 1) + (\lambda_1 + 1)(-1)^{\lambda_2 + \lambda_3 + S' + T}]$, with a similar replacement in eq. (27).

REFERENCES

[1] D. J. Rowe, J. Math. Phys. **25**, 2662 (1984) D. J. Rowe, G. Rosensteel, and R. Gilmore, J. Math. Phys. **26**, 2787 (1985); R. LeBlanc and D. J. Rowe, J. Phys. A: Math. Gen. **18**, 1891, 1905 (1985); and **19**, 1083 (1986).

[2] J. Deenen and C. Quesne, J. Math. Phys. **25**, 1638, 2354 (1984); C. Quesne, J. Math Phys. **27**, 869 (1986).

[3] K. T. Hecht, *The vector coherent state method and its applications to problems of higher symmetry*, Lecture Notes in Physics Vol. 290 (Springer, New York 1987).

[4] D. J. Rowe, R. LeBlanc, and J. Repka, J. Phys. A: Math. Gen. **22**, L309 (1989).

[5] K. T. Hecht, J. Phys. A: Math Gen. **27**, 3445 (1994).

[6] D. J. Rowe, J. Math. Phys. **35**, 3163 (1994).

[7] D. J. Rowe and K. T. Hecht, to be publ.

[8] K. T. Hecht in *Proc. of the Second Workshop on Harmonic Oscillators*, Cocoyoc, Mexico, K. B. Wolf, and D. Han, eds. NASA Conf. Publication (1994).

[9] J. P. Elliott, Proc. Roy. Soc. **A245**, 128 (1958).

[10] J. P. Elliott, Proc. Roy. Soc. **A245**, 562 (1958).

[11] J. P. Draayer, J. Math. Phys. **11**, 3225 (1970).

[12] S. A. Williams and D. L. Pursey, J. Math. Phys. **9** 1230 (1968).

[13] K. T. Hecht and Sing Chin Pang, J. Math. Phys. **10**, 1571 (1969).

[14] K. T. Hecht in *Symmetries in Science VI*, B. Gruber, ed., Plenum Press p. 299 (1994).

[15] M. Moshinsky, J. Math. Phys. **4**, 1128 (1963).

[16] G. E. Baird and L. C. Biedenharn, J. Math. Phys. **4**, 1449 (1963).

[17] K. T. Hecht, J. Phys. A: Math. Gen. **23**, 407 (1990).

ALGEBRAIC THEORY OF THE THREE-BODY PROBLEM

F. Iachello

Center for Theoretical Physics, Sloane Physics Laboratory
Yale University, New Haven, CT 06520-8120, USA

1. INTRODUCTION

The three-body problem appears in many branches of physics (Fig.1). Many techniques have been developed to solve the non-relativistic quantum mechanical problem for 3 particles interacting through a two-body or three-body force. These techniques are designed to solve the differential or integro-differential Schrödinger equation. In this article, an alternative formulation of the three-body is given, in terms of Lie algebras. This formulation provides a framework for detailed calculations and, most importantly, allows one to classify all solutions that can be obtained in closed form (exactly solvable problems).

2. ALGEBRAIC THEORY

Algebraic theory[1] is a map of a quantum mechanical system onto an algebraic structure $\mathcal{G}$. The logic scheme of algebraic theory is:

$$\text{Quantum mechanical system}$$
$$\Downarrow$$
$$\text{Algebraic Structure}$$

$$\left\{ \begin{array}{l} \text{Lie Algebras} \\ \text{Graded Lie Algebras} \\ \text{Infinite Dimensional (Kac-Moody) Algebras} \\ \text{q-deformed (Hopf) Algebras} \\ \cdots \end{array} \right.$$

$$\Downarrow$$
$$\text{Observables}$$

$$\left\{ \begin{array}{l} \text{Energy Spectra} \\ \text{Transition Rates} \\ \cdots \end{array} \right.$$

Symmetries in Science VII, Edited by
B. Gruber, Plenum Press, New York, 1995

⇓

Experiment

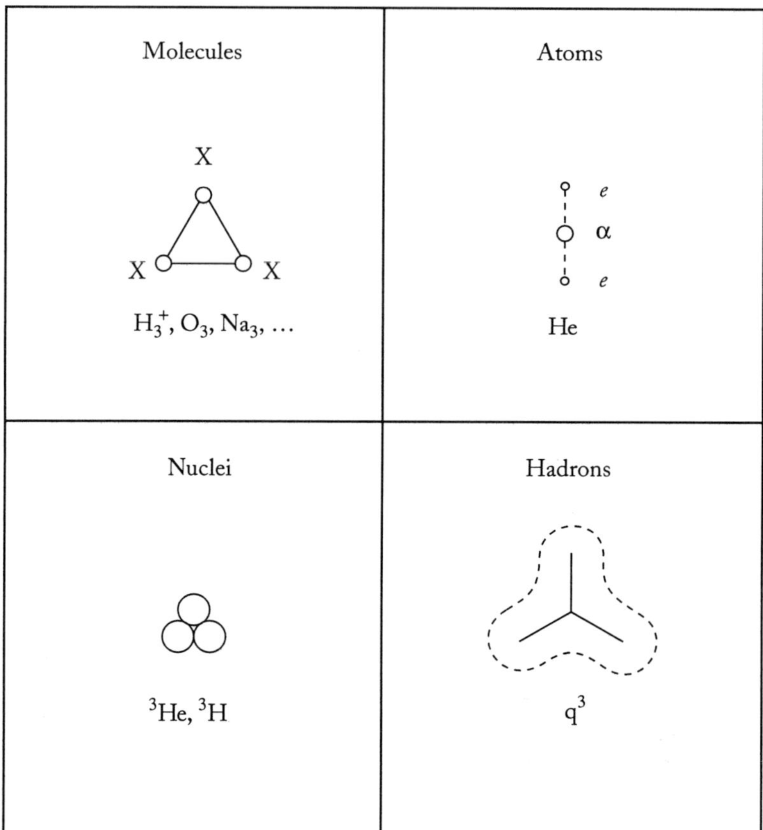

Figure 1. Some examples of three-body problems in physics: molecules, atoms, nuclei, hadrons.

All physical operators are expanded onto elements $G_\alpha \epsilon \mathcal{G}$. Usually this expansion is a polynomial in the G_α 's. For example, the Hamiltonian operator is written as

$$H = E_o + \sum_\alpha \epsilon_\alpha G_\alpha + \sum_{\alpha\beta} u_{\alpha\beta} G_\alpha G_\beta + \ldots \quad , \qquad G_\alpha \epsilon \mathcal{G} \quad . \tag{1}$$

Thus H is in the envelopping algebra of $\mathcal{G}$. The algebra $\mathcal{G}$ is called the spectrum generating algebra (SGA). It has been suggested[1] that, for non-relativistic quantum mechanical problems with no spin in ν space dimensions, one can always take the unitary Lie algebra $u(\nu+1)$ (or its complex extensions or its contractions) as spectrum generating algebra, and the totally symmetric representations [N] of $u(\nu+1)$ (or their appropriate modifications) as the corresponding quantum mechanical space of states. The introduction of a unitary algebra related to the number of dimensions is obvious, since there are, in a quantum mechanical problem, ν coordinates and momenta. The novel and most important aspect is the embedding onto a unitary algebra with <u>one additional</u> (complex) <u>degree of freedom</u>, which allows one to describe all states of the system within a single irreducible representation of $\mathcal{G}$. Alternative embeddings have been suggested in the past[2], most notably the symplectic real $sp(2\nu, R)$ for the harmonic oscillator problem in ν dimensions, and the orthogonal $o(\nu + 1, 2)$ for the Coulomb problem in ν dimensions. $u(\nu + 1)$ (and its variations) covers these cases, as well as all others.

The calculational scheme of algebraic theory is straightforward:

$$\begin{array}{c}
\text{Assign H (i.e. the coefficients } \epsilon_\alpha, u_{\alpha\beta}, \cdots \text{ in (1))} \\
\Downarrow \\
\text{Diagonalize H in the basis [N] obtaining the eigenstates} \\
\mid \psi_i > \text{ and eigenenergies } E_i \\
\Downarrow \\
\text{Evaluate matrix elements of other operators of interest,} < \psi_j \mid T \mid \psi_i > .
\end{array} \qquad (2)$$

Since the Hamiltonian H and the operators T

$$T = t_0 + \sum_\alpha t_\alpha G_\alpha + \cdots \quad , \qquad (3)$$

are all in the enveloping algebra of $\mathcal{G}$, the calculation of any observable quantity becomes a purely algebraic problem, hence the name algebraic theory given to it.

In some special cases, the Hamiltonian H does not contain all the elements of $\mathcal{G}$, but only those combinations which form invariant (Casimir) operators, $\mathcal{C}$, of $\mathcal{G}$ and of its subalgebras $\mathcal{G} \supset \mathcal{G}' \supset \mathcal{G}'' \supset \cdots$,

$$H = \alpha\mathcal{C}(\mathcal{G}) + \alpha'\mathcal{C}(\mathcal{G}') + \alpha''\mathcal{C}(\mathcal{G}'') + \cdots \quad . \qquad (4)$$

These situations are called dynamic symmetries (DS), since the eigenvalue problem for H can be solved in closed form

$$E = \alpha < \mathcal{C}(\mathcal{G}) > + \alpha' < \mathcal{C}(\mathcal{G}') > + \alpha'' < \mathcal{C}(\mathcal{G}'') > + \cdots \quad , \qquad (5)$$

where $< \mathcal{C}(\mathcal{G}) >$ is the expectation value of $\mathcal{C}$ in the appropriate representation of $\mathcal{G}$. These special cases, are also called exactly solvable problems. A consequence of the suggestion[1] is that all exactly solvable problems in ν space dimensions, can be found by studying the branchings of $u(\nu + 1)$ or its complex extensions or its contractions (subject to some conditions, if any).

3. BRIEF REVIEW OF ALGEBRAIC THEORY OF THE TWO-BODY PROBLEM

In the two-body problem, Fig.2, after removal of the center of mass coordinate, $\vec{R}$, there remain 3 degrees of freedom, the relative vector coordinate $\vec{r} = \vec{r}_2 - \vec{r}_1$ (and its associated momentum, $\vec{p}$). Thus, here $\nu=3$, and the spectrum generating algebras is $u(4)$ [3] (or its contraction $h(4)$, or its complex extension, $u(3,1)$). In this article, only some aspects of $u(4)$ will be briefly discussed. First of all, in order to do explicit calculations, it is convenient to introduce a realization of the algebra of $u(4)$ in terms of boson operators. Three of these boson operators can be related to the usual coordinates and momenta,[1]

---

[1]The relation between coordinates and momenta and boson operators is, in general, more complex than that in Eq.(6), but it becomes simple in the limit $N \to \infty$. See Chapt.7 of [15] for details.

$$b_m^\dagger = \frac{1}{\sqrt{2}}(r_m - ip_m) \quad , \qquad m = 0, \pm 1 \quad ,$$

$$b_m = \frac{1}{\sqrt{2}}(r_m + ip_m) \quad , \qquad m = 0, \pm 1 \quad . \tag{6}$$

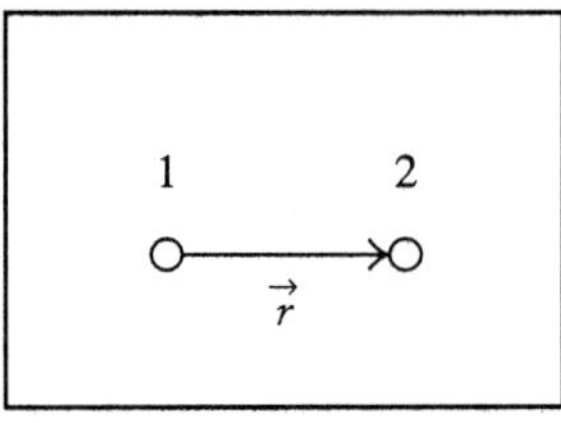

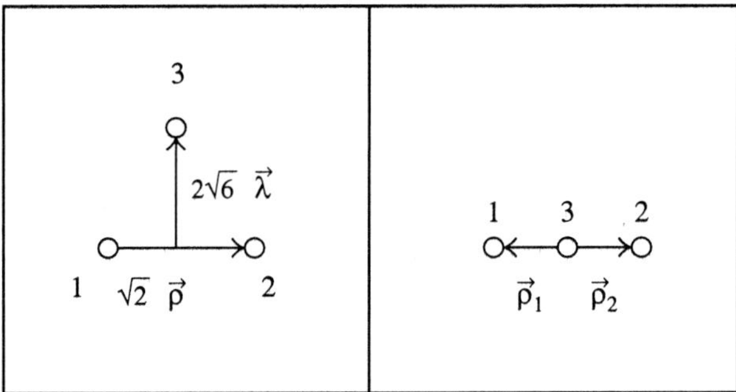

Figure 2. Coordinates used in the description of the two-body and three-body (planar and linear) problems.

The fourth boson operator $s^\dagger(s)$ is an auxiliary boson which is introduced in order to be able to place all the bound states of a generic two-body problem into a single unitary representation [of $u(4)$]. The bilinear products

$$G_{\alpha\alpha'} = b_\alpha^\dagger b_{\alpha'} \qquad (\alpha, \alpha' = 1, \cdots, 4) \quad , \tag{7}$$

of the boson operators $b_\alpha^\dagger \equiv (b_m^\dagger, s^\dagger)$ and $b_\alpha \equiv (b_m, s)$, $\alpha = 1, \cdots, 4$, generate the Lie algebra $u(4)$. For applications to problems with rotational invariance, it is convenient to introduce tensor operators with respect to $o(3)$. The creation operators $b_m^\dagger$ and $s^\dagger$ transform as a vector and a scalar under $o(3)$. In order to construct tensor operators with annihilation operators, one introduces the operators $\tilde{b}_m = (-)^{1-m} b_{-m}$ and $\tilde{s} = s$.

The 16 elements of the algebra $u(4)$ can then be written in their Racah form:

$$\begin{aligned}
G_0^{(0)}(ss) &= (s^\dagger \times \tilde{s})_0^{(0)} \\
G_\kappa^{(1)}(bs) &= (b^\dagger \times \tilde{s})_\kappa^{(1)} \\
G_\kappa^{(1)}(sb) &= (s^\dagger \times \tilde{b})_\kappa^{(1)} \\
G_0^{(0)}(bb) &= (b^\dagger \times \tilde{b})_0^{(0)} \\
G_\kappa^{(1)}(bb) &= (b^\dagger \times \tilde{b})_\kappa^{(1)} \\
G_\kappa^{(2)}(bb) &= (b^\dagger \times \tilde{b})_\kappa^{(2)} \quad .
\end{aligned} \tag{8}$$

An alternative form, which is more useful when studying subalgebras of $u(4)$ is:

$$
\begin{aligned}
\hat{n}_s &= (s^\dagger \times \tilde{s})^{(0)} \\
\hat{D}_\kappa &= (b^\dagger \times \tilde{s} - s^\dagger \times \tilde{b})^{(1)}_\kappa \\
\hat{A}_\kappa &= i(b^\dagger \times \tilde{s} + s^\dagger \times \tilde{b})^{(1)}_\kappa \\
\hat{Q} &= (b^\dagger \times \tilde{b})^{(2)}_\kappa \\
\hat{L} &= (b^\dagger \times \tilde{b})^{(1)}_\kappa \\
\hat{n}_b &= (b^\dagger \times \tilde{b})^{(0)} \quad .
\end{aligned}
\tag{9}
$$

The crosses here and in Eq.(8) denote tensor products with respect to $o(3)$. An important property of the boson operators is also how they transform under permutation of the two bodies. The permutation group S_2 is a two element discrete group, isomorphic to C_2 (rotation of $180°$) and P(parity). In order to characterize the transformation properties of the operators (and the states) one can use labels appropriate to S_2, C_2 or P:

$$
\begin{array}{ccc}
S_2 & C_2 & P \\[4pt]
\square\square & A & + \\[10pt]
\begin{array}{c}\square\\\square\end{array} & B & -
\end{array}
\tag{10}
$$

The operator $b^\dagger$ is odd under permutation P(12),

$$
P(12)b^\dagger_m = (-)b^\dagger_m \quad ,
\tag{11}
$$

while the operator $s^\dagger$ is even. The parity label $(\pm)$ is that which is often used.

3.1 Dynamic symmetries of the two-body problem

The dynamic symmetries of the two-body problem can be studied by considering all subalgebra chains of $u(4)$. If one insists on rotational invariance (i.e. that $o(3)$ be contained in the chain) there are two possible chains:[2]

$$
u(4)
\begin{cases}
\nearrow & u(3) \supset o(3) \supset o(2) \quad (I) \\
\\
\searrow & o(4) \supset o(3) \supset o(2) \quad (II)
\end{cases}
\tag{12}
$$

<u>Chain (I)</u>

States in this chain are characterized by the quantum numbers

---

[2]Since from the algebraic point of view there is no difference between $o(n)$ and $so(n)$, the notation $o(n)$ and $so(n)$ will be used in the following formulas in an interchangeable way.

$$\left| \begin{array}{cccc} u(4) & \supset & u(3) & \supset & o(3) & \supset & o(2) \\ \downarrow & & \downarrow & & \downarrow & & \downarrow \\ N & & n & & L & & M_L \end{array} \right\rangle \qquad . \tag{13}$$

The quantum numbers n, L, M_L contained in a given representation $[N] \equiv [N,0,0,0]$ of $u(4)$ are

$$\begin{aligned} n &= N, N-1, \cdots, 0 \quad , \\ L &= n, n-2, \cdots, 1 \text{ or } 0 \quad (n = \text{odd or even}) \quad , \\ -L &\leq M_L \leq +L \quad . \end{aligned} \tag{14}$$

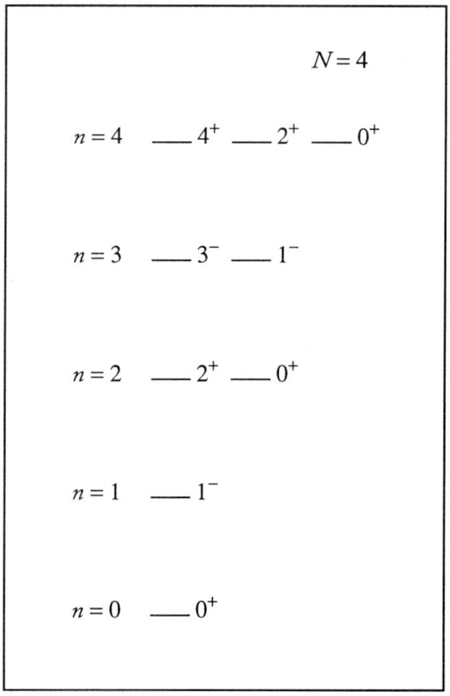

Figure 3. Schematic representations of the spectrum of a two-body problem with $u(3)$ symmetry, Eq.(16) with $E_o = 0, \alpha = 0, \beta = 0$. (Harmonic oscillator in $\nu = 3$ dimensions).

The Hamiltonian H with dynamic symmetry I, can be written, up to terms quadratic in the generators, as

$$H^{(I)} = E_0 + \epsilon \mathcal{C}_1(u3) + \alpha \mathcal{C}_2(u3) + \beta \mathcal{C}_2(o3) \quad , \tag{15}$$

where $\mathcal{C}_1(\mathcal{G})$ and $\mathcal{C}_2(\mathcal{G})$ denote linear and quadratic invariants of $\mathcal{G}$. The eigenvalues of (15) are

$$E^{(I)}(N, n, L, M_L) = E_o + \epsilon n + \alpha n(n+2) + \beta L(L+1) \quad . \tag{16}$$

The corresponding spectrum of states is shown in Fig.3. A special case of (16) is the harmonic oscillator ($\alpha = 0, \beta = 0$).

Chain (II)

States in this chain are characterized by the quantum numbers

$$\left| \begin{array}{cccc} u(4) & \supset & o(4) & \supset & o(3) & \supset & o(2) \\ \downarrow & & \downarrow & & \downarrow & & \downarrow \\ N & & \omega & & L & & M_L \end{array} \right\rangle \tag{17}$$

The quantum numbers ω, L, M_L contained in a given representation [N] of $u(4)$ are given by:

$$\begin{aligned} \omega &= N, N-2, \cdots, 1 \text{ or } 0 \quad (N = \text{odd or even}), \\ L &= \omega, \omega - 1, \cdots, 0 \\ -L &\leq M_L \leq +L \quad . \end{aligned} \tag{18}$$

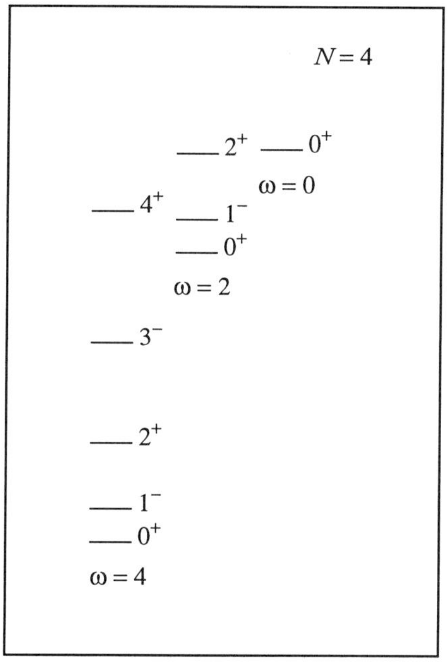

Figure 4. Schematic representation of the spectrum of a two-body problem with $o(4)$ symmetry, Eq.(20) with $E_o = -AN(N+2), A < 0, B > 0$. (Rotovibrator in $\nu = 3$ dimensions.)

The Hamiltonian H with dynamic symmetry, II, can be written as

$$H^{(II)} = E_0 + AC_2(o4) + \beta C_2(o3) \quad , \tag{19}$$

with eigenvalues

$$E^{(II)}(N,\omega,L,M_L) = E_0 + A\omega(\omega+2) + BL(L+1) \quad . \tag{20}$$

The corresponding spectrum of states is shown in Fig.4. This is the spectrum of a 3 dimensional rotovibrator.

The analysis of this subsection shows that in $\nu = 3$ dimensions, there are only two classes of exactly solvable problems (corresponding to $u(3)$ and $o(4)$ symmetry respectively). This situation is summarized in the following diagram, called a lattice of algebras,

$$
\begin{array}{ccc}
 & u(4) & \\
\nearrow & & \searrow \\
u(3) & & o(4) \\
\searrow & & \nearrow \\
 & o(3) & \\
 & \downarrow & \\
 & o(2) &
\end{array}
\tag{21}
$$

[It should be mentioned that the Coulomb problem, which is also exactly solvable in $\nu = 3$ dimensions, belongs to the second class, $o(4)$ symmetry. However, in the Coulomb case, $o(4)$ is not realized bilinearly in the coordinates and momenta, as in (8), and the Hamiltonian is not a polynomial in the $G'_\alpha s$, as in (1), but rather $H = -A/2\,(C_2(o4)+1)$. The dynamic symmetries of both the harmonic oscillator and of the Coulomb problem are discussed in many textbooks[2].]

4. ALGEBRAIC THEORY OF THE THREE-BODY PROBLEM

4.1 The Planar Case

In a generic three-body problem, Fig.2, the three particles lie on a plane. After removal of the center of mass coordinate, $\vec{R}$, there remain two vector coordinates, which can be taken as the Jacobi coordinates (and their associated momenta)

$$
\begin{aligned}
\vec{\rho} &= \frac{1}{\sqrt{2}}(\vec{r}_2 - \vec{r}_1) \quad , \quad \vec{p}_\rho \quad , \\
\vec{\lambda} &= \frac{1}{\sqrt{6}}(2\vec{r}_3 - \vec{r}_1 - \vec{r}_2) \quad , \quad \vec{p}_\lambda \quad .
\end{aligned}
\tag{22}
$$

Since here $\nu = 3+3 = 6$, the spectrum generating algebra is $u(7)$[4]. A bosonic realization of $u(7)$ can be simply done by introducing the boson operators

$$
\begin{aligned}
b^\dagger_{\rho,m} &= \frac{1}{\sqrt{2}}(\rho_m - ip_{\rho,m}), \\
b_{\rho,m} &= \frac{1}{\sqrt{2}}(\rho_m + ip_{\rho,m}), \\
b^\dagger_{\lambda,m} &= \frac{1}{\sqrt{2}}(\lambda_m - ip_{\lambda,m}),
\end{aligned}
$$

$$b_{\lambda,m} = \frac{1}{\sqrt{2}}(\lambda_m + i p_{\lambda,m}),$$
$$m = 0, \pm 1 \quad , \tag{23}$$

together with an auxiliary boson $s^\dagger(s)$. If one denotes the boson operators generically by $b_\alpha^\dagger \equiv (b_{\rho,m}^\dagger, b_{\lambda,m}^\dagger, s^\dagger)$ and $b_\alpha \equiv (b_{\rho,m}, b_{\lambda,m}, s^\dagger), (\alpha = 1, \cdots, 7)$, the 49 bilinear products

$$G_{\alpha\alpha'} = b_\alpha^\dagger b_{\alpha'} \quad , \tag{24}$$

generate the Lie algebra $u(7)$.

The Racah form of $u(7)$ can be written as[4]

$$
\begin{aligned}
\hat{n}_s &= (s^\dagger \times \tilde{s})^{(0)} \\
\hat{D}_{\rho,\mu} &= (b_\rho^\dagger \times \tilde{s} - s^\dagger \times \tilde{b}_\rho)_\mu^{(1)} \\
\hat{D}_{\lambda,\mu} &= (b_\lambda^\dagger \times \tilde{s} - s^\dagger \times \tilde{b}_\lambda)_\mu^{(1)} \\
\hat{A}_{\rho,\mu} &= i(b_\rho^\dagger \times \tilde{s} + s^\dagger \times \tilde{b}_\rho)_\mu^{(1)} \\
\hat{A}_{\lambda,\mu} &= i(b_\lambda^\dagger \times \tilde{s} + s^\dagger \times \tilde{b}_\lambda)_\mu^{(1)} \\
\hat{G}_{S,\mu}^{(\ell)} &= (b_\rho^\dagger \times \tilde{b}_\rho + b_\lambda^\dagger \times \tilde{b}_\lambda)_\mu^{(\ell)} \\
\hat{G}_{\lambda,\mu}^{(\ell)} &= (b_\rho^\dagger \times \tilde{b}_\rho - b_\lambda^\dagger \times \tilde{b}_\lambda)_\mu^{(\ell)} \\
\hat{G}_{\rho,\mu}^{(\ell)} &= (b_\rho^\dagger \times \tilde{b}_\lambda + b_\lambda^\dagger \times \tilde{b}_\rho)_\mu^{(\ell)} \\
\hat{G}_{A,\mu}^{(\ell)} &= i(b_\rho^\dagger \times \tilde{b}_\lambda - b_\lambda^\dagger \times \tilde{b}_\rho)_\mu^{(\ell)}
\end{aligned}
\quad , \tag{25}
$$

with $\ell = 0,1,2$.

An important property of the boson operators is how they transform under permutations of the three bodies. The permutation group S_3 is a six element discrete group which is isomorphic to the dihedral group D_3. In order to characterize the transformation properties of the operators (and the states) one can use the label of either group

$$
\begin{array}{cc}
S_3 & D_3 \\[4pt]
\square\square\square & A_1 \\[4pt]
\begin{matrix}\square\square \\ \square\end{matrix} & E \\[4pt]
\begin{matrix}\square \\ \square \\ \square\end{matrix} & A_2
\end{array}
\quad . \tag{26}
$$

The three creation operators of (23) transform under the transposition P(12) and the cyclic permutation P(123) as

$$
P(12)\begin{pmatrix} s^\dagger \\ b_{\rho,m}^\dagger \\ b_{\lambda,m}^\dagger \end{pmatrix} = \begin{pmatrix} 1 & 0 & 0 \\ 0 & -1 & 0 \\ 0 & 0 & 1 \end{pmatrix}\begin{pmatrix} s^\dagger \\ b_{\rho,m}^\dagger \\ b_{\lambda,m}^\dagger \end{pmatrix} \quad ,
$$

$$
P(123) \begin{pmatrix} s^\dagger \\ b^\dagger_{\rho,m} \\ b^\dagger_{\lambda,m} \end{pmatrix} = \begin{pmatrix} 1 & 0 & 0 \\ 0 & \cos(2\pi/3) & \sin(2\pi/3) \\ 0 & -\sin(2\pi/3) & \cos(2\pi/3) \end{pmatrix} \begin{pmatrix} s^\dagger \\ b^\dagger_{\rho,m} \\ b^\dagger_{\lambda,m} \end{pmatrix} . \tag{27}
$$

In general, one must also add the label that characterizes the transformation properties under parity. The two operators $b^\dagger_{\rho,m}$ and $b^\dagger_{\lambda,m}$ are odd under parity, while the operator $s^\dagger$ is even. The total discrete group is the 12 element group $D_{3h} \sim S_3 \times P$. The parity label is $(\pm)$ as usual.

4.2 Dynamic symmetries of the three-body problem

Since $u(7)$ has a relatively large number of dimensions, the three-body problem has a rather complex set of dynamic symmetires. It is convenient here to introduce the concept of lattice of algebras. In Figs.5-7 all pathways that start with $u(7)$ and end with $o(3)$ are shown. The inclusion of $o(3)$ implies rotational invariance. Each pathway in Figs.5-7 is a chain of subalgebras of $u(7)$.

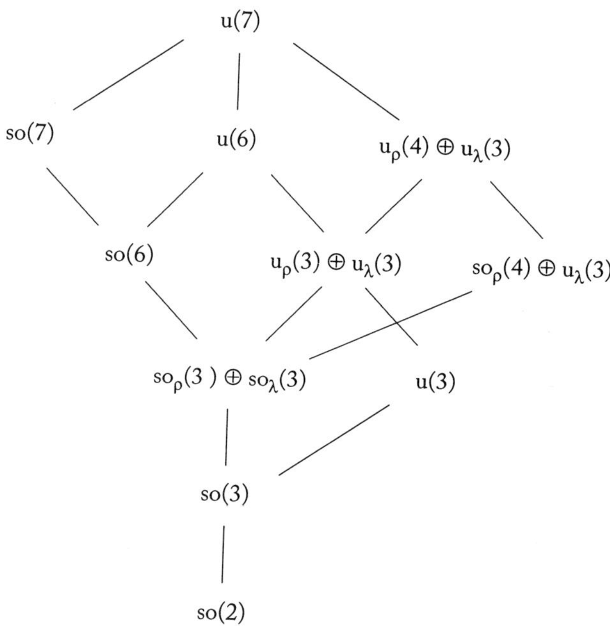

Figure 5. A portion of the lattice of algebras $u(7) \rightarrow so(3)$. The two chains of Eq.(28) are contained in this figure.

A complete study of all chains is too long to be reported here, and I will therefore discuss only a small, selected, number of chains. I begin with chains with hyperrotational, $o(6)$, and rotational, $o(3)$, invariance. These are a generalization to $\nu=6$ of the chains of the three-dimensional problem:

$$u(7) \quad\nearrow\quad u(6) \;\supset\; o(6) \;\supset\; o_\rho(3) \oplus o_\lambda(3) \;\supset\; o(3) \;\supset\; o(2) \qquad (I)$$

$$\searrow\quad o(7) \;\supset\; o(6) \;\supset\; o_\rho(3) \oplus o_\lambda(3) \;\supset\; o(3) \;\supset\; o(2) \qquad (II).$$

$$(28)$$

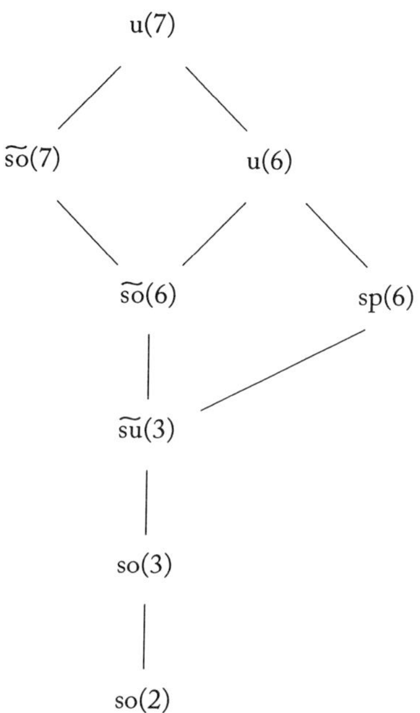

Figure 6. A portion of the lattice of algebras $u(7) \to so(3)$. The algebras denoted with a tilde, for example $s\tilde{o}(6)$, are isomorphic to those of Figure 6, but have different elements. (A different linear combination of the elements of Eq. (25).)

It is of interest to write down explicitly the elements of at least one of the chains in (28) in terms of the elements of $u(7)$ given in (25):

$u(6)$:

$$\hat{L}^{(1)}_{\rho,\mu}; \quad \hat{L}^{(1)}_{\lambda,\mu}; \quad \hat{Q}^{(2)}_{\rho,\mu}; \quad \hat{Q}^{(2)}_{\lambda,\mu}; \quad (b^\dagger_\rho \times \tilde{b}_\lambda)^{(\ell)}, (b^\dagger_\lambda \times \tilde{b}_\rho)^{(\ell)}, l = 0,1,2;$$
$$\hat{n}_\rho; \hat{n}_\lambda \quad .$$

$so(6)$:

$$\hat{L}^{(1)}_{\rho,\mu}; \quad \hat{L}^{(1)}_{\lambda,\mu}; \quad i\left[(b^\dagger_\rho \times \tilde{b}_\lambda) - (b^\dagger_\lambda \times \tilde{b}_\rho)\right]^{(0)};$$
$$\left[(b^\dagger_\rho \times \tilde{b}_\lambda) + (b^\dagger_\lambda \times \tilde{b}_\rho)\right]^{(1)}_\mu; \quad i\left[(b^\dagger_\rho \times \tilde{b}_\lambda) - (b^\dagger_\lambda \times \tilde{b}_\rho)\right]^{(2)}_\mu \quad .$$

$so_\rho(3) \oplus so_\lambda(3)$:

$$\hat{L}^{(1)}_{\rho,\mu}; \hat{L}^{(1)}_{\lambda,\mu}$$

$so(3)$:

$$\hat{L}^{(1)}_{\rho,\mu} + \hat{L}^{(1)}_{\lambda,\mu}$$

$so(2)$:

$$\hat{L}^{(1)}_{\rho,o} + \hat{L}^{(1)}_{\lambda,o}$$

$$(29)$$

The operators $\hat{L}, \hat{Q}, \hat{n}$ are defined in (9). As one can see from (29), the total $o(3)$ algebra is the direct sum of the two algebras $o_\rho(3)$ and $o_\lambda(3)$. The elements of the other chain can be written down in a similar fashion. One can then study the dynamic symmetries associated with (I) and (II).

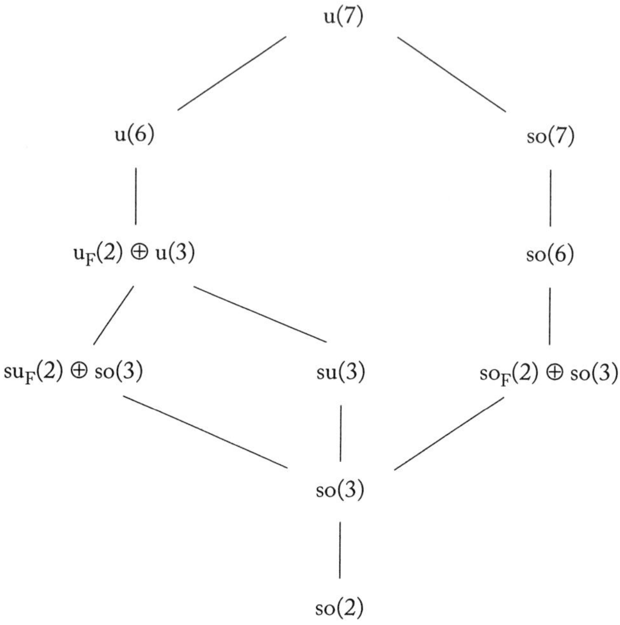

Figure 7. A portion of the lattice of algebras $u(7) \to o(3)$. This figure lists a set of algebras called F-spin algebras. F-spin is discussed in Chapt.5 of [16] and in this case refers to the two indices λ and ρ.

<u>Chain (I)</u>

States in this chain are characterized by the quantum numbers [3]

$$\left|\begin{array}{ccccccc} u(7) & \supset & u(6) & \supset & o(6) & \supset & o_\rho(3) & \oplus & o_\lambda(3) & \supset & o(3) & \supset & o(2) \\ \downarrow & & \downarrow & & \downarrow & & \downarrow & & \downarrow & & \downarrow & & \downarrow \\ N & & n & & \gamma & & L_\rho & & L_\lambda & & L & & M_L \end{array}\right\rangle . \tag{30}$$

The quantum numbers $n, \gamma, L_\rho, L_\lambda, L, M_L$ contained in a given representation $[N] \equiv [N,0,0,0,0,0,0]$ of $u(7)$ are given by:

$$n = N, N-1, \cdots, 0 \quad ;$$
$$\gamma = n, n-2, \cdots, 1 \text{ or } 0 \quad ;$$
$$L_\rho \text{ and } L_\lambda \text{ obtained by partitioning } \gamma \text{ as}$$
$$\gamma = 2\nu + L_\rho + L_\lambda, \quad \nu = 0, 1, \cdots \quad ;$$
$$\mid L_\rho - L_\lambda \mid \leq L \leq \mid L_\rho + L_\lambda \mid \quad ,$$
$$-L \leq M_L \leq +L \quad . \tag{31}$$

---

[3] In this article, the notation appropriate to algebras, i.e. lowercase letters and $\oplus$ signs will be used.

The Hamiltonian H with dynamic symmetry (I) can be written, up to quadratic terms, as

$$H^{(I)} = E_o + \epsilon C_1(u6) + \alpha C_2(u6) + \beta C_2(o6) +$$
$$+\eta_\rho C_2(o_\rho 3) + \eta_\lambda C_2(o_\lambda 3) + \eta C_2(o3) \quad , \tag{32}$$

with eigenvalues

$$E^{(I)}(N,n,\gamma,L_\rho,L_\gamma,L,M_L) = E_0 + \epsilon n + \alpha n(n+5) + \beta\gamma(\gamma+4) +$$
$$+\eta_\rho L_\rho(L_\rho+1) + \eta_\lambda L_\lambda(L_\lambda+1) + \eta L(L+1). \tag{33}$$

If one insists on problems with S_3 symmetry (three identical particles), then one must have $\eta_\rho = \eta_\lambda = 0$. Contrary to the case of the two-body problem, where the transformation properties of the states under S_2 are simply given by $(-)^L$, the construction of states which transform as representations of S_3 in the three-body problem is a very difficult problem. This is due to the fact that the Jacobi coordinates $\vec{\rho}$ and $\vec{\lambda}$ do not contain the particle coordinates in a symmetric way. The chains (I) and (II) are not particularly well suited for this construction. Alternative chains, such as the F-spin chains of Fig.7, are best suited, as discussed by Dragt[6], Kramer and Moshinsky[7], and Bowler et al[8]. Transforming back to the states of chain I and using (32), one

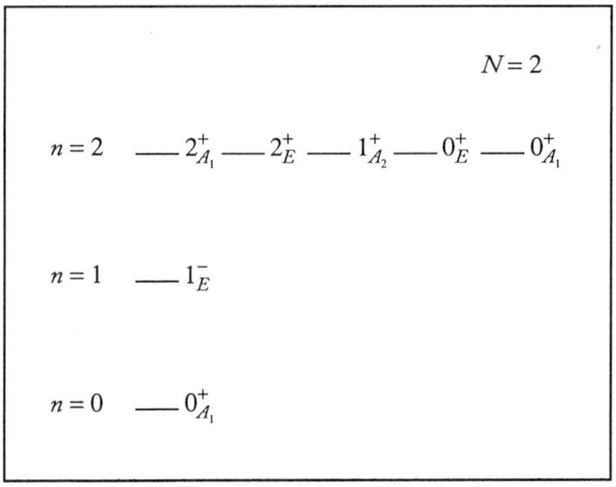

Figure 8. Schematic representation of the spectrum of the 3+3 dimensional harmonic oscillator with S_3 symmetry ($\alpha = \beta = \eta_\rho = \eta_\lambda = \eta = 0$). The angular momentum, parity, and S_3 species are indicated next to each level. The $u(7)$ representation, N, is shown on the upper right hand corner.

obtains the spectrum shown in Fig.8. This figure shows the special case of the 3+3 dimensional harmonic oscillator with S_3 symmetry ($\alpha = \beta = \eta_\rho = \eta_\lambda = \eta = 0$). The role of the algebra $o(6)$, as a subalgebra of $u(6)$, in the context of the three-body problem has also been emphasized by Cutkosky and Hendrick[9].

Chain (II)

States in this chain are characterized by the quantum numbers

$$\left| \begin{array}{ccccccc} u(7) & \supset & o(7) & \supset & o(6) & \supset & o_\rho(3) & \oplus & o_\lambda(3) & \supset & o(3) & \supset & o(2) \\ \downarrow & & \downarrow & & \downarrow & & \downarrow & & \downarrow & & \downarrow & & \downarrow \\ N & & \omega & & \gamma & & L_\rho & & L_\lambda & & L & & M_L \end{array} \right\rangle \quad . \tag{34}$$

The quantum numbers $\omega, \gamma, L_\rho, L_\lambda, L, M_L$ contained in a given representation $[N]$ of $u(7)$ are given by:

$$
\begin{aligned}
&\omega = N, N-2, \cdots, 1 \text{ or } 0(N = \text{odd or even}); \\
&\gamma = \omega, \omega - 1, \cdots, 0; \\
&L_\rho \text{ and } L_\lambda \text{ obtained by partitioning } \gamma \text{ as} \\
&\gamma = 2\nu + L_\rho + L_\gamma \quad , \quad \nu = 0, 1, \cdots \quad ; \\
&\mid L_\rho - L_\lambda \mid \le L \le \mid L_\rho + L_\lambda \mid \quad ; \\
&-L \le M_L \le +L \quad .
\end{aligned}
\tag{35}
$$

The Hamiltonian H with dynamic symmetry (II) can be written, up to quadratic terms, as

$$H^{(II)} = E_0 + A\mathcal{C}_2(o7) + B\mathcal{C}_2(o6) + C_\rho\mathcal{C}_2(o_\rho 3) + C_\lambda\mathcal{C}_2(o_\lambda 3) + C\mathcal{C}(o3), \tag{36}$$

with eigenvalues

$$
\begin{aligned}
E^{(II)}(N, \omega, \gamma, L_\rho, L_\lambda, L, M_L) = {} & E_o + A\omega(\omega + 5) + B\gamma(\gamma + 4) \\
& + C_\rho L_\rho(L_\rho + 1) + C_\lambda L_\lambda(L_\lambda + 1) + CL(L+1) \quad .
\end{aligned}
\tag{37}
$$

If one insists on problems with S_3 symmetry, then $C_\rho = C_\lambda = 0$. The corresponding spectrum is shown in Fig.9.

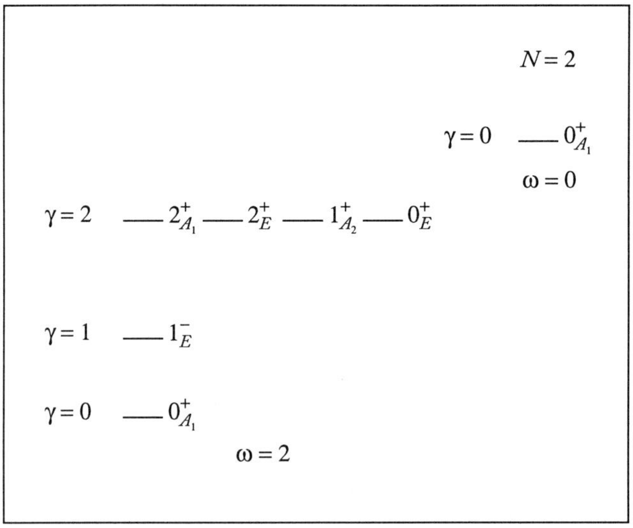

Figure 9. Schematic representation of the spectrum of the 3+3 dimensional rotovibrator with S_3 symmetry $C_\rho = C_\lambda = 0$. In this figure $A < 0, B > 0$ and $C = 0$. The notation is as in Fig.8.

This completes the classification of the two main branches of exactly solvable problems with hyperrotational, $o(6)$, invariance.

[It should be mentioned at this stage that the Coulomb problem, which is also exactly solvable in $\nu = 6$ dimensions, belongs to class II, $o(7)$ symmetry[10]. However, as in the 3 dimensional case, $o(7)$ is not realized bilinearly in the coordinates and momenta, and the Hamiltonian is not a polynomial, but rather $H = -A/2(C_2(o7) + \frac{25}{4})$. The Coulomb problem in 3+3 dimensions is discussed in the accompanying article by Santopinto[11].]

Another point worth mentioning is that, the dynamic symmetry (II) has degeneracies which are reminiscent of those of the experimentally observed spectrum of baryons, if one takes $A > 0, B \approx C \approx 0$ in (37). This spectrum is shown in Fig.10.

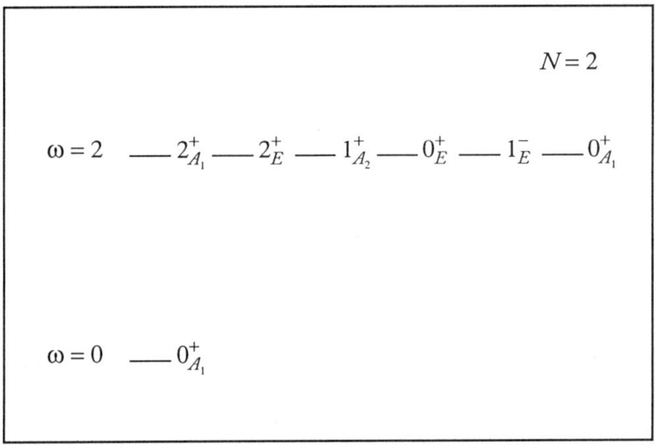

$$\omega = 2 \quad \underline{\quad} 2^+_{A_1} \underline{\quad} 2^+_E \underline{\quad} 1^+_{A_2} \underline{\quad} 0^+_E \underline{\quad} 1^-_E \underline{\quad} 0^+_{A_1} \qquad\qquad N = 2$$

$$\omega = 0 \quad \underline{\quad} 0^+_{A_1}$$

Figure 10. Schematic representation of a spectrum with $o(7)$ symmetry and $A > 0$ (B=C=0).

I come next to the case in which there is no hyperrotational invariance. As one can see from Figs.5-7, the number of possibilities here is larger than in the previous case. I consider here only one chain:

$$u(7) \supset u(6) \supset u_\rho(3) \oplus u_\lambda(3) \supset o_\rho(3) \oplus o_\lambda(3) \supset o(3) \supset o(2) \quad (I_a). \qquad (38)$$

This chain has been extensively used in the treatment of the three-body problem in nuclear[7] and hadronic[12] physics.

Chain I_a

States in this chain are characterized by the quantum numbers

$$\left|
\begin{array}{ccccccccc}
u(7) & \supset & u(6) & \supset & u_\rho(3) & \oplus & u_\lambda(3) & \supset & o_\rho(3) & \oplus & o_\lambda(3) & \supset & o(3) & \supset & o(2) \\
\downarrow & & \downarrow & & \downarrow & & \downarrow & & \downarrow & & \downarrow & & \downarrow & & \downarrow \\
N & & (n) & & n_\rho & & n_\lambda & & L_\rho & & L_\lambda & & L & & M_L
\end{array}
\right\rangle \quad (I_a). \qquad (39)$$

The quantum number n of $u(6)$ is redundant since $n = n_\rho + n_\lambda$. The values of the quantum numbers contained in a given representation [N] of $u(7)$ are:

$$n = N, N - 1, \cdots, 0 \quad ; \quad n_\rho = n, n - 1, \cdots, 0 \quad ; \quad n_\lambda = n - n_\rho \quad ;$$
$$L_\rho = n_\rho, n_\rho - 2, \cdots, 1 \text{ or } 0(n_\rho = \text{odd or even}) \quad ;$$
$$L_\lambda = n_\lambda, n_\lambda - 2, \cdots, 1 \text{ or } 0(n_\lambda = \text{odd or even}) \quad ;$$
$$| L_\rho - L_\lambda | \leq L \leq | L_\rho + L_\lambda | \quad ;$$
$$-L \leq M_L \leq +L \quad . \tag{40}$$

The Hamiltonian H with dynamic symmetry I_a can be written as

$$\begin{aligned}
H^{(I_a)} \; = \;\; & E_0 + \epsilon C_1(u6) + \alpha C_2(u6) + \epsilon_\rho C_1(u_\rho 3) + \alpha_\rho C_2(u_\rho 3) + \\
& + \epsilon_\lambda C_1(u_\lambda 3) + \alpha_\lambda C_2(u_\lambda 3) + \alpha_{\rho\lambda} C_1(u_\rho 3) C_1(u_\lambda 3) + \\
& + \beta_\rho C_2(o_\rho 3) + \beta_\lambda C_2(o_\lambda 3) + \beta C_2(O3) \quad ,
\end{aligned} \tag{41}$$

with eigenvalues

$$\begin{aligned}
E^{(I_a)} = \; & E_0 + \epsilon n + \alpha n(n + 5) + \epsilon_\rho n_\rho + \alpha_\rho n_\rho(n_\rho + 2) + \epsilon_\lambda n_\lambda + \alpha_\lambda n_\lambda(n_\lambda + 2) \\
& + \alpha_{\rho\lambda} n_\rho n_\lambda + \beta_\rho L_\rho(L_\rho + 1) + \beta_\lambda L_\lambda(L_\lambda + 1) + \beta L(L + 1) \quad .
\end{aligned} \tag{42}$$

However, two terms in (41) can be eliminated since $n = n_\rho + n_\lambda$. Chain I_a corresponds to two coupled harmonic oscillators. The spectrum of states is shown in Fig.11. This spectrum should be compared with that of Fig.8. When no S_3 symmetry is present, the two-dimensional representation E of S_3 splits into two separate pieces, corresponding to oscillations in the ρ or λ degree of freedom.

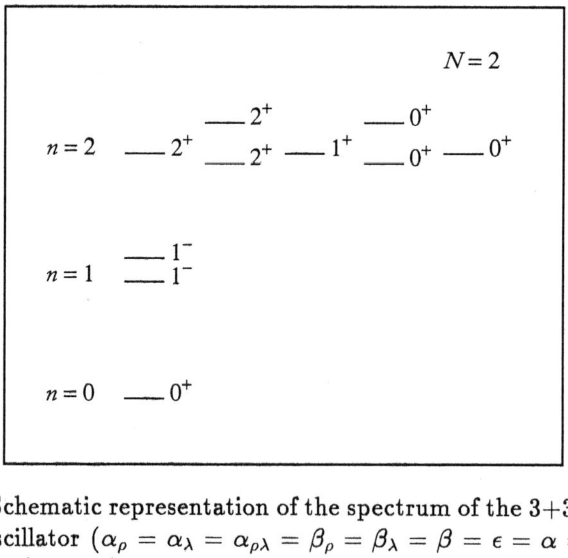

Figure 11. Schematic representation of the spectrum of the 3+3 dimensional harmonic oscillator $(\alpha_\rho = \alpha_\lambda = \alpha_{\rho\lambda} = \beta_\rho = \beta_\lambda = \beta = \epsilon = \alpha = 0)$ without S_3 symmetry $(\epsilon_\rho \neq \epsilon_\lambda)$.

4.3 Hyperspherical coordinates

It is interesting to note that for problems with hyperrotational invariance, it is convenient to introduce, instead of the individual Jacobi variables, $\vec{\rho}$ and $\vec{\lambda}$, a new set of coordinates, the hyperspherical coordinates[13]:

$$x = (\vec{\rho}^2 + \vec{\lambda}^2)^{1/2}, \qquad \xi = arctg(\rho/\lambda), \qquad \Omega_\rho, \Omega_\lambda \quad . \tag{43}$$

The basis states can then be written, in a coordinate representation, as

$$\Psi(x, \xi, \Omega_\rho, \Omega_\lambda) = f_{[\gamma]}(x) Y^{[\gamma]}_{L_\rho, L_\gamma, L, M_L}(\xi, \Omega_\rho, \Omega_\lambda) \quad , \tag{44}$$

where $Y^{[\gamma]}$ are the hyperspherical harmonics. The Hamiltonian

$$H^{(I)} = E_0 + \epsilon \mathcal{C}_1(u6) \quad , \tag{45}$$

(cfr.32) leads to the differential equation

$$\left[\frac{d^2}{dx^2} + \frac{5}{x}\frac{d}{dx} - \frac{\gamma(\gamma+4)}{x^2}\right]\Psi(x) = \left(\frac{2m}{\hbar^2}\right)\left[E - \frac{3}{2}kx^2\right]\psi(x) \quad , \tag{46}$$

with eigenvalues

$$E^{(I)} = E_0 + \epsilon n \quad , \tag{47}$$

as in (33), and $\epsilon = \sqrt{\frac{3k}{m}}$.

4.4 The Linear Case

In some cases, most notably linear triatomic molecules, the 3 particles lie on a line, Fig.2. In this case, it is convenient to use a different set of coordinates, the bond coordinates $\vec{\rho}_1, \vec{\rho}_2$ of Fig.2.

$$\begin{aligned}
\vec{\rho}_1 &= \vec{r}_1 - \vec{r}_3 \quad , \quad \vec{p}_{\rho_1} \quad , \\
\vec{\rho}_2 &= \vec{r}_2 - \vec{r}_3 \quad , \quad \vec{p}_{\rho_2} \quad .
\end{aligned} \tag{48}$$

The coordinates $\vec{\rho}_1$ and $\vec{\rho}_2$ are then quantized separately, leading to $u_1(4) \oplus u_2(4)$ as spectrum generating algebra. A bosonic realization of this algebra is:

$$\begin{aligned}
b^\dagger_{1,m} &= \frac{1}{\sqrt{2}}(\rho_{1m} - ip_{\rho 1m}) \\
b_{1,m} &= \frac{1}{\sqrt{2}}(\rho_{1m} + ip_{\rho 1m}) \\
b^\dagger_{2,m} &= \frac{1}{\sqrt{2}}(\rho_{2m} - ip_{\rho 2m}) \\
b_{2,m} &= \frac{1}{\sqrt{2}}(\rho_{2m} + ip_{\rho 2m}) \quad , m = 0, \pm 1 \quad ,
\end{aligned} \tag{49}$$

together with the auxiliary bosons $s_1^\dagger(s_1)$ and $s_2^\dagger(s_2)$. The dynamic symmetries of $u_1(4)\oplus u_2(4)$ have been classified completely [14]. They are:

$$
\begin{array}{l}
u_1(4) \ \oplus\ u_2(4) \ \supset\ u_1(3) \ \oplus\ u_2(3) \ \supset\ o_1(3) \ \oplus\ o_2(3) \ \supset\ o(3) \ \supset\ o(2)\quad I_a\\
\supset\ u_1(3) \ \oplus\ o_2(4) \ \supset\ o_1(3) \ \oplus\ o_2(3) \ \supset\ o(3) \ \supset\ o(2)\quad I_b\\
\supset\ o_1(4) \ \oplus\ o_2(4) \ \supset\ o_1(3) \ \oplus\ o_2(3) \ \supset\ o(3) \ \supset\ o(2)\quad I_c\\
\supset\ u_1(3) \ \oplus\ u_2(3) \ \supset\ u(3) \ \supset\ o(3) \ \supset\ o(2)\qquad\quad II_a\\
\supset\ o_1(4) \ \oplus\ o_2(4) \ \supset\ o(4) \ \supset\ o(3) \ \supset\ o(2)\qquad\quad II_b\\
\supset\ u(4) \ \supset\ u(3) \ \supset\ o(3) \ \supset\ o(2)\qquad\qquad\qquad III_a\\
\supset\ u(4) \ \supset\ o(4) \ \supset\ o(3) \ \supset\ o(2)\qquad\qquad\qquad III_b
\end{array}
$$

$$\tag{50}$$

Detailed descriptions of these chains are given in [14]. Here, I discuss briefly only chain II_b which is of particular relevance for linear diatomic molecules.

Chain II_b

States in this chain are characterized by the quantum numbers

$$
\left|
\begin{array}{ccccccc}
u_1(4) & \oplus\ u_2(4) & \supset\ o_1(4) & \oplus\ o_2(4) & \supset\ o(4) & \supset\ o(3) & \supset\ o(2)\\
\downarrow & \downarrow & \downarrow & \downarrow & \downarrow & \downarrow & \downarrow\\
N_1 & N_2 & \omega_1 & \omega_2 & \tau_1,\tau_2 & L & M_L
\end{array}
\right\rangle \ . \tag{51}
$$

The values of the quantum numbers contained in a given representation $[N_1]\otimes[N_2]$ of $u_1(4)\oplus u_2(4)$ are:

$$
\begin{aligned}
\omega_1 &= N_1, N_1-2, \cdots, 1 \text{ or } 0 \ (N_1 = \text{odd or even});\\
\omega_2 &= N_2, N_2-2, \cdots, 1 \text{ or } 0 \ (N_2 = \text{odd or even});\\
&(\tau_1,\tau_2) \text{ obtained from the direct product } (\omega_1,0)\otimes(\omega_2,0)\\
&\quad \tau_1 = \omega_1+\omega_2-\mu-\nu, \quad \tau_2 = \mu-\nu \ ;\\
&\mu = 0,1\cdots, \ min\,(\omega_1,\omega_2), \quad \nu = 0,1,\cdots,\mu \ .\\
L^P &= 0^+, 1^-, \cdots, (\tau_1^+ \text{or} \tau_1^-), \quad \text{when } \tau_2 = 0 \text{ and } \tau_1 = \text{even or odd} \ ;\\
L^P &= \tau_2^\pm, (\tau_2+1)^\pm, \cdots, \tau_1^\pm \ , \quad \text{when } \tau_2 \neq 0 \ ;\\
&\qquad\qquad -L \leq M_L \leq +L \ . \tag{52}
\end{aligned}
$$

The parity label P is necessary here to distinguish the states. The Hamiltonian H with dynamic symmetry II_b is

$$
H^{(II_b)} = E_0 + A_1 C_2(o_1 4) + A_2 C_2(o_2 4) + A_{12} C_2(o4) + B C_2(o3) \ , \tag{53}
$$

with eigenvalues

$$
E^{(II_b)} = E_0 + A_1\omega_1(\omega+2) + A_2\omega_2(\omega_2+2) + A_{12}\left[\tau_1(\tau_1+2)+\tau_2^2\right] + BL(L+1) \ . \tag{54}
$$

The spectrum corresponding to (54) is shown in Fig.4 of Ref.[14]. It describes linear triatomic molecules.

It should be noted that both in the chains of Figs.5-7 and in those of (50), there are some automorphisms, arising from sign and phase changes among the elements of $\mathcal{G}$, that increase the number of possible chains. For example, the automorphism

$$
\begin{aligned}
\mathcal{A}(\hat{n}_s) &= -\hat{n}_s \quad, \quad \mathcal{A}(\hat{n}_b) = -\hat{n}_b \quad, \\
\mathcal{A}(\hat{D}) &= \hat{D} \quad, \quad \mathcal{A}(\hat{L}) = \hat{L} \quad, \\
\mathcal{A}(\hat{A}) &= -\hat{A} \quad, \quad \mathcal{A}(\hat{Q}) = -\hat{Q} \quad,
\end{aligned}
\tag{55}
$$

leaves the commutation relations of $u(4)$ unchanged. For a single algebra, this is not relevant, because the phase changes can be reabsorbed in the coefficients in front of the operators. For coupled algebras, the sign is relevant leading to other chains. For example,

$$
u_1(4) \oplus \bar{u}_2(4) \supset u^*(4) \supset o(4) \supset o(3) \supset o(2) \quad, \qquad (III_c) \tag{56}
$$

where $\bar{u}(4)$ denotes the algebra (9), and $u^*(4)$ that obtained by combining $u_1(4)$ of (9) with $\bar{u}_2(4)$ of Eq.(55). This problem is discussed in [14].

5. CONCLUSIONS

In this article, an alternative formulation of the three-body problem in terms of algebraic structures has been given. The ingredients of this formulation are: a spectrum generating algebra (SGA) + a space on which it acts (irrep). It has been suggested[4, 14] to use:

$$
\begin{array}{ccc}
 & SGA & Irrep \\
\text{Planar case} & u(7) & [N] \\
\text{Linear case} & u_1(4) \oplus u_2(4) & [N_1] \otimes [N_2]
\end{array}
\tag{57}
$$

The algebraic formulation of the three-body problem is particularly useful when the interactions between the constituent particles are not well known and must be modeled. This situation occurs in molecules[15] and hadrons[5].

Since the number of space degrees of freedom of the three-body problem is $\nu = 6$, this problem has a rich algebraic structure, as evident from Figs.5-7 and Eq.(50). In this article, only the classification scheme of the algebras $u(7)$ and $u_1(4) \oplus u_2(4)$ has been briefly discussed.

However, the most important aspect of algebraic theory is the possibility to do realistic calculations, in which the dynamic symmetries are used as a basis to diagonalize a Hamiltonian and to compute matric elements of operators. In this respect, much work remains to be done. Outstanding problems are: (i) the explicit construction of transformation brackets from the $o(7)$ chain II to the $u(6)$ chain I; this will allow one to use chain II as an alternative to the harmonic oscillator basis, for problems where the $o(7)$ symmetry is relevant; (ii) the explicit construction of the transformation brackets from $o(6)$ to $u_\rho(3) \oplus u_\lambda(3)$; this will allow one to use the hyperspherical basis for problems with hyperspherical symmetry; (iii) the embedding of the appropriate discrete group, for example S_3, onto the chains of Figs.5-7. This problem has only partially been solved (for the harmonic oscillator).

6. ACKNOWLEDGEMENTS

This work has been supported in part by D.O.E. Grant DE-FG02-91ER40608. I wish to thank Roelof Bijker and Amiran Leviatan for discussions on the algebraic structure of $u(7)$.

REFERENCES

[1] F. Iachello, *Nucl. Phys.* **A560**, 23 (1993); F. Iachello, in "Lie Algebras, Cohomologies and New Applications of Quantum Mechanics", Contemporary Mathematics, AMS, Vol. 160 (1994), p.151-171.

[2] B.W. Wybourne, Classical Groups for Physicists, *J. Wiley and Sons*, N.Y. (1976), **Chapts. 20 and 21.**

[3] F. Iachello, *Chem. Phys. Lett.* **78**, 581 (1981); F. Iachello and R.D. Levine, *J. Chem. Phys.* **77**, 3046 (1982).

[4] R. Bijker and A. Leviatan, in "Symmetries in Science VII: Spectrum Generating Algebras and Dynamic Symmetries in Physics", B. Gruber and T. Otsuka, eds., Plenum Press, New York (1994), p.87.

[5] R. Bijker, F. Iachello, and A. Leviatan, *Ann. Phys.* (N.Y.) 1994, in press.

[6] A.J. Dragt, J. Math. Phys. **6**, 533 (1965).

[7] P. Kramer and M. Moshinsky, *Nucl. Phys.* **82**, 241 (1966).

[8] K.C. Bowler, P.J. Corvi, A.J.G. Hey, P.D. Jarvis and R.C. King, *Phys. Rev.* **D24**, 197 (1981).

[9] R.E. Cutkosky and R.E. Hendrik, *Phys. Rev.* **D16**, 793 (1977).

[10] A.O.Barut and Y. Kitagawara, *J. Phys.* **A14**, 2581 (1981); **A15**, 117 (1982).

[11] E. Santopinto, These Proceedings.

[12] N. Isgur and G.Karl, *Phys. Rev.* **D18**, 4187 (1978); **D19**, 2653 (1979), **D20**, 1191 (1979).

[13] G. Morpurgo, *Nuovo Cimento* **9**, 461 (1952); J.L. Ballot and M. Fabre de La Ripelle, *Ann. Phys.* (N.Y.) **127**, 62 (1980); M. Giannini, *Nuovo Cimento* **A76**, 455 (1983), and references therein.

[14] O.S. van Roosmalen, A.E.L. Dieperink, and F. Iachello, *Chem. Phys. Lett.* **85**, 32 (1982); O.S. van Roosmalen, F. Iachello, R.D. Levine, and A.E.L. Dieperink, *J. Chem. Phys.* **79**, 2515 (1983).

[15] F. Iachello and R.D. Levine, "Algebraic Theory of Molecules", *Oxford University Press*, Oxford (1994).

[16] F. Iachello and A. Arima, "The Interacting Boson Model", *Cambridge University Press*, Cambridge (1987).

QUANTUM EFFECT OF NONLINEAR BORN–INFELD FIELD

Masahiko Kanenaga, Mikio Namiki and Hiroshi Hotta

Department of Physics
Waseda University
Tokyo 160, Japan

INTRODUCTION

Many years ago Born and Infeld [1] presented a nonlinear electromagnetic field with a non-polynomial action including the so-called *universal length*. One of the most important characteristics of the Born-Infeld field is found in its static solution which has no infra-red divergence. Many physicists expected that this might be an example of divergence-free field theory. However, no one could succeed to quantize the field, by means of the standard canonical quantization method, because of the complicated nonlinearity. Even the path-integral quantization can hardly be applied to this field, because we cannot easily manipulate such a non-polynomial action. We have to invent a new quantization method if we want to quantize the Born-Infeld field.

About ten years ago, Parisi and Wu [2] proposed a new quantization method, called *stochastic quantization*, by introducing a hypothetical stochastic process with respect to a new (fictitious) time, say t, other than ordinary time, say x_0. The stochastic process is so designed as to yield quantum mechanics as thermal equilibrium limit for very large t. This theory starts from a hypothetical langevin equation for the stochastic process by adding the fictitious-time derivative and the random source to the classical equation of motion. That is, the stochastic quantization can be formulated only on the basis of classical field equation, without resorting to canonical formalism.

BRIEF REVIEW OF STATIC BORN-INFELD FIELD

The ordinary electromagnetic field is described by the following Lagrangian density

$$\mathcal{L} = -\frac{1}{4}F_{\mu\nu}F^{\mu\nu} = \frac{1}{2}(\mathbf{E}^2 - \mathbf{H}^2) \,, \tag{1}$$

where $F_{\mu\nu} = \partial_\mu A_\nu - \partial_\nu A_\mu$. The corresponding action is given by

$$\mathcal{S} = \int d^4x\,\mathcal{L} \,. \tag{2}$$

Symmetries in Science VII, Edited by
B. Gruber, Plenum Press, New York, 1995

Here we have followed the usual notation. Note that we keep the Minkowski metric.

In the case of spherically symmetric static electric field, we can put

$$\mathbf{E} = -\nabla A_0(r), \qquad \mathbf{H} = 0 , \tag{3}$$

and obtain

$$A_0 = \frac{e}{r} \tag{4}$$

for a point charge e. That this field becomes ∞ for $r \to 0$ is a well-known fact.

Let us introduce the action functional of the Born-Infeld field [1]:

$$\mathcal{S}_B = \int d^4x \left[-b^2\sqrt{1 - \frac{2}{b^2}\mathcal{L}} + b^2 \right] \tag{5}$$

where $1/\sqrt{b}$ is a sort of universal constant called *universal length*, and has the dimension of length in natural unit $\hbar = c = 1$. We can easily see that $\mathcal{S}_B \to \mathcal{S}$, that is, the Born-Infeld will become the ordinary electromagnetic field as b tends to ∞.

As is well known, all physical quantities can be written only in terms of the dimension of length in natural unit $\hbar = c = 1$. Many years ago Heisenberg anticipated that we could formulate a finite field theory, free from field-theoretical divergences, if we could bring a sort of *universal length* into physics in an appropriate way. Following his idea, Born and Infeld [1] proposed to use the above field given by action (5). Unfortunately, however, the Heisenberg's anticipation was not accomplished yet even now.

For the spherically symmetric static field $A_0(r)$ (3), the Born-Infeld action (5) yields the following equation

$$\frac{\partial}{\partial r} \left[r^2 \frac{\frac{\partial}{\partial r} A_0(r)}{\sqrt{1 - \frac{1}{b^2}(\frac{\partial}{\partial r} A_0(r))^2}} \right] = 0 , \tag{6}$$

whose solution is given by

$$A_0 = \frac{e}{r_0} \int_{r/r_0}^{\infty} d\xi \frac{1}{\sqrt{1+\xi^4}} \to \left\{ \begin{array}{ll} (e/r) & \text{for } r \gg r_0 , \\ 1.8541 \cdot (e/r_0) & \text{for } r \to 0 , \end{array} \right. \tag{7}$$

where $r_0 = \sqrt{|e|/b}$. We surely realize that the Born-Infeld field has finite static self-energy. Of course, the self-energy goes back to the original infinity as $b \to \infty$.

WAVE GAUGE

We are not interested in static field but in wave field propagating to remote places. The Euler-Lagrange equation of $\mathcal{S}_B$ is generally written as

$$\frac{\delta \mathcal{S}_B}{\delta A_\nu(x)} = \partial_\mu \left[\frac{F^{\mu\nu}}{\sqrt{1 + \frac{1}{2b^2}F^2}} \right] = \partial_\mu \left[\frac{F^{\mu\nu}}{\chi} \right] = 0 , \tag{8}$$

or

$$\partial_\mu F^{\mu\nu} = (\partial_\mu \ln \chi) F^{\mu\nu} , \tag{9}$$

where $F^2 = F_{\mu\nu}F^{\mu\nu}, \chi = \sqrt{1 + \frac{1}{2b^2}F^2}$.

The right-hand side of this equation is proportional to $1/b^2$, and can be expanded in a power series of $1/b^2$. Needless to say, its unperturbed one (for $1/b = 0$) is nothing other than the free Maxwell equation

$$\partial_\mu F^{\mu\nu} = 0 , \tag{10}$$

234

which allows us to use the wave gauge given by

$$A_0 = 0 \; , \quad \nabla \cdot \mathbf{A} = 0 \; . \tag{11}$$

Consequently, one may naively expect to have the perturbative theory based on (9) and (11). In this case, however, we can hardly develop this kind of the perturbative approach to the quantized Born-Infeld field, because the interaction part includes higher powers of derivative terms. This is the reason why we attempt to develop an unperturbative approach to the quantized Born-Infeld field by means of *stochastic quantization* [2] in the present paper.

STOCHASTIC QUANTIZATION OF BORN-INFELD FIELD

As is well-known, it is convenient to use the Euclidean action $\mathcal{S}_E$ derived by the Wick rotation $(x_0 \to -ix_0)$ from the original Minkowski action, for the purpose of carrying out the Parisi-Wu stochastic quantization. In order to perform *stochastic quantization* of the Born-Infeld field A_μ, we have to introduce the additional dependence on fictitious-time t (other than ordinary-time x_0) into the field quantities and then to set the basic Langevin equation

$$\frac{\partial}{\partial t} A_\mu(x,t) = -\frac{\delta \mathcal{S}_E}{\delta A_\mu}\Big|_{A=A(x,t)} + \eta_\mu(x,t) \tag{12}$$

for a hypothetical stochastic process of $A_\mu(x,t)$ with respect to t [2]. Here $\eta_\mu(x,t)$ is the Gaussian white-noise field subject to

$$< \eta_\mu(x,t) > \; = \; 0 \; , \tag{13}$$

$$< \eta_\mu(x,t)\eta_\nu(x',t') > \; = \; 2\delta_{\mu\nu}\delta(x-x')\delta(t-t') \; , \tag{14}$$

where we have put $\hbar = 1$ for simplicity.

According to the prescription of stochastic quantization, we can derive the field-theoretical propagators through the well-known formula

$$D^A_{\mu\nu}(x,x') = \lim_{t\to\infty}\{< A_\mu(x,t)A_\nu(x',t) > - < A_\mu(x,t) >< A_\nu(x',t) >\} \; , \tag{15}$$

where $A_\mu(x,t)$ as a function of η_μ is to be obtained by solving (12).

Let us decompose A_μ and η_μ into their longitudinal and transverse components, A^L_μ and A^T_μ, and η^L_μ and η^T_μ, given by

$$A^L_\mu \;\; = \;\; \frac{1}{\Box}\partial_\mu\partial_\nu A_\nu \tag{16}$$

$$A^T_\mu \;\; = \;\; (\delta_{\mu\nu} - \frac{1}{\Box}\partial_\mu\partial_\nu)A_\nu \; ; \quad \partial_\mu A^T_\mu = 0 \; , \tag{17}$$

and similar ones for η's. Therefore, we can decompose the basic Langevin equation (12) as follows;

$$\frac{\partial}{\partial t} A^L_\mu(x,t) \;\; = \;\; 0 + \eta^L_\mu(x,t) \; , \tag{18}$$

$$\frac{\partial}{\partial t} A^T_\mu(x,t) \;\; = \;\; -\frac{\delta \mathcal{S}_E}{\delta A^T_\mu}\Big|_{A^T = A^T(x,t)} + \eta^T_\mu(x,t) \; . \tag{19}$$

The absence of drift force in (18) is an important reflection of the gauge invariance that $\mathcal{S}_E$ does not depend on A^L_μ. Consequently, the longitudinal component $A^L_\mu(x,t)$

makes a random walk around its initial value $A_\mu^L(x,0) = \frac{1}{\Box}\partial_\mu\phi(x)$, $\phi(x)$ being a scalar field. As was discussed in detail in the case of non-Abelian gauge field [3], we must introduce *gauge parameter* α by taking average of ϕ over random fluctuations around zero $(\overline{\phi(x)\phi(x')} = \alpha\delta(x - x'))$. Thus we obtain

$$D_{\mu\nu}^{A^L}(x, x') = \frac{1}{\Box}\partial_\mu\partial_\nu(\alpha\frac{1}{\Box} + 2t)\delta(x - x') \, . \tag{20}$$

Of course, we know that the longitudinal components never appear in gauge-invariant (physical) quantities.

We should also notice that the transverse components and their propagators are completely decoupled with the longitudinal ones in the Abelian gauge field case. This is an important point quite different from the non-Abelian gauge field case. Thus we can safely discard A_μ^L, and use the wave gauge (11), even in this case, for the purpose of deriving propagators of the transverse components.

Considering wave propagation along the $z(= x_3)$-axis, we put

$$A_0 \;=\; 0 \, , \qquad A_3 = 0 \, , \tag{21}$$
$$A_1 \;=\; A_1(x_0, x_3) \, , \qquad A_2 = A_2(x_0, x_3) \, , \tag{22}$$

and

$$\eta_0 \;=\; 0 \, , \qquad \eta_3 = 0 \, , \tag{23}$$
$$\eta_1 \;=\; \eta_1(x_0, x_3) \, , \qquad \eta_2 = \eta_2(x_0, x_3) \, . \tag{24}$$

Note that (11) for A_μ and the similar ones for η_μ are automatically satisfied.

In this case, the Euclid action becomes

$$S_E = \int dx_0 dx_3 \left[b^2\sqrt{1 + \frac{1}{2b^2}F_E^2} - b^2 \right] \, , \tag{25}$$

where $F_E^2 = 2\{(\partial_0 A_1)^2 + (\partial_3 A_2)^2 + (\partial_0 A_2)^2 + (\partial_3 A_1)^2\}$. The corresponding classical field equation is given by

$$\frac{\delta S_E}{\delta A_i(x)} = -\partial_0\left[\frac{\partial_0 A_i}{\sqrt{1 + \frac{1}{2b^2}F_E^2}}\right] - \partial_3\left[\frac{\partial_3 A_i}{\sqrt{1 + \frac{1}{2b^2}F_E^2}}\right] = 0, \quad (i = 1, 2) \, . \tag{26}$$

Note that the dimensions of the field and b are different from the original ones.

Based on the classical equation, we can set up the basic Langevin equation for stochastic quantization of the Born-Infeld field as follows;

$$\frac{\partial}{\partial t}A_i(x_0, x_3, t) \;=\; \partial_0\left[\frac{\partial_0 A_i}{\sqrt{1 + \frac{1}{2b^2}F_E^2}}\right] + \partial_3\left[\frac{\partial_3 A_i}{\sqrt{1 + \frac{1}{2b^2}F_E^2}}\right] + \eta_i(x_0, x_3, t) \, ,$$
$$(i = 1, 2) \, , \tag{27}$$

where the fluctuating source-field η_i should have the statistical properties

$$< \eta_i(x_0, x_3, t) >_\eta \;=\; 0 \, , \tag{28}$$
$$< \eta_i(x_0, x_3, t)\eta_j(x_0', x_3', t') >_\eta \;=\; 2\delta_{ij}\delta(x_0 - x_0')\delta(x_3 - x_3')\delta(t - t') \, . \tag{29}$$

236

Consequently, we have to obtain A_i as a function (or functional) of η_i, by solving (27), and then to calculate expectation values of physical quantities, by making use of (28) and (29). For example, the field-theoretical propagator of A_i is given by the formula

$$\Delta_{ij}^A(x_0 - x_0', x_3 - x_3') \equiv \lim_{t \to \infty} \{ < A_i(x_0, x_3, t) A_j(x_0', x_3', t) >$$
$$- < A_i(x_0, x_3, t) A_j(x_0', x_3', t) > \} \ . \tag{30}$$

For conventional fields, we can extract real information about the particle mass, $\mathcal{M}$ and $\mathcal{M}'$, associated with the field (or the first energy gap) from the asymptotic formulas

$$\Delta_{ii}^A(0, x_3) \xrightarrow{|x_3| \to \infty} \text{const.} \exp[-\mathcal{M}|x_3|] \ , \tag{31}$$

$$\Delta_{ii}^A(x_0, 0) \xrightarrow{|x_0| \to \infty} \text{const.} \exp[-\mathcal{M}'|x_0|] \tag{32}$$

(no summation), in which we can put $\mathcal{M} = \mathcal{M}'$ for the Euclidean symmetry in space-time. Unfortunately in the Born-Infeld case, however, we have no reliable theory to justify this scheme. Despite of this situation, we intend to follow the conventional approach to the "particle mass" associated with the (transverse) Born-Infeld field, based on (31) and/or (32).

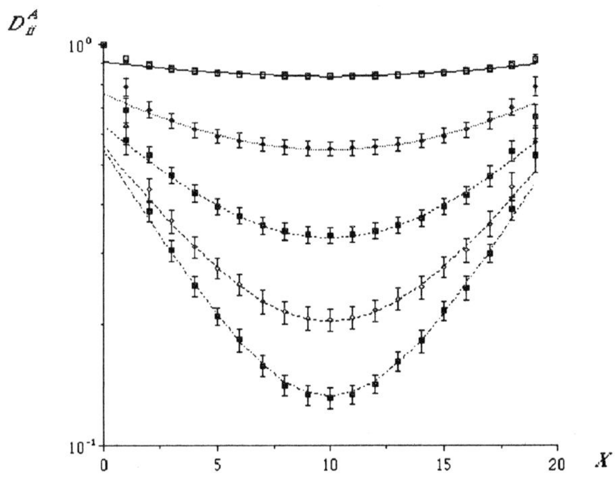

Figure 1. This figure shows the field-theoretical propagators for several values of b^{-1}. As the value of b^{-1} comes up from 10 to 50, the line expressing the propagator comes down.

NUMERICAL SIMULATION AND PARTICLE MASS

Of course, we know that it is very difficult to solve (27) analytically, so that we have discretized the equation and then solved it numerically, under the periodic boundary condition

$$A_i(x_0 + 2x_c, x_3) = A_i(x_0, x_3) \ , \quad A_i(x_0, x_3 + 2x_c) = A_i(x_0, x_3) \tag{33}$$

with period $2x_c$, in order to obtain the field-theoretical propagator $\Delta_{ij}(x, y)$. Practically, we have used the Langevin-source method (for example, see [2]). We choose a discretized fictitious time step of $\Delta t = 0.01$, perform 5.4×10^6 iterations for a lattice of 20×20 sites, and use the 2.0×10^5 iterations to calculate the field-theoretical propagator.

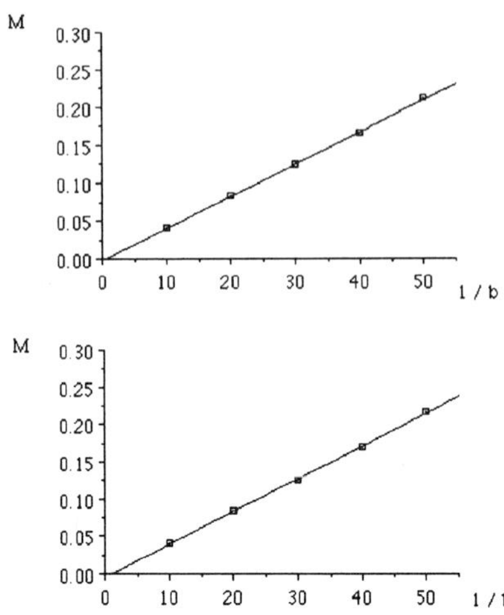

Figure 2. These figures plot the "particle mass", $\mathcal{M}$, as a function of b^{-1}. The "particle mass", $\mathcal{M}$, of the upper figure is taken from the propagator $< A_1 A_1 >$. The "particle mass", $\mathcal{M}$, of the lower figure is taken from the propagator $< A_2 A_2 >$.

Figure 1 shows the field-theoretical propagators for $b^{-1} = 10, 20, 30, 40$ and 50, from which we can estimate the "particle mass", $\mathcal{M}$, associated with the Born-Infeld field, by making use of the asymptotic formula

$$D_{ii}^{A}(X,0) = D_{ii}^{A}(0,X) = C\frac{\cosh \mathcal{M}|X - x_c|}{\cosh \mathcal{M}|x_c|} \quad \text{(no summation)}, \tag{34}$$

which is the substitute of (31) and/or (32) under the boundary condition (33). Here we have chosen x_c as the center of the correlation function i.e. $x_c = 10$, and put $C = D_{ii}^{A}(0,0) = < A_i^2 >$ (no summation). Note that C is independent of i due to the space-time uniformity, and that C is gauge-invariant because the correlation function of A_μ^T is gauge-invariant.

Rigorously speaking, we can only assert that the above $\mathcal{M}$ is proportional to the "particle mass" associated with the (transverse) Born-Infeld field, because we have no renormalization group theory to give the scaling formula in the case of Born-Infeld field. The problem is, of course, still open to questions. In this paper, however, we want to tell the "particle mass" by $\mathcal{M}$.

By making use of

$$\mathbf{E} = -\frac{\partial \mathbf{A}}{\partial x_0} \ , \quad \mathbf{B} = \nabla \times \mathbf{A} \ , \tag{35}$$

we can also obtain the field-theoretical propagators of $\mathbf{E}$ and $\mathbf{B}$ as follows;

$$\begin{aligned} D_{ii}^{E}(X,0) &= D_{ii}^{E}(0,X) = D_{ii}^{B}(X,0) = D_{ii}^{B}(0,X) \\ &= C\mathcal{M}^2 \frac{\cosh \mathcal{M}|X - x_c|}{\cosh \mathcal{M}|x_c|} \quad \text{(no summation)}. \end{aligned} \tag{36}$$

Of course, all the propagators have the same form characterized by $\mathcal{M}$.

Figures 2 and 3 plot the "particle mass", $\mathcal{M}$, as a function of b^{-1}. It seems that the particle mass is proportional to b^{-1}, but unfortunately, we do not know what kind of physical implications this fact suggests.

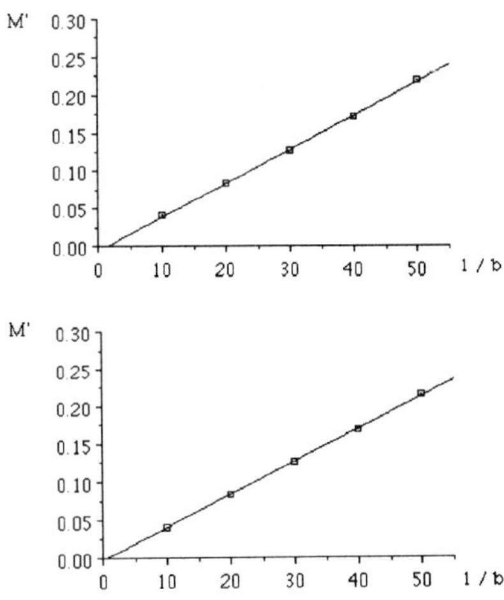

Figure 3. These figures plot the "particle mass", $\mathcal{M}'$, as a function of b^{-1}. The "particle mass", $\mathcal{M}'$, of the upper figure is taken from the propagator $< A_1 A_1 >$. The "particle mass", $\mathcal{M}'$, of the lower figure is taken from the propagator $< A_2 A_2 >$.

The particle mass seems vanishing, in the case of $b^{-1} = 0$, as expected from the fact that the Born-Infeld field must go back to the free Maxwell field in this limit.

Summarizing, we have stochastically quantized the Born-Infeld field, characterized by the so-called *universal length*, which cannot be dealt with by means of the conventional quantization methods. Even though we can hardly justify the whole procedure theoretically, we have derived the particle mass associated with the (transverse) Born-Infeld field, as a function of the universal length, through the conventional formulas to give them.

It would be interesting to observe that we have derived the "particle mass" from a perfectly gauge-invariant field theory. Of course, we can guess that the "particle mass" is produced by introducing the *universal length* b^{-1} having the dimension of length.

The authors are indebted to Drs. I. Ohba, Y. Yamanaka, K. Okano and Zeng Bo for many discussions and suggestions.

References

[1] M. Born and L. Infeld, Proc. Roy. Soc. **150** (1934) 141;
M. Born and L. Infeld, Proc. Roy. Soc. **147** (1934) 522;
M. Born, Proc. Roy. Soc. **143** (1934) 410.

[2] G. Parisi and S.Y. Wu, Sci. Sin. **24** (1981) 483; For review articles, see, M. Namiki, *Stochastic Quantization*, Springer, Heidelberg, 1992; M. Namiki and K. Okano ed. *Stochastic Quantization*, Prog. Theor. Phys. Supplement No.111, Kyoto, 1993.

[3] M. Namiki, I. Ohba, K. Okano and Y. Yamanaka, Prog. Theor. Phys. **69** (1983) 1580; and see the above review articles.

SOME ASPECTS OF q- AND qp-BOSON CALCULUS

M.R. Kibler, R.M. Asherova,[1] and Yu.F. Smirnov[2]

Institut de Physique Nucléaire de Lyon
IN2P3-CNRS et Université Claude Bernard
43 Boulevard du 11 Novembre 1918
F-69622 Villeurbanne Cedex
France

1. PRELIMINARIES

The aim of the present paper is to continue the program of extending in the framework of q-deformations the main results of the work in ref. 1 on the SU_2 unit tensor or Wigner operator (the matrix elements of which are coupling coefficients or $3 - jm$ symbols). A first part of this program was published in the proceedings of *Symmetries in Science VI* (see ref. 2) where the q-deformed Schwinger algebra was defined and where an algorithm, based on the method of complementary q-deformed algebras, was given for obtaining three- and four-term recursion relations for the Clebsch-Gordan coefficients (CGc's) of $U_q(\mathrm{su}_2)$ and $U_q(\mathrm{su}_{1,1})$. The algorithm was fully exploited in ref. 3 where the complementary of three quantum algebras in a q-deformation of the symplectic Lie algebra $\mathrm{sp}(8, \mathbb{R})$ was used for producing 32 recursion relations.

This paper is organized as follows. In section 2, we derive 12 explicit forms for the CGc's of $U_q(\mathrm{su}_2)$. They are q-deformations of the most usual formulas displayed in the literature. [In the following, we shall use the terminology X *form* which means that the corresponding formula can be identified with the one originally derived by the author(s) X (X = Wigner,[4] van der Waerden,[5] Racah,[6] Majumdar,[7] etc.) in the limiting case where $q = 1$.] From each of the 12 X forms, it is possible to derive, as explained in section 3, a q-boson realization of the $U_q(\mathrm{su}_2)$ unit tensor. Finally,

---

[1] Permanent address : Physics and Power Engineering Institute, Obninsk, Kaluga region, Russia.

[2] Present address : Instituto de Física, Universidad Nacional Autonoma de México, México D.F. 001, México. On leave of absence from : Institute of Nuclear Physics, Moscow State University, 119899 Moscow, Russia.

Symmetries in Science VII, Edited by
B. Gruber, Plenum Press, New York, 1995

section 4 deals with the two-parameter Hopf algebra $U_{qp}(\mathrm{u}_2)$ and it is sketched there how to transcribe to $U_{qp}(\mathrm{u}_2)$ the results obtained in section 2 for $U_q(\mathrm{su}_2)$.

Some words about the notation are in order. In section 4, we shall use the notation

$$[[x]]_{qp} := \frac{q^x - p^x}{q - p} \tag{1}$$

while in sections 2 and 3 we shall use the abbreviation $[x]$ to denote

$$[x] \equiv [x]_q := [[x]]_{qq^{-1}} = \frac{q^x - q^{-x}}{q - q^{-1}} \tag{2}$$

where x may stand for an operator or a (real) number. The other notations to be used concern the q- and qp-deformed factorials

$$
\begin{aligned}
&[n]! \text{ or } [n]_q! := [1][2] \cdots [n] \equiv [1]_q [2]_q \cdots [n]_q && [0]! \text{ or } [0]_q! := 1 \\
&[[n]]_{qp}! := [[1]]_{qp} [[2]]_{qp} \cdots [[n]]_{qp} && [[0]]_{qp}! := 1
\end{aligned}
\tag{3}
$$

where n is a positive integer.

We shall take the commutation relations of the quantum algebra $U_q(\mathrm{su}_2)$ in the usual (Kulish-Reshetikhin-Drinfeld-Jimbo) form, viz.,

$$[J_3, J_-] = -J_- \qquad [J_3, J_+] = +J_+ \qquad [J_+, J_-] = [2J_3]_q \tag{4}$$

The co-product of the Hopf algebra $U_q(\mathrm{su}_2)$ corresponds to eq. (41) below with $p = q^{-1}$. The extension of eq. (4) to the quantum algebra $U_{qp}(\mathrm{u}_2)$ is given in section 4.

For Hermitean conjugation requirements [more precisely, to insure that $(J_+)^\dagger = J_-$], the values of the parameters q and p must be restricted to the following domains: either (i) $q \in \mathbb{R}$ and $p \in \mathbb{R}$ or (ii) $q \in \mathbb{C}$ and $p \in \mathbb{C}$ with $p = q^*$ (the $*$ indicates complex conjugation). In the special case where $p = q^{-1}$, we may take either (i) $q \in \mathbb{R}$ or (ii) $q = e^{i\beta}$ with $0 \le \beta < 2\pi$. Therefore, in all cases the product qp is real.

2. ANALYTICAL EXPRESSIONS FOR q-DEFORMED CGC'S

2.1. The Philosophy

In this section, our aim is to derive analytical expressions for the $U_q(\mathrm{su}_2)$ CGc's from a given (fully checked) formula. This can be done with various means including: (i) the resummation procedure, (ii) the use of ordinary symmetry properties (corresponding to the 12 simple ordinary symmetries of the Regge array) for the $U_q(\mathrm{su}_2)$ CGc's, (iii) the use of pure Regge symmetries (corresponding to the $72 - 12 = 60$ nonordinary symmetries of the Regge array) for the $U_q(\mathrm{su}_2)$ CGc's, (iv) the use of the mirror reflection symmetry for the $U_q(\mathrm{su}_2)$ CGc's, and (v) the transition from the given formula to its expression in terms of the q-deformed hypergeometric function $_3F_2$ and the use of symmetry properties of this function.

The resummation procedure (i) to be used below is an adaptation, in the framework of q-deformations, of the procedure described by Jucys and Bandzaitis[8] and applied to the standard SU_2 CGc's. It amounts to introduce

$$\delta(a, b) = \sum_s (-1)^{s-a} \frac{q^{s(a-b-1)+a}}{[a-s]![s-b]!} \tag{5}$$

in the sum occurring in a given formula for the $U_q(su_2)$ CGc's. The resummation of the so-obtained expression may then be achieved by using some of the following summation identities (or q-factorial sums)

$$\sum_s q^{-as} \frac{1}{[s]![b-s]![c-s]![a-b-c+s]!} = q^{-bc} \frac{[a]!}{[b]![c]![a-b]![a-c]!} \qquad (6)$$

$$\sum_s (-1)^s q^{s(b+c-a-1)} \frac{[a-s]!}{[s]![b-s]![c-s]!} = (-1)^c q^{bc} \frac{[a-c]![b+c-a-1]!}{[b]![c]![b-a-1]!}, \qquad (7)$$
$$b > a \geq c$$

$$\sum_s (-1)^s q^{s(b+c-a-1)} \frac{[a-s]!}{[s]![b-s]![c-s]!} = q^{bc} \frac{[a-b]![a-c]!}{[b]![c]![a-b-c]!}, \quad a \geq b,\ a \geq c \qquad (8)$$

$$\sum_s q^{s(b+c-a+2)} \frac{[b-s]![c+s]!}{s![a-s]!} = q^{a(c+1)} \frac{[b-a]![c]![b+c+1]!}{[a]![b+c-a+1]!} \qquad (9)$$

where, as in eq. (5), the factorials are q-factorials. The identities (6)-(9) coincide with the well-known factorials sums given in ref. 8 in the $q = 1$ limit.

Among the ordinary symmetry properties (ii), we shall use the following relations

$$(j_1 j_2 m_1 m_2 | j m)_q = (-1)^{j_2-j-m_1}\, q^{m_1} \sqrt{\frac{[2j+1]}{[2j_2+1]}}\, (j j_1, -m m_1 | j_2, -m_2)_q \qquad (10)$$

$$(j_1 j_2 m_1 m_2 | j m)_q = (-1)^{j_1-j+m_2}\, q^{-m_2} \sqrt{\frac{[2j+1]}{[2j_1+1]}}\, (j j_2 m, -m_2 | j_1 m_1)_q \qquad (11)$$

$$(j_1 j_2 m_1 m_2 | j m)_q = (-1)^{j_1-j+m_2}\, q^{-m_2} \sqrt{\frac{[2j+1]}{[2j_1+1]}}\, (j_2 j m_2, -m | j_1, -m_1)_q \qquad (12)$$

We shall also use

$$(j_1 j_2 m_1 m_2 | j m)_q = (-1)^{j_2+m_2}\, q^{-m_2} \sqrt{\frac{[2j+1]}{[2j_1+1]}}\, (j_2 j, -m_2 m | j_1 m_1)_{q^{-1}} \qquad (13)$$

$$(j_1 j_2 m_1 m_2 | j m)_q = (-1)^{j_1-m_1}\, q^{m_1} \sqrt{\frac{[2j+1]}{[2j_2+1]}}\, (j_1 j m_1, -m | j_2, -m_2)_{q^{-1}} \qquad (14)$$

$$(j_1 j_2 m_1 m_2 | j m)_q = (-1)^{j_2+m_2}\, q^{-m_2} \sqrt{\frac{[2j+1]}{[2j_1+1]}}\, (j j_2, -m m_2 | j_1, -m_1)_{q^{-1}} \qquad (15)$$

where it should be observed that, in contradistinction with eqs. (10)-(12), the index of the CGc in the right-hand-sides of eqs. (13)-(15) is $q^{-1} = 1/q$. In the limiting situation where $q = 1$, eqs. (10)-(15) reduce to 6 of the 12 ordinary symmetry properties for the CGc's of SU_2 in an $SU_2 \supset U_1$ basis.

In the following, we shall be mainly concerned with the points (i) and (ii). However, the points (iii) (i.e., the Regge symmetries) and (iv) (i.e., the mirror reflection symmetry: m_1, m_2, m unchanged; $j_i \mapsto -j_i - 1$ for $i = 1, 2$ and $j \mapsto -j - 1$) were used to check certain forms given below.

2.2. The Expressions

The problem of finding analytical expressions for the $U_q(\mathrm{su}_2)$ CGc's and for the corresponding Wigner (unit tensor) operator was attacked by numerous authors (see refs. 9 to 14 for a nonexhaustive list of works). Generally speaking, the methods valid for the ordinary CGc's (for which $q = 1$) can be extended to q-deformed CGc's.

In the limiting case where $q = 1$, a useful form for the CGc's of the nondeformed chain $\mathrm{su}_2 \supset \mathrm{u}_1$ was derived by Shapiro[15] by making use of the (Löwdin) method of projection operators (see also the work by Calais[16]). Such a method was adapted and applied by Smirnov, Tolstoĭ and Kharitonov[13] to the q-deformed chain $U_q(\mathrm{su}_2) > \mathrm{u}_1$. Our starting point is

The Shapiro (Smirnov-Tolstoĭ-Kharitonov) form (1st form) :

$$(j_1 j_2 m_1 m_2 | j m)_q = (-1)^{j_1 + j_2 - j} q^{-\frac{1}{2}(j_1 + j_2 - j)(j_1 + j_2 + j + 1) + j_1 m_2 - j_2 m_1}$$

$$\times \left([2j + 1] \frac{[j_1 - j_2 + j]![j_1 + j_2 - j]![j_1 + j_2 + j + 1]![j_2 - m_2]![j + m]!}{[j - j_1 + j_2]![j_2 + m_2]![j - m]![j_1 - m_1]![j_1 + m_1]!} \right)^{\frac{1}{2}} \tag{16}$$

$$\times \sum_z (-1)^z q^{z(j_1 + m_1)} \frac{[2j_2 - z]![j_1 + j_2 - m - z]!}{[z]![j_1 + j_2 - j - z]![j_2 - m_2 - z]![j_1 + j_2 + j + 1 - z]!}$$

which is eq. (5.17) of ref. 13. In eq. (16), as well as in eqs. (17)-(27), it is assumed that $m = m_1 + m_2$. [If $m \neq m_1 + m_2$, the right-hand-sides of (16)-(27) should be replaced by 0.] The substitution $z \mapsto j_1 + j_2 - j - z$ allows us to rewrite eq. (16) in an alternative form. We thus get

The Shapiro (Smirnov-Tolstoĭ-Kharitonov) form (2nd form) :

$$(j_1 j_2 m_1 m_2 | j m)_q = q^{-\frac{1}{2}(j_1 + j_2 - j)(j - j_1 + j_2 - 2m_1 + 1) + j_1 m_2 - j_2 m_1}$$

$$\times \left([2j + 1] \frac{[j + m]![j_2 - m_2]![j_1 - j_2 + j]![j_1 + j_2 - j]![j_1 + j_2 + j + 1]!}{[j - m]![j_2 + m_2]![j_1 - m_1]![j_1 + m_1]![j - j_1 + j_2]!} \right)^{\frac{1}{2}} \tag{17}$$

$$\times \sum_z (-1)^z q^{-z(j_1 + m_1)} \frac{[j - m + z]![j - j_1 + j_2 + z]!}{[z]![2j + 1 + z]![j_1 + j_2 - j - z]![j - j_1 - m_2 + z]!}$$

From the latter form, we can derive a useful intermediate form by using the resummation procedure. Indeed, by introducing eq. (5) into the right-hand-side of eq. (17) and then by using successively eq. (8) and eq. (7), a long but straightforward calculation leads to

The intermediate form :

$$(j_1 j_2 m_1 m_2 | j m)_q = q^{-\frac{1}{2}(j_1 + j_2 - j)(j_2 - j_1 - j + 2m_2 - 1) + j_1 m_2 - j_2 m_1}$$

$$\times \left([2j + 1] \frac{[j_2 - m_2]![j_2 + m_2]![j + j_1 - j_2]![j - j_1 + j_2]![j_1 + j_2 - j]!}{[j_1 - m_1]![j_1 + m_1]![j - m]![j + m]![j + j_1 + j_2 + 1]!} \right)^{\frac{1}{2}}$$

$$\times \sum_z (-1)^z q^{-z(j + j_1 - j_2 + 1)} \frac{[j - m + z]![j_1 + j_2 + m - z]!}{[z]![j_2 + m_2 - z]![j - j_1 - m_2 + z]![j_1 + j_2 - j - z]!}$$

$$\tag{18}$$

which constitutes in turn an initial point for spanning other analytical expressions of the $U_q(\mathrm{su}_2)$ CCc's.

By starting from the intermediate form (18), it is possible to derive, still in the context of the resummation procedure, the q-analog of the van der Waerden[5] (symmetrical) formula. As a net result, we have found

The van der Waerden form :

$$(j_1 j_2 m_1 m_2 | j m)_q = q^{\frac{1}{2}(j_1+j_2-j)(j_1+j_2+j+1)+j_1 m_2 - j_2 m_1}$$

$$\times \left([2j+1] \frac{[j-j_1+j_2]![j+j_1-j_2]![j_1+j_2-j]!}{[j_1+j_2+j+1]!} \right)^{\frac{1}{2}}$$

$$\times \left([j_1-m_1]![j_1+m_1]![j_2-m_2]![j_2+m_2]![j-m]![j+m]! \right)^{\frac{1}{2}} \tag{19}$$

$$\times \sum_z (-1)^z q^{-z(j_1+j_2+j+1)} \frac{1}{[z]![j_1-m_1-z]![j_2+m_2-z]!}$$

$$\times \frac{1}{[j-j_1-m_2+z]![j-j_2+m_1+z]![j_1+j_2-j-z]!}$$

From the van der Waerden form (19), we can obtain the q-analog of the Racah[6] formula by using the resummation procedure together with a repeated application of the summation identity (6). This yields

The Racah form (1st form) :

$$(j_1 j_2 m_1 m_2 | j m)_q = (-1)^{j_1-m_1} \, q^{-\frac{1}{2}[j_1(j_1+1)-j_2(j_2+1)+j(j+1)]+m_1(m+1)}$$

$$\times \left([2j+1] \frac{[j_1+j_2-j]![j-m]![j+m]![j_1-m_1]![j_2-m_2]!}{[j+j_1-j_2]![j-j_1+j_2]![j_1+j_2+j+1]![j_1+m_1]![j_2+m_2]!} \right)^{\frac{1}{2}} \tag{20}$$

$$\times \sum_z (-1)^z \, q^{z(j+m+1)} \frac{[j_1+m_1+z]![j+j_2-m_1-z]!}{[z]![j-m-z]![j_1-m_1-z]![j_2-j+m_1+z]!}$$

Then, from the Racah form (20), we can derive another useful form, viz., the q-analog of the formula (13.1||) by Jucys and Bandzaitis,[8] again through the resummation procedure. As a matter of fact, we have obtained

The Jucys-Bandzaitis form (2nd form) :

$$(j_1 j_2 m_1 m_2 | j m)_q = (-1)^{j_1-m_1} q^{-\frac{1}{2}[j_1(j_1+1)-j_2(j_2+1)+j(j+1)]-j_1 j+j_1 m+j m_1+m_1}$$

$$\times \left(\frac{[j+j_1+j_2+1]!}{[-j+j_1+j_2]![j-j_1+j_2]![j+j_1-j_2]!} \right)^{\frac{1}{2}}$$

$$\times \left([2j+1] \frac{[j_1-m_1]![j_1+m_1]![j-m]![j+m]!}{[j_2-m_2]![j_2+m_2]!} \right)^{\frac{1}{2}} \tag{21}$$

$$\times \sum_z (-1)^z q^{z(j_1-j_2+j)} \frac{[j_2+j-m_1-z]![j_1+j_2-m-z]!}{z![j_1-m_1-z]![j-m-z]![j_1+j_2+j+1-z]!}$$

Going back to the van der Waerden form, by making the substitution $z \mapsto j_2 + m_2 - z$ in eq. (19) and by applying the resummation procedure to the so-obtained relation, we arrive at

The Majumdar form (1st form) :

$$(j_1 j_2 m_1 m_2 | j m)_q = (-1)^{j_2+m_2} \, q^{-\frac{1}{2}(j-j_1+j_2)(j_1+j_2-j+1)+m_1 j-m j_1-m_2}$$

$$\times \left([2j+1] \frac{[j-j_1+j_2]![j_1-m_1]![j_1+m_1]![j_2-m_2]![j+m]!}{[j+j_1-j_2]![j_1+j_2-j]![j_1+j_2+j+1]![j_2+m_2]![j-m]!} \right)^{\frac{1}{2}} \tag{22}$$

$$\times \sum_z (-1)^z \, q^{z(j_1-m_1+1)} \frac{[2j-z]![j_1+j_2-j+z]!}{[z]![j+m-z]![j_1-j-m_2+z]![j-j_1+j_2-z]!}$$

Finally, the substitution $z \mapsto j + m - z$ in eq. (22) produces

The Majumdar form (2nd form) :

$$(j_1 j_2 m_1 m_2 | jm)_q = (-1)^{j-j_2+m_1} q^{\frac{1}{2}[j_1(j_1+1)-j_2(j_2+1)+j(j+1)]-m_1(m-1)}$$

$$\times \left([2j+1] \frac{[j-j_1+j_2]![j_1-m_1]![j_1+m_1]![j_2-m_2]![j+m]!}{[j+j_1-j_2]![j_1+j_2-j]![j_1+j_2+j+1]![j_2+m_2]![j-m]!} \right)^{\frac{1}{2}} \quad (23)$$

$$\times \sum_z (-1)^z \, q^{-z(j_1-m_1+1)} \frac{[j-m+z]![j_1+j_2+m-z]!}{[z]![j+m-z]![j_1+m_1-z]![j_2-j_1-m+z]!}$$

Equation (22) is the q-analog of a formula originally derived by Majumdar[7] while eq. (23) is a simple consequence of (22).

Other analytical formulas for the CGc's of $U_q(\mathrm{su}_2) > \mathrm{u}_1$ may be spanned from the just obtained forms owing to the symmetry properties (10-15). For example, by applying the symmetry property (13) to the intermediate form (18) we obtain

The Wigner form :

$$(j_1 j_2 m_1 m_2 | jm)_q = (-1)^{j_2+m_2} \, q^{\frac{1}{2}(j-j_1+j_2)(j-j_1-j_2+2m-1)-j_2 m-(j+1)m_2}$$

$$\times \left([2j+1] \frac{[j_1+j_2-j]![j-j_1+j_2]![j+j_1-j_2]![j-m]![j+m]!}{[j_1+j_2+j+1]![j_1-m_1]![j_1+m_1]![j_2-m_2]![j_2+m_2]!} \right)^{\frac{1}{2}} \quad (24)$$

$$\times \sum_z (-1)^z \, q^{z(j_1+j_2-j+1)} \frac{[j_1-m_1+z]![j+j_2+m_1-z]!}{[z]![j+m-z]![j_1-j_2-m+z]![j-j_1+j_2-z]!}$$

[In other words, by introducing the intermediate form (18) for the CGc in the right-hand-side of (13), then the Wigner form (24) is obtained as the left-hand-side of (13).] Similarly, the introduction of the intermediate form (18) in the right-hand-side of the symmetry property (11) generates

The Zukauskas-Mauza form :

$$(j_1 j_2 m_1 m_2 | jm)_q = (-1)^{j_1-j+m_2} \, q^{-\frac{1}{2}(j-j_1+j_2)(j_2-j_1-j-2m_2-1)-(j+1)m_2-j_2 m}$$

$$\times \left([2j+1] \frac{[j_1+j_2-j]![j-j_1+j_2]![j+j_1-j_2]![j_2-m_2]![j_2+m_2]!}{[j_1+j_2+j+1]![j_1-m_1]![j_1+m_1]![j-m]![j+m]!} \right)^{\frac{1}{2}}$$

$$\times \sum_z (-1)^z \, q^{-z(j_1-j_2+j+1)} \frac{[j_1-m_1+z]![j+j_2+m_1-z]!}{[z]![j_2-m_2-z]![j_1-j+m_2+z]![j-j_1+j_2-z]!}$$

$$(25)$$

The q-analog of the Racah form used by Kibler and Grenet in ref. 1 [at the level of their eq. (92)] can be deduced from the Racah form (20) : it is sufficient to introduce (20) into the symmetry property (14); this leads to

The Racah form (2nd form) :

$$(j_1 j_2 m_1 m_2 | jm)_q = q^{\frac{1}{2}[j_1(j_1+1)+j_2(j_2+1)-j(j+1)]+m_1(m_2-1)}$$

$$\times \left([2j+1] \frac{[j+j_1-j_2]![j_1-m_1]![j_2-m_2]![j_2+m_2]![j+m]!}{[j_1+j_2-j]![j-j_1+j_2]![j_1+j_2+j+1]![j_1+m_1]![j-m]!} \right)^{\frac{1}{2}} \quad (26)$$

$$\times \sum_z (-1)^z q^{-z(j_2-m_2+1)} \frac{[j_1+m_1+z]![j+j_2-m_1-z]!}{[z]![j_1-m_1-z]![j_2+m_2-z]![j-j_2+m_1+z]!}$$

Finally, by putting the Shapiro form (1st form) [eq. (16)] in the right-hand-side of the symmetry property (12), we obtain

The Jucys-Bandzaitis form (1st form) :

$$(j_1 j_2 m_1 m_2 | jm)_q = (-1)^{j_2+m_2} \, q^{\frac{1}{2}(j_1-j_2-j)(j_1+j_2+j+1)-mj_2-m_2 j-m_2}$$

$$\times \left(\frac{[-j_1+j_2+j]![j_1+j_2-j]![j_1+j_2+j+1]!}{[j_1-j_2+j]!} \right)^{\frac{1}{2}}$$

$$\times \left([2j+1] \, \frac{[j_1-m_1]![j+m]!}{[j_1+m_1]![j_2-m_2]![j_2+m_2]![j-m]!} \right)^{\frac{1}{2}} \tag{27}$$

$$\times \sum_z (-1)^z \, q^{z(j_2+m_2)} \, \frac{[2j-z]![j+j_2+m_1-z]!}{[z]![j+m-z]![j+j_2-j_1-z]![j_1+j_2+j+1-z]!}$$

which is the q-analog of the relation (13.1B) obtained by Jucys and Bandzaitis and listed in ref. 8.

3. TOWARDS q-BOSON REALIZATIONS OF THE $U_q(\mathrm{su}_2)$ CGC'S

3.1. The Philosophy

Following the approach of ref. 1, we are now in a position to find q-boson realizations of the $U_q(\mathrm{su}_2)$ unit tensor. The components $t[q : k\rho\Delta]$ of such a tensor operator are defined by[1,2]

$$\langle j'm'|t[q : k\rho\Delta]|jm\rangle := \delta(j',j+\Delta)\delta(m',m+\rho)(-1)^{2k} \left([2j'+1]_q\right)^{-\frac{1}{2}} (jkm\rho|j'm')_q \tag{28}$$

where $(jkm\rho|j'm')_q$ is a CGc for $U_q(\mathrm{su}_2)$. The operator $t[q : k\rho\Delta]$ constitutes a q-deformation of the operator $t_{kq\alpha} \equiv t[1 : kq\alpha]$ worked out by Kibler and Grenet.[1]

It was shown in ref. 1 that the operator $t[1 : kq\alpha]$ can be expressed in the enveloping algebra of the Schwinger algebra (isomorphic with $\mathrm{so}_{2,3}$). The extension to the case where $q \neq 1$ is trivial so that it is possible to find a realization of the operator $t[q : k\rho\Delta]$ in terms of q-boson operators defined in a two-particle Fock space (with two sets $\{a_+, a_+^+\}$ and $\{a_-, a_-^+\}$ of q-bosons). As in ref. 2, we take Macfarlane[17] and Biedenharn[18] q-bosons corresponding to

$$a_+ |n_1 n_2\rangle = \sqrt{[n_1]_q} \, |n_1-1, n_2\rangle$$

$$a_+^+ |n_1 n_2\rangle = \sqrt{[n_1+1]_q} \, |n_1+1, n_2\rangle$$

$$a_- |n_1 n_2\rangle = \sqrt{[n_2]_q} \, |n_1, n_2-1\rangle \tag{29}$$

$$a_-^+ |n_1 n_2\rangle = \sqrt{[n_2+1]_q} \, |n_1, n_2+1\rangle$$

$$N_i |n_1 n_2\rangle = n_i |n_1 n_2\rangle \quad (i=1,2)$$

[from which it is clear that $a_+^+ = (a_+)^\dagger$ and $a_-^+ = (a_-)^\dagger$ when $q \in \mathbb{R}$ or $q \in S^1$]. Then, from the transformation

$$|jm\rangle \equiv |n_1 n_2\rangle \qquad j := \frac{1}{2}(n_1+n_2) \qquad m := \frac{1}{2}(n_1-n_2) \tag{30}$$

we have

$$a_\pm |jm\rangle = \sqrt{[j \pm m]_q} \, |j-\frac{1}{2}, m \mp \frac{1}{2}\rangle$$

$$a_\pm^+ |jm\rangle = \sqrt{[j \pm m + 1]_q} \, |j+\frac{1}{2}, m \pm \frac{1}{2}\rangle \tag{31}$$

which is at the root of the (Jordan-Schwinger) realization of $U_q(\mathrm{su}_2)$. A simple iteration of (31) yields

$$
\begin{aligned}
(a_+)^n \, |jm\rangle &= \left(\frac{[j+m]_q!}{[j+m-n]_q!} \right)^{1/2} |j - \frac{n}{2}, m - \frac{n}{2}\rangle \\[2mm]
(a_-)^n \, |jm\rangle &= \left(\frac{[j-m]_q!}{[j-m-n]_q!} \right)^{1/2} |j - \frac{n}{2}, m + \frac{n}{2}\rangle \\[2mm]
(a_+^+)^n \, |jm\rangle &= \left(\frac{[j+m+n]_q!}{[j+m]_q!} \right)^{1/2} |j + \frac{n}{2}, m + \frac{n}{2}\rangle \\[2mm]
(a_-^+)^n \, |jm\rangle &= \left(\frac{[j-m+n]_q!}{[j-m]_q!} \right)^{1/2} |j + \frac{n}{2}, m - \frac{n}{2}\rangle
\end{aligned}
\tag{32}
$$

for $n \in \mathbb{N}$. Thus, it is possible to understand why $t[q : k\rho\Delta]$ can be developed in the polynomial form

$$
t[q : k\rho\Delta] = \sum_{\alpha,\beta,\gamma,\delta} C_{\alpha\beta\gamma\delta}(q,k,\rho,\Delta) \, (a_+^+)^\alpha \, (a_+)^\beta \, (a_-^+)^\gamma \, (a_-)^\delta
\tag{33}
$$

where the sum on α, β, γ, and δ is in fact a sum on a single variable z [likewise in eqs. (16)-(27)].

3.2. Some q-Bosonized Expressions

As an example, the q-boson realization of the operator $t[q : k\rho\Delta]$ corresponding to eq. (18) gives

The intermediate realization :

$$
\begin{aligned}
t[q : k\rho\Delta] =\ & (-1)^{2k} \, q^{\frac{1}{2}(k-\Delta)(2j+\Delta-k+1-2\rho)+j\rho-km} \\[2mm]
& \times \left(\frac{[k-\Delta]![k+\Delta]![k-\rho]![k+\rho]![2j+\Delta-k]!}{[2j+\Delta+k+1]!} \right)^{\frac{1}{2}} \\[2mm]
& \times \sum_z (-1)^z \, q^{-z(2j+\Delta-k+1)} \, \frac{(a_+)^{k-\Delta-z}(a_+^+)^{k+\rho-z}(a_-)^z(a_-^+)^{\Delta-\rho+z}}{[k-\Delta-z]! \, [k+\rho-z]! \, [z]! \, [\Delta-\rho+z]!}
\end{aligned}
\tag{34}
$$

where, as in section 2, the abbreviation $[x]$ stands for $[x]_q$. The correctness of eq. (34) can be verified by taking the $j'm'$-jm matrix element of the right-hand-side of (34): then, by using (32) and (28), it can be checked that we obtain the CGc $(jkm\rho|j'm')_q$ in the intermediate form (18). It is to be observed that, in equations of type (34), the q-factorials in the denominators of the sum over z give a guaranty that the powers of all q-boson operators are nonnegative integers.

In a similar way, we have obtained the q-analogs of eqs. (89)-(92) of ref. 1. They are given by

The van der Waerden realization :

$$
\begin{aligned}
t[q : k\rho\Delta] =\ & (-1)^{k+\Delta} \, q^{\frac{1}{2}(k-\Delta)(2j+\Delta+k+1)+j\rho-km} \\[2mm]
& \times \left(\frac{[k+\rho]![k-\rho]![k+\Delta]![k-\Delta]![2j+\Delta-k]!}{[2j+\Delta+k+1]!} \right)^{1/2} \\[2mm]
& \times \sum_z (-1)^z \, q^{-z(2j+\Delta+k+1)} \, \frac{(a_-^+)^{k-\rho-z}(a_-)^{k-\Delta-z}(a_+^+)^{\rho+\Delta+z}(a_+)^z}{[k-\rho-z]! \, [k-\Delta-z]! \, [\rho+\Delta+z]! \, [z]!}
\end{aligned}
\tag{35}
$$

The Zukauskas-Mauza realization :

$$t[q:k\rho\Delta] = (-1)^{\rho+\Delta}\, q^{\frac{1}{2}(k+\Delta)(2j+\Delta-k+1+2\rho)-(j+\Delta+k+1)\rho-km}$$

$$\times \left(\frac{[k+\rho]![k-\rho]![k+\Delta]![k-\Delta]![2j+\Delta-k]!}{[2j+\Delta+k+1]!}\right)^{1/2}$$

$$\times \sum_z (-1)^z\, q^{-z(2j+\Delta-k+1)}\,\frac{(a_+)^{k-\rho-z}(a_+^+)^{k+\Delta-z}(a_-)^{\rho-\Delta+z}(a_-^+)^z}{[k-\rho-z]!\,[k+\Delta-z]!\,[\rho-\Delta+z]!\,[z]!}$$

(36)

The Wigner realization :

$$t[q:k\rho\Delta] = (-1)^{k-\rho}\, q^{-\frac{1}{2}(k+\Delta)(\Delta+k+1-2m-2\rho)-(j+\Delta+k+1)\rho-km}$$

$$\times \left(\frac{[k+\Delta]![k-\Delta]![2j+\Delta-k]!}{[k+\rho]![k-\rho]![2j+\Delta+k+1]!}\right)^{1/2}$$

$$\times \sum_z (-1)^z\, q^{z(k-\Delta+1)}\,\frac{(a_+^+)^z(a_+)^{k-\rho}(a_+^+)^{k+\Delta-z}(a_-^+)^{k+\Delta-z}(a_-)^{k+\rho}(a_-^+)^z}{[k+\Delta-z]!\,[z]!}$$

(37)

The Racah (2nd) realization :

$$t[q:k\rho\Delta] = (-1)^{2k}\, q^{\frac{1}{2}[k(k+1)-\Delta(2j+\Delta+1)]+m(\rho-1)}$$

$$\times \left(\frac{[k+\rho]![k-\rho]![2j+\Delta-k]!}{[k+\Delta]![k-\Delta]![2j+\Delta+k+1]!}\right)^{1/2}$$

$$\times \sum_z (-1)^z\, q^{-z(k-\rho+1)}\,\frac{(a_+^+)^{k+\rho-z}(a_+)^{k-\Delta}(a_+^+)^z(a_-)^{k+\rho-z}(a_-^+)^{k+\Delta}(a_-)^z}{[k+\rho-z]!\,[z]!}$$

(38)

Note that in eqs. (34)-(38), it may be appropriate to replace the eigenvalues j and m *in terms of* the operators $(1/2)(N_1+N_2)$ and $(1/2)(N_1-N_2)$, respectively, in order to have basis independent operators.

The remaining forms of section 2 can be q-bosonized too. The detailed results shall be given elsewhere. Remark that the Hermitean conjugation property

$$t[q:k\rho\Delta]^\dagger = (-1)^{\alpha-\rho}t[q:k-\rho-\Delta]$$

(39)

[that is connected with the permutation of j and j' in eq. (28)] may be exploited to pass from one realization to another or to produce other q-boson realizations.

4. EXTENSION TO THE QUANTUM ALGEBRA $U_{qp}(\mathrm{u}_2)$

The question of extending the one-parameter algebra $U_q(\mathrm{su}_2)$ to a two-parameter algebra was addressed by several authors.[19-30] Indeed, the qp-quantized universal enveloping algebra $U_{qp}(\mathrm{su}_2)$ can be seen to be amenable to the one-parameter algebra $U_q(\mathrm{su}_2)$. To get a truly two-parameter algebra, it is necessary to qp-deform u_2 rather than su_2. We follow here the presentation of ref. 27 (see also ref. 2): the two-parameter quantum algebra $U_{qp}(\mathrm{u}_2)$ is spanned by the four generators J_α (with $\alpha = 0, 3, +, -$) which satisfy the following commutation relations

$$[J_3, J_\pm] = \pm J_\pm \qquad [J_+, J_-] = (qp)^{J_0-J_3}\,[[2J_3]]_{qp} \qquad [J_0, J_\alpha] = 0,\ \alpha = 0, 3, +, -$$

(40)

[Observe that the two first commutation relations in (40) reduce to (4) when $p = q^{-1}$.]
In order to endow $U_{qp}(\mathrm{u}_2)$ with a Hopf algebraic structure, it is necessary to introduce
a co-product Δ_{qp}. It is defined by the application $\Delta_{qp} : U_{qp}(\mathrm{u}_2) \otimes U_{qp}(\mathrm{u}_2) \to U_{qp}(\mathrm{u}_2)$
such that

$$\Delta_{qp}(J_0) := J_0 \otimes 1 + 1 \otimes J_0$$
$$\Delta_{qp}(J_3) := J_3 \otimes 1 + 1 \otimes J_3 \tag{41}$$
$$\Delta_{qp}(J_\pm) := J_\pm \otimes (qp)^{\frac{1}{2}J_0}\,(qp^{-1})^{+\frac{1}{2}J_3} + (qp)^{\frac{1}{2}J_0}\,(qp^{-1})^{-\frac{1}{2}J_3} \otimes J_\pm$$

[Note that with the constraint $p = q^*$, the co-product satisfies the Hermitean conjugation property $(\Delta_{qp}(J_\pm))^\dagger = \Delta_{pq}(J_\mp)$ and is compatible with the commutation
relations for the four operators $\Delta_{qp}(J_\alpha)$ with $\alpha = 0, 3, +, -$.] The universal $\mathcal{R}$-matrix
associated to the co-product Δ_{qp} reads

$$\mathcal{R}_{pq} = \begin{pmatrix} p & 0 & 0 & 0 \\ 0 & \sqrt{pq} & 0 & 0 \\ 0 & p-q & \sqrt{pq} & 0 \\ 0 & 0 & 0 & p \end{pmatrix} \tag{42}$$

and it can be proved that $\mathcal{R}_{pq}$ verifies the so-called Yang-Baxter equation. The counit and anti-pode required for the Hopf algebraic structure of $U_{qp}(\mathrm{u}_2)$ are given in
ref. 27.

The operator defined by[27]

$$C_2(U_{qp}(\mathrm{u}_2)) := \frac{1}{2}(J_+ J_- + J_- J_+) + \frac{1}{2}\,[[2]]_{qp}\,(qp)^{J_0 - J_3}\,([[J_3]]_{qp})^2 \tag{43}$$

is an invariant of the quantum algebra $U_{qp}(\mathrm{u}_2)$. The latter invariant gives back the
well-known invariant of the quantum algebra $U_q(\mathrm{su}_2)$ when $p = q^{-1}$.

In the case where neither q nor p are roots of unity, the representation the-
ory of $U_{qp}(\mathrm{u}_2)$ easily follows from the one of the Lie algebra u_2. An irreducible
representation of the quantum algebra $U_{qp}(\mathrm{u}_2)$ is characterized by a doublet (j_0, j)
where $j_0 \in \mathbb{R}$ and $2j \in \mathbb{N}$. Such a representation is associated to a subspace
$\mathcal{E}(j_0, j) = \{|j_0 j m\rangle : m = -j, -j+1, \cdots, j\}$. The generic basis vector $|j_0 j m\rangle$ of
$\mathcal{E}(j_0, j)$ can be obtained from the highest weight vector $|j_0 j j\rangle$ owing to

$$|j_0 j m\rangle = (qp)^{-\frac{1}{2}(j_0 - j)(j - m)} \sqrt{\frac{[[j+m]]_{qp}!}{[[2j]]_{qp}![[j-m]]_{qp}!}}\,(J_-)^{j-m}\,|j_0 j j\rangle \tag{44}$$

Then, the action of the generators J_α (with $\alpha = 0, 3, +, -$) on the subspace $\mathcal{E}(j_0, j)$
is given by

$$J_0\,|j_0 j m\rangle = j_0\,|j_0 j m\rangle$$
$$J_3\,|j_0 j m\rangle = m\,|j_0 j m\rangle$$
$$J_+\,|j_0 j m\rangle = (qp)^{\frac{1}{2}(j_0 - j)}\sqrt{[[j-m]]_{qp}[[j+m+1]]_{qp}}\,|j_0 j m + 1\rangle \tag{45}$$
$$J_-\,|j_0 j m\rangle = (qp)^{\frac{1}{2}(j_0 - j)}\sqrt{[[j+m]]_{qp}[[j-m+1]]_{qp}}\,|j_0 j m - 1\rangle$$

The eigenvalues of the invariant operator $C_2(U_{qp}(\mathrm{u}_2))$ on $\mathcal{E}(j_0, j)$ read

$$\frac{q^{j+j_0+1}p^{j_0-j} - q^{j_0+1}p^{j_0} - q^{j_0}p^{j_0+1} + q^{j_0-j}p^{j+j_0+1}}{(q-p)^2} = (qp)^{(j_0-j)}\,[[j]]_{qp}\,[[j+1]]_{qp} \tag{46}$$

in the general case where $j_0 \neq j$. In the particular case where $j_0 = j$, the eigenvalues (46) are equal to $[[j]]_{qp} [[j+1]]_{qp}$, a fact that was used as a basic ingredient for the qp-rotor model developed in refs. 31 and 32.

The quantum algebra $U_{qp}(u_2)$ clearly depends on the two parameters q and p. It is however interesting to show its relation with the well-known one-parameter algebra $U_q(su_2)$. In this respect, eq. (41) suggests the following change of parameters

$$Q := (qp^{-1})^{\frac{1}{2}} \qquad P := (qp)^{\frac{1}{2}} \tag{47}$$

In terms of the parameters Q and P, we have

$$[[x]]_{qp} = P^{x-1} [x]_Q \qquad [[x]]_{qp}! = P^{\frac{1}{2}x(x-1)} [x]_Q! \tag{48}$$

Then, by introducing the generators A_α (with $\alpha = 0,3,+,-$)

$$A_0 := J_0 \qquad A_3 := J_3 \qquad A_\pm := (qp)^{-\frac{1}{2}(J_0 - \frac{1}{2})} J_\pm \tag{49}$$

it can be shown that the two-parameter quantum algebra $U_{qp}(u_2)$ is isomorphic to the central extension

$$U_{qp}(u_2) = u_1 \otimes U_Q(su_2) \tag{50}$$

where u_1 is spanned by the operator A_0 and $U_Q(su_2)$ by the set $\{A_3, A_+, A_-\}$. The Q-deformation $U_Q(su_2)$ of the Lie algebra su_2 corresponds to the usual commutation relations [cf. eq. (4)]

$$[A_3, A_\pm] = \pm A_\pm \qquad [A_+, A_-] = [2A_3]_Q \tag{51}$$

(Of course, we have $[A_0, A_\alpha] = 0$ for $\alpha = 0,3,+,-$.) Furthermore, the co-product relations (41) leads to

$$\begin{aligned}
\Delta_{qp}(J_\pm) &= \left(P^{A_0 - \frac{1}{2}} \otimes P^{A_0} \right) \Delta_Q(A_\pm) \\
&= P^{\Delta_Q(A_0) - \frac{1}{2}} \Delta_Q(A_\pm)
\end{aligned} \tag{52}$$

where the co-product $\Delta_Q : U_Q(su_2) \otimes U_Q(su_2) \to U_Q(su_2)$ is given via

$$\begin{aligned}
\Delta_Q(A_0) &:= A_0 \otimes 1 + 1 \otimes A_0 \\
\Delta_Q(A_3) &:= A_3 \otimes 1 + 1 \otimes A_3 \\
\Delta_Q(A_\pm) &:= A_\pm \otimes Q^{+A_3} + Q^{-A_3} \otimes A_\pm
\end{aligned} \tag{53}$$

The invariant $C_2(U_{qp}(u_2))$ may be transcribed in terms of the two parameters Q and P. In fact, eq. (43) can be rewritten as

$$C_2(U_{qp}(u_2)) = P^{2A_0 - 1} C_2(U_Q(su_2)) \tag{54}$$

where

$$C_2(U_Q(su_2)) := \frac{1}{2}(A_+ A_- + A_- A_+) + \frac{1}{2} [2]_Q ([A_3]_Q)^2 \tag{55}$$

so that the eigenvalues (46) of $C_2(U_{qp}(u_2))$ on the subspace $\mathcal{E}(j_0, j)$ can be rewritten as $P^{2j_0 - 1} [j]_Q [j+1]_Q$.

From the point of view of the representation theory, eqs. (47)-(55) strongly suggest that the coupling (i.e., CGc's and $3-jm$ symbols) and recoupling (i.e., $6-j$ and

$9 - j$ symbols) coefficients of $U_{qp}(u_2)$ may be deduced from the corresponding ones of $U_Q(su_2)$. This intuition is further reinforced by the fact that the CGc's

$$(j_{01}j_{02}j_1j_2m_1m_2|j_0jm)_{qp} \equiv (m_1m_2|m)_{qp} \tag{56}$$

of $U_{qp}(u_2)$ are easily seen to satisfy the three-term recursion relations

$$\sqrt{[j \mp m]_Q\,[j \pm m + 1)]_Q}\;(m_1m_2|m \pm 1)_{qp}$$
$$= Q^{+m_2}\,\sqrt{[j_1 \pm m_1]_Q\,[j_1 \mp m_1 + 1]_Q}\;(m_1 \mp 1, m_2|m)_{qp}$$
$$+ Q^{-m_1}\,\sqrt{[j_2 \pm m_2]_Q\,[j_2 \mp m_2 + 1]_Q}\;(m_1, m_2 \mp 1|m)_{qp} \tag{57}$$

which are identical to the ones satisfied by the CGc's $(j_1j_2m_1m_2|jm)_Q$ of the quantum algebra $U_Q(su_2)$ (see ref. 3). Therefore, there exists a proportionality relation between the qp-CGc's and the Q-CGc's. Reality and normalization conditions can be used to justify that the proportionality constant is equal to 1. Indeed, this may be checked by direct calculation: by adapting to the qp-deformation $U_{qp}(u_2)$ the method of projection operators used for su_2 (in ref. 15) and for $U_q(su_2)$ (in ref. 13), we can show that we have the connecting formula:

$$(j_{01}j_{02}j_1j_2m_1m_2|j_0jm)_{qp} = \delta(j_0, j_{01} + j_{02})\,(j_1j_2m_1m_2|jm)_Q \tag{58}$$

a result to be compared with the ones in refs. 28 and 29 for $U_{qp}(su_2)$. Therefore, all formulas of subsection 2.2 may be adapted to the case of the two-parameter quantum algebra $U_{qp}(u_2)$.

5. CLOSING REMARKS

We obtained in this work various analytical forms for the $U_q(su_2)$ CGc's. A special effort was put on the compatibility between the different forms discussed. Indeed, all the obtained forms were deduced from one single form, namely, the so-called Shapiro form, either by applying the resummation procedure or by using ordinary symmetry properties of the CGc's. Some other checks (not reported in the present paper) were also achieved by means of Regge symmetries and of the mirror reflection symmetry extended to the $U_q(su_2) > u_1$ chain. In addition, the forms were compared to existing formulas when possible.

In the classical limit where $q = 1$, the various q-dependent forms reduce to well-known expressions (like the formulas derived by Wigner, van der Waerden, and Racah in the early days of what is refered to as Wigner-Racah algebra of the rotation group) and to less-known expressions for the CGc's of the group SU_2 in an $SU_2 \supset U_1$ basis. For a given form, the passage from $q = 1$ to $q \neq 1$ manifests itself by: (i) the replacement of $2j + 1$ by $[2j + 1]_q$ and of ordinary factorials by q-factorials, (ii) the introduction of an *internal* q-factor which depends on a summation index z, and (iii) the introduction of an *external* q-factor (z-independent).

It is remarkable that the internal q-factor (which is z-dependent) assumes the form $q^{\pm z(N-D+1)}$, where N and D stand respectively for the numerator and the denominator of the fraction involved in the sum over z. The occurrence of this heuristic rule might be rationalized on the basis of some properties of the q-deformation of the hypergeometric function $_3F_2$ in term of which it is possible to express the q-deformed CGc's of SU_2 [see ref. 33 for the use of $_3F_2(abc, de; 1)$ in connection with SU_2 CGc's].

The various forms of the $U_q(\mathrm{su}_2)$ CGc's fall in three families: the Wigner, van der Waerden, and Racah families. Each family is characterized by a given distribution of the q-factorials $[j_1 \pm m_1]_q!$, $[j_2 \pm m_2]_q!$, and $[j \pm m]_q!$ in the numerator of the factor in front of the sum over z. The van der Waerden family has the six q-factorials in the numerator. The Wigner family [including the Shapiro forms, the intermediate form, the Zukauskas-Mauza form, and the Jucys-Bandzaitis (1st) form] presents two q-factorials in the numerator while the Racah family [including the Majumdar forms and the the Jucys-Bandzaitis (2nd) form] has four q-factorials in the numerator. In principle, the parents in a family may be obtained from any member of the family by applying ordinary symmetry properties of the CGc's.

Some preliminary results on the q-bosonization of the CGc's of $U_q(\mathrm{su}_2)$ have been reported. Various expressions have been given for the unit tensor operator $t[q:k\rho\Delta]$. Not all the possible forms of the operator $t[q:k\rho\Delta]$ have been described. In particular, the Majumdar realization of $t[q:k\rho\Delta]$ has been omited since it presents some peculiarities as in the $q=1$ case. A more complete listing of q-boson realizations of the operator $t[q:k\rho\Delta]$ shall be published elsewhere.

Finally, we have extended to a two-parameter deformation of u_2 the various expressions for the q-deformed CGc's of the chain $\mathrm{SU}_2 \supset \mathrm{U}_1$. The derivation of a true (nontrivial) two-parameter deformation of su_2 has been questioned by various authors. Although, the CGc's for $U_{qp}(\mathrm{u}_2)$ may be obtained from the ones for $U_Q(\mathrm{su}_2)$ by means of a formula where the two parameters q and p are unified in a single one [i.e., $Q=(qp)^{\frac{1}{2}}$], the two parameters q and p [or, equivalently, Q and $P=(qp^{-1})^{\frac{1}{2}}$] really appear in the Casimir(s) of the Hopf algebra $U_{qp}(\mathrm{u}_2)$. For the practitioner, interested in putting some numbers on the (real) world, the two parameters q and p may have some interest when dealing with a comparison between theory and experiment (via fitting procedures for example). The latter point was applied to the derivation of a qp-rotor model for describing rotational bands of (superdeformed) nuclei.[31,32]

To complete this work, it would be interesting to study the $U_q(\mathrm{su}_2)$ CGc's in terms of the q-deformed hypergeometric function ${}_3F_2$ and to examine the contraction of ${}_3F_2$ into a ${}_2F_1$ operator-valued function when going from the CGc to the Wigner operator. Also, to find q-boson realizations for the Racah operator (the matrix elements of which are q-deformed recoupling coefficients or $6-j$ symbols) is an interesting problem. In addition, a treatment via symbolic programming languages (like Maple or Mathematica) of the formulas in this paper is presently under study. [A programme is obtainable[34] for calculating any $U_q(\mathrm{su}_2)$ CGc.] We hope to return on these matters in a future work.

ACKNOWLEDGEMENTS

Part of this work was presented (together with some application to rotational spectroscopy) by one of the authors (M.R. K.) to the Symposium "Symmetries in Science VIII". The authors acknowledge Prof. B. Gruber for giving them the opportunity to present their results to this beautiful symposium.

6. REFERENCES

1. M. Kibler and G. Grenet, J. Math. Phys. 21:422 (1980).
2. Yu.F. Smirnov and M.R. Kibler, Some aspects of q-boson calculus, in: Symmetries in Science VI: From the rotation group to quantum algebras, B. Gruber, ed., Plenum Press, New York (1993), p. 691.
3. M. Kibler, C. Campigotto and Yu.F. Smirnov, Recursion relations for Clebsch-

Gordan coefficients of $U_q(su_2)$ and $U_q(su_{1,1})$, *in*: Proceedings of the International Workshop "Symmetry Methods in Physics, in Memory of Professor Ya.A. Smorodinsky", A.N. Sissakian, G.S. Pogosyan and S.I. Vinitsky, eds., J.I.N.R., Dubna, Russia (1994), p. 246.

4. E. Wigner, Gruppentheorie und ihre Anwendungen auf die Quantenmechanik der Atomspektren, Vieweg & Sohn, Braunschweig (1931).

5. B.L. van der Waerden, Die Gruppentheoretische Methode in der Quantenmechanik, Springer, Berlin (1932).

6. G. Racah, Phys. Rev. 62:438 (1942).

7. S.D. Majumdar, Prog. Theor. Phys. 20:798 (1958).

8. A.P. Jucys and A.A. Bandzaitis, Theory of angular momentum in quantum mechanics, Mintis, Vilnius (1965).

9. L.C. Biedenharn and M. Tarlini, Lett. Math. Phys. 20:271 (1990).

10. M. Nomura, J. Phys. Soc. Jpn. 59:1954 (1990).

11. M. Nomura, J. Phys. Soc. Jpn. 59:2345 (1990).

12. V.A. Groza, I.I. Kachurik and A.U. Klimyk, J. Math. Phys. 31:2769 (1990).

13. Yu.F. Smirnov, V.N. Tolstoĭ and Yu.I. Kharitonov, Sov. J. Nucl. Phys. 53:593 (1991).

14. C. Quesne, q-bosons and irreducible tensors for q-algebras, *in*: Symmetries in Science VI: From the rotation group to quantum algebras, B. Gruber, ed., Plenum Press, New York (1993).

15. J. Shapiro, J. Math. Phys. 6:1680 (1965).

16. J.-L. Calais, Int. J. Quantum Chem. 2:715 (1968).

17. A.J. Macfarlane, J. Phys. A 22:4581 (1989).

18. L.C. Biedenharn, J. Phys. A 22:L873 (1989).

19. A. Sudbery, J. Phys. A: Math Gen. 23:L697 (1990).

20. N. Reshetikhin, Lett. Math. Phys. 20:331 (1990).

21. D.B. Fairlie and C.K. Zachos, Phys. Lett. B 256:43 (1991).

22. A. Schirrmacher, J. Wess and B. Zumino, Z. Phys. C 49:317 (1991).

23. S.T. Vokos, J. Math. Phys. 32:2979 (1991).

24. R. Chakrabarti and R. Jagannathan, J. Phys. A: Math. Gen. 24:L711 (1991).

25. V.K. Dobrev, J. Math. Phys. 33:3419 (1992).

26. Yu.F. Smirnov and R.F. Wehrhahn, J. Phys. A: Math. Gen. 25:5563 (1992).

27. M.R. Kibler, Introduction to quantum algebras, *in*: Symmetry and Structural Properties of Condensed Matter, W. Florek, D. Lipiński and T. Lulek, eds., World Scientific, Singapore (1993), p. 445.

28. S. Meljanac and M. Mileković, J. Phys. A: Math. Gen. 26:5177 (1993).

29. C. Quesne, Phys. Lett. A 174:19 (1993).

30. R. Chakrabarti and R. Jagannathan, J. Phys. A: Math. Gen. 27:2023 (1994).

31. R. Barbier, J. Meyer and M. Kibler, J. Phys. G: Nucl. Part. Phys. 17:L67 (1994).

32. R. Barbier, J. Meyer and M. Kibler, A qp-rotor model for rotational bands of superdeformed nuclei, Preprint LYCEN 9437, IPNL (1994).

33. Ya.A. Smorodinskiĭ and L.A. Shelepin, Usp. Fiz. Nauk 15:1 (1972).

34. M. Kibler, G.-H. Lamot and Yu.F. Smirnov, to be published.

QUARKS AND PARTONS AS TWO DIFFERENT MANIFESTATIONS OF ONE COVARIANT ENTITY

Y. S. Kim

Department of Physics, University of Maryland
College Park, Maryland 20742, U.S.A.

1 Introduction

Unlike classical physics, modern physics depends heavily on observer's state of mind or environment. Quantum mechanics depends on how we measure physical quantities, and this issue has not yet been completely settled. In relativity, observers in different Lorentz frames see the same physical system differently. The importance of the observer's subjective viewpoint was emphasized by Immanuel Kant in his book entitled *Kritik der reinen Vernunft* whose first and second editions were published in 1781 and 1787 respectively. However, using his own logic, he ended up with a conclusion that there must be an absolute inertial frame, and that we only see the frames dictated by our subjectivity.

Einstein's special relativity was developed along Kant's line of thinking: things depend on the frame from which you make observations. However, there is one big difference. Instead of the absolute frame, Einstein introduced an extra dimension. Let us illustrate this using a CocaCola can. It appears like a circle if you look at it from the top, while it appears as a rectangle from the side. The real thing is a three-dimensional circular cylinder.

I was fortunate enough to be close to Eugene Wigner, and enjoyed the privilege of asking him many questions. I was of course able to do this because I read carefully many of his papers and wrote two books on the research lines initiated by him. I once asked him whether he thinks like Immanuel Kant. He said Yes. I then asked him whether Einstein was a Kantianist in his opinion. Wigner said very firmly Yes. I then asked him whether he studied the philosophy of Kant while he was in college. He said No, and said that he realized he had been a Kantianist after writing so many papers in physics. He added that philosophers do not dictate people how to think, but their job is to describe systematically how people think. Wigner told me that I was the only one who asked him this question, and asked me how I knew the Kantian way of reasoning was working in his mind. I gave him the following answer.

I never had any formal education in oriental philosophy, but I know that my frame of thinking is affected by my Korean background. One important aspect is that Immanuel

Kant's name is known to every high-school graduate in Korea, while he is unknown to Americans, particularly to American physicists. The question then is whether there is in oriental culture a "natural frequency" which can resonate with one of the frequencies radiated from Kantianism developed in Europe.

I would like to answer this question in the following way. Koreans absorbed a bulk of Chinese culture during the period of the Tang dynasty (618-907 AD). At that time, China was the center of the world as the United States is today. This dynasty's intellectual life was based on Taoism which tells us, among others, that everything in this universe has to be balanced between its plus (or bright) side and its minus (or dark) side. This way of thinking forces us to look at things from two different or opposite directions. This aspect of Taoism could constitute a "natural frequency" which can be tuned to the Kantian view of the world where things depend how they are observed.

I would like to point that Hideki Yukawa was quite fond of Taoism and studied systematically the books of Laotse and Chuangtse who were the founding fathers of Taoism [1]. Both Laotse and Chuangtse lived before the time of Confucius. It is interesting to note that Kantianism is also popular is Japan, and it is my assumption that Kant's books were translated into Japanese by Japanese philosophers first, and Koreans of my father's age learned about Kant by reading the translated versions.

My publication record will indicate that I studied Yukawa's papers before becoming seriously interested in Wignerism. Indeed, I picked up a signal of possible connection between Kantianism and Taoism while reading Yukawa's papers carefully, and this led to my bold venture to ask Wigner whether he is a Kantianist..

	Massive Slow	between	Massless Fast
Energy Momentum	$E = \dfrac{p^2}{2m}$	Einstein's $E = \sqrt{m^2+p^2}$	$E = p$
Spin, Gauge Helicity	S_3 S_1 S_2	Wigner's Little Group	S_3 Gauge Trans.

BUILD YOUR OWN HOUSE

Figure 1. Further implications of Einstein's $E = mc^2$. Massive and massless particles have different energy-momentum relations. Einstein's special relativity gives one relation for both. Winger's little group unifies the internal space-time symmetries for massive and massless particles which are locally isomorphic to $O(3)$ and $E(2)$ respectively. It is a great challenge for us to find another unification. In this note, we present a unified picture of the quark and parton models which are applicable to slow and ultra-fast hadrons respectively.

Since I have made a confession about my intellectual background, I can explain to you how I do my physics. When I look for problems in physics, I always look for a gap between two observations of the same event. Let us look at the energy-momentum relation. If a particle is very slow, its energy-momentum relation is $E = p^2/2m$, while it is $E = cp$ for massless particles or those moving with speed close to that of light. I was very unhappy about this until I learned that Einstein's mass-energy settles this question. The first row of Fig. 1 illustrates this point.

I told Wigner further that, when I was a graduate student, I was not completely happy with his 1939 paper on the representations of the Poincaré group because the massive and massless particles have different little groups which dictate their internal space-time symmetries [2]. Wigner told me that he was trying to do something about this, and this is why he started writing papers with E. Inonu on group contractions [3]. Indeed, this problem has been solved and is summarized in Fig. 1.

There is another problem which makes physicists unhappy. These days, hadrons are regarded as bound states of quarks. This model works well when they move slowly or are at rest. However, when they move with speed close to that of light, they appear as a collection of infinite-number of partons. This picture was formulated by Feynman, and is called Feynman's parton model [4]. The problem is that the same physical object looks quite differently to observers in the two different Lorentz frames. If we settle this quarrel between these two observers, we will build a "new house" as is indicated in Fig. 1.

In order to solve this problem, we need wave a set of functions which can be Lorentz boosted. How can we then construct such a set? In constructing wave functions for any purpose in quantum mechanics, the standard procedure is to try first harmonic oscillator wave functions. In studying the Lorentz boost, the standard language is the Lorentz group. Thus the first step to construct covariant wave functions is to work out representations of the Lorentz group using harmonic oscillators [5, 6, 7]

With the wave function which can be Lorentz boosted, we resolve the mentioned quark-parton puzzle. In Sec. 2, we review Wigner's little groups of the Poincaré group. In Sec. 3, it is pointed out that the formalism of covariant harmonic oscillators is a representation of the $O(3)$-like little group for massive particles. In this formalism, harmonic oscillator wave functions become squeezed under the Lorentz boost. In Sec. 4, it is shown that the squeeze transformation leads to Feynman's parton picture of hadrons.

2   Little Groups of the Poincaré Group

The Poincaré group is the group of inhomogeneous Lorentz transformations, namely Lorentz transformations followed by space-time translations. In order to study this group, we have to understand first the group of Lorentz transformations, the group of translations, and how these two groups are combined to form the Poincaré group. The Poincaré group is a semi-direct product of the Lorentz and translation groups. The two Casimir operators of this group correspond to the (mass)^2 and (spin)^2 of of a given particle.

The question then is how to construct the representations of the Lorentz group which are relevant to physics. For this purpose, Wigner in 1939 studied the subgroups of the Lorentz group whose transformations leave the four-momentum of a given free particle [2]. Indeed, the maximal subgroup of the Lorentz group which leaves the four-momentum invariant is called the little group.

Since the little group leaves the four-momentum invariant, it governs the internal space-time symmetries of relativistic particles. Wigner shows in his paper that the internal space-time symmetries of massive and massless particles are dictated by the $O(3)$-like and $E(2)$-like little groups respectively. However, this paper was not fully appreciated until recently for the following reasons.

First, electrons obey the Dirac equation which constitutes a non-unitary representation of the Lorentz group. On the other hand, the readers of Wigner's 1939 papers were not flexible enough to extend Wigner's idea to non-unitary representations such as the Dirac equation.

Second, in order to understand full implications of the E(2)-like symmetry of the massless particle, the readers of Wigner's paper did not want to examine the physical implications of the translation-like degrees of freedom for the E(2)-like little group. This is the reason why it used to be "impossible " to connect Wigner's representation theory with Maxwell's equations. These days, it is widely accepted that the translation-like degrees of freedom are gauge degrees of freedom. This point was not mentioned in Winger's original paper, and I am not the one who discovered this. However, I am not able to give an exact historical account of this important addition to Wigner's work, and I hope to do this in the future. I am however the first one to Wigner this story in 1985, and he was pleased.

Another important development along this line of research is the application of group contractions to the unifications of the two different little groups for massive and massless particles. The Lorentz group is generated by three rotation generators J_i and three boost generators K_i. If a massless particle is at rest, the little group is the three-dimensional rotation group generated by J_1, J_2 and J_3. The four-momentum is not affected by this rotation, but the spin variable changes its direction. For a massless particle moving along the z direction, Wigner observed that the little group is generated by J_3, N_1 and N_2, where

$$N_1 = J_1 + K_2, \qquad N_2 = J_2 - K_1, \tag{1}$$

and that these generators satisfy the Lie algebra for the two-dimensional Euclidean group. Here, J_3 is like the rotation generator, while N_1 and N_2 are like translation generators in the two-dimensional Euclidean plane.

We always associate the three-dimensional rotation group with a spherical surface. Let us consider a circular area of radius 1 kilometer centered at the north pole of the earth. Since the radius of the earth is more than 6,450 times longer, the circular region appears flat. Thus, within this region, we use the E(2) symmetry group for this region. The validity of this approximation depends on the ratio of the two radii.

Indeed, in 1953, Inonu and Wigner formulated this problem as the contraction O(3) to E(2). How about then the little groups which are isomorphic to O(3) and E(2)? It is reasonable to expect that the E(2)-like little group be obtained as a limiting case for of the O(3)-like little group for massless particles. Indeed, in 1981, it was observed by Ferrara and Savoy that this limiting process is the Lorentz boost [8]. In 1983, using the same limiting process as that of Ferrara and Savoy, Han *et al* showed that transverse rotation generators become the generators of gauge transformations in the limit of infinite momentum and/or vanishing mass [9]. In 1987, Kim and Wigner showed that the little group for massless particles is the cylindrical group which is isomorphic to the E(2) group [10]. This completes the second raw in Fig. 1, where Wigner's little group unifies the internal space-time symmetries of massive and massless particles.

We are now interested in constructing the third row in Fig. 1. As we promised in Sec. 1, we will be dealing with hadrons which are bound states of quarks with space-time extension. For this purpose, we need the covariant harmonic oscillator formalism in which wave functions carry a covariant probability interpretation.

3 Covariant Harmonic Oscillators

If we construct a representation of the Lorentz group using normalizable harmonic oscillator wave functions, the result is the covariant harmonic oscillator formalism [7]. The formalism constitutes a representation of Wigner's O(3)-like little group for a massive particle with internal space-time structure. This oscillator formalism has been shown to be effective in explaining the basic phenomenological features of relativistic extended hadrons observed in high-energy laboratories. In particular, the formalism shows that the quark model and Feynman's parton picture are two different manifestations of one relativistic entity [7, 11].

The essential feature of the covariant harmonic oscillator formalism is that Lorentz boosts are squeeze transformations [12]. In the light-cone coordinate system, the boost transformation expands one coordinate while contracting the other so as to preserve the product of these two coordinate remains constant. We shall show that the parton picture emerges from this squeeze effect.

The covariant harmonic oscillator formalism has been discussed exhaustively in the literature, and it is not necessary to give another full-fledged treatment in the present paper. We shall discuss here one of the most puzzling problems in high-energy physics, namely whether quarks are partons. It is now a well-accepted view that hadrons are bound states of quarks. This view is correct if the hadron is at rest or nearly at rest. On the other hand, it appears as a collection of partons when it moves with a speed very close to that of light. This is called Feynman's parton picture [4].

Let us consider a bound state of two particles. For convenience, we shall call the bound state the hadron, and call its constituents quarks. Then there is a Bohr-like radius measuring the space-like separation between the quarks. There is also a time-like separation between the quarks, and this variable becomes mixed with the longitudinal spatial separation as the hadron moves with a relativistic speed. There are no quantum excitations along the time-like direction. On the other hand, there is the time-energy uncertainty relation which allows quantum transitions. It is possible to accommodate these aspect within the framework of the present form of quantum mechanics. The uncertainty relation between the time and energy variables is the c-number relation, which does not allow excitations along the time-like coordinate. We shall see that the covariant harmonic oscillator formalism accommodates this narrow window in the present form of quantum mechanics.

Let us consider a hadron consisting of two quarks. If the space-time position of two quarks are specified by x_a and x_b respectively, the system can be described by the variables

$$X = (x_a + x_b)/2, \qquad x = (x_a - x_b)/2\sqrt{2}. \tag{2}$$

The four-vector X specifies where the hadron is located in space and time, while the variable x measures the space-time separation between the quarks. In the convention of Feynman *et al.* [13], the internal motion of the quarks bound by a harmonic oscillator potential of unit strength can be described by the Lorentz-invariant equation

$$\frac{1}{2}\left\{ x_\mu^2 - \frac{\partial^2}{\partial x_\mu^2} \right\} \psi(x) = \lambda \psi(x). \tag{3}$$

We use here the space-favored metric: $x^\mu = (x, y, z, t)$.

It is possible to construct a representation of the Poincaré group from the solutions of the above differential equation [7]. If the hadron is at rest, the solution should take the form

$$\psi(x, y, z, t) = \psi(x, y, z) \left(\frac{1}{\pi}\right)^{1/4} \exp\left(-t^2/2\right), \tag{4}$$

where $\psi(x,y,z)$ is the wave function for the three-dimensional oscillator with appropriate angular momentum quantum numbers. Indeed, the above wave function constitutes a representation of Wigner's $O(3)$-like little group for a massive particle [7].

We note in the above expression that there are no time-like excitations, and this is consistent with our observation. It was Dirac who noted first this space-time asymmetry in quantum mechanics [14]. However, this asymmetry is quite consistent with the $O(3)$ symmetry of the little group for hadrons. The left portion of Fig. 2 illustrates the uncertainty relations along the space-like and time-like directions.

Since the three-dimensional oscillator differential equation is separable in both spherical and Cartesian coordinate systems, $\psi(x,y,z)$ consists of Hermite polynomials of $x,y,$ and z. If the Lorentz boost is made along the z direction, the x and y coordinates are not affected, and can be dropped from the wave function. The wave function of interest can be written as

$$\psi^n(z,t) = \left(\tfrac{1}{\pi}\right)^{1/4} \exp\left(-t^2/2\right)\psi_n(z), \tag{5}$$

with

$$\psi^n(z) = \left(\frac{1}{\pi n!2^n}\right)^{1/2} H_n(z)\exp(-z^2/2), \tag{6}$$

where $\psi^n(z)$ is for the n-th excited oscillator state. The full wave function $\psi^n(z,t)$ is

$$\psi_0^n(z,t) = \left(\frac{1}{\pi n!2^n}\right)^{1/2} H_n(z)\exp\left\{-\frac{1}{2}\left(z^2+t^2\right)\right\}. \tag{7}$$

The subscript 0 means that the wave function is for the hadron at rest. The above expression is not Lorentz-invariant, and its localization undergoes a Lorentz squeeze as the hadron moves along the z direction [7].

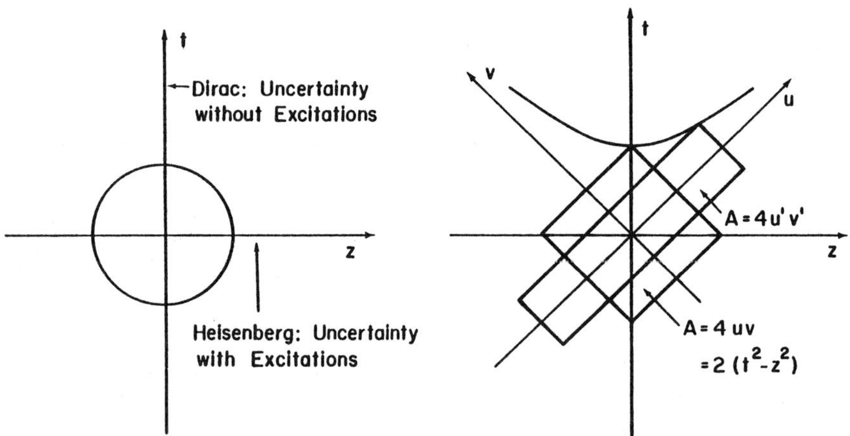

Figure 2. Quantum mechanics and special relativity. The present form of quantum mechanics allows quantum excitations along the space-like directions, but does not allow excitations along the time-like direction even though there is an uncertainty relation between the time and energy variables. According to special relativity, a space-time point traces a hyperbola when it is boosted. It is also important to note that the same Lorentz boost squeezes the square in this figure into a rectangle. The Lorentz boost is indeed a squeeze transformation.

It is convenient to use the light-cone variables to describe Lorentz boosts. The light-cone coordinate variables are

$$u = (z + t)/\sqrt{2}, \qquad v = (z - t)/\sqrt{2}. \tag{8}$$

In terms of these variables, the Lorentz boost along the z direction,

$$\begin{pmatrix} z' \\ t' \end{pmatrix} = \begin{pmatrix} \cosh\eta & \sinh\eta \\ \sinh\eta & \cosh\eta \end{pmatrix} \begin{pmatrix} z \\ t \end{pmatrix}, \tag{9}$$

takes the simple form

$$u' = e^{\eta}u, \qquad v' = e^{-\eta}v, \tag{10}$$

where η is the boost parameter and is $\tanh^{-1}(v/c)$. As we can see in Fig. 2, this is a boost transformation.

The wave function of Eq.(7) can be written as

$$\psi_o^n(z,t) = \psi_0^n(z,t) = \left(\frac{1}{\pi n! 2^n}\right)^{1/2} H_n\left((u+v)/\sqrt{2}\right) \exp\left\{-\frac{1}{2}(u^2 + v^2)\right\}. \tag{11}$$

If the system is boosted, the wave function becomes

$$\psi_\eta^n(z,t) = \left(\frac{1}{\pi n! 2^n}\right)^{1/2} H_n\left((e^{-\eta}u + e^{\eta}v)/\sqrt{2}\right) \times \exp\left\{-\frac{1}{2}\left(e^{-2\eta}u^2 + e^{2\eta}v^2\right)\right\}. \tag{12}$$

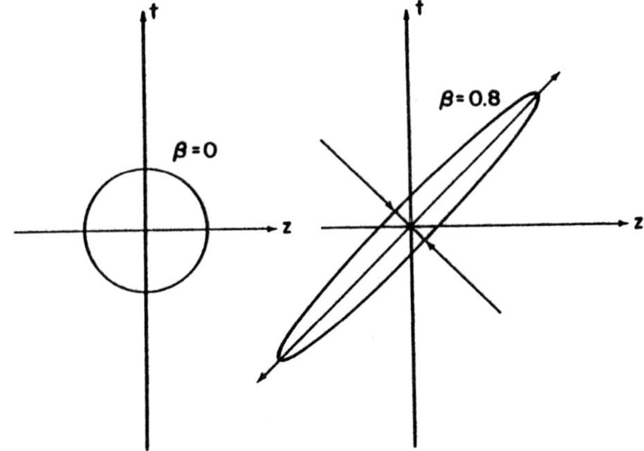

Figure 3. Relativistic quantum mechanics. If quantum mechanics described in Fig. 2 is combined with special relativity in the same figure, the result will be the circle being squeezed into an ellipse.

In both Eqs. (11) and (12), the localization property of the wave function in the uv plane is determined by the Gaussian factor, and it is sufficient to study the ground state only for the essential feature of the boundary condition. The wave functions in Eq.(11) and Eq.(12) then respectively become

$$\psi_0(z,t) = \left(\frac{1}{\pi}\right)^{1/2} \exp\left\{-\frac{1}{2}(u^2 + v^2)\right\}. \tag{13}$$

If the system is boosted, the wave function becomes

$$\psi_\eta(z,t) = \left(\frac{1}{\pi}\right)^{1/2} \exp\left\{-\frac{1}{2}\left(e^{-2\eta}u^2 + e^{2\eta}v^2\right)\right\}. \tag{14}$$

We note here that the transition from Eq.(13) to Eq.(14) is a squeeze transformation. The wave function of Eq.(13) is distributed within a circular region in the uv plane, and thus in the zt plane. On the other hand, the wave function of Eq.(14) is distributed in an elliptic region. This ellipse is a "squeezed" circle with the same area as the circle, as is illustrated in Fig. 3.

4　Feynman's Parton Picture

It is safe to believe that hadrons are quantum bound states of quarks having localized probability distribution. As in all bound-state cases, this localization condition is responsible for the existence of discrete mass spectra. The most convincing evidence for this bound-state picture is the hadronic mass spectra which are observed in high-energy laboratories [7, 13]. However, this picture of bound states is applicable only to observers in the Lorentz frame in which the hadron is at rest. How would the hadrons appear to observers in other Lorentz frames? More specifically, can we use the picture of Lorentz-squeezed hadrons discussed in Sec. 3.

Proton's radius is 10^{-5} of that of the hydrogen atom. Therefore, it is not unnatural to assume that the proton has a point charge in atomic physics. However, while carrying out experiments on electron scattering from proton targets, Hofstadter in 1955 observed that the proton charge is spread out. In this experiment, an electron emits a virtual photon, which then interacts with the proton. If the proton consists of quarks distributed within a finite space-time region, the virtual photon will interact with quarks which carry fractional charges. The scattering amplitude will depend on the way in which quarks are distributed within the proton. The portion of the scattering amplitude which describes the interaction between the virtual photon and the proton is called the form factor.

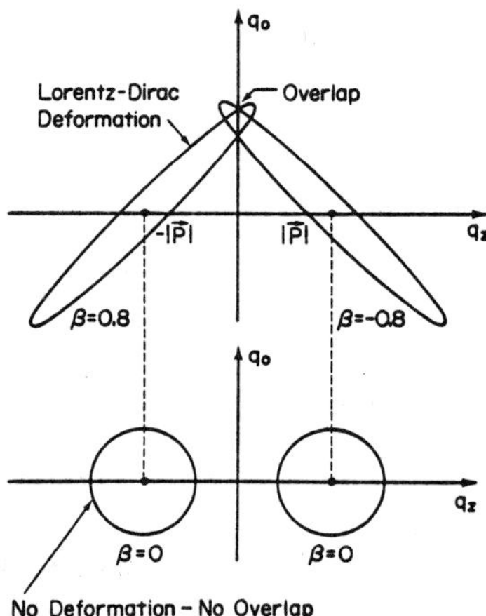

Figure 4. Two overlapping wave functions in the form factor calculation. This figure explains the form factor calculation of Fujimura *et al.* in terms of the hadronic momentum-energy wave function [15]. Without the squeeze, the overlap between the initial and final wave functions becomes negligible, while the squeezed wave functions have an overlapping region. This is why the form factor has a polynomial decrease instead of an exponential cutoff as the momentum transfer variable increases. This overlap picture is explained in detail in Ref. [7].

Although there have been many attempts to explain this phenomenon within the framework of quantum field theory, it is quite natural to expect that the wave function in the quark model will describe the charge distribution. In high-energy experiments, we are dealing with the situation in which the momentum transfer in the scattering process is large. Indeed, the Lorentz-squeezed wave functions lead to the correct behavior of the hadronic form factor for large values of the momentum transfer [15]. Figure 4 illustrates how the squeeze property of the wave function leads to the behavior of the form factor.

While the form factor is the quantity which can be extracted from the elastic scattering, it is important to realize that in high-energy processes, many particles are produced in the final state. They are called inelastic processes. While the elastic process is described by the total energy and momentum transfer in the center-of-mass coordinate system, there is, in addition, the energy transfer in inelastic scattering. Therefore, we would expect that the scattering cross section would depend on the energy, momentum transfer, and energy transfer. However, one prominent feature in inelastic scattering is that the cross section remains nearly constant for a fixed value of the momentum-transfer/energy-transfer ratio. This phenomenon is called "scaling"[16]

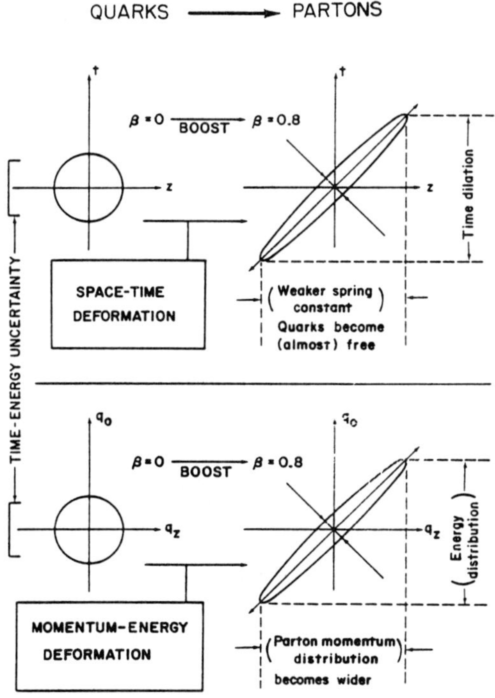

Figure 5. Lorentz-squeezed space-time and momentum-energy wave functions. As the hadron's speed approaches that of light, both wave function become concentrated along their respective positive light-cone axes. These light-cone concentrations lead to Feynman's parton picture.

In order to explain the scaling behavior in inelastic scattering, Feynman in 1969 observed that a fast-moving hadron can be regarded as a collection of many "partons" whose properties do not appear to be identical to those of quarks [4]. For example, the number of quarks inside a static proton is three, while the number of partons in a rapidly moving proton appears to be infinite. The question then is how the proton looking like a bound state of quarks to one observer can appear different to an observer in a different Lorentz frame? Feynman made the following systematic observations.

a). The picture is valid only for hadrons moving with velocity close to that of light.

b). The interaction time between the quarks becomes dilated, and partons behave as free independent particles.

c). The momentum distribution of partons becomes widespread as the hadron moves fast.

d). The number of partons seems to be infinite or much larger than that of quarks. Because the hadron is believed to be a bound state of two or three quarks, each of the above phenomena appears as a paradox, particularly b) and c) together. We would like to resolve this paradox using the covariant harmonic oscillator formalism.

For this purpose, we need a momentum-energy wave function. If the quarks have the four-momenta p_a and p_b, we can construct two independent four-momentum variables [13]

$$P = p_a + p_b, \qquad q = \sqrt{2}(p_a - p_b). \tag{15}$$

The four-momentum P is the total four-momentum and is thus the hadronic four-momentum. q measures the four-momentum separation between the quarks.

We expect to get the momentum-energy wave function by taking the Fourier transformation of Eq.(14):

$$\phi_\eta(q_z, q_0) = \left(\frac{1}{2\pi}\right) \int \psi_\eta(z, t) \exp\left\{-i(q_z z - q_0 t)\right\} dx \, dt. \tag{16}$$

Let us now define the momentum-energy variables in the light-cone coordinate system as

$$q_u = (q_0 - q_z)/\sqrt{2}, \qquad q_v = (q_0 + q_z)/\sqrt{2}. \tag{17}$$

In terms of these variables, the Fourier transformation of Eq.(16) can be written as

$$\phi_\eta(q_z, q_0) = \left(\frac{1}{2\pi}\right) \int \psi_\eta(z, t) \exp\left\{-i(q_u u + q_v v)\right\} du \, dv. \tag{18}$$

The resulting momentum-energy wave function is

$$\phi_\eta(q_z, q_0) = \left(\frac{1}{\pi}\right)^{1/2} \exp\left\{-\frac{1}{2}\left(e^{-2\eta} q_u^2 + e^{2\eta} q_v^2\right)\right\}. \tag{19}$$

Because we are using here the harmonic oscillator, the mathematical form of the above momentum-energy wave function is identical to that of the space-time wave function. The Lorentz squeeze properties of these wave functions are also the same, as is indicated in Fig. 5.

When the hadron is at rest with $\eta = 0$, both wave functions behave like those for the static bound state of quarks. As η increases, the wave functions become continuously squeezed until they become concentrated along their respective positive light-cone axes. Let us look at the z-axis projection of the space-time wave function. Indeed, the width of the quark distribution increases as the hadronic speed approaches that of the speed of light. The position of each quark appears widespread to the observer in the laboratory frame, and the quarks appear like free particles.

Furthermore, interaction time of the quarks among themselves become dilated. Because the wave function becomes wide-spread, the distance between one end of the harmonic oscillator well and the other end increases as is indicated in Fig. 5. This effect, first noted by Feynman [4], is universally observed in high-energy hadronic experiments.

The period is oscillation is increases like e^η. On the other hand the interaction time
with the external signal, since it is moving in the direction opposite to the direction
of the hadron, it travels along the negative light-cone axis. If the hadron contracts
along the negative light-cone axis, the interaction time decreases by $e^{-\eta}$. The ratio of
the interaction time to the oscillator period becomes $e^{-2\eta}$. The energy of each proton
coming out of the Fermilab accelerator is $900 GeV$. This leads the ratio to 10^{-6}. This
is indeed a small number. The external signal is not able to sense the interaction of
the quarks among themselves inside the hadron.

The momentum-energy wave function is just like the space-time wave function.
The longitudinal momentum distribution becomes wide-spread as the hadronic speed
approaches the velocity of light. This is in contradiction with our expectation from
nonrelativistic quantum mechanics that the width of the momentum distribution is
inversely proportional to that of the position wave function. Our expectation is that if
the quarks are free, they must have their sharply defined momenta, not a wide-spread
distribution. This apparent contradiction presents to us the following two fundamental
questions:

a). If both the spatial and momentum distributions become widespread as the
hadron moves, and if we insist on Heisenberg's uncertainty relation, is Planck's constant
dependent on the hadronic velocity?

b). Is this apparent contradiction related to another apparent contradiction that
the number of partons is infinite while there are only two or three quarks inside the
hadron?

The answer to the first question is "No", and that for the second question is "Yes".
Let us answer the first question which is related to the Lorentz invariance of Planck's
constant. If we take the product of the width of the longitudinal momentum distribution
and that of the spatial distribution, we end up with the relation

$$< z^2 >< q_z^2 >= (1/4)[\cosh(2\eta)]^2. \tag{20}$$

The right-hand side increases as the velocity parameter increases. This could lead us to
an erroneous conclusion that Planck's constant becomes dependent on velocity. This
is not correct, because the longitudinal momentum variable q_z is no longer conjugate
to the longitudinal position variable when the hadron moves.

In order to maintain the Lorentz-invariance of the uncertainty product, we have to
work with a conjugate pair of variables whose product does not depend on the velocity
parameter. Let us go back to Eq.(17) and Eq.(18). It is quite clear that the light-
cone variable u and v are conjugate to q_u and q_v respectively. It is also clear that the
distribution along the q_u axis shrinks as the u-axis distribution expands. The exact
calculation leads to

$$< u^2 >< q_u^2 >= 1/4, \qquad < v^2 >< q_v^2 >= 1/4. \tag{21}$$

Planck's constant is indeed Lorentz-invariant.

Let us next resolve the puzzle of why the number of partons appears to be infinite
while there are only a finite number of quarks inside the hadron. As the hadronic speed
approaches the speed of light, both the x and q distributions become concentrated along
the positive light-cone axis. This means that the quarks also move with velocity very
close to that of light. Quarks in this case behave like massless particles.

We then know from statistical mechanics that the number of massless particles is
not a conserved quantity. For instance, in black-body radiation, free light-like particles
have a widespread momentum distribution. However, this does not contradict the

known principles of quantum mechanics, because the massless photons can be divided into infinitely many massless particles with a continuous momentum distribution.

Likewise, in the parton picture, massless free quarks have a wide-spread momentum distribution. They can appear as a distribution of an infinite number of free particles. These free massless particles are the partons. It is possible to measure this distribution in high-energy laboratories, and it is also possible to calcuate it using the covariant harmonic oscillator formalism. We are thus forced to compare these two results [18]. Figure 6 shows the result.

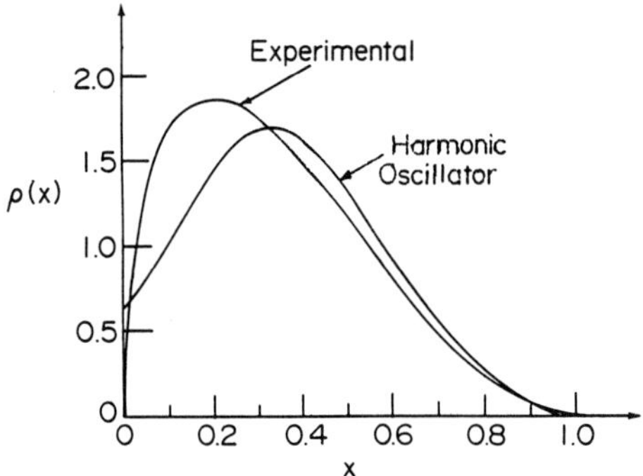

Figure 6. Calculation of the parton distribution based on the harmonic oscillator wave function. It is possible to construct the covariant harmonic oscillator wave functions for the three-quark system, and compare the parton distribution function with experiment. This graph shows a good agreement between the oscillator-based theory and the observed experimental data.

Acknowledgments

I am grateful to Prof. Bruno Gruber for inviting me to the 8th Symposium on Symmetries in Science. I was indeed fortunate enough to meet a number of physicists who had a deeper knowledge of Immanuel Kant. Prof. Walter Schempp kindly explained to me the scope of Kant's publications. Prof. Yoshio Ohnuki for telling me about the status of Kantianism in Japan.

During the coffee break after my talk, Prof. Peter Kramer suggested that I add an explanation of where the quark mass stands in the covariant oscillator formalism. This is an important question because this mass used to be the stumbling block against the development of relativistic quantum mechanics of bound systems, as Dirac pointed out in his 1949 paper [17]. First, let me give a mathematical explanation. In constructing the oscillator representation of the Poincaré group, the hadronic (mass)^2 is one of the Casimir operators, but the quark mass is not. Thus, the quark mass is not necessarily a Lorentz-invariant constant. Then, is this explanation consistent with our present understanding of physics. The answer is Yes. If quarks are inside the hadron, they do not have to be on their mass shell. This is perfectly consistent with the present form of quantum field theory based on Feynman diagrams [19]. Indeed, if we remove the mass-shell condition on the constituent particles, the difficulty mentioned in Dirac's 1949 paper disappears.

References

[1] Y. Tanikawa, *Hideki Yukawa: Scientific Works* (Iwanami Shoten, Tokyo, 1979).

[2] E. P. Wigner, Ann. Math. **40**, 149 (1939); V. Bargmann and E. P. Wigner, Proc. Natl. Acad. Scie (U.S.A.) **34**, 211 (1948); E. P. Wigner, in *Theoretical Physics* edited by A. Salam (Int'l Atomic Energy Agency, Vienna, 1963).

[3] E. Inonu and E. P. Wigner, Proc. Natl. Acad. Sci. (U.S.) **39**, 510 (1953).

[4] R. P. Feynman, in *High Energy Collisions*, Proceedings of the Third International Conference, Stony Brook, New York, edited by C. N. Yang *et al.* (Gordon and Breach, New York, 1969).

[5] P. A. M. Dirac, Proc. Roy. Soc. (London) **A183**, 284 (1945).

[6] H. Yukawa, Phys. Rev. *91*, 415 (1953).

[7] Y. S. Kim and M. E. Noz, *Theory and Applications of the Poincaré Group* (Reidel, Dordrecht, 1986).

[8] S. Ferrara and C. Savoy, in *Supergravity 1981*, S. Ferrara and J. G. Taylor eds. (Cambridge Univ. Press, Cambridge, 1982), p. 151. See also P. Kwon and M. Villasante, J. Math. Phys. **29**, 560 (1988); *ibid.* **30**, 201 (1989).

[9] D. Han, Y. S. Kim, and D. Son, Phys. Lett. B **131**, 327 (1983). See also D. Han, Y. S. Kim, M. E. Noz, and D. Son, Am. J. Phys. **52**, 1037 (1984).

[10] Y. S. Kim and E. P. Wigner, J. Math. Phys. **28**, 1175 (1987) and **32**, 1998 (1991).

[11] Y. S. Kim, Phys. Rev. Lett. **63**, 348-351 (1989).

[12] Y. S. Kim and M. E. Noz, Phys. Rev. D **8**, 3521 (1973).

[13] R. P. Feynman, M. Kislinger, and F. Ravndal, Phys. Rev. D **3**, 2706 (1971).

[14] P. A. M. Dirac, Proc. Roy. Soc. (London) **A114**, 243 and 710 (1927).

[15] K. Fujimura, T. Kobayashi, and M. Namiki, Prog. Theor. Phys. **43**, 73 (1970).

[16] J. D. Bjorken and E. A. Paschos, Phys. Rev. **185**, 1975 (1969).

[17] P. A. M. Dirac, Rev. Mod. Phys. **21**, 392 (1949).

[18] P. E. Hussar, Phys. Rev. D **23**, 2781 (1981).

[19] D. Han and Y. S. Kim, Prog. Theor. Phys. **64**, 1852 (1980).

SPECTRA AND EIGENFUNCTIONS OF REPRESENTATION OPERATORS FOR QUANTUM GROUPS AND q-OSCILLATORS

A. U. Klimyk

Institute for Theoretical Physics
Ukrainian National Academy of Sciences
Kiev 252143, Ukraine

1. INTRODUCTION

Properties of operators of irreducible representations of quantum algebras (q-deformed universal enveloping algebras of Lie algebras) very often differ from these of representation operators for Lie algebras. The main differences are:

(a) discrete spectra of operators of representations of finite dimensional representations are mostly non-equidistant;

(b) closures of unbounded symmetric operators of representations of infinite dimensional irreducible representations are not mostly selfadjoint (in these cases, they have equal deficiency indices and therefore we can construct their selfadjoint extensions).

These differences show that the theory of spectra of operators of irreducible representations of quantum algebras must be constructed, especially, in the case of unbounded representation operators. At present, we have no such theory. The corresponding theory of unbounded operators of infinite dimensional irreducible infinitesimally unitary representations of universal enveloping algebras of Lie algebras is well developed and is good exposed in Chapter 11 of the book by Barut and Raczka [1]. In the case of quantum algebras, only some concrete operators of irreducible representations of quantum algebras of a low rank and of simplest classes of irreducible representations of quantum algebras of a high rank are studied [2–5]. Spectra of symmetric operators of representations of q-oscillator algebras are studied in [6]. In the present paper we give a survey of results on spectra of representation operators of quantum algebras and q-oscillator algebras, discuss the problem of selfadjoint extensions of closures of symmetric unbounded operators of infinite dimensional representations of these algebras, consider overlap functions (coefficients of transition from a basis with respect to which an operator is not diagonal to a basis which diagonalizes this operator). So, we concern here the following problems:

(a) properties of self-adjointness of unbounded symmetric operators of infinite dimensional irreducible representations of quantum algebras and of q-oscillator algebras;

(b) spectra and eigenfunctions of bounded operators of infinite dimensional irreducible representations of quantum algebras and of q-oscillator algebras;

(c) spectra and eigenfunctions of operators of finite dimensional irreducible representations of "compact" quantum algebras;

(d) applications of spectra and eigenvectors of operators of irreducible representations of quantum algebras.

In this paper we are not interested in the structure of a Hopf algebra (comultiplication, counit, antipode) for quantum algebras because we do not need this structure for our consideration. Quantum algebras and q-deformed algebras are considered here as associative algebras generated by finite number of elements. Below we suppose (if the opposite is not indicated) that $0 < q < 1$.

2. THE CASE OF LIE GROUPS AND LIE ALGEBRAS

Let G be a noncompact real Lie group and let g be its Lie algebra. Let T be an infinite dimensional irreducible representation of G and let dT be the corresponding representation of g. We denote by U the universal enveloping algebra of the Lie algebra g. Then a representation dT determines a corresponding representation of U. We denote it by the same symbol dT.

Symmetric elements of the universal enveloping algebra U are defined [1]. One of the main results of the theory of representations of Lie groups and Lie algebras states that if T is a unitary representation of G, then closures of operators corresponding to symmetric elements of U are selfadjoint operators. In this case, there is a dense subspace of analytical vectors for the representation T of the group G and for the representation dT of g. Recall that a vector $\mathbf{x}$ of the carrier space H of the representation dT is analytical if it is analytical for every operator of representation dT. A vector $\mathbf{x}$ of H is analytical for a symmetric (skew-symmetric) operator A if and only if the series

$$\sum_{n=0}^{\infty} \frac{1}{n!} \|A^n \mathbf{x}\| s^n$$

converges for some numbers $s > 0$.

Conversely, if there exists a dense subspace of analytical vectors of the representation dT of g, then this representation can be integrated to a unitary representation of G.

These assertions are used in the theory of representations of Lie groups and Lie algebras. For example, in 1965, Gel'fand and Graev [7] constructed representations of discrete series of the Lie algebra $u(p,q)$ by symmetric operators. But they have no information about self-adjointness of closures of these operators. Therefore, they could not state that these representations can be integrated to unitary representations of the Lie group $U(p,q)$. Only in 1973, Mickelson and Niederle [8] proved that there exists an everywhere dense subspace of analytical vectors for the representations, constructed by Gel'fand and Graev. This means that these representations of $u(p,q)$ can be integrated to unitary representations of $U(p,q)$.

When we explicitly diagonalize a selfadjoint operator, then we actually construct coefficients which give a transition from the original basis to the basis with respect to which this operator is diagonal. These coefficients are called overlap coefficients or overlap functions.

Overlap coefficients for two bases of carrier spaces of irreducible representations of Lie groups and Lie algebras are of great importance for physics and mathematics. If we interpret infinitesimal operators as physical observables, then overlap coefficients are connected with probabilities of observable values. Overlap coefficients for representations of quantum algebras can be also used in physics. In particular, if we connect representation operators with dynamical operators or with observables.

Overlap coefficients are evaluated for many representation operators for infinite dimensional representations of simplest Lie groups (for the groups $SL(2, \mathbf{C})$, $SO_0(3, 1)$) and for infinite dimensional class 1 representations of high dimension Lie groups [9–12]. They are used for investigation of special functions.

In this paper we give overlap coefficients for some representation spaces of q-deformed algebras. They are expressed in terms of q-orthogonal polynomials [13].

3. DIFFERENCE OPERATORS OF THE SECOND ORDER

Let V be the Hilbert space with the orthonormal basis $|n\rangle$, $n = 0, 1, 2, \cdots$. Let L be a linear operator on V acting upon basis elements as

$$L|n\rangle = a_n|n+1\rangle + b_n|n\rangle + c_n|n-1\rangle, \tag{1}$$

and let

$$|z\rangle = \sum_{n=0}^{\infty} p_n(z)|n\rangle \tag{2}$$

be an eigenvector of L with an eigenvalue z: $L|z\rangle = z|z\rangle$. Then

$$L|z\rangle = \sum_{n=0}^{\infty}(p_n(z)a_n|n+1\rangle + p_n(z)b_n|n\rangle + p_n(z)c_n|n-1\rangle) = z\sum_{n=0}^{\infty} p_n(z)|n\rangle.$$

Equating coefficients at the vector $|n\rangle$ we obtain the recurrence relation for the coefficients from (2):

$$c_{n+1}p_{n+1}(z) + b_n p_n(z) + a_{n-1}p_{n-1}(z) = zp_n(z). \tag{3}$$

Since $p_{-1}(z) \equiv 0$ then setting $p_0(z) \equiv 1$ we have that this relation completely determines the coefficients $p_n(z)$. Moreover, $p_n(z)$ are polynomials in z of degree n.

Sometimes, vectors $\mathbf{v} = \sum_{n=0}^{\infty} v_n|n\rangle$ of V are represented as numerical sequences $(v_0, v_1, \cdots)$. In this case formula (1) can be written as

$$(L\mathbf{v})_n = a_{n-1}v_{n-1} + b_n v_n + c_{n+1}v_{n+1}.$$

Because of this, such operators L are called *second order difference operators*.

Let now L be a symmetric operator. Then formula (1) is of the form

$$L|n\rangle = a_n|n+1\rangle + b_n|n\rangle + a_{n-1}|n-1\rangle \tag{4}$$

and equation (3), determining eigenvectors, is reduced to the recurrence relation

$$a_n p_{n+1}(z) + b_n p_n(z) + a_{n-1}p_{n-1}(z) = zp_n(z). \tag{5}$$

If the coefficients a_n and b_n are real, then all coefficients of the polynomials $p_n(z)$ are real.

We suppose that a_n and b_n are real and $a_n > 0$. If the operator L is unbounded, then we denote the closure of L by $\bar{L}$. The operator $\bar{L}$ may be not selfadjoint. In this case $\bar{L}$ has nonzero deficiency indices (m, k) which determine dimensions of deficiency subspaces. (The definitions of deficiency indices and deficiency subspaces, as well as their properties, can be found in [14].) If $(m, k) \neq (0, 0)$, then to every complex number z, $\mathrm{Im}\, z \neq 0$, there correspond its deficiency subspaces. The following statements may be used for studying operators L:

(a) Deficiency indices of the operator $\bar{L}$ are coinciding. Moreover, these indices are $(0, 0)$ or $(1, 1)$. In the first case the operator $\bar{L}$ is selfadjoint. In the second case $\bar{L}$ is not selfadjoint, however it has selfadjoint extensions.

(b) Deficiency indices of $\bar{L}$ are $(0, 0)$ if and only if the series $\sum_{n=0}^{\infty} |p_n(z)|^2$ diverges for all complex z, $\mathrm{Im}\, z \neq 0$, where $p_n(z)$ are polynomials from (5). If deficiency indices are $(1, 1)$, then this series converges for all complex z, $\mathrm{Im}\, z \neq 0$.

(c) If deficiency indices are $(1, 1)$, then deficiency subspaces are one-dimensional. The deficiency subspace $N_{\bar{z}}$ corresponding to a complex number $\bar{z}$ is spanned by the vector

$$\sum_{n=0}^{\infty} p_n(z)|n\rangle,$$

where $p_n(z)$ are taken from formula (5).

To find whether or not the operator $\bar{L}$ is selfadjoint, we may use the following statements [15]:

(a) If the coefficients a_n and b_n from (4) are bounded, then the operator $\bar{L}$ is bounded and, therefore, selfadjoint.

(b) If b_n are any real numbers and a_n are such that $\sum_{n=0}^{\infty} (a_n)^{-1} = \infty$, then the operator $\bar{L}$ is selfadjoint.

(c) Let $|b_n| \leq C$, $n = 0, 1, 2, \cdots$, and let begining with some positive integer j we have $a_{n-1} a_{n+1} \leq a_n^2$, $n \geq j$. If

$$\sum_{n=0}^{\infty} \frac{1}{a_n} < \infty \tag{6}$$

then the operator $\bar{L}$ is not selfadjoint.

If $\bar{L}$ is not selfadjoint operator, then it has selfadjoint extensions. There are infinitely many selfadjoint extensions of $\bar{L}$. We refer the reader to the books [14, 15] for a more detailed discussion of selfadjoint extensions.

We considered the case of operators L on the Hilbert space V with the orthonormal basis $|n\rangle$, $n = 0, 1, 2, \cdots$. Let now V have the orthonormal basis $|n\rangle$, $n = 0, \pm 1, \pm 2, \cdots$ and let a symmetric operator L act upon this basis by formula (4). Then for eigenvectors (2) of $\bar{L}$ we have the recurrence relation (5) where $n = 0, \pm 1, \pm 2, \cdots$. According to results of the paper [16], in order to find deficiency indices of $\bar{L}$ we have to divide $\bar{L}$ into two operators L_1 and L_2: one acting upon the basis vectors $|m\rangle$, where $m > 0$, and the second acting upon $|m\rangle$ with $m < 0$. The deficiency indices of $\bar{L}$ are equal to the sum of the corresponding deficiency indices of $\bar{L}_1$ and $\bar{L}_2$. For example, if the deficiency indices of $\bar{L}_1$ and $\bar{L}_2$ are $(1, 1)$, then those for $\bar{L}$ are $(2, 2)$. If the deficiency indices of $\bar{L}_1$ are $(1, 1)$ and L_2 is a bounded operator, then the deficiency indices of $\bar{L}$ are $(1, 1)$.

4. SPECTRA OF REPRESENTATION OPERATORS FOR $U_q(\mathrm{su}_{1,1})$

The quantum algebra $U_q(\mathrm{sl}_2)$ is the associative algebra generated by the elements E_+, E_-, H that satisfy the commutation relations

$$[H, E_\pm] = \pm E_\pm, \quad [E_+, E_-] = \frac{q^H - q^{-H}}{q^{1/2} - q^{-1/2}} = \frac{\sinh hH}{\sinh(h/2)}, \tag{7}$$

where $q = \exp h$. Introducing into $U_q(\mathrm{sl}_2)$ the involution, defined by the relations $E_\pm^* = -E_\mp$, $H^* = H$, we obtain the real quantum algebra $U_q(\mathrm{su}_{1,1})$.

The representations T_l^+ of the discrete series of the algebra $U_q(\mathrm{su}_{1,1})$ are given by a positive number l and act on the Hilbert spaces V_l with the orthonormal bases $|m\rangle$, $m = l+1, l+2, \cdots$. The operators $T_l^+(E_\pm)$ and $T_l^+(H)$ act upon basis elements $|m\rangle$ by the formulas

$$T_l^+(H)|m\rangle = m|m\rangle, \quad T_l^+(E_+)|m\rangle = ([-l+m][l+m+1])^{1/2}|m+1\rangle \tag{8}$$

$$T_l^+(E_-)|m\rangle = -([-l+m-1][l+m])^{1/2}|m-1\rangle, \tag{9}$$

where $[a]$ is a q-number defined by the formula

$$[a] = (q^{a/2} - q^{-a/2})(q^{1/2} - q^{-1/2})^{-1}.$$

The elements H, $E_+ - E_-$, $\mathrm{i}(E_+ + E_-)$ are symmetric with respect to the involution. It is seen from formula (8) that the operator $T_l^+(H)$ is unbounded. Since it is diagonal with respect to the basis $\{|m\rangle\}$, then its closure is a selfadjoint operator. It follows from formulas (8) and (9) that the operators $L' = T_l^+(\mathrm{i}E_+ + \mathrm{i}E_-)$ and $L = T_l^+(E_+ - E_-)$ are also unbounded. It is easy to show that when passing from the basis $|m\rangle$, $m = l+1, l+2, \cdots$, to the basis $|m\rangle' = \mathrm{i}^m|m\rangle$, $m = l+1, l+2, \cdots$, we go over from the matrix of the operator L' to the matrix of the operator L. Because of this the closures $\bar{L}$ and $\bar{L}'$ of the operators L and L' are simultaneously selfadjoint or not and their deficiency indices are coinciding. We consider only the operator L. We have

$$L|m\rangle = a_m|m+1\rangle + a_{m-1}|m-1\rangle, \tag{10}$$

where $a_m = ([-l+m][l+m+1])^{1/2}$.

The operator $\bar{L}$ is not selfadjoint. Really, direct evaluations show that for a real number a the inequality

$$[a+m-1][a+m+1] \le [a+m]^2$$

reduces to the inequality $q + q^{-1} \ge 2$ which is fulfilled for all real positive q and the equality has place only for $q = 1$. This means that for the coefficients a_m from (10) we have $a_{m-1}a_{m+1} \le a_m^2$. The series (6) converges for the coefficients a_m since it can be majorized by the convergent integral

$$\int_0^\infty \frac{q^{x/2}dx}{(q^{x-l} - 1)^{1/2}(q^{x+l+1} - 1)^{1/2}}.$$

These arguments show that the operator $\bar{L}$ is not selfadjoint. It has selfadjoint extensions. The deficiency indices of $\bar{L}$ are $(1,1)$. Therefore, deficiency subspaces

N_z and $N_{\bar{z}}$ are one-dimensional. Moreover, deficiency subspaces $N_{\bar{z}}$, Im $z \neq 0$, are generated by the vectors

$$|z\rangle = \sum_{k=0}^{\infty} P_k(z)|l + k + 1\rangle,$$

where polynomials $P_k(z)$, $k = 0, 1, 2, \cdots$, satisfy the recurrence relation

$$([k + 1][k + 2l + 2])^{1/2} P_{k+1}(z) + ([k][k + 2l + 1])^{1/2} P_{k-1}(z) = z P_k(z). \tag{11}$$

and the initial conditions $P_{-1}(z) = 0$, $P_0(z) = 1$.

Spectral properties of operators of representations of the negative discrete series can be considered completely in the same way. The quantum algebra $U_q(\mathrm{su}_{1,1})$ has other series of unitary representations (the principal unitary series, the supplementary series, the strange series; see, for example, [17]). The operators $L = T(E_+ - E_-)$ in these representations are given by formula (10), where coefficients a_m have other expressions (they can be found in [17]) and m changes from $-\infty$ to $+\infty$. It is easy to verify that the operator L is symmetric and unbounded for all these unitary representations. The deficiency indices of the operator $\bar{L}$ in any unitary irreducible representation with m changing from $-\infty$ to $+\infty$ is equal to $(2,2)$. Thus, this operator has selfadjoint extensions.

If $q = \exp ih$, $h \in \mathbf{R}$, and is not a root of unity, then the operators $L' = T(iE_+ + iE_-)$ and $L = T(E_+ - E_-)$ are bounded for all unitary series (see [17]). Therefore, their closures are selfadjoint operators.

We could not solve the recurrence relation (11). Instead of $T_l^+(E_+ - E_-)$ we now consider the symmetric operator

$$B = T_l^+(q^{H/4}(E_+ - E_-)q^{H/4}).$$

We have

$$B|m\rangle = b_m|m + 1\rangle + b_{m-1}|m - 1\rangle, \quad b_m = ([k + 1][k + 2l + 2])^{1/2} q^{m/2} q^{1/4},$$

where $k = m - l - 1$. Since $\lim_{m \to \infty} b_m = (q^{-1/2} - q^{1/2})^{-1}$, then the operator B is bounded. Therefore, its closure is a bounded selfadjoint operator. A generalized vector

$$|y\rangle = \sum_{k=0}^{\infty} P_k(y)|k + l + 1\rangle \tag{12}$$

is an eigenvector of B corresponding to an eigenvalue y if $P_k(y)$, $k = 0, 1, 2, \cdots$, satisfy the recurrence relation

$$q^{k/2} q^{1/4} ([k + 1][k + 2l + 2])^{1/2} P_{k+1}(y) + q^{k/2} q^{-1/4} ([k][k + 2l + 1])^{1/2} P_{k-1}(y)$$

$$= y P_k(y) q^{-(l+1)/2}$$

and the initial conditions $P_{-1}(y) = 0$, $P_0(y) = 1$. The substitution

$$P_k(y) = \{(q; q)_k (q^{2l+2}; q)_k\}^{-1/2} P'_k(y) \tag{13}$$

reduce this relation to the form

$$P'_{k+1}(y) + (1 - q^k)(1 - q^{k+2l+1}) P'_{k-1}(y) = y(q^{-1/2} - q^{1/2}) P'_k(y).$$

Replacing $q^{-1/2}(1-q)y$ by $2x$ and $P'_k(2x(q^{-1/2}-q^{1/2})^{-1})$ by $P''_k(x)$, we obtain the relation

$$P''_{k+1}(x) + (1-q^k)(1-q^{k+2l+1})P''_{k-1}(x) = 2xP''_k(x). \tag{14}$$

To solve this recurrence relation we consider the orthogonal polynomials [18]

$$p_n(\cos(\theta+\phi); a, c|q) = a^{-n}e^{-in\phi}(ace^{2i\phi}; q)_n(a^2; q)_n(ac; q)_n$$
$$\times {}_4\varphi_3\left(\begin{array}{c} q^{-n}, q^{n-1}a^2c^2, ae^{2i\phi}e^{i\theta}, ae^{-i\theta} \\ ace^{2i\phi}, \quad a^2, \quad ac \end{array}; q, q\right), \tag{15}$$

where ${}_4\varphi_3$ is the basic hypergeometric function (the definition of this function see in [13]) and $(d; q)_n = (1-d)(1-dq)\cdots(1-dq^{n-1})$. The recurrence relation for these polynomials is of the form [18]

$$2xp_n(x) = A_np_{n+1}(x) + B_np_n(x) + C_np_{n-1}(x) \tag{16}$$

where $x = \cos(\theta+\phi)$ and expressions for A_n, B_n, C_n are given in [18]. direct evaluation shows that relation (14) coincides with recurrence relation (16) for $\phi = \pi/2$, $c = 0$, $a = q^{l+1}$, $x = \cos(\theta + \frac{\pi}{2}) = -\sin\theta$. Therefore,

$$P''_k(x) = p_k(\cos(\theta + \frac{\pi}{2}); q^{l+1}, 0|q) = (iq^{l+1})^{-k}(q^{2l+2}; q)_k$$
$$\times {}_3\varphi_2\left(\begin{array}{c} q^{-k}, e^{-i\theta}q^{l+1}, -e^{i\theta}q^{l+1} \\ q^{2l+2}, \quad 0 \end{array}; q, q\right).$$

Passing on to the polynomials $P_k(y)$, normalized by the condition $P_0(y) = 1$, we obtain from (13) that in formula (12) we have

$$P_k(y) = (iq^{l+1})^{-k}\left(\frac{(q^{2l+1}; q)_k}{(q; q)_k}\right)^{1/2} {}_3\varphi_2\left(\begin{array}{c} q^{-k}, e^{-i\theta}q^{l+1}, -e^{i\theta}q^{l+1} \\ q^{2l+2}, \quad 0 \end{array}; q, q\right),$$

where

$$y = \frac{2}{q^{-1/2} - q^{1/2}}\cos\left(\theta + \frac{\pi}{2}\right) = -i\frac{e^{-i\theta} - e^{i\theta}}{q^{-1/2} - q^{1/2}} \equiv -i[i\theta/2]. \tag{18}$$

Using the orthogonality relation for the polynomials $p_n(x; a, c|q)$ from [18] we find that the orthogonality relation for the polynomials $P_k(y)$ is of the form

$$\int_{-b}^{b} P_n(y)P_k(y)w(y)dy = \delta_{kn}, \tag{19}$$

where $b = 2/(q^{-1/2} - q^{1/2})$ and

$$w(y) = \frac{q^{-1/2} - q^{1/2}}{4\pi\cos\theta}(q; q)_\infty(q^{2l+2}; q)_\infty\left|\frac{(-e^{2i\theta}; q)_\infty}{(e^{2i\theta}q^{2l+2}; q^2)_\infty}\right|^2.$$

This relation means that the spectrum of the operator B is simple and covers exactly the interval $(-2/(q^{-1/2} - q^{1/2}), 2/(q^{-1/2} - q^{1/2}))$. The spectral measure of this operator coincides with the measure $d\sigma(y)=w(y)dy$. When $q \to 1$ then the spectrum turns into the real line and polynomials $P_n(y)$ tend to the corresponding Meixner-Pollaczek polynomials. This agrees with results for the classical Lie group $SU(1,1)$ (see [9], Chapter 7).

5. SPECTRA OF REPRESENTATION OPERATORS FOR $U_q(\mathrm{u}_n)$

We considered spectra of representation operators for the 'noncompact' quantum algebra $U_q(\mathrm{su}_{1,1})$. Let us consider spectra of representation operators for the 'compact' quantum algebra $U_q(\mathrm{u}_n)$ which is a real form of the quantum algebra $U_q(\mathrm{gl}(n,\mathbf{C}))$.

The quantum algebra $U_q(\mathrm{gl}(n,\mathbf{C}))$ is generated by the elements $E_{i,i-1}$, $E_{i-1,i}$, $i = 2,3,\cdots,n$, and E_{ii}, $i = 1,2,\cdots,n$, satisfying the certain relations (see, for example, [19]). The involution in $U_q(\mathrm{gl}(n,\mathbf{C}))$, defined by the relations

$$E^*_{i,i-1} = E_{i-1,i}, \quad E^*_{ii} = E_{ii},$$

determines the 'compact' quantum algebra $U_q(\mathrm{u}_n)$.

We consider irreducible finite dimensional representations T_l of the algebra $U_q(\mathrm{u}_n)$ with highest weights $(l,0,\cdots,0)$, $l \geq 0$. They act on the spaces V_l with the Gel'fand-Tsetlin bases. The basis elements of V_l are labelled by

$$|m,j,k,\cdots,r\rangle, \quad l \geq m \geq j \geq k \geq \cdots \geq |r|.$$

We denote these elements by $|m,M\rangle$, where M denotes the collection of indices $j,k,\cdots,r$. The representation T_l is given by the formulas

$$T_l(E_{n,n-1})|m,M\rangle = ([l-m][m-j+1])^{1/2}|m+1,M\rangle, \tag{20}$$

$$T_l(E_{n-1,n})|m,M\rangle = ([l-m+1][m-j])^{1/2}|m-1,M\rangle, \tag{21}$$

$$T_l(E_{nn})|m,M\rangle = (m-l)|m,M\rangle, \tag{22}$$

$$T_l(E_{n-1,n-1})|m,M\rangle = (j-m)|m,M\rangle \tag{23}$$

and by the analogous formulas for other operators $T_l(E_{j,j-1})$, $T_l(E_{j-1,j})$, $T_l(E_{jj})$.

The space V_l can be decomposed into the orthogonal sum

$$V_l = \sum_M \oplus\, V_M, \quad M = (j,k,\cdots,r).$$

The operators (20)–(23) leave the subspaces V_M invariant. We shall evaluate the spectrum and the eigenvectors

$$|x,j,k,\cdots\rangle = \sum_{m=j}^{l} P_{m-j}(x)|m,j,k,\cdots\rangle \tag{24}$$

of the operator

$$L_j = T_l((q^{1/4}E_{n,n-1} + q^{-1/4}E_{n-1,n})q^{(E_{nn}-E_{n-1,n-1})/4})$$

on the $l-j+1$-dimensional subspace $V_M = V_{(j,k,\cdots)}$. We suppose that q is real and $L_j|x,j,k,\cdots\rangle = [x]|x,j,k,\cdots\rangle$.

The vector (24) is an eigenvector for the operator L_j with the eigenvalue $[x]$ if $P_{m-j}(x)$ satisfy the recurrence relation

$$q^{1/4}([N-n][n+1])^{1/2}P_{n+1}(x) + q^{-1/4}([N-n+1][n])^{1/2}P_{n-1}(x)$$

$$= q^{N/4}q^{-n/2}[x]P_n(x),$$

where $n = m - j$, $N = l - j$. Setting here

$$P_n(x) = \left(\frac{q^{n(N-1)/2}[N]!}{[n]![N-n]!} \right)^{1/2} P'_n(x)$$

we reduce this relation to the form

$$q^n(1 - q^{N-n})P'_{n+1}(x) + (1 - q^n)P'_{n-1}(x) = (q^{-1/2} - q^{1/2})q^{N/2}[x]P'_n(x).$$

It coincides with the recurrence relation for the dual q-Krawtchouk polynomials [13]

$$k_n(\lambda(y); c, N|q) = {}_3\varphi_2 \left(\begin{matrix} q^{-y}, -cq^{y+1}, q^{-n} \\ 0, \quad q^{-N} \end{matrix} ; q, q \right),$$

where $\lambda(y) = q^{-y} - cq^{y+1}$ if $c = q^{-N-1}$, and

$$(q^{-1/2} - q^{1/2})[x]q^{N/2} = q^y - q^{N-y}.$$

Thus,

$$P_n(x) = \left(\frac{q^{(N-1)n/2}[N]!}{[n]![N-n]!} \right)^{1/2} k_n(\lambda(y); q^{-N-1}, N|q), \tag{25}$$

where $[N]! = [1][2] \cdots [N]$, $[x] = [N - 2y]$, and y takes the values $0, 1, 2, \cdots, N$. It follows from here and from the orthogonality relation for dual q-Krawtchouk polynomials [13] that the spectrum of L_j coincides with the set of points

$$[k], \quad k = -(l - j), \ -(l-j)+2, \ -(l-j)+4, \ \cdots, \ l - j. \tag{26}$$

The corresponding eigenvectors are determined by formulas (24) and (25). The orthogonality relation for the polynomials $P_n(x)$ are of the form

$$\sum_{y=0}^{N} P_n(x)P_m(x)W(x) = \delta_{mn}.$$

Here

$$W(x) = \frac{q^{y(N-y)-N(N-1)/4}[N]![4x - 2N]}{2[2y]!![2N - 2y]!![2x - N]}$$

where $[2k]!! = [2k][2k - 2][2k - 4] \cdots [2]$ and y is related to x as in formula (25). The vectors

$$|x, j, k, \cdots\rangle' = W(x)^{1/2} \sum_{m=j}^{l} P_{m-j}(x)|m, j, k, \cdots\rangle$$

are orthonormalized and

$$|m, j, k, \cdots\rangle = \sum_{x} W(x)^{1/2} P_{m-j}(x)|x, j, k, \cdots\rangle'.$$

The spectrum of the operator

$$L = T_l((q^{1/4} E_{n,n-1} + q^{-1/4} E_{n-1,n})q^{(E_{nn} - E_{n-1,n-1})/4})$$

is obtained by union of spectra (26) for $j = 0, 1, 2, \cdots, l$.

Spectra of operators of other irreducible representations of the quantum algebra $U_q(\mathbf{u}_n)$ are not studied.

6. THE q-DEFORMED ALGEBRA $U_q(\mathrm{so}(n, \mathbf{C}))$

Drinfeld [20] and Jimbo [21] defined q-deformed (quantum) algebras $U_q(g)$ for all simple complex Lie algebras g by means of Cartan subalgebras and root subspaces. However, these approaches do not give a satisfactory presentation of the quantum algebra $U_q(\mathrm{so}(n, \mathbf{C}))$ from point of view of some problems of quantum physics and representation theory. In fact, they admit the inclusion

$$U_q(\mathrm{so}(n, \mathbf{C})) \supset U_q(\mathrm{so}(n - 2, \mathbf{C}))$$

and do not admit

$$U_q(\mathrm{so}(n, \mathbf{C})) \supset U_q(\mathrm{so}(n - 1, \mathbf{C})). \tag{27}$$

This is why we cannot construct the quantum algebra $U_q(\mathrm{so}_{n,1})$ in the frame of these approaches. In order to obtain inclusion (27) we proposed in [22] (see also [23]) another q-deformation of the classical universal enveloping algebra $U(\mathrm{so}(n, \mathbf{C}))$. The algebra $U(\mathrm{so}(n, \mathbf{C}))$ is generated by the elements $I_{i,i-1}$, $i = 2, 3, \cdots, n$, that satisfy the relations

$$I_{i,i-1} I_{i+1,i}^2 - 2 I_{i+1,i} I_{i,i-1} I_{i+1,i} + I_{i+1,i}^2 I_{i,i-1} = -I_{i,i-1}, \tag{28}$$

$$I_{i,i-1}^2 I_{i+1,i} - 2 I_{i,i-1} I_{i+1,i} I_{i,i-1} + I_{i+1,i} I_{i,i-1}^2 = -I_{i+1,i}, \tag{29}$$

$$[I_{i,i-1}, I_{j,j-1}] = 0, \quad |i - j| > 1 \tag{30}$$

(they follow from the well-known commutation relations for the generators I_{ij} of the Lie algebra $\mathrm{so}(n, \mathbf{C})$).

In our approach to the q-deformed orthogonal algebra we define q-deformation of the associative algebra $U(\mathrm{so}(n, \mathbf{C}))$ by deforming relations (28)–(30). The q-deformed relations are of the form

$$I_{i,i-1} I_{i+1,i}^2 - a I_{i+1,i} I_{i,i-1} I_{i+1,i} + I_{i+1,i}^2 I_{i,i-1} = -I_{i,i-1}, \tag{31}$$

$$I_{i,i-1}^2 I_{i+1,i} - a I_{i,i-1} I_{i+1,i} I_{i,i-1} + I_{i+1,i} I_{i,i-1}^2 = -I_{i+1,i}, \tag{32}$$

$$[I_{i,i-1}, I_{j,j-1}] = 0, \quad |i - j| > 1, \tag{33}$$

where $a = q^{1/2} + q^{-1/2}$ and $[\cdot, \cdot]$ denotes the usual commutator. Obviously, in the limit $q \to 1$ formulas (31)–(33) give relations (28)–(30). Remark that relations (31) and (32) differ from the q-deformed Serre relations in the approach of Jimbo and Drinfeld to quantum orthogonal algebras by appearance of nonzero right hand side and by possibility of reduction (27). Below by the algebra $U_q(\mathrm{so}(n, \mathbf{C}))$ we mean the q-deformed algebra defined by formulas (31)–(33).

The compact real form $U_q(\mathrm{so}_n)$ of the algebra $U_q(\mathrm{so}(n, \mathbf{C}))$ is defined by the involution given as

$$I_{i,i-1}^* = -I_{i,i-1}, \quad i = 2, 3, \cdots, n. \tag{34}$$

The noncompact real form $U_q(\mathrm{so}_{n-1,1})$ of $U_q(\mathrm{so}(n, \mathbf{C}))$ is determined by the involution

$$I_{i,i-1}^* = -I_{i,i-1}, \ i \neq n, \quad I_{n,n-1}^* = I_{n,n-1}. \tag{35}$$

The q-deformed algebras $U_q(\mathrm{so}_{n-1,1})$ and $U_q(\mathrm{so}_n)$ contain the subalgebra $U_q(\mathrm{so}_{n-1})$. This fact allows us to consider Gel'fand-Tsetlin bases of carrier spaces of representations of $U_q(\mathrm{so}_{n,1})$ and $U_q(\mathrm{so}_n)$.

It was shown by Noumi *et al.* [24] that the algebra $U_q(\mathrm{so}(n,\mathbf{C}))$ can be embedded into $U_q(\mathrm{sl}(n,\mathbf{C}))$. In particular, we have the embedding $U_q(\mathrm{so}(3,\mathbf{C})) \subset U_q(\mathrm{sl}(3,\mathbf{C}))$ important from the point of view of nuclear physics. It was shown by Noumi [25] that the algebra $U_q(\mathrm{so}(n,\mathbf{C}))$ allows to define quantum analogues of the homogeneous spaces $GL(n)/SO(n)$.

As in the classical case, the q-algebras $U_q(\mathrm{so}(3,\mathbf{C}))$ and $U_q(\mathrm{so}(4,\mathbf{C}))$ can also be described in terms of bilinear relations (q-commutators). In fact, defining the algebra $U_q(\mathrm{so}(3,\mathbf{C}))$ by relations (31)–(33) we have only two generators I_{21} and I_{32}. However, we can define the third element I_{31} according to the formula [23]

$$I_{31} = q^{1/4} I_{21} I_{32} - q^{-1/4} I_{32} I_{21}. \tag{35}$$

Then by the algebra $U_q(\mathrm{so}(3,\mathbf{C}))$ we mean the associative algebra generated by the elements I_{21}, I_{32} and I_{31} which satisfy the relations

$$q^{1/4} I_{21} I_{32} - q^{-1/4} I_{32} I_{21} = I_{31}, \tag{36}$$
$$q^{1/4} I_{31} I_{21} - q^{-1/4} I_{21} I_{31} = I_{32}, \tag{37}$$
$$q^{1/4} I_{32} I_{31} - q^{-1/4} I_{31} I_{32} = I_{21}. \tag{38}$$

It is clear that if the generators I_{21}, I_{32} and I_{31} satisfy relations (36)–(38), then the pair I_{21} and I_{32} satisfies the trilinear relations (31) and (32). Remark that the algebra given by formulas (36)–(38) coincides with the cyclically symmetric Fairlie algebra [26].

The q-deformed algebra $U_q(\mathrm{so}(4,\mathbf{C}))$ is generated by the generators I_{21}, I_{32} and I_{43}. Moreover, for the first two generators everything, said above concerning $U_q(\mathrm{so}(3,\mathbf{C}))$, is true. Thus, the inclusion

$$U_q(\mathrm{so}(3,\mathbf{C})) \subset U_q(\mathrm{so}(4,\mathbf{C}))$$

takes place. The generators I_{21} and I_{43} mutually commute and the pair I_{32}, I_{43} in turn must satisfy relations (31) and (32). Again, $U_q(\mathrm{so}(4,\mathbf{C}))$ can be also given in terms of bilinear q-commutators. Namely, we can add to the triple of generators I_{21}, I_{32} and I_{43} the element I_{31} from (35) and the elements I_{42}, I_{41} defined as

$$I_{42} = q^{1/4} I_{32} I_{43} - q^{-1/4} I_{43} I_{32}, \tag{39}$$

$$I_{41} = q^{1/4} I_{31} I_{43} - q^{-1/4} I_{43} I_{31} = q^{1/4} I_{21} I_{42} - q^{-1/4} I_{42} I_{21}. \tag{40}$$

Contrary to the case of Lie algebra $\mathrm{so}(4,\mathbf{C})$, the quantum algebra $U_q(\mathrm{so}(4,\mathbf{C}))$ cannot be represented as a direct sum (or a direct product) of two algebras $U_q(\mathrm{so}(3,\mathbf{C}))$.

7. SPECTRA OF REPRESENTATION OPERATORS FOR $U_q(\mathrm{so}_4)$

Finite dimensional irreducible representations of the algebra $U_q(\mathrm{so}_3)$ are given by integral or half-integral nonnegative number l. We denote these representations by T_l. The carrier space of the representation T_l has the orthonormal basis $\{|m\rangle,\ m = l, l-1, \cdots, -l\}$, and the operators $T_l(I_{21})$ and $T_l(I_{32})$ act upon this basis as

$$T_l(I_{21})|m\rangle = \mathrm{i}[m]|m\rangle, \tag{41}$$

$$T_l(I_{32})|m\rangle = d(m)([l-m][l+m+1])^{\frac{1}{2}}|m+1\rangle - d(m-1)([l-m+1][l+m])^{\frac{1}{2}}|m-1\rangle, \tag{42}$$

where

$$d(m) = ([m][m+1]/[2m][2m+2])^{1/2}$$

and $[a]$ denotes a q-number.

As in the case of the Lie group $SO(4)$, finite dimensional irreducible representations T_{rs} of the q-deformed algebra $U_q(\mathrm{so}_4)$ are given by two integral or half-integral numbers r and s such that $r \geq |s| \geq 0$ (see [27]). Restriction of T_{rs} onto the subalgebra $U_q(\mathrm{so}_3)$ decomposes into the sum of irreducible representations T_l of this subalgebra for which $l = |s|, |s|+1, \cdots, r$. Uniting bases of subspaces of irreducible representations T_l of $U_q(\mathrm{so}_3)$ we obtain a basis of the carrier space V_{rs} of the representation T_{rs} of $U_q(\mathrm{so}_4)$. Thus, the corresponding orthonormal basis of V_{rs} consists of the vectors

$$|l,m\rangle, \quad |s| \leq l \leq r, \quad m = -l, -l+1, \cdots, l.$$

The operator $T_{rs}(I_{43})$ acts upon these vectors by the formula [27]

$$T_{rs}(I_{43})|l,m\rangle = \mathrm{i}\frac{[r+1][s][m]}{[l][l+1]}|l,m\rangle$$

$$+ \left(\frac{[r-l][l+s+1][l-s+1][l+m+1][l-m+1]}{[r+l+2]^{-1}[l+1]^2[2l+1][2l+3]}\right)^{1/2}|l+1,m\rangle$$

$$- \left(\frac{[r+l+1][l+s][l-s][l+m][l-m]}{[r-l+1][l]^2[2l-1][2l+1]}\right)^{1/2}|l-1,m\rangle, \tag{43}$$

where numbers in the square brackets are q-numbers. The operators $T_{rs}(I_{21})$ and $T_{rs}(I_{32})$ act upon the basis vectors by formulas (41) and (42). Formulas (41)–(43) completely determine the representation T_{rs}.

Let us find the spectrum of the operator $L = -\mathrm{i}T_{rs}(I_{43})$, $\mathrm{i} = \sqrt{-1}$. It is selfadjoint. Replacing the vectors $|l,m\rangle$ by $|l,m\rangle' = \mathrm{i}^{-l}|l,m\rangle$ we obtain that L acts upon the vectors $|l,m\rangle'$ by formula (43) in which the sign $-$ of the third summand is replaced by $+$ and the first summand is multipled by $-\mathrm{i}$.

The space V_{rs} can be decomposed into the sum

$$V_{rs} = \sum_{m=-r}^{r} \oplus\, V_m,$$

where V_m is spanned by the vectors $|l,m\rangle$ with fixed m. Let us find the spectrum and the eigenvectors

$$|x,m\rangle' = \sum_{l=k}^{r} P_{l-k}(x)|l,m\rangle, \quad k = \max\,(|m|, |s|) \tag{44}$$

of the operator L on the subspace V_m:

$$L|x,m\rangle' = [x]|x,m\rangle', \tag{45}$$

where $[x]$ is a q-number. Formula (43) is symmetric with respect to permutation of s and m and to change of signs at m and s. Therefore, we may assume, without loss of generality, that s and m are positive and that $s \geq m$.

Substituting expression (44) for $|x, m\rangle'$ into (45) and acting by L upon $|l, m\rangle$ we easily find that vector (44) is an eigenvector of L with the eigenvalue $[x]$ if P_{l-k} satisfy the recurrence relation

$$\left(\frac{[u][n + 2s + 1][n + 1][n + s + m + 1][n + s - m + 1]}{[r + n + s + 2]^{-1}[n + s + 1]^2[2n + 2s + 1][2n + 2s + 3]} \right)^{1/2} P_{n+1}(x)$$

$$+ \left(\frac{[r + n + s + 1][r - n - s + 1][n + 2s][n][n + s + m]}{[n + s - m]^{-1}[n + s]^2[2n + 2s - 1][2n + 2s + 1]} \right)^{1/2} P_{n-1}(x)$$

$$+ \frac{[r + 1][s][m]}{[n + s][n + s + 1]}P_n(x) = [x]P_n(x) \tag{46}$$

(here $u = r - n - s$, $n = l - k$) and the initial conditions $P_0(x) = 1$, $P_{-1}(x) = 0$.

Making in (46) the substitution

$$P_n(x) = -q^c \left(\frac{[n + 2s]![n + s + m]![2n + 2s + 1]}{[n]![n + s - m]![r - n - s]![r + n + s + 1]!} \right)^{1/2} P_n'(x)$$

where $c = (s + m - r)/2$, we obtain recurrence relation (46) in a transformed form. Comparing it with recurrence relation (7.5.2) from [13] for q-Racah polynomials

$$R_n(\mu(y); \; \alpha, \beta, \gamma, \delta | q) = {}_4\varphi_3 \left(\begin{matrix} q^{-y}, q^{y+1}\gamma\delta, q^{-n}, q^{n+1}\alpha\beta \\ \alpha q, \quad \beta\delta q, \quad \gamma q \end{matrix} ; \; q, q \right)$$

at

$$\alpha = \beta = -q^s, \; \gamma = q^{s+m}, \; \delta = -q^{-r-1}, \tag{47}$$

after cumbersome transformations we conclude that

$$P_n'(x) = R_n(\mu(y); \; \alpha, \beta, \gamma, \delta | q),$$

where α, β, γ, δ are given by formulas (47) and $x = (r - s - m) - 2y$. Thus, the polynomials $P_n(x)$ from (46) normalized by the condition $P_0(x) = 1$ are of the form

$$P_n(x) = N^{1/2} R_n(\mu(y); \; -q^s, -q^s, q^{s+m}, -q^{-r-1} | q), \tag{48}$$

$$N = \frac{[n + 2s]![n + s + m]![2n + 2s + 1][s - m]![r - s]![r + s + 1]!}{[n]![n + s - m]![r - n - s]![r + n + s + 1]![2s]![s + m]![2s + 1]},$$

where $x = (r - s - m) - 2y$. The variable y takes the values $0, 1, 2, \cdots, r - s$. Therefore, the spectrum of L on the subspace V_m consists of the points

$$[r - s - m], \; [r - s - 2 - m], \; [r - s - 4 - m], \; \cdots, \; [-(r - s) - m]. \tag{49}$$

The corresponding eigenvectors are determined by formulas (44) and (48). The orthogonality relation for the polynomials $P_n(x)$ follows from the orthogonality of q-Racah polynomials [13] and is of the form

$$\sum_{y=0}^{r-s} P_n(x)P_k(x)W(x) = \delta_{nk}. \tag{50}$$

Here $W(x)$ is equal to the expression

$$\frac{[4y + 2k - 2r][2y + 2k - 2r - 2]!![2y + 2s]!![2r - 2y]!![r - m - y]!}{[2y + 2k - 2r][y + k - r - 1]![y + s]![2y + 2m]!![r - y]![r - s - y]![y]!}$$

$$\times [y + m]![k + y]![2s + 1]!!([s]!)^2[r - s]([2s]!![s - m]![k]![r + s + 1]!)^{-1},$$

where $k = s + m$, $[n]! = [n][n - 1] \cdots [1]$ and $[n]!! = [n][n - 2][n - 4] \cdots [1]$ or $[2]$.

Formula (50) shows that vectors (44) are not normalized. The vectors

$$|x, m\rangle = W(x)^{1/2}|x, m\rangle'$$

are normal and due to formula (45) we have

$$T_{rs}(I_{43})|x, m\rangle = \mathrm{i}[x]|x, m\rangle. \tag{51}$$

Joining spectra (49) for all subspaces V_m, we obtain the spectrum of the operator T_{rs}, and therefore the spectrum of the operator $T_{rs}(I_{43})$.

In analogous way, spectra and eigenvectors of operators of irreducible infinite dimensional representations of the quantum algebras $U_q(\mathrm{so}_{2,1})$ and $U_q(\mathrm{so}_{3,1})$ are found in [4]. Spectra and eigenvectors of operators of class 1 irreducible representations of the quantum algebras $U_q(\mathrm{so}_n)$ and $U_q(\mathrm{so}_{n,1})$ are evaluated in [3].

8. REPRESENTATIONS T_{rs} OF $U_q(\mathrm{so}_4)$ IN THE BASIS $|x, m\rangle$

Spectra and eigenvectors of operators of irreducible representations of quantum algebras can be used for studying representations of these algebras. Here, as example, we show [28] how eigenvectors from Section 7 are used to derive formulas for operators of representations T_{rs} of the algebra $U_q(\mathrm{so}_4)$ with respect to the basis $|x, m\rangle$ which corresponds to the reduction of $U_q(\mathrm{so}_4)$ onto the subalgebra $U_q(\mathrm{so}_2) \times U_q(\mathrm{so}_2)$.

The operator $T_{rs}(I_{43})$ acts upon the basis vectors $|x, m\rangle$ by formula (51). It is clear from formulas (41) and (44) that

$$T_{rs}(I_{21})|x, m\rangle = \mathrm{i}[m]|x, m\rangle. \tag{52}$$

Thus, to have the representation T_{rs} in the basis $|x, m\rangle$, we must find the action formula for the operator $T_{rs}(I_{32})$ upon this basis.

Since

$$|x, m\rangle = \sum_{l=s}^{r} P_{l-s}^m(x)|l, m\rangle, \tag{53}$$

with $P_{l-s}^m(x) = W(x)^{1/2}P_{l-s}(x)$, then due to formula (17) we have

$$T_{rs}(I_{32})|x, m\rangle = d(m) \sum_{l=s}^{r} P_{l-s}^m(x)([l - m][l + m + 1])^{1/2}|l, m + 1\rangle$$

$$- d(m - 1) \sum_{l=s}^{r} P_{l-s}^m(x)([l - m + 1][l + m])^{1/2}|l, m - 1\rangle. \tag{54}$$

Applying to $([l - m][l + m + 1])^{1/2}P_{l-s}^m(x)$ recurrence relation (7.2.14) from [13] with

$$a = q^{m-r-1}, \quad b = -q^s, \quad c = d = -q^m, \quad n = (r - s - m - x)/2, \quad j = l - m$$

and using the equalities

$$[2x]/[x] = q^{x/2} + q^{-x/2}, \quad (q^{(a+b)/2} \pm q^{-(a+b)/2})(q^{(a-b)/2} \mp q^{-(a-b)/2}) = [2a] \mp [2b],$$

after some calculations we obtain for the first summand of the right hand side of (54) the expression

$$d(m)d(x-1)\{([r+1]+[s-m+x-1])([r+1]+[s+m-x+1])\}^{\frac{1}{2}} \sum_{l=s}^{r} P_{l-s}^{m+1}(x-1)|l,m+1\rangle$$

$$-d(m)d(x)\{([r+1]-[s+m+x+1])([r+1]-[s-m-x-1])\}^{\frac{1}{2}} \sum_{l=s}^{r} P_{l-s}^{m+1}(x+1)|l,m+1\rangle.$$

$$(55)$$

To transform the second summand on the right hand side of (54) we apply to the basic hypergeometric function $_4\varphi_3$ from the expression for $P_{l-s}^m(x)$ the transformation

$$_4\varphi_3 \left(\begin{array}{c} q^{-N}, \ \alpha, \ \beta, \ \gamma \\ \delta, \ \sigma, \ \rho \end{array} ; \ q,q \right) =$$

$$\frac{(\sigma/\alpha;q)_N(\rho/\alpha;q)_N}{(\sigma;q)_N(\rho;q)_N} \alpha^N \ _4\varphi_3 \left(\begin{array}{c} q^{-N}, \alpha, \ \delta/\beta, \ \delta/\gamma \\ \delta, \ \alpha q^{1-N}/\sigma, \ \alpha q^{1-N}/\rho \end{array} ; \ q,q \right)$$

where $N = l - s$ and $\alpha = q^{l+s+1}$, $\beta = -q^{(s-r+m-x)/2}$, $\gamma = q^{(s-r+m+x)/2}$, $\delta = q^{s-r}$, $\sigma = -q^{s+1}$, and $\rho = q^{s+m+1}$. Now we apply to $([l-m+1][l+m])^{1/2}P_{l-s}^m(x)$ the same recurrence relation (7.2.14) from [13] with

$$a = q^{-(r+m+1)}, \quad b = -q^{-s}, \quad c = d = -q^m, \quad n = (r-s+m+x)/2, \quad j = l+m.$$

Then the second summand of the right hand side of (54) takes the form

$$d(m-1)d(x)\{([r+1]+[s+m-x-1])([r+1]+[s-m+x+1])\}^{\frac{1}{2}} \sum_{l=s}^{r} P_{l-s}^{m-1}(x+1)|l,m-1\rangle$$

$$-d(m-1)d(x-1)\{([r+1]-[s-m-x+1])([r+1]-[s+m+x-1])\}^{\frac{1}{2}} \sum_{l=s}^{r} P_{l-s}^{m-1}(x-1)|l,m-1\rangle.$$

$$(56)$$

We substitute expressions (55) and (56) into (54) and take into account formula (53). As a result, we find that the operator $T_{rs}(I_{32})$ acts upon the vectors $|x,m\rangle$ as

$$T_{rs}(I_{32})|x,m\rangle =$$

$$d(m)d(x-1)\{([r+1]+[s-m+x-1])([r+1]+[s+m-x+1])\}^{\frac{1}{2}}|x-1,m+1\rangle$$

$$-d(m)d(x)\{([r+1]-[s+m+x+1])([r+1]-[s-m-x-1])\}^{\frac{1}{2}}|x+1,m+1\rangle$$

$$+d(m-1)d(x-1)\{([r+1]-[s-m-x+1])([r+1]-[s+m+x-1])\}^{\frac{1}{2}}|x-1,m-1\rangle$$

$$-d(m-1)d(x)\{([r+1]+[s+m-x-1])([r+1]+[s-m+x+1])\}^{\frac{1}{2}}|x+1,m-1\rangle. \quad (57)$$

Now we completely determined representations T_{rs} of $U_q(\mathrm{so}_4)$ with respect to the basis corresponding to reduction onto the subalgebra $U_q(\mathrm{so}_2) \times U_q(\mathrm{so}_2)$.

9. REPRESENTATIONS OF THE q-DEFORMED ALGEBRA $U_q(\mathrm{so}_{2,2})$

As in the case of representations of compact and noncompact real Lie groups, by making use of analytical continuation in parameters giving representations we can obtain [28] infinite dimensional representations of the q-deformed algebra $U_q(\mathrm{so}_{2,2})$ from the representations T_{rs} of $U_q(\mathrm{so}_4)$. In this way, the representations $T_{\sigma\tau}^\epsilon$, $\sigma \in \mathbf{C}$, $\tau \in \mathbf{C}$, $\epsilon \in \{0,1\}$, of $U_q(\mathrm{so}_{2,2})$ which act on the Hilbert spaces H_ϵ with the orthonormal basis

$$|x,m\rangle, \quad x \in \mathbf{Z}, \quad m \in \mathbf{Z}, \quad x + m \equiv \epsilon \ (\mathrm{mod}\ 2)$$

are obtained. The operators $T_{\sigma\tau}^\epsilon(I_{21})$ and $T_{\sigma\tau}^\epsilon(I_{43})$ act upon these basis vectors by formulas (51) and (52). For the operator $T_{\sigma\tau}^\epsilon(I_{32})$ we have

$$T_{\sigma\tau}^\epsilon(I_{32})|x,m\rangle =$$

$$d(m)d(x-1)\{([\sigma+1]+[\tau-m+x-1])([\sigma+1]+[\tau+m-x+1])\}^{\frac{1}{2}}|x-1,m+1\rangle$$

$$- d(m)d(x)\{([\sigma+1]-[\tau+m+x+1])([\sigma+1]-[\tau-m-x-1])\}^{\frac{1}{2}}|x+1,m+1\rangle$$

$$+ d(m-1)d(x-1)\{([\sigma+1]-[\tau-m-x+1])([\sigma+1]-[\tau+m+x-1])\}^{\frac{1}{2}}|x-1,m-1\rangle$$

$$- d(m-1)d(x)\{([\sigma+1]+[\tau+m-x-1])([\sigma+1]+[\tau-m+x+1])\}^{\frac{1}{2}}|x+1,m-1\rangle.$$

This formula can be transformed to the form

$$T_{\sigma\tau}^\epsilon(I_{32})|x,m\rangle =$$

$$\left(\frac{[\sigma-\tau+m-x+2][\sigma-\tau-m+x][(\sigma+\tau+m-x+2)/2]}{[(\sigma-\tau+m-x+2)/2][(\sigma-\tau-m+x)/2][(\sigma+\tau-m+x)/2]^{-1}}\right)^{1/2}$$
$$\times\, d(m)d(x-1)|x-1,m+1\rangle$$

$$-\left(\frac{[\sigma+\tau+m+x+2][\sigma+\tau-m-x][(\sigma-\tau+m+x+2)/2]}{[(\sigma+\tau+m+x+2)/2][(\sigma+\tau-m-x)/2][(\sigma-\tau-m-x)/2]^{-1}}\right)^{1/2}$$
$$\times\, d(m)d(x)|x+1,m+1\rangle$$

$$+\left(\frac{[\sigma+\tau-m-x+2][\sigma+\tau+m+x][(\sigma-\tau-m-x+2)/2]}{[(\sigma+\tau-m-x+2)/2][(\sigma+\tau+m+x)/2][(\sigma-\tau+m+x)/2]^{-1}}\right)^{1/2}$$
$$\times\, d(m-1)d(x-1)|x-1,m-1\rangle$$

$$-\left(\frac{[\sigma-\tau-m+x+2][\sigma-\tau+m-x][(\sigma+\tau-m+x+2)/2]}{[(\sigma-\tau-m+x+2)/2][(\sigma-\tau+m-x)/2][(\sigma+\tau+m-x)/2]^{-1}}\right)^{1/2}$$
$$\times\, d(m-1)d(x)|x+1,m-1\rangle. \tag{58}$$

In every summand here there are two expressions of the form

$$[\sigma-\tau-m+x]/[(\sigma-\tau-m+x)/2].$$

This expression is equal to

$$q^{(\sigma-\tau-m+x)/4} + q^{-(\sigma-\tau-m+x)/4}.$$

Irreducibility of the representations $T_{\sigma\tau}^\epsilon$ is studied in the same way as in the case of the q-deformed algebras $U_q(\mathrm{so}_{2,1})$ and $U_q(\mathrm{so}_{3,1})$ (see [23]). Namely, invariant subspaces in the representation space appear because of vanishing of some coefficients in formula (58). This studying leads to the following result: A representation $T_{\sigma\tau}^\epsilon$ of the algebra $U_q(\mathrm{so}_{2,2})$ is irreducible if and only if $\sigma + \tau \not\equiv \epsilon \ (\mathrm{mod}\ 2)$ and $\sigma - \tau \not\equiv \epsilon \ (\mathrm{mod}\ 2)$.

10. SPECTRA OF OPERATORS FOR q-OSCILLATORS

The q-oscillator algebra is generated by the elements a^+, a^-, N satisfying the relations

$$[a^-, a^+]_q \equiv a^- a^+ - q a^+ a^- = q^{-N},$$

$$[N, a^+] = a^+, \quad [N, a^-] = -a^-.$$

Putting

$$a^- = q^{N/2} \alpha^-, \quad a^+ = \alpha^+ q^{N/2}$$

and replacing q^{-2} by q we obtain the associative algebra generated by the relations

$$[\alpha^-, \alpha^+] = q^N, \quad [N, \alpha^+] = \alpha^+, \quad [N, \alpha^-] = -\alpha^-.$$

In the same way, putting

$$a^+ = (1-q)^{-1} b^+ q^{-N/2}, \quad a^- = (1-q)^{-1} q^{-N/2} b^-$$

and replacing q by $q^{1/2}$ we obtain the associative algebra generated by the relations

$$[b^-, b^+]_q = 1 - q, \quad [N, b^+] = b^+, \quad [N, b^-] = -b^-.$$

The Fock representations T of these q-oscillator algebras act in the Hilbert space with the orthonormal basis $|n\rangle$, $n = 0, 1, 2, \cdots$. They are given by the formulas [29]

$$T(a^-)|n\rangle = [n]^{1/2}|n-1\rangle, \quad T(a^+)|n\rangle = [n+1]^{1/2}|n+1\rangle,$$
$$T(\alpha^-)|n\rangle = \{n\}^{1/2}|n-1\rangle, \quad T(\alpha^+)|n\rangle = \{n+1\}^{1/2}|n+1\rangle,$$
$$T(b^-)|n\rangle = (1-q^n)^{1/2}|n-1\rangle, \quad T(b^+)|n\rangle = (1-q^{n+1})^{1/2}|n+1\rangle,$$

where $[n]$ is a q-number and $\{n\} = (1 - q^n)/(1 - q)$. For all cases we have $T(N)|n\rangle = n|n\rangle$.

Here we consider spectral properties for the symmetric operators

$$Q_\alpha = T(\alpha^+) + T(\alpha^-), \quad Q_a = T(a^+) + T(a^-), \quad Q_b = T(b^+) + T(b^-) \tag{59}$$

corresponding to q-coordinate operators in the q-oscillator algebras. The q-momentum operators

$$P_\alpha = i(T(\alpha^+) - T(\alpha^-)), \quad P_a = i(T(a^+) - T(a^-)), \quad P_b = i(T(b^+) - T(b^-)) \tag{60}$$

(they are also symmetric) have the same spectral properties as the corresponding operators (59). Namely, it is easy to show that pairs of operators from (59) and (60) (or their closures) are simultaneously selfadjoint or not selfadjoint. They simultaneously have a selfadjoint extension or not.

For the operator Q_α we have

$$Q_\alpha|n\rangle = d_n|n+1\rangle + d_{n-1}|n-1\rangle, \quad d_n = \{n+1\}^{1/2}. \tag{61}$$

If $n \to \infty$ then $\{n\} \to -\infty$ if $q > 1$ and $\{n\} \to (1-q)^{-1}$ if $q < 1$. Therefore, the operator Q_α is bounded for $q < 1$ and unbounded for $q > 1$. In the case $q < 1$ it is selfadjoint.

Let us consider the operator Q_α for $q > 1$. According to the criterion of Section 3, if the coefficients d_n in (61) satisfy the conditions

$$d_n \geq 0, \quad d_{n-1}d_{n+1} \leq d_n^2, \quad \sum_{n=0}^{\infty} \frac{1}{d_n} < \infty, \tag{62}$$

then the closure $\bar{Q}_\alpha$ of Q_α is not a selfadjoint operator. Substituting $d_n = \{n+1\}^{1/2}$ into $d_{n-1}d_{n+1} \leq d_n^2$, we reduce this condition to the inequality $q + q^{-1} \geq 2$. This inequality is fulfilled for all real positive q and the equality is achieved at $q = 1$. Up to a multiplier, the series from formula (62) is majorized by the integral

$$\int_a^{\infty} \frac{dx}{(q^x - 1)^{1/2}},$$

which is convergent for $q > 1$. Thus, the operator $\bar{Q}_\alpha$ is not selfadjoint [6]. The deficiency indices of $\bar{Q}_\alpha$ are $(1,1)$. This means that $\bar{Q}_\alpha$ has a selfadjoint extension and deficiency subspaces are one-dimensional. Moreover, the deficiency subspaces $N_{\bar{z}}$, $\mathrm{Im}\, z \neq 0$, are defined by the generalized vectors $|z\rangle = \sum_n P_n(z)|n\rangle$ such that

$$\{n\}^{1/2}P_{n-1}(z) + \{n+1\}^{1/2}P_{n+1}(z) = zP_n(z) \tag{63}$$

and the initial conditions $P_{-1}(z) = 0$ and $P_0(z) \equiv 1$ are satisfied.

Since $q > 1$ then the solution of relation (63) with the given initial conditions is

$$P_n(z) = (q;q)_n^{-1/2} h_n\left(\tfrac{1}{2}(q-1)z|q\right), \tag{64}$$

where $h_n(w|q)$, $q > 1$, is the q-Hermite polynomial studied by Askey [30]. Really, substituting expression (64) for $P_n(z)$ into formula (63), after some transformations, we obtain the recurrence relation (1.7) from [30] for q-Hermite polynomials $h_n(w|q)$.

Thus, the operator $\bar{Q}_\alpha$ can be extended to be a selfadjoint operator defined on the subspace $D(\bar{Q}_\alpha) \oplus N_z \oplus N_{\bar{z}}$, where $D(\bar{Q}_\alpha)$ is the domain of definition of $\bar{Q}_\alpha$ and z is any fixed complex number. There are many selfadjoint extensions of the operator $\bar{Q}_\alpha$.

For real z, the polynomials $P_n(z)$ are the coefficients of transition from the basis $|n\rangle$, $n = 0, 1, 2, \cdots$, to the continual basis consisting of generalized eigenvectors $|x\rangle$ of the operator $\bar{Q}_\alpha$: $\bar{Q}_\alpha|x\rangle = x|x\rangle$. Namely, $|x\rangle = \sum_{n=0}^{\infty} P_n(x)|x\rangle$. The polynomials $P_n(x)$ satisfy the orthogonality relations

$$\int_{\infty}^{\infty} P_n\left(\frac{2\sinh u}{q-1}\right) P_m\left(\frac{2\sinh u}{q-1}\right) d\sigma(u) = \delta_{mn}, \tag{65}$$

where the measure $d\sigma(u)$ is of the form

$$d\sigma(u) = \frac{(-1)^n\, du}{(-q^{-1}e^{2u};q^{-1})_\infty(-q^{-1}e^{-2u};q^{-1})_\infty(q^{-1};q^{-1})_\infty \log q}.$$

This orthogonality defines the orthogonality relation for the generalized vectors $|x\rangle$.

Let now $q < 1$. In this case the operator Q_α is bounded and selfadjoint. The generalized vector $|x\rangle = \sum_{n=0}^{\infty} P_n(x)|n\rangle$ is an eigenvector for Q_α with an eigenvalue x, $x \in \mathbf{R}$, if $P_n(x)$ satisfies recurrence relation (63) and the initial condition $P_0(x) = 1$. Since $q < 1$ then the solution is the polynomial

$$P_n(x) = (q;q)_n^{-1/2} H_n\left(\tfrac{1}{2}(1-q)x|q\right), \tag{66}$$

where $H_n(x|q)$ is the continuous q-Hermite polynomials [31]. It follows from formula (6.6) in [31] that polynomials (66) satisfy the orthogonality relation

$$\int_{-2/(1-q)}^{2/(1-q)} P_n(x)P_m(x)d\sigma(x) = \delta_{mn},$$

where the measure is given by the formula

$$d\sigma(x) = \frac{(1-q)^2(q;q)}{4\pi((1-q)^2 - 4x^2)^{1/2}} \prod_{k=0}^{\infty} \left(1 - 2\left(\frac{8x^2}{(1-q)^2} - 1\right)q^k + q^{2k}\right) dx. \tag{67}$$

Therefore, the spectrum of the operator Q_α is the interval $(-2/(1-q), 2/(1-q))$ and the spectral measure is given by formula (67). The spectrum turns into the real line when $q \to 1$.

For the operator Q_a we have

$$Q_a|n\rangle = c_n|n+1\rangle + c_{n-1}|n-1\rangle, \qquad c_n = [n+1]^{1/2}.$$

Since $[n] \to \infty$ if $n \to \infty$ then Q_a is unbounded symmetric operator for $q > 1$ and for $q < 1$. Repeating the arguments of the first case, we conclude that the closure $\bar{Q}_a$ of Q_a is not a selfadjoint operator. Its deficiency indices are $(1,1)$. This means that $\bar{Q}_a$ has selfadjoint extensions and the deficiency subspaces N_z and $N_{\bar{z}}$ are one-dimensional. The subspace $N_{\bar{z}}$ is spanned by the vector $\sum_{n=0}^{\infty} p_n(z)|n\rangle$, where the components $p_n(z)$ satisfy the recurrence relation

$$zp_n(z) = [n]^{1/2}p_{n-1}(z) + [n+1]^{1/2}p_{n+1}(z)$$

and the initial conditions $p_{-1}(z) = 0$ and $p_0(z) \equiv 1$. The polynomials

$$p_n(z) = (q;q)_n^{-1/2} H_n^q\left(\tfrac{1}{2}(q - q^{-1})z\right)$$

are solutions of this recurrence relation, where $H_n^q(z)$ are q-Hermite polynomials defined in [32]. The orthogonality relation for these polynomials is not known.

Since for the operator Q_b we have

$$Q_b|n\rangle = f_n|n+1\rangle + f_{n-1}|n-1\rangle, \qquad f_n = (1 - q^{n+1})^{1/2},$$

then Q_b is a symmetric operator unbounded for $q > 1$ and bounded for $q < 1$. Repeating the arguments of the first case, we derive that the closure $\bar{Q}_b$ of Q_b is not a selfadjoint operator. It has the deficiency numbers $(1,1)$. Thus, the operator $\bar{Q}_b$ admits selfadjoint extensions and the deficiency subspaces N_z and $N_{\bar{z}}$ are one-dimensional. The subspace $N_{\bar{z}}$ is spanned by the vectors $\sum_n R_n(z)|n\rangle$, where polynomials $R_n(z)$ satisfy the recurrence relation

$$(1 - q^n)R_{n-1}(z) + (1 - q^{n+1})R_{n+1}(z) = zR_n(z)$$

with the initial conditions $R_{-1}(z) \equiv 0$ and $R_0(z) \equiv 1$. Therefore, we have

$$R_n(z) = (q;q)_n^{-1/2} h_n(z/2|q),$$

where $h_n(w|q)$, $q > 1$, are such as in (64). The orthogonality relation for polynomials $R_n(z)$ at real z is obtained from (65) replacing $(2\sinh u)/(q-1)$ by $2\sinh v$. If

$q < 1$ then the operator Q_b is bounded and selfadjoint. A generalized vector $|x\rangle = \sum_n R_n(x)|n\rangle$ is an eigenvector of Q_b with the eigenvalue x if

$$R_n(x) = (q;q)_n^{-1/2} H_n(x/2|q), \tag{68}$$

where $H_n(y|q)$ are such as in (66). The orthogonality relation for polynomials (68) is

$$\int_{-2}^{2} R_n(x)R_m(x)d\tau(x) = \delta_{mn},$$

where

$$d\tau(x) = \frac{(q;q)_\infty}{4\pi(1-4x^2)^{-1/2}} \prod_{k=0}^{\infty} \left(1 - 2(8x^2 - 1)q^k + q^{2k}\right). \tag{69}$$

Thus, in this case the spectrum of Q_b is the interval $(-2, 2)$ and the spectral measure is given by formula (69).

ACKNOWLEDGEMENTS

The research described in this publication was made possible in part by Grant No. U4J000 from the International Science Foundation.

REFERENCES

1. A. Barut, R. Raczka, *Theory of Group Representations and Applications* (PWN, Warszawa, 1977).
2. I. I. Kachurik, A. U. Klimyk, Preprint ITP–93–37E, Kiev (1993).
3. I. I. Kachurik, A. U. Klimyk, *Algebras, Groups and Geometries*, **11** (1994).
4. I. I. Kachurik, A. U. Klimyk, *Comm. Math. Phys.* (submitted for publication).
5. A. Hebecker *et al*, Preprint MPI–Ph/93–45, Munich (1993).
6. I. M. Burban, A. U. Klimyk, *Lett. Math. Phys.* **29**, 13 (1993).
7. I. M. Gel'fand, M. I. Graev, Izv. Akad. Nauk SSSR, Ser. Mat. **29**, 1329 (1965).
8. J. Mickelsson, J. Niederle, *Ann. Inst. H. Poincare*, **19**, 171 (1973).
9. N. Ja. Vilenkin, A. U. Klimyk, *Representation of Lie Groups and Special Functions* (Kluwer, Dordrecht, vol. 1, 1991; vol. 2, 1993), Chapters 7 and 10.
10. D. Basu, K. B. Wolf, *J. Math. Phys.* **23**, 189 (1982).
11. A. V. Rozemblyum, L. V. Rozemblyum, *Multivariate Special Functions in the Theory of Group Representations* (Universitetskoye, Minsk, 1992).
12. N. M. Atakishiev, S. K. Suslov, *J. Phys. A*, **18**, 1583 (1985).
13. G. Gasper, M. Rahman, *Basic Hypergeometric Functions* (Cambridge Univ. Press, Cambridge, 1991).
14. N. I. Ahiezer, I. M. Glazman, *The Theory of Linear Operators in Hilbert Spaces* (Ungar, New York, 1961).
15. Ju. M. Berezanskij, *Expansions in Eigenfunctions of Selfadjoint Operators* (Amer. Math. Soc., Providence, R. I., 1968).
16. F. G. Maksudov, B. P. Allakhverdiev, *Dokl. Russian Akad. Nauk*, **328**, 654 (1993).
17. I. M. Burban, A. U. Klimyk, *J. Phys. A*, **26**, 2139 (1993).
18. R. Askey, J. Wilson, *Memoirs of Amer. Math. Soc.* **54**, 1 (1985).
19. M. Noumi, K. Mimachi, H. Yamada, *Jap. J. Math.* **19**, 31 (1993).

20. V. G. Drinfeld, *Sov. Math. Dokl.* **32**, 254 (1985).

21. M. Jimbo, *Lett. Math. Phys.* **10**, 63 (1985).

22. A. M. Gavrilik, I. I. Kachurik, A. U. Klimyk, Preprint ITP–90–26E, Kiev (1990).

23. A. M. Gavrilik, A. U. Klimyk, *J. Math. Phys.* **35**, No 7 (1994).

24. M. Noumi, T. Umeda, M. Wakayama, *Lett. Math. Phys.* (in press).

25. M. Noumi, *Adv. in Math.* (in press).

26. D. Fairlie, *J. Phys. A*, **32**, L183 (1990).

27. A. M. Gavrilik, *Teoret. i Matem. Fizika*, **95**, 251 (1993).

28. I. I. Kachurik, A. U. Klimyk, *J. Phys. A* (submitted for publication).

29. E. V. Damaskinsky, P. P. Kulish, *Zap. Nauchn. Sem. LOMI*, **189**, 37 (1991).

30. R. Askey, in *q-Series and Partitions*, ed. D. Stanton (Springer, New York, 1989), pp. 151–158.

31. R. Askey, M. Ismail, in *Studies in Pure Mathematics*, ed. P. Erdös (Birkhäuser, Basel, 1983), pp. 55–58.

32. E. M. Damaskinsky, P. P. Kulish, *Zap. Nauchn. Sem. LOMI*, **199**, 81 (1992).

GEOMETRY OF AUTOMORPHISMS FOR FREE GROUPS

Peter Kramer[1]

[1]Institut für Theoretische Physik der Universität
D 72076 Tübingen, Germany

ABSTRACT

New generators, relations and subgroups are described for the group $Aut(F_3)$ of automorphisms of the free group F_3. A non-commutative crystallography is described with finite point groups, non-commutative translations, and corresponding space groups. A geometric setting is given in terms of affine transformations.

1 INTRODUCTION

The group of automorphisms $Aut(F_n)$ of the free group F_n was studied first by Nielsen [12, 11]. There are two important compatible homomorphisms which map the pair $(F_n, Aut(F_n))$ to the pair $(Z^n, Aut(Z^n) = Gl(n, Z))$:

$$\begin{aligned} hom_1: \quad & F_n & \to \quad & Z^n, \\ hom_2: \quad & Aut(F_3) & \to \quad & Gl(n, Z). \end{aligned} \qquad (1)$$

The first one is just the abelianization of F_n, the second one assigns to an automorphism, given as a map of the generators $< x_1, x_2 \ldots x_n >$, its substitution matrix which has entries from Z^{n^2} and, because of the invertibility of the automorphism, determinant $det = \pm 1$.

These general relations motivated the notion of a non-commutative (NC) crystallography based on $Aut(F_3)$, [5]. Various general and specific results and applications were given in [1, 2, 3, 6, 7]. The group $Aut(F_2)$ is analyzed in [8], and its interpretation in terms of NC crystallography in two dimensions is given in [10]. The group $Aut(F_3)$ has a much richer structure, it is analyzed in [9]. In the following sections we give an account of these results and interpret them algebraically as the basis for a NC crystallography in three dimensions. The algebraic results are then represented by geometric actions on R^3.

Symmetries in Science VII, Edited by
B. Gruber, Plenum Press, New York, 1995

2 NEW GENERATORS AND RELATIONS FOR
$Aut\,(F_3)$

In this section we give a new set of generators and relations for the group of automorphisms $Aut(F_3)$ of the free group $F_3 = <x_1, x_2, x_3>$ with three generators derived in [9].

1.Prop.: The group $Aut(F_3)$ is defined in terms of five generators and 20 relations Q_i given below by:

$$Aut(F_3) := <c_{12}, c_{23}, c_{34}, \sigma_1, c_2 | \; Q_1 \ldots Q_{20}> . \tag{2}$$

where the first 12 relations are

$$(c_{ii+1})^2 = e, \; i = 1, 2, 3, \; (Q_1, Q_2, Q_3), \tag{3}$$
$$(c_{12}c_{23})^3 = e, \; (c_{23}c_{34})^3 = e, \; (Q_4, Q_5),$$
$$c_{12} \rightleftharpoons c_{34}, \; (Q_6).$$
$$(\sigma_1)^2 = e, \; (Q_7)$$
$$(c_{13}\sigma_1)^4 = e, \; (Q_8)$$
$$c_{34} \rightleftharpoons \sigma_1, \; (Q_9)$$
$$\sigma_1 \rightleftharpoons c_{23}(\sigma_1 c_{13})^2 c_{23}, \; (Q_{10})$$
$$\sigma_1 \rightleftharpoons c_{14}c_{23}\sigma_1 c_{23}c_{14}, \; (Q_{11})$$
$$(c_2)^2 = e, \; (Q_{12}).$$

The next relations are given in terms of the elements

$$X_1 := c_2 c_3 c_{13} c_2, \; X_2 := c_2 c_{23}, \; c_3 := c_{12} c_2 c_{12}. \tag{4}$$

For the two elements X_1, X_2 we require the following transformations under conjugation $(g, X) \to X^g := gXg^{-1}$ with $<c_2, c_3, \sigma_1>$:

$$
\begin{aligned}
(X_1)^{c_2} &= X_1 X_2, & (X_2)^{c_2} &= (X_2)^{-1} & (Q_{13}, Q_{14}) \\
(X_1)^{c_3} &= (X_2)^{-1} & (X_2)^{c_3} &= (X_1)^{-1} & (Q_{15}, Q_{16}) \\
(X_1)^{\sigma_1} &= (X_1)^{-1}, & (X_2)^{\sigma_1} &= X_2, & (Q_{17}, Q_{18})
\end{aligned}
\tag{5}
$$

For the conjugation of the generator σ_1 with elements from the Coxeter group A_3 defined in Eq.(9) below we introduce the notation $\sigma_{212} := \sigma_1$. In view of the stability Q_9 of σ_{212} under c_{34} there are altogether 12 conjugates of σ_1 under A_3 which we denote by

$$i \neq j : \; \sigma_{iji} := (\sigma_{212})^{c_{j1}c_{i2}} \tag{6}$$

The remaining relations are

$$
\begin{aligned}
c_2 &\rightleftharpoons \sigma_{343} = (\sigma_1)^{c_{41}c_{32}}, \; (Q_{19}) \\
c_{23} &= c_2 \sigma_2 q, \; (Q_{20}) \\
q &:= c_{23} c_{14} c_2 c_{14} c_{23}, \\
\sigma_2 &:= c_3 \sigma_1 c_3.
\end{aligned}
\tag{7}
$$

2.Prop.: The generators of $Aut(F_3)$ act on the free group F_3 according to

$$
\begin{array}{llll}
\epsilon: & x_1 & x_2 & x_3 \\
c_{12}: & (x_1)^{-1} & x_1 x_2 & x_3 \\
c_{23}: & x_1 x_2 & (x_2)^{-1} & x_2 x_3 \\
c_{34}: & x_1 & x_2 x_3 & (x_3)^{-1} \\
\sigma_1: & (x_1)^{-1} & x_2 & x_3 \\
c_2: & x_1 x_2 & (x_2)^{-1} & x_3
\end{array}
\tag{8}
$$

3 SUBGROUPS OF *(Aut F₃)*

The first set of 6 relations determines a Coxeter group [4]

$$A_3 := < c_{12}, c_{23}, c_{34} | Q_1 \ldots Q_6 > \tag{9}$$

A second set of relations derived from $Q_1 \ldots Q_9$ determines a Coxeter group

$$\begin{aligned}
B_3 &:= \quad < c_{34}, c_{13} = c_{12}c_{23}c_{12}, \sigma_1 >, \\
(c_{34})^2 &= \quad (c_{13})^2 = (\sigma_1)^2 = (c_{34}c_{13})^3 = (c_{13}\sigma_1)^4 = e.
\end{aligned} \tag{10}$$

These finite groups will serve as point groups in the application to crystallography.

3.Def.: The group $\Psi_4 < Aut(F_3)$ is the subgroup of $Aut(F_3)$ given by

$$\Psi_4 := < c_{12}, c_{23}, c_{34}, \sigma_1 | Q_1 \ldots Q_{11} > \tag{11}$$

This subgroup has only the first four generators of $Aut(F_3)$ and all the relations between them.

4.Prop.:In the group Ψ_4, the elements

$$S_1 := \sigma_{121}\sigma_{131}\sigma_{141}, \ S_2 := \sigma_{212}\sigma_{232}\sigma_{242}, \ S_3 := \sigma_{313}\sigma_{323}\sigma_{343}, \ S_4 := \sigma_{414}\sigma_{424}\sigma_{434}, \tag{12}$$

are involutive and generate a normal subgroup $\mathcal{X}_4$. The conjugation properties of the elements Eq.(12) under $p \in A_3$ and σ_1 are

$$\begin{aligned}
(S_i)^p &= \quad S_{p(i)}, \tag{13} \\
(S_1)^{\sigma_1} &= \quad S_2 S_1 S_2, \ (S_2)^{\sigma_1} = S_2, \ (S_3)^{\sigma_1} = S_3, (S_4)^{\sigma_1} = S_4 \tag{14}
\end{aligned}$$

In terms of this group we pass to

5.Def.: The subgroup $\mathcal{T}$ is defined by

$$\begin{aligned}
\mathcal{T} &:= \quad < T_1, T_2, T_3 >, \tag{15} \\
T_1 &= \quad S_1 S_2, \ T_2 = S_2 S_3, \ T_3 = S_3 S_4.
\end{aligned}$$

Written in terms of the involutive generators of $\mathcal{X}_4$, $\mathcal{T}$ is the normal subgroup $Inn(F_3) \lhd Aut(F_3)$ of inner automorphisms in Φ_3. This group in turn is isomorphic to F_3 itself. Under conjugation with the generators of $Aut(F_3)$, the transformation laws are

$$\begin{array}{llll}
T_i : & T_1 & T_2 & T_3 \\
(T_i)^{c_{12}} : & (T_1)^{-1} & T_1 T_2 & T_3 \\
(T_i)^{c_{23}} : & T_1 T_2 & (T_2)^{-1} & T_2 T_3 \\
(T_i)^{c_{34}} : & T_1 & T_2 T_3 & (T_3)^{-1} \\
(T_i)^{\sigma_1} : & (T_1)^{-1} & T_2 & T_3 \\
(T_i)^{c_2} : & T_1 T_2 & (T_2)^{-1} & T_3
\end{array} \tag{16}$$

These conjugation properties assure the normal property of $\mathcal{T}$. Moreover the generators $< T_1, T_2, T_3 >$ are the images of $< x_1, x_2, x_3 >$ under the isomorphism. The transformations Eq.(16) correspond exactly to the action Eq.(8).

The occurrence of the group F_3 as a normal subgroup of $Aut(F_3)$ extends from $n = 3$ to arbitrary n and will be used to define non-commutative translations in section 5.

4 POINT GROUPS

The finite Coxeter groups in $Aut(F_3)$ determined in section 3 can serve as *point groups* in a frame for NC crystallography in three dimensions. For comparison we put them into standard crystallographic notation:

$$
\begin{aligned}
A_3 &= T_d = \overline{4}3m, \\
B_3 &= O_h = m3m.
\end{aligned}
\tag{17}
$$

The relation between these groups must be taken from their embedding in $Aut(F_3)$: Their intersection is $A_3 \cap B_3 = B_2$, and their union $A_3 \cup B_3$ generates the infinite group Ψ_4. For the present purpose we restrict the discussion to these finite subgroups.

5 NON-COMMUTATIVE TRANSLATIONS

We have seen in section 3 that there is a normal subgroup $\mathcal{T} := < T_1, T_2, T_3 >$ of $Aut(F_3)$, isomorphic to F_3 and with the isomorphism $T_i \sim x_i$, $i = 1, 2, 3$. In view of the abelianization $hom_1 : F_3 \to Z^3$ it is justified to call T the *non-commutative* (NC) *translation group*. The non-commutative property can be seen in the following way: Consider an element $h = f(x_1, x_2, x_3) \in F_3$ of the kernel $ker(hom_1) \lhd F_3$. Its isomorphic image in $\mathcal{T}$ is $H = f(T_1, T_2, T_3)$. All the different NC translations isomorphic to elements of $ker(hom_1)$ collapse under abelianization into the identity of the commutative translation group.

Note that these non-commutative translations form a subgroup of $Aut(F_3)$, but belong to the kernel $ker(hom_2)$ and hence have no counterpart within $Gl(3, Z)$. This will allow us to introduce the notion of NC space groups in terms of subgroups of $Aut(F_3)$.

6 NON-COMMUTATIVE SPACE GROUPS

Since $\mathcal{T}$ is normal in $Aut(F_3)$, we can look for subgroups which form semidirect products. In particular we are interested in NC counterparts of symmorphic space groups.
6.Prop.: The group $Aut(F_3)$ admits as subgroups the two semidirect products

$$
\mathcal{T} \times_s A_3, \ \mathcal{T} \times_s B_3,
\tag{18}
$$

which we call NC *symmorphic space groups*.

Proof: The first factor in these products is normal since $\mathcal{T}$ is normal in $Aut(F_3)$. The intersection between the factors reduces to the identity. Then it is easy to verify that the conditions for a semidirect product formed from two subgroups of a group are fulfilled $\square$.

Other NC symmorphic subgroups can be found by restricting A_3, B_3 to one of their subgroups.

The NC space groups differ in several respects from the space groups of standard crystallography. From space group elements in standard crystallography we can obtain new finite order elements with fix-points which are not points of the lattice. This construction in general does not extend to NC space groups.

7 THE AFFINE GEOMETRIC REPRESENTATION

We shall discuss now an algorithm which yields a representation of the subgroup $\Psi_4 < Aut(F_3)$ by linear affine transformations. This infinite subgroup of $Aut(F_3)$ encompasses already the finite point groups A_3, B_3 and the NC translation group.

We introduce [10] the following *graph algorithm* for $Aut(F_3)$:

$$
\begin{aligned}
x_1, x_2 & & & : & \textit{directed straight path } x_1, x_2, \\
x_i & \rightarrow & x_i^{-1} & : & \textit{invert direction of } x_i, \\
x_i, x_j & \rightarrow & x_i x_j & : & \textit{concatenate path } x_i, \textit{ path } x_j \textit{ from right to left}, \\
x_i x_i^m & = & x_i^m x_i & : & \textit{allow for discrete shifts of } x_i \textit{ on its line}.
\end{aligned}
\tag{19}
$$

We take three linear independent vectors in R^3 for $< x_1, x_2, x_3 >$ and arrange the corresponding three lines as follows: We define a tetrahedron or simplex in R^3 with vertices $4, 3, 2, 1$ by the prescription: The directed path x_i points from vertex $i+1$ to vertex i for $i = 3, 2, 1$. It follows that the concatenated path $x_1 x_2 x_3$ forms a *Hamilton line* on the simplex, i.e. a connected sequence of edges which passes once through each vertex. Next we give rules for the representation of the generators of Ψ_4 by affine reflections. An affine reflection $(\phi, \mathbf{r})$ in R^3 is a map which preserves a fixed plane and reverses all vector components parallel to a fixed vector $\mathbf{r}$, not in the plane. We specify affine reflections as follows:

(a) The action corresponding to the transposition $c_{i,i+1}$ preserves the plane through the midpoint of the edge $(i+1, i)$ and the opposite edge. The vector is $\mathbf{r} = x_i$ for $i = 1, 2, 3$.

(b) The action corresponding to the generator σ_1 preserves the plane with vertices $2, 3, 4$. The vector is $\mathbf{r} = x_1$.

These generating actions are shown in Fig.1.

Note that these affine reflections act on the simplex, but also globally on R^3. Any sequence of these operations maps the original simplex into an image, so that the image of the Hamilton line can be constructed. It can now be verified that the actions for the generators are in line with the algebraic expressions of Eq.(8). Whereas the transpositions map the simplex into itself, the generator σ_1 propagates the simplex into an image which shares only one face with its preimage. With this affine representation, we can now give a geometric description of the various subgroups of Ψ_4, considered so far only in algebraic terms:

(1) The images of the vertices of the simplex are on a lattice $\frac{1}{2}\Gamma$. The points of this lattice are on the three lines determined by the vectors $< x_1, x_2, x_3 >$ with a spacing corresponding to twice the length of these vectors, and on all their parallels obtained by discrete shifts along the two other not collinear vectors.

(2) The two Coxeter groups each have a fixed point: A_3 leaves fixed the midpoint, B_3 leaves fixed the vertex 2 of the simplex.

(3) The subgroup $\mathcal{X}_4$ is generated by the 4 reflections S_i of the simplex in its vertices. The action corresponding to S_4 is shown in Fig.2.

(4) The normal subgroup $\mathcal{T}$, which algebraically corresponds to $Inn(F_3)$, is generated by translations T_i of the simplex along three of its edges, by a distance corresponding to twice the length of x_i, and in the direction opposite to this vector. The vertex positions obtained in this way form the lattice Γ. Each of the basic translations is a product of two different reflections from $\mathcal{X}_4$.

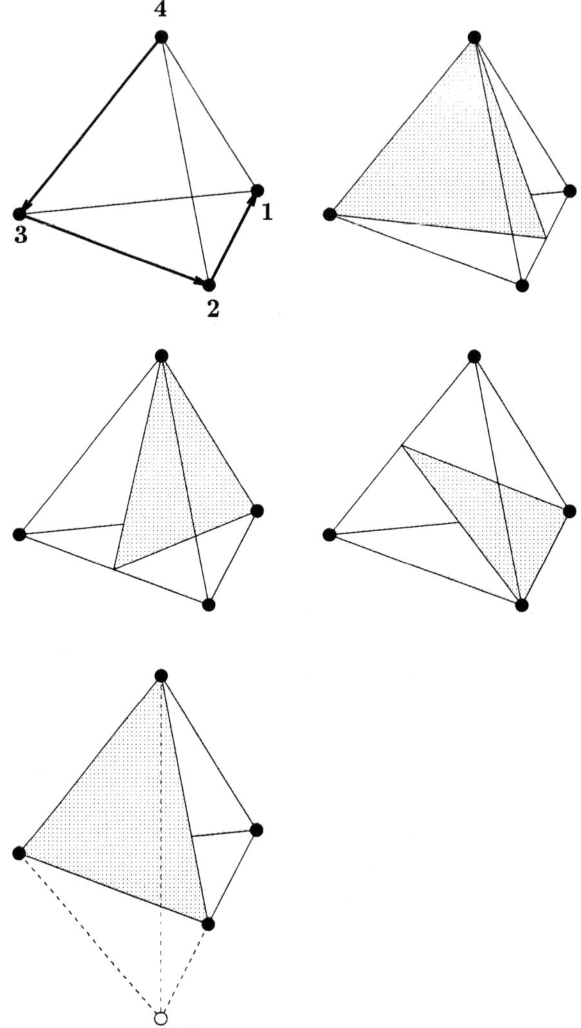

Figure 1. The simplex with vertices $1, 2, 3, 4$ is given by the Hamilton line formed from the three vectors corresponding to $< x_1, x_2, x_3 >$. The generators $< c_{12}, c_{23}, c_{34} >$ correspond to three affine mirror planes passing through an edge and through the midpoint of the opposite edge. The generator σ_1 corresponds to an affine mirror plane passing through a face of the simplex and an affine reflection of the vertex 1.

The general elements of $\mathcal{T}$ are in one-to-one correspondence to the elements of F_3 and can be associated with a unique path on the lattice Γ.

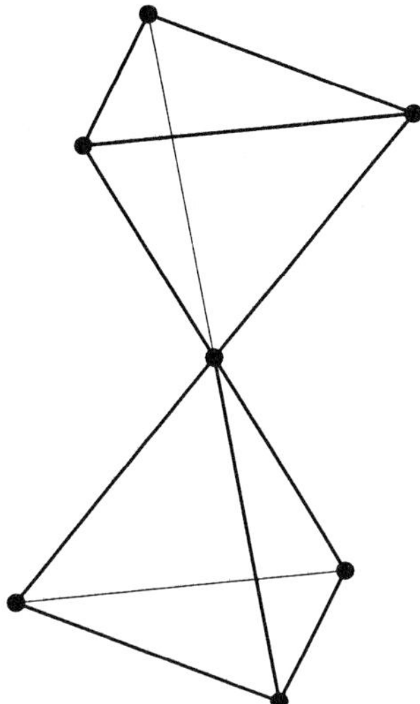

Figure 2. The element S_4 reflects the simplex in its vertex 4.

In this affine geometry, the action of the non-commutative translation groups must be interpreted as follows: At any lattice position, there are infinitely many different copies of a given "motive" on top of one another. Their number is given by the cardinality of the group $ker(hom_1)$.

The NC crystallography provides a large frame for many possible applications. In particular the hope is to find in the group $Aut(F_n)$ classes of automorphisms which select the subsets of points appropriate for the description of quasicrystals, compare [10, 2].

REFERENCES

[1] J Garcia-Escudero and P Kramer 1993, Anales de Fisica, Monografias **1**, vol.1, 339-42, Madrid

[2] J Garcia-Escudero and P Kramer 1993, J Phys **A26** L1029-35

[3] J Garcia-Escudero and P Kramer 1993, Proc. Int. Wigner Symposium, Oxford

[4] J E Humphreys 1990, *Reflection Groups and Coxeter Groups*, Cambridge University Press

[5] P Kramer 1993, Anales de Fisica, Monografias **1**, vol.2, 370-3, Madrid

[6] P Kramer 1993, J Phys **A26** 213-228

[7] P Kramer 1993, J Phys Lett **A26** L245-L250

[8] P Kramer 1994, J Phys **A27** 2011-22

[9] P Kramer 1994,
Generators and subgroups for $Aut(F_3)$,
submitted for publication

[10] P Kramer 1994,
Non-commutative geometry for quasicrystals,
submitted for publication

[11] W Magnus, A Karras and D Solitar 1976, *Combinatorial Group Theory*, Dover,
New York

[12] J Nielsen 1924, Math. Ann. **91** 169-209

QUANTUM CENTRAL LIMIT THEOREMS

Romuald Lenczewski

Hugo Steinhaus Center for Stochastic Methods
Instytut Matematyki, Politechnika Wroclawska
50-370 Wroclaw, Poland
e-mail lenczew@graf.im.pwr.wroc.pl

1. INTRODUCTION

This work is a review of certain quantum limit theorems that may be viewed as non-commutative versions of the classical central limit theorem. Since various kinds of independences can be introduced in quantum probability, there are many quantum versions of this fundamental result in classical probability. We concentrate here only on various algebraic approaches and even within this scope we do not give a complete survey of the vast literature on the subject.

Quantum stochastic processes are usually understood in the sense of [AFL], where the role of random variables is played by observables or their complex linear combinations and measure is replaced by a state, i.e. a normalized positive functional. An algebraic setting is often more general, for example the object of study may be a coalgebra and instead of states arbitrary functionals may be considered. It is often assumed that there is an involution defined on the algebra (thus making it into a *-algebra), or even that the *-algebra is a C^*-algebra.

The aim of the quantum analogs of the central limit theorem that we shall present is to study the limits when $N \to \infty$ of the (quantum) moments of collective variables

$$\frac{X_{N1} + \ldots + X_{Nk_N}}{\alpha_N}$$

for an infinite triangular array X_{ij} of operators from some associative algebras such that $X_{N1}, \ldots, X_{Nk_N}$ are for fixed N independent in some sense (with respect to the functionals ϕ^N) and $\alpha_N = \sqrt{N}$ (most cases), or $\alpha_N = \sqrt{[N]}$ (quantum groups). In the qclt's we study one can take one functional $\hat{\phi}$ by an appropriate embedding.

The investigations concerning quantum central limit theorems (qclt), i.e. quantum analogs of the classical central limit theorem with classical independent random variables replaced by independent (in some sense) operators started with the works of Cushen and Hudson [CuH] and Giri and von Waldenfels [GvW] for commuting observables and Hudson [Hud] and von Waldenfels [vWa] for anticommuting observables.

Since we want to concentrate on the algebraic approach, the first two types of qclt's presented (Section 2) will be the commuting and anticommuting independences as studied in [GvW] and [vWa]. The independent variables take here the standard second quantization form. As far as the combinatorics is concerned, all pair-partitions survive in the limit. A realization of the limit can be given as the CCR algebra and the CAR algebra, respectively.

Another kind of independence is the so-called free independence introduced by Voiculescu [Voi1, Voi2] related to the reduced free product of C^*-algebras. It has a different factorization law for moments of collective variables. This leads to a different quantum limit theorem (see [Spe1,Spe2]) in which only the so-called non-crossing (or admissible, in the terminology of [Spe1,Spe2]) pair-partitions survive in the limit. Note that in the tensor product case, where the tensor product (and not the reduced free product) separates independent operators, all pair-partitions appear in the limit. An outline of the free version of qclt is given in Section 3.

In Section 4 the generalization of the approach to coalgebras is given (see [Sch1, Sch2, Sch3]). Here, the succesive iterations of the coproduct represent addition of independent quantum variables. The limit functional can be represented in the convolution exponential form. In particular, if the coalgebra is the tensor algebra with the symmetric or antisymmetric product and the coproduct takes the standard second quantization form, we obtain commuting or anticommuting independence, respectively. In [Sch2] the theorem has been applied to left and right q-bialgebras (see also [Sch3]) and realizations of the limit states in terms of Azema quantum martingales (for the left q-bialgebras) and the Boson-Fermion interpolations (for the right q-bialgebra) were given.

Other approaches to twisted versions of the qclt and twisted white noise were also proposed ([BSp1, BSp2, LiP, SpB]). In Section 5 we concentrate on the qclt's given in [BSp1] and [SpB].

In [LPo, Len1, Len2, Len3] we studied a q-analog of the qclt for quantum groups with scaling different from the standard one. From the probabilistic point of view, in the qclt we study the independent variables as they appear in quantum groups (summands in the coproduct) are not identically distributed. Independent observables represented by quantum groups have to be added to give observables of the same kind (with preserved symmetry), for example the sum of two twisted spins has to represent a spin, etc. For states like the vacuum state that forces a different scaling of the sums, namely the classical $\sqrt{N}$ is replaced by its q-analog, namely $\sqrt{[N]}$. That leads to a different analog of the quantum central limit theorem which we call its q-analog. The combinatorics of this approach is different in the sense that all (ordered) partitions survive in the limit. The result is more general and can be extended to free *-algebras [Len4]. Then the GNS representation of the limit state can be given in terms of generalized creation and annihilation operators. This q-analog of qclt is summarized in Section 6.

We do not discuss the analytical aspects of qclt's as discussed, for instance for the Boson case, in [GVV].

2. BOSON AND FERMION INDEPENDENCE

Consider a *-algebra $\mathcal{C}$ generated by $\mathcal{V}_+$ and let $\mathcal{V}_- = \mathcal{V}_+^*$, $\mathcal{V} = \mathcal{V}_+ \cup \mathcal{V}_-$. Let $\phi \in \mathcal{C}^*$ be a normalized functional, i.e. $\phi(1) = 1$. For each $N \in \mathbf{N}$, $1 \leq j \leq N$ and each $v \in \mathcal{V}$ we define a canonical injection $j_{i,N} : \mathcal{C} \to \mathcal{C}^{\otimes N}$ by

$$j_{i,N}(v) = 1^{\otimes(i-1)} \otimes v \otimes 1^{\otimes(N-i)}.$$

Thus, $j_{i,N}$ are the canonical injections of $\mathcal{C}$ into $\mathcal{C}^{\otimes N}$ that form an infinite triangular array. From the quantum probabilistic point of view, for given $N \in \mathbf{N}$, $v \in \mathcal{V}$ and $i \neq k$, $j_{i,N}(v)$ and $j_{k,N}(v)$ represent independent quantum variables. They commute, namely

$$j_{i,N}(v)j_{k,N}(w) = j_{k,N}(w)j_{i,N}(v)$$

where $v, w \in \mathcal{V}$ if only $i \neq k$. The sum S_N of such independent variables will be given by

$$S_N(v) = \sum_{i=1}^{N} j_{i,N}(v)$$

which can be also viewed as a collective (second quantization) variable associated with variable v.

It is easy to show that

$$\phi^{\otimes N}(S_N(v_1)\ldots S_N(v_p)) = \sum_{r=1}^{p} \binom{N}{r} \sum_{S \in \mathcal{P}_r^{ord}\{1,\ldots,p\}} \phi(v_{S_1})\ldots\phi(v_{S_r})$$

$$= \sum_{r=1}^{p} A_{r,N} \sum_{S \in \mathcal{P}_r\{1,\ldots,p\}} \phi(v_{S_1})\ldots\phi(v_{S_r})$$

where $v_1,\ldots,v_p \in \mathcal{V}$ and $A_{r,N} = N(N-1)\ldots(N-r+1)$. Further, we use the following notation concerning partitions: $\mathcal{P}_r^{ord}\{1,\ldots p\}$ denotes the set of all *ordered* partitions of $\{1,\ldots,p\}$ into r disjoint subsets, i.e. tuples $(S_1,\ldots,S_r)$, whereas $\mathcal{P}_r\{1,\ldots,p\}$ stands for the set of all partitions in the usual sense, i.e. collections of disjoint subsets $\{S_1,\ldots,S_r\}$. Thus, for each partition $S = \{S_1,\ldots,S_r\}$ there are $r!$ ordered partitions associated with it. We also adopt here and in the sequel the following notation: for any given partition $S = \{S_1,\ldots,S_r\} \in \mathcal{P}_r\{1,\ldots,p\}$ we denote $v_{S_j} = \prod_{i \in S_j} v_i$ where the product is taken in the ascending order of indices.

The qclt for this kind of independence was formulated and proved by Giri and von Waldenfels [GvW]. The method of moments presented in [GvW] consists in evaluating

$$\lim_{N \to \infty} \phi^{\otimes N}(S_N(v_1^N)\ldots S_N(v_p^N))$$

where $v_i^N = 1/\sqrt{N}v_i$ and $v_i \in \mathcal{V}$, $i = 1,\ldots,p$. In order to formulate the qclt of [GvW], let us define the functional obtained in the limit. Thus a *Gaussian functional* Φ_G on $\mathcal{C}$ is the normalized functional given by

$$\Phi_G(v_1\ldots v_p) = \begin{cases} \sum_{S \in \mathcal{P}_{pair}\{1,\ldots,2k\}} \phi(v_{S_1})\ldots\phi(v_{S_k}) & \text{if } p = 2k \\ 0 & \text{if } p \text{ odd} \end{cases}$$

where $\mathcal{P}_{pair}\{1,\ldots,2k\}$ denotes the set of all pair-partitions of the set $\{1,\ldots,2k\}$.

We are now able to formulate the following **Boson-type qclt**.

Theorem 2.1. *Let $v_1,\ldots,v_p \in \mathcal{V}$ and assume that $\phi \in \mathcal{C}^*$ has vanishing first moments on $\mathcal{V}$, i.e. $\phi(v) = 0$ for all $v \in \mathcal{V}$. Then*

$$\lim_{N \to \infty} \phi^{\otimes N}(S_N(v_1^N)\ldots S_N(v_p^N)) = \Phi_G(v_1\ldots v_p)$$

where Φ_G denotes the Gaussian functional.

The proof is of combinatorial nature and follows directly from the formulas given above.

It rests on the fact that partitions which are not pair-partitions do not contribute to the limit since they either have a singleton, i.e. S_i consists of one element only for some i (but then $\phi(v_{S_i}) = 0$), or they have fewer than k sets, say r, where $r < k$, but then

$$\lim_{N \to \infty} \frac{A_{r,N}}{N^{p/2}} = 0$$

and that finishes the proof. A sample calculation is given below.

$$\lim_{N \to \infty} \phi^{\otimes N}(v_1^N v_2^N v_3^N v_4^N)$$

$$= \lim_{N \to \infty} \left(\frac{1}{N} \phi(v_1 v_2 v_3 v_4) + \frac{N-1}{N} (\phi(v_1 v_2)\phi(v_3 v_4) + \phi(v_1 v_3)\phi(v_2 v_4) + \phi(v_1 v_4)\phi(v_2 v_3)) \right)$$

$$= \phi(v_1 v_2)\phi(v_3 v_4) + \phi(v_1 v_3)\phi(v_2 v_4) + \phi(v_1 v_4)\phi(v_2 v_3)$$

The structure of the Gaussian functional gives more information on the limit algebra, namely gives a connection between the commuting independence and Boson symmetry. The easiest way to see it is probably by finding a Fock-space realization of the Gaussian functional for ϕ a state on $\mathcal{C}$. By GNS construction we get (π, ω), a *-representation on a Hilbert space $\mathcal{H}$. Take the symmetric (Boson) Fock space

$$\Gamma_s(\mathcal{H}) = \bigoplus_{n=0}^{\infty} \hat{S}(\mathcal{H}^{\otimes n})$$

with vacuum Ω and the canonical creation and annihilation operators, namely

$$a^*(f)\hat{S}(h_1 \otimes \ldots \otimes h_n) = \hat{S}(f \otimes h_1 \otimes \ldots \otimes h_n)$$

$$a(f)\hat{S}(h_1 \otimes \ldots \otimes h_n) = \sum_{k=1}^{n} < f, h_k > \hat{S}(h_1 \otimes \ldots \otimes h_{k-1} \otimes h_{k+1} \otimes \ldots \otimes h_n)$$

for $f, h_1, \ldots, h_n \in \mathcal{H}$, where $\hat{S}$ denotes the standard symmetrizer. The creation and annihilation operators satisfy the canonical commutation relations (CCR)[HuP]

$$[a(f), a^*(g)] = a(f)a^*(g) - a^*(g)a(f) = < f, g >$$

and one can show that they give the same combinatorics as the one of the Gaussian state. For convenience, take hermitian elements $v_1, \ldots, v_p \in \mathcal{V}_+ \oplus \mathcal{V}_-$. Then each moment of the limit functional can be represented on the Fock space as follows

$$\Phi_G(v_1 \ldots v_p) = < \Omega, c(f_1) \ldots c(f_p)\Omega >$$

where $c(f_i) = a(f_i) + a^*(f_i)$ and $f_i = \pi(v_i)\omega$, since $\phi(v_i v_j) = < f_i, f_j >$. Thus, one can conclude that if ϕ is a state, in the commuting independence case the limit state is a quasi-free Boson state. In particular, $\mathcal{C}$ may be a free *-algebra. Hence, one can conclude that (Boson) symmetry arises as a consequence of the limit theorem, not any algebraic structure on $\mathcal{C}$. For details on this approach (formulated for associative algebras) see [GvW].

A similar approach was applied by von Waldenfels to anticommuting (Fermion) independence [vWa], or even more generally, to the $\mathbf{Z}_2$-graded (supersymmetric) case. Let $\mathcal{V} = \mathcal{V}_0 \cup \mathcal{V}_1$, where $\mathcal{V}_0$ represents even generators and $\mathcal{V}_1$ odd generators. That defines the canonical $\mathbf{Z}_2$ gradation on $\mathcal{C} = \mathcal{C}_0 \oplus \mathcal{C}_1$ and we denote by $d(v)$ the corresponding degree of any homogenous $v \in \mathcal{C}$. Consider now $\mathcal{C}^{\hat{\otimes} N}$ with a graded product given by

$$(v_1 \otimes \ldots \otimes v_N)(w_1 \otimes \ldots \otimes w_N) = (-1)^{\sum_{i<j} d(v_j)d(b_i)} v_1 w_1 \otimes \ldots \otimes v_N w_N$$

and in this case

$$j_{i,N}(v)j_{k,N}(w) = (-1)^{d(v)d(w)}j_{k,N}(w)j_{i,N}(v)$$

where $i \neq k, N \in \mathbf{N}$ and v, w are homogenous. In particular, if $v, w \in \mathcal{V}_1$, then the injections anticommute. Again, it is easy to show that if $v_1, \ldots, v_p$ are odd generators, then

$$\phi^{\otimes N}(S_N(v_1)\ldots S_N(v_p)) = \sum_{r=1}^{p} \binom{N}{r} \sum_{S \in \mathcal{P}_r^{ord}\{1,\ldots,p\}} (-1)^{\#W_S} \phi(v_{S_1})\ldots\phi(v_{S_r})$$

where W_S is the set of inversions in S and $\#$ indicates cardinality. By an inversion of an ordered partition $S = (S_1, \ldots, S_r) \in \mathcal{P}_r^{ord}\{1, \ldots, p\}$ we understand any pair (i, j), $i \in S_{k_i}, j \in S_{k_j}$, such that $i < j$ and $k_i > k_j$. In turn, by the set of inversions of a pair-partition $S = \{S_1, \ldots, S_k\} \in \mathcal{P}_{pair}\{1, \ldots, 2k\}$, where $S_i = (e_i, z_i)$ we shall understand the set $I_S = \{(i,j)|e_i < e_j < z_i < z_j\}$.

The limit functional in the anticommuting case (when all generators ore odd) is the *antisymmetric Gaussian functional* Φ_{AG} defined as a normalized functional such that

$$\Phi_{AG}(v_1 \ldots v_p) = \begin{cases} \sum_{S \in \mathcal{P}_{pair}\{1,\ldots,2k\}} (-1)^{\#I_S} \phi(v_{S_1})\ldots\phi(v_{S_k}) & \text{if } p = 2k \\ 0 & \text{if } p \text{ odd} \end{cases}$$

Then the following **Fermion-type qclt** holds (proof is similar, for details see [vWa]).

Theorem 2.2. *Assume that graded product is taken on $\mathcal{C}^{\hat{\otimes}N}$ and let $v_1, \ldots, v_p$ be odd generators. Further, assume that first moments of ϕ vanish. Then*

$$\lim_{N \to \infty} \phi^{\otimes N}(S_N(v_1^N)\ldots S_N(v_p^N)) = \Phi_{AG}(v_1 \ldots v_p)$$

where $v_i^N = 1/\sqrt{N}v_i, i \in \{1, \ldots, p\}$.

A sample calculation gives

$$\lim_{N \to \infty}(S_N(v_1^N)S_N(v_2^N)S_N(v_3^N)S_N(v_4^N)) = \phi(v_1v_2)\phi(v_3v_4) - \phi(v_1v_3)\phi(v_2v_4) + \phi(v_1v_4)\phi(v_2v_3).$$

Now, to find a realization of the limit state (if ϕ is a state on $\mathcal{C}$) on a Fock-type space, one needs to take the antisymmetric Fock space $\Gamma_a(\mathcal{H})$, where $\mathcal{H}$ is the GNS representation Hilbert space for $(\mathcal{C}, \phi)$. The situation is completely analogous to the Boson case, except that instead of the symmetrizer $\hat{S}$ we have the antisymmetrizer $\hat{A}$ and the creation and annihilation operators satisfy the canonical anticommutation relations (CAR). The limit states are called quasi-free Fermion states. The theorem can be extended to all generators, not necessarily only the odd ones (supersymmetric case).

Another approach was presented in [CuH, Hud, PQu, Qua], where the starting point is a CCR- or CAR algebra.

3. FREE INDEPENDENCE

Instead of the tensor or graded tensor product that is used in the case of bosons or fermions, Voiculescu [Voi1,Voi2] proposed the so-called *reduced free product* of C^* algebras. For the purposes of the central limit theorem we define the reduced free product of the same algebra. Thus, let J be an index set and for all $i \in J$ let $\mathcal{C}_i$ be a copy of a C^* algebra $\mathcal{C}$ and let ϕ be a state on $\mathcal{C}$. Furthermore, let a C^* algebra $\hat{\mathcal{C}}$ with unity and a state $\hat{\phi}$ on $\hat{\mathcal{C}}$ be given. Then $(\hat{\mathcal{C}}, \hat{\phi})$ is a reduced free product of $(\mathcal{C}_i, \phi)$ if:

1. there exist unital *-homomorphisms $j_i : \mathcal{C} \to \hat{\mathcal{C}}$ such that $\hat{\mathcal{C}}$ is generated by all $j_i(\mathcal{C})$,

2. $\hat{\phi} \circ j_i = \phi_i$ for all $i \in J$,

3. for $p \in \mathbf{N}$, $k_i \in J$ with $k_1 \neq k_2 \neq \ldots \neq k_m$ (consecutive indices are distinct) and $v_i \in \mathcal{C}$ and $\phi(v_i) = 0$ we have

$$\hat{\phi}(j_{k_1}(v_1) \ldots j_{k_p}(v_p)) = 0$$

4. the GNS construction applied to $(\hat{\mathcal{C}}, \hat{\phi})$ yields a faithful representation of $\hat{\mathcal{C}}$.

Condition (3) is the condition of *free independence* and it allows us to calculate all moments of $j_{k_1}(v_1), \ldots, j_{k_p}(v_p)$ with respect to $\hat{\phi}$ if all moments of $v_1, \ldots, v_p$ with respect to ϕ are known.

The combinatorics of the limit distribution (functional) is different from the tensor case. The so-called non-crossing (or, admissible) partitions are needed. Thus, the partition $S = \{S_1, \ldots, S_r\} \in \mathcal{P}_r\{1, \ldots, p\}$ is *non-crossing* if at least one of the sets S_i is a segment of $\{1, \ldots, p\}$, i.e. has the form $S_i = (k, k+1, \ldots, k+m)$ and $\{S_1, \ldots, S_{i-1}, S_{i+1}, \ldots, S_r\}$ is a non-crossing partition of $\{1, \ldots, p\} - S_i$. Otherwise S is called *crossing* (or, non-admissible). If we build bridges by connecting in the sequence $1, 2, \ldots, p$ the numbers belonging to the same S_i, then the partition is non-crossing if it is possible to build the corresponding bridge in such a way that the lines do not cross. The set of all non-crossing partitions of $\{1, \ldots, p\}$ consisting of r sets will be denoted by $\mathcal{P}_r^{nc}\{1, \ldots, p\}$ and the set of non-crossing pair-partitions will be $\mathcal{P}_{pair}^{nc}\{1, \ldots, p\}$.

When calculating moments, contributions from the crossing and non-crossing partitions take different forms. Namely, only for non-crossing partitions the moment $\hat{\phi}(j_1(v_1) \ldots j_p(v_p))$ factorizes, i.e.

$$\hat{\phi}(j_1(v_1) \ldots j_p(v_p)) = \phi(v_{S_1}) \ldots \phi(v_{S_p})$$

where S is the partition in which to the same S_i belong all indices of $v_1, \ldots, v_p$ that are injected in the same way, i.e. $k, m \in S_i$ iff $j_k = j_m$ (see [Spe1]). If S is a crossing partition, then $\hat{\phi}$ can be expressed as a sum of products of moments, where each summand contains at least $r + 1$ factors (see [Spe1]). The following two examples illustrate the difference.

First, let $v_1, v_2, v_3 \in \mathcal{C}$. For simplicity denote $V_1 = j_1(v_1), V_2 = j_2(v_2), V_3 = j_1(v_3)$, and assume that $j_1 \neq j_2$. That corresponds to a non-crossing partition $S = \{S_1, S_2\}$, where $S_1 = \{1, 3\}, S_2 = \{2\}$. We write

$$V_i = V_i^0 + \hat{\phi}(V_i) = V_i^0 + \phi(v_i).$$

Thus, $\hat{\phi}(V_i^0) = 0$ and we get

$$\hat{\phi}(V_1 V_2 V_3) = \phi(v_1)\hat{\phi}(V_2^0 V_3^0) + \phi(v_2)\hat{\phi}(V_1^0 V_3^0) + \phi(v_3)\hat{\phi}(V_1^0 V_2^0)$$

$$+ \phi(v_1)\phi(v_2)\hat{\phi}(V_3^0) + \phi(v_1)\phi(v_3)\hat{\phi}(V_2^0) + \phi(v_2)\phi(v_1)\hat{\phi}(V_3^0) + \phi(v_1)\phi(v_2)\phi(v_3)$$

$$= \phi(v_2)\hat{\phi}(V_1^0 V_3^0) + \phi(v_1)\phi(v_2)\phi(v_3) = \phi(v_1 v_3)\phi(v_2)$$

In a similar way we can calculate the contribution from crossing partitions, but in this case the moment does not factorize. For instance, let $v_1, v_2, v_3, v_4 \in \mathcal{C}$ and assume that $j_1 = j_3$ and $j_2 = j_4$ and $j_1 \neq j_2$. Then we obtain

$$\hat{\phi}(V_1 V_2 V_3 V_4) = \phi(v_1 v_3)\phi(v_2)\phi(v_4) + \phi(v_2 v_4)\phi(v_1)\phi(v_3) - \phi(v_1)\phi(v_2)\phi(v_3)\phi(v_4),$$

each summand containing more than 2 factors. It can also be seen that in both cases the final expression does not depend on the particular values of j_i's, thus leading to natural equivalence classes.

The collective variable associated with $v \in \mathcal{C}$ will be again (as in the tensor case)

$$S_N(v) = \sum_{i=1}^{N} j_i(v)$$

except that the injected variables are multiplied differently (the reduced free product is used) and we get

$$\hat{\phi}(S_N(v_1)\ldots S_N(v_p)) = \sum_{r=1}^{p} A_{r,N} \sum_{S \in \mathcal{P}_r\{1,\ldots,p\}} \hat{\phi}(S)$$

where $\hat{\phi}(S)$ denotes the contribution from the partition $\{S_1, \ldots, S_r\}$. Here, we cannot write it in the factorized form yet since crossing partitions are still present. However, in the central limit theorem they will disappear.

By analogy with the commuting independence one can define the *free Gaussian functional* Φ_{FG} as a normalized functional for which it holds

$$\Phi_{\mathrm{FG}}(v_1 \ldots v_p) = \begin{cases} \sum_{S \in \mathcal{P}^{nc}_{pair}\{1,\ldots,2k\}} \phi(v_{S_1}) \ldots \phi(v_{S_k}) & \text{if } p = 2k \\ 0 & \text{if } p \text{ odd} \end{cases}$$

Then we obtain the following **qclt for free independence**.

Theorem 3.1. *Let $v_1, \ldots, v_p \in \mathcal{C}$ and $\phi \in \mathcal{C}^*$, with vanishing first moments,.i.e. $\phi(v_i) = 0$ for all $i = 1, \ldots, p$. Then*

$$\lim_{N \to \infty} \hat{\phi}(S_N(v_1^N) \ldots S_N(v_p^N)) = \Phi_{FG}(v_1 \ldots v_p)$$

where $v_i^N = \frac{1}{\sqrt{N}} v_i$.

The proof differs from the tensor case only for crossing pair-partitions (when $p = 2k$). Namely, if S is a crossing pair-partition then its contribution is expressed in terms of summands with more than k factors, hence each of those summands must have a singleton and hence must vanish.

For instance, in the case of $\hat{\phi}(S_N(v_1)\ldots S_N(v_4))$ the contribution of the crossing partitions is

$$N(N-1)(\phi(v_1 v_3)\phi(v_2)\phi(v_4) + \phi(v_2 v_4)\phi(v_1)\phi(v_3) - \phi(v_1)\phi(v_2)\phi(v_3)\phi(v_4))$$

and vanishes since first moments vanish. Thus, also they will not appear in the limit and we obtain

$$\lim_{N \to \infty} \hat{\phi}(S_N(v_1^N) \ldots S_N(v_4^N)) = \phi(v_1 v_2)\phi(v_3 v_4) + \phi(v_1 v_4)\phi(v_2 v_3)$$

which differs from the tensor case.

The aim is now to find an explicit realization of the limit state and its moments in terms of some creation and annihilation operators acting on a Hilbert space. Here it is the *free Fock space*

$$\Gamma_{fr} = \bigoplus_{n=0}^{\infty} \mathcal{H}^{\otimes n}$$

where $\mathcal{H}$ is the Hilbert space obtained from the GNS construction applied to ϕ and $\mathcal{C}$. For simplicity we write the representation for the hermitian elements of the algebra $\mathcal{C}$. Thus, let ω be the cyclic vector and π the GNS representation. For $v_1, \ldots, v_p$, hermitian elements of $\mathcal{C}$, we put $f_i = \pi(v_i)\omega$. For any $h_1, \ldots, h_n \in \mathcal{H}$ we define free creation and annihilation operators as follows:

$$a^*(f) \, h_1 \otimes \ldots \otimes h_n = f \otimes h_1 \otimes \ldots \otimes h_n$$

$$a(f) \, h_1 \otimes \ldots \otimes h_n = <f, h_1> \, h_2 \otimes \ldots \otimes h_n$$

and $a(f)\Omega = f$, $a^*(f)\Omega = 0$. Note that we have the following relation

$$a^*(f)a(g) = <f, g>$$

and the following representation can be proved:

$$\Phi_{\mathrm{FG}}(v_1 \ldots v_p) = < \Omega, c(f_1) \ldots c(f_p)\Omega >$$

where $c(f_i) = a(f_i) + a^*(f_i)$. This realization is the Cuntz algebra.

4. COALGEBRAS

The way commuting and anticommuting independences are defined leads to a natural generalization to coalgebras where the tensor product also appears as the source of independence [Sch1]. A coalgebra over $\mathbf{C}$ is a complex vector space $\mathcal{C}$ over $\mathbf{C}$ equipped with a coproduct $\Delta : \mathcal{C} \to \mathcal{C} \otimes \mathcal{C}$ and a counit $\delta : \mathcal{C} \to \mathbf{C}$ such that

$$(\mathrm{Id} \otimes \Delta) \circ \Delta = (\Delta \otimes \mathrm{Id}) \circ \Delta, \quad (\mathrm{Id} \otimes \delta) \circ \Delta = (\delta \otimes \mathrm{Id}) \circ \Delta = \mathrm{Id}.$$

The convolution product of $\phi, \psi \in \mathcal{C}^*$ is given by

$$\phi * \psi = (\phi \otimes \psi) \circ \Delta$$

and let ϕ^{*N} denote the n-th convolution of ϕ. Note that here Δ_{N-1}, the $N-1$-th iteration of the coproduct replaces S_N used so far and represents the sum of N independent random variables. Besides, $\phi^{*N} = \phi^{\otimes} \circ \Delta_{N-1}$, thus the object of interest in the qclt is

$$\phi^{*N}(v_1^N \ldots v_p^N)$$

In the limit we obtain the convolution exponential

$$\exp_* \phi = \sum_{N=0}^{\infty} \frac{\phi^{*N}}{N!}$$

which converges pointwise (see [Sch1]).

Further, assume that $\mathcal{C}$ is $\mathbf{N}$-graded, i.e. $\mathcal{C} = \bigoplus_{j=0}^{\infty} \mathcal{C}^{(j)}$ where $\Delta \mathcal{C}^{(j)} \subset \bigoplus_{j_1+j_2=j} \mathcal{C}^{(j_1)} \otimes \mathcal{C}^{(j_2)}$. For a given $v \in \mathcal{C}$ we denote by $\deg(c)$ its degree. If we let

$$v^N = (\frac{1}{\sqrt{N}})^{\deg(v)} v$$

then we obtain the following **qclt on coalgebras**.

Theorem 4.1. *Let C be an $\mathbf{N}$-graded coalgebra. Assume that $\phi \in C^*$ vanishes on $C^{(1)}$ and agrees with δ on $C^{(0)}$. Then*

$$\lim_{N \to \infty} \phi^{*N}(v_1^N \ldots v_p^N) = \Phi_{\mathrm{conv}}(v_1 \ldots v_p)$$

where $\Phi_{\mathrm{conv}} = \exp_* \mathrm{d}_\phi$ and d_ϕ is the linear functional on C which vanishes on $C^{(j)}$, $j \neq 2$, and agrees with ϕ on $C^{(2)}$.

The proof can be found in [Sch1]. It is easy to notice that if C is the tensor algebra $\mathcal{T}(V)$ (of a vector space V) then the commuting independence becomes a special case. Thus, let the canonical gradation on $\mathcal{T}(V)$ be introduced, namely

$$\mathcal{T}(V) = \bigoplus_{j=0}^{\infty} \mathcal{T}(V)^{(j)}$$

where

$$\mathcal{T}(V)^{(j)} = \mathrm{span}\{v_1 \otimes \ldots \otimes v_j | v_1, \ldots, v_j \in V\}$$

for $j \geq 1$ and $\mathcal{T}(V)^{(0)} = \mathbf{C}$. Then $\mathcal{T}(V)$ becomes a coalgebra (in addition to its algebra structure, thus a bialgebra) if the coproduct $\Delta : \mathcal{T}(V) \to \mathcal{T}(V) \otimes \mathcal{T}(V)$ is a homomorphic extension of

$$\Delta(v) = v \otimes 1 + 1 \otimes v, \quad \Delta(1) = 1 \otimes 1$$

and the counit δ is given by $\delta(v) = 0$, $\delta(1) = 1$, where $v \in V$. Here, the multiplication in $\mathcal{T}(V) \otimes \mathcal{T}(V)$ is standard and Φ_{conv} becomes the Gaussian functional Φ_{G}.

In turn, if $\Delta : \mathcal{T}(V) \to \mathcal{T}(V) \hat{\otimes} \mathcal{T}(V)$ is a homomorphic extension of the above formulas ($\hat{\otimes}$ signifies graded multiplication in the tensor product of graded algebras), then Φ_{conv} becomes the antisymmetric Gaussian functional Φ_{AG}.

In a few other cases realizations of the limit functional have been given (when C is a left or right q-bialgebra [Sch2]).

5. TWISTED QUANTUM CENTRAL LIMIT THEOREMS

In this section we will concentrate on twisted quantum central limit theorems by which we understand limit theorems with limit functionals depending on a parameter q (in general, complex-valued, although in many models the range of q is restricted to get positive limit functionals). They lead to certain interpolations between Bosonic and Fermionic relations and have been studied in many works [BSp1,BSp2,LiP,Sch2,SpB,Spe2]. We start with a certain generalization of the qclt presented in [SpB] that can be traced back to [Spe1,Spe2].

Namely, let $\hat{C}$ be a unital *-algebra equipped with a state $\hat{\Phi}$. It is more convenient to formulate the theorem using selfadjoint elements. Thus, consider elements $V_i = V_i^*$ $\in \hat{C}$, $i \in \mathbf{N}$ which satisfy the following assumptions:

1. $\hat{\phi}(V_{i_1} \ldots V_{i_p}) = 0$ if one of the i_k's is different from all others, i.e. there exists $1 \leq k \leq p$ such that $i_k \neq i_l$ for $l \neq k$

2. $\hat{\phi}(V_{i_1} \ldots V_{i_p}) = \hat{\phi}(V_{\pi(i_1)} \ldots V_{\pi(i_p)})$ for any permutation $\pi \in S_\infty$ of $\mathbf{N}$.

for all $i_1, \ldots, i_p \in \mathbf{N}$, $p \in \mathbf{N}$. Thus, the first condition is a general independence condition and the second signifies the invariance with respect to permutations of $\mathbf{N}$.

This invariance defines an equivalence relation on the set of p-tuples of natural numbers. Namely, $(i_1, \ldots, i_p) \sim (j_1, \ldots, j_p)$ iff ($i_k = i_l$ iff $j_k = j_l$). Thus, for instance, $(3, 2, 7, 3) \sim (5, 1, 4, 5)$. With each equivalence class of p-tuples with r distinct numbers ($r < p$) can be naturally associated a partition $S \in \mathcal{P}_r\{1, \ldots, p\}$, such that $i_k = i_l$ iff k, l belong to the same S_i. For instance, with $(3, 2, 7, 3)$ we associate $S = \{S_1, S_2, S_3\}$, where $S_1 = \{1, 4\}$, $S_2 = \{2\}$, $S_3 = \{3\}$.

Now, the scaled sum of "independent" variables (actually, the definition of independence still has to be given and there are various possibilities) is given by

$$S_N = \frac{V_1 + \ldots + V_N}{\sqrt{N}}$$

and the following familiar-looking formula can be derived using only the S_∞-invariance:

$$\hat{\phi}((S_N)^p) = \sum_{r=1}^{p} A_{r,N} \sum_{S \in \mathcal{P}_r\{1,\ldots,p\}} \hat{\phi}(S)$$

where $\hat{\phi}(S) = \hat{\phi}(V_{i_1} \ldots V_{i_p})$ for any p-tuple associated with S (it is identical on the whole equivalence class). We can see that the essential elements of the algebraic approach to qclt's are preserved and thus the proof of the qclt given below goes along the same lines, except that the limit is written in a more general form since no specific factorization law has been given yet. Let us call it a **generalized qclt**.

Theorem 5.1. *Let $\hat{\mathcal{C}}$ be a unital *-algebra and let $\hat{\phi} \in \hat{\mathcal{C}}^*$ be a state. Suppose that assumptions (1) and (2) are satisfied. Then*

$$\lim_{N \to \infty} \hat{\phi}((S_N)^p) = \begin{cases} \sum_{S \in \mathcal{P}_{pair}\{1,\ldots,2k\}} \hat{\phi}(S) & \text{if } p = 2k \\ 0 & \text{if } p \text{ odd} \end{cases}$$

where $\hat{\phi}(S) = \hat{\phi}(V_{i_1} \ldots V_{i_p})$ for any p-tuple $i_1, \ldots, i_p$ associated with S.

The approach given in [BSp1,BSp2,SpB] is to add some additional conditions on independence, namely the so-called *pyramidal factorization* [Küm], which is equivalent to $\hat{\phi}$ being *multiplicative*, namely

$$\hat{\phi}(S_1 \cup S_2) = \hat{\phi}(S_1)\hat{\phi}(S_2)$$

for any $S_1 \in \mathcal{P}_{pair}\{1, \ldots, k, l+1, \ldots, n\}$ and $S_2 \in \mathcal{P}_{pair}\{k+1, \ldots, l\}$, where $k, l, n \in \mathbf{N}$ are arbitrary. Moreover, it is required that multiplicative functions be positive definite so that they give positive limit functionals.

Two examples of positive definite multiplicative functions were given by Bożejko and Speicher. In [BSp1] $\hat{\phi}(S)$ was defined in terms of the number of inversions in S, and in [SpB] it was defined with the help of the number of *connected components* of S. Let us concentrate here on the first of those cases which gives canonical twisted commutation relations. Thus, let $\hat{\phi}(S) = \phi(v_{S_1}) \ldots \phi(v_{S_k}) q^{\#I_S}$. This function is multiplicative and nonnegative definite for $q \in [-1, 1]$ (see [BSp1, BSp2]), thus we get a restriction on q. The aim is now to find a realization of the limit state in terms of some creation and annihilation operators on a Hilbert space.

Let ω be the cyclic vector and π the $*$-representation of of $\mathcal{C}$ on a Hilbert space $\mathcal{H}$. Define the following (twisted) creation and annihilation operators on Γ, the set of finite linear combinations of product vectors:

$$a_q^*(f)\, h_1 \otimes \ldots \otimes h_n = f \otimes h_1 \otimes \ldots \otimes h_n$$

$$a_q(f)\, h_1 \otimes \ldots h_n = \sum_{k=1}^{n} q^{k-1} < f, h_k > h_1 \otimes \ldots \otimes h_{k-1} \otimes \ldots \otimes h_n$$

and $a_q(f)\Omega = f$, $a_q^*(f)\Omega = 0$. Here, the creation and annihilation operators are not adjoint with respect to the canonical scalar product $< .,. >$ on $\Gamma_{fr}(\mathcal{H})$ but they are adjoint with respect to a twisted scalar product given by

$$< g_1 \otimes \ldots \otimes g_n, h_1 \otimes \ldots \otimes h_m >_q = \delta_{nm} < \Omega, a_q(g_n) \ldots a_q(g_1) h_1 \otimes \ldots \otimes h_n >$$

which, for $q \in (-1,1)$ is positive definite and for $q \in \{-1,1\}$ is nonnegative definite (see [BSp1]). Thus, a twisted Fock space $\Gamma_q(\mathcal{H})$ is defined as the completion of Γ (when $q \in \{1,-1\}$ we first have to divide out the kernel). We have the following commutation relations

$$a_q(f)a_q^*(g) - qa_q^*(g)a_q(f) =< f, g >$$

and the following representation can be given (again we write it for $v_1, \ldots v_p$ hermitian elements of $\mathcal{C}$):

$$\Phi(v_1 \ldots v_p) =< \Omega, c(f_1) \ldots c(f_p)\Omega >$$

where $c(f_i) = a(f_i) + a^*(f_i)$.

6. TWISTED FREE *-BIALGEBRAS AND QUANTUM GROUPS

In this section we present still another approach to twisted qclt's. It is motivated by quantum groups and started with the q-analog of the qclt obtained for the fundamental representation of $SU_q(2)$ in [LPo] for q positive (for q complex with $|q| \neq 0,1$ see [Len1]). Here we present the essential elements of our approach developed first for $SU_q(2)$ ([Len2]) and then for Jimbo-Drinfel'd quantum groups [Jim, Dri] in [Len3] and free *-bialgebras in [Len4].

Assume $\mathcal{C}$ to be a free *-algebra over $\mathbf{C}$ generated (as a *-algebra) by the set $\mathcal{V}_+$ and let set $\mathcal{V} = \mathcal{V}_+ \cup \mathcal{V}_-$, where $\mathcal{V}_- = \{v^*|v \in \mathcal{V}_+\}$. On $\mathcal{C}$ we define a gradation that counts the generators by $d_1(v) = 1$, $v \in \mathcal{V}$, $d_1(\mathbf{C}) = 0$, extended to all free products in the only possible way. We denote the direct sum decomposition resulting from this $\mathbf{N}$-gradation with superscripts:

$$\mathcal{C} = \mathcal{C}^{(0)} \oplus \mathcal{C}^{(1)} \oplus \ldots$$

The algebra $\mathcal{C}$ can be twisted by adding hermitian generators t, t^{-1} subject to $tt^{-1} = t^{-1}t = 1$ and $tv = q^2vt$, $tw = q^{-2}wt$ where $q \in \mathbf{R}^+$ and $v \in \mathcal{V}_+$, $w \in \mathcal{V}_-$. The twisted *-algebra thus obtained will be denoted by $\mathcal{C}_q$. More general twistings on *-algebras were studied in [Len3, Len4, Sch2].

Now, we introduce a coproduct on $\mathcal{C}_q$, inspired by the coproduct in quantum groups. Thus let

$$\Delta : \mathcal{C}_q \to \mathcal{C}_q \otimes \mathcal{C}_q$$

be the homomorphic extension of

$$\Delta(1) = 1 \otimes 1, \quad \Delta(t) = t \otimes t, \quad \Delta(t^{-1}) = t^{-1} \otimes t^{-1}, \quad \Delta(v) = t^{-1} \otimes v + v \otimes t$$

where $v \in \mathcal{V}$. Moreover, $\mathcal{C}_q$ can be equipped with a counit δ. Namely, let δ be the homomorphism defined by $\delta(1)=\delta(t)=\delta(t^{-1})=1$, $\delta(v)=0$, $v \in \mathcal{V}$.

Here, the N-th iteration of the coproduct is given by

$$\Delta_{N-1}(v) = \sum_{i=1}^{N} j_{i,N}(v),$$

$$\Delta_{N-1}(t) = t^{\otimes N}, \quad \Delta_{N-1}(t^{-1}) = (t^{-1})^{\otimes N}$$

where $v \in \mathcal{V}$ and

$$j_{i,N}(v) = (t^{-1})^{\otimes(i-1)} \otimes v \otimes (t)^{\otimes(N-i)}$$

define canonical embeddings of $\mathcal{V}$ into $\otimes^N \mathcal{C}_q$. They have the following commutation rules:

$$j_{i,N}(v)j_{k,N}(w) = \epsilon(v,w)j_{k,N}(w)j_{i,N}(v)$$

for $i > k$, where

$$\epsilon(v,w) = \begin{cases} q^4 & \text{if } v,w \in \mathcal{V}_+ \\ q^{-4} & \text{if } v,w \in \mathcal{V}_- \\ 1 & \text{if } v \in \mathcal{V}_-, \ w \in \mathcal{V}_+ \ \text{or} \ v \in \mathcal{V}_+, \ w \in \mathcal{V}_- \end{cases}$$

Now, let $S \in \mathcal{P}_r^{ord}\{1,\ldots,p\}$. By the *signature* γ_S of S we understand the r-tuple

$$(\gamma_{S_1}, \ldots, \gamma_{S_r}) = (\#(S_1), \ldots, \#(S_r))$$

where $\#(S_j)$ denotes the number of elements in S_j. Partitions for which $\#(S_j)$ is even for all $j = 1, \ldots, r$ are called *even* and the set of all ordered even partitions of $\{1,\ldots,p\}$ will be denoted by $\mathcal{P}_e^{ord}\{1,\ldots,p\}$. Further, let

$$S_j^+ = \{k \in S_j | v_k \in \mathcal{V}_+\} \quad S_j^- = \{k \in S_j | v_k \in \mathcal{V}_-\},$$

and, $\gamma_{S_j^+} = \#(S_j^+)$, $\gamma_{S_j^-} = \#(S_j^-)$. We also assume that $\phi \in \mathcal{C}_q$ is *left and right* $\mathbf{C}[t,t^{-1}]$-*independent* , i.e.

$$\phi(scr) = \phi(s)\phi(c)\phi(r)$$

where $s, r \in \mathbf{C}[t,t^{-1}]$, $c \in \mathcal{C}_q$. Further, let $\phi(t) = a \in \mathbf{R}^+ - \{1\}$ and write $a = q^\alpha$, $\alpha > 0$ (in [Sch2] it is assumed that $a = 1$, whereas in the case of the fundamental representation of $SU_q(2)$ we have $a = q$ [LPo], which motivated our assumption in our work). Thus, in order to obtain finite and nontrivial limits for the moments, the usual $1/\sqrt{N}$ is not appropriate and we need to introduce its twisted analog, namely

$$v^N = 1/\sqrt{[N]_{a^2}}v$$

where $v \in \mathcal{V}$ and $[N]_x = (x^N - x^{-N})/(x - x^{-1})$ is "twisted" N.

We obtain a quantum limit theorem that we call a **q-analog of the qclt.**

Theorem 6.1. *Let* $\mu = q^{-2} \in \mathbf{R}^+$, $v_1, \ldots, v_p \in \mathcal{V}_- \cup \mathcal{V}_+$. *Let* $\phi \in \mathcal{C}_q^*$ *be left and right* $\mathbf{C}[t,t^{-1}]$-*independent with* $\phi(t) = a = \mu^{-\alpha/2} \in \mathbf{R}^+$. *Assume also that* $\phi(\mathcal{C}^{(j)}) = 0$ *for* j *odd. Then, if* $p = 2k$,

$$\lim_{N \to \infty} \phi_N^*(v_1^N \ldots v_p^N) = \sum_{r=1}^{p} \sum_{S=(S_1,\ldots,S_r) \in \mathcal{P}_e^{ord}(I)} Q(S)D(S|\mu) \prod_{j=1}^{r} \phi(v_{S_j})$$

where

$$D(S|\mu) = D_{k,r}(\alpha_{S_1}, \ldots, \alpha_{S_r}|\mu) = \frac{(\mu^{2\alpha} - 1)^{p/2}}{(\mu^{\alpha_{S_1}} - 1)(\mu^{\alpha_{S_1} + \alpha_{S_2}} - 1)\ldots(\mu^{\alpha_{S_1} + \ldots + \alpha_{S_r}} - 1)}$$

for $1 < \mu < \infty$ *,* $\alpha_i = \alpha\gamma_{S_i}$*, and* $D(S|\mu) = D_{k,r}(\alpha_{S_r}, \ldots, \alpha_{S_1}|\mu^{-1})$ *for* $0 < \mu < 1$*, and* $Q(S)$ *denotes the combinatorial twisted commutation factor:*

$$Q(S) = \prod_{(i,j) \in W_S} \epsilon(v_i, v_j) = \prod_{(i,j) \in W_S} \epsilon(t, v_i)\epsilon(t, v_j).$$

If p is odd, then the above limit vanishes.

This qclt is a twisted Boson-type qclt in the following sense. It holds

$$\lim_{\mu \to 1+} D(S|\mu) = \begin{cases} \frac{1}{k!} & r = k \\ 0 & r < k \end{cases}$$

and $\lim_{N \to \infty} Q(S) = 1$. Thus, in the limit we obtain the Boson symmetry and Boson-type qclt with only pair-partitions present. In general, when no other relations are assumed on $\mathcal{C}$, all partitions survive in the limit. Let us also remark that the limit functional can be written in terms of the usual partitions, however it is more natural and convenient to use ordered partitions.

Let us assume that ϕ is a state. In order to define generalized creation and annihilation operators we start with the GNS representation (π, ω) of $(\mathcal{C}, \phi)$ on a Hilbert space $\mathcal{H}$. Then we construct the *incomplete direct product space* (IDPS) relative to $\Omega = \omega^{\otimes\infty}$. It is spanned by

$$\{\pi(c_1)\omega \otimes \ldots \otimes \pi(c_n)\omega \otimes \omega^{\otimes\infty} | c_1, \ldots, c_n \in \mathcal{C}\}.$$

Now, to given $v_1, \ldots, v_p \in \mathcal{V}$ and ordered partition $S = (S_1, \ldots, S_r)$, or to given $v_{S_1}, \ldots, v_{S_r}$, we associate the element

$$v_S = v_{S_1} \otimes \ldots \otimes v_{S_r} \otimes 1^{\otimes\infty} \in \mathcal{C}^{\otimes\infty}$$

in the infinite tensor product of $\mathcal{C}$ and

$$v_S^\pi = \pi(v_{S_1})\omega \otimes \ldots \otimes \pi(v_{S_r})\omega \otimes \omega^{\otimes\infty} \in \mathcal{H}^{\otimes^\omega \infty}.$$

in the IDPS. Let ω be the vacuum vector, which implies that $\pi(v)\omega = 0$ for $v \in \mathcal{V}_-$. Then, for $v \in \mathcal{V}_+$, we define the following one-cluster operators

$$L_j(v)v_{S_{\text{in}}} = \epsilon(v, v_{S_1} \ldots v_{S_{j-1}})t(S_{\text{in}}, S_{\text{out}})v_{S_1} \otimes \ldots \otimes v_{S_{j-1}} \otimes vv_{S_j} \otimes v_{S_{j+1}} \otimes \ldots \otimes v_{S_r} \otimes 1^{\otimes\infty}$$

$$A_j(v)v_{S_{\text{in}}} = \epsilon(v, v_{S_1} \ldots v_{S_{j-1}})t(S_{\text{in}}, S_{\text{out}})v_{S_1} \otimes \ldots \otimes v_{S_{j-1}} \otimes v \otimes v_{S_j} \otimes \ldots \otimes v_{S_r} \otimes 1^{\otimes\infty}$$

where

$$\epsilon(v, v_{S_1} \ldots v_{S_k}) = \prod_{i=1}^{k} \epsilon(v, v_{S_i}),$$

$$t(S, T) = \frac{D^*(T|\mu)}{D^*(S|\mu)}, \quad D^*(S|\mu) = D_{k,r}(2\alpha_{S_1^+}, \ldots, 2\alpha_{S_r^+}|\mu)$$

and $\alpha_{S_j^+} = \alpha\gamma_j$, $k = \sum_{j=1}^{r} 2\gamma_{S_j^+}$. We assume that $D(\emptyset|\mu) = 1$ and S_{out} denotes the partition obtained on the right hand side. For $v \in \mathcal{V}_+$ one can interpret $L_j(v)$ as the operator creating v at the existing cluster at site j, and $A_j(v)$ as the operator creating v in a new cluster at site j. Similar definitions can be given on the IDPS.

In turn, for $w \in \mathcal{V}_-$ we define

$$L_j(w)v_S = \epsilon(w, v_{S_1}, \ldots, v_{S_r})v_{S_1} \otimes \ldots \otimes v_{S_{j-1}} \otimes wv_{S_j} \otimes v_{S_{j+1}} \otimes \ldots \otimes v_{S_r} \otimes 1^{\otimes\infty}$$

as the operator creating w at the existing cluster at site j.

The *generalized creation and annihilation operators* are sums of one-cluster operators:

$$A(v)v_S = \sum_{j=1}^{r} L_j(v)v_S + \sum_{j=1}^{r+1} A_j(v)v_S$$

$$A(w)v_S = \sum_{j=1}^{r} L_j(w)v_S$$

where $S = (S_1, \ldots, S_r)$, $v \in \mathcal{V}_+$, $w \in \mathcal{V}_-$. Thus, one can say that the generalized creation operator $A(v)$ is the sum of operators creating v in all existing clusters and in new clusters that could be created at different sites, whereas the generalized annihilation operator is the sum of operators annihilating v in all existing clusters. We obtain the following **representation theorem**.

Theorem 6.2. *Let $v_1, \ldots, v_p \in \mathcal{V}_- \cup \mathcal{V}_+$, $\mu \in [1, \infty)$ and let $\phi \in \mathcal{C}_\mu^*$ be the vacuum state. Then, for a given cyclic representation (π, ω) of $(\mathcal{C}, \phi)$ on Hilbert space $\mathcal{H}$ we have, for p even*

$$\lim_{N\to\infty} \phi_N^*(v_1^N \ldots v_p^N) = <\Omega, A(v_1)\ldots A(v_p)\Omega>$$

where $A(v)$ are generalized creation or annihilation operators and the scalar product is the canonical scalar product on the incomplete direct product space $\mathcal{H}^{\otimes^\omega \infty}$ relative to $\Omega = \omega^{\otimes\infty}$. For p odd, the above limit vanishes.

This gives in fact the GNS representation for the limit states but it takes a very specific form, reminding the Fock space representation. For details see [Len4].

Let

$$\Gamma = \text{span}\{A(v_1)\ldots A(v_p)\Omega | v_1, \ldots, v_p \in \mathcal{V}\}.$$

and define the twisted semi-inner product on Γ by linear extension of:

$$< A(v_1)\ldots A(v_p)\Omega, A(w_1)\ldots A(w_r)\Omega >_\mu = <\Omega, A(v_p^*)\ldots A(v_1^*)A(w_1)\ldots A(w_r)\Omega >.$$

It becomes inner product on the quotient space $\Gamma/\text{Ker}||.||$, where

$$\text{Ker}||.|| = \{x \in D_{\otimes\pi}| \, ||x||_\mu = 0\}$$

and $||x||_\mu^2 = <x, x>_\mu$. Its completion Γ_D is a Hilbert space and the carrier of the *-representation of $\mathcal{C}$ with vacuum Ω.

Let $v_S = v_{S_1} \otimes \ldots \otimes v_{S_r}$. The following commutation rule between the generalized creation and annihilation operators can be derived:

$$[A(v), A(w)]v_S = \sum_{j=1}^{r} L_j[v, w]v_S + \sum_{j=1}^{r+1} A_j(vw)v_S$$

where $L_j[v, w] \equiv [L_j(v), L_j(w)]$, $A_j(vw) \equiv L_j(v)A_j(w)$, and $v \in \mathcal{V}_-$, $w \in \mathcal{V}_+$. Thus, the bracket is expressed in terms of the bracket in $\mathcal{C}$ and this is as far as one can get in the case of free *-algebras.

However, one can apply the above results to algebras which have some additional relations (as long as Δ is a homomorphism). In particular, let $\mathcal{C}_q = SU_q(2)$, $q > 1$, be generated by J_+, J_-, t, t^{-1} subject to

$$[J_+, J_-] = \frac{t^2 - t^{-2}}{q^2 - q^{-2}}, \quad tJ_+ = q^2 J_+ t, \quad tJ_- = q^{-2} J_- t$$

with the coproduct

$$\Delta(J_+) = J_+ \otimes t + t^{-1} \otimes J_+, \quad \Delta(J_-) = J_- \otimes t + t^{-1} \otimes J_-, \quad \Delta(t) = t \otimes t, \quad \Delta(t) = t^{-1} \otimes t^{-1}$$

and the counit $\delta(J_+) = \delta(J_-) = 0$, $\delta(t) = \delta(t^{-1}) = 1$. When we apply the q-analog of qclt we obtain in the limit the q-oscillator algebra generated by $a_q, a_q^*, r_q, r_q^{-1}$ subject to

$$[a_q, a_q^*] = r_q^{-2}, \quad a_q r_q = q^2 r_q a_q, \quad a_q^* r_q = q^{-2} r_q a_q^*$$

or, $[a_q, a_q^*] = q^{-4\hat{N}}$, $[\hat{N}, a_q^*] = a_q^*$, $[\hat{N}, a_q] = -a_q$ (the correspondence is given by $r_q = q^{-2\hat{N}}$). Moreover, the q-analog qclt corresponds to the qroup contraction:

$$\frac{\Delta_{N-1}(J_-)}{\sqrt{[N]_{q^2}}} \to a_q^* , \quad \frac{\Delta_{N-1}(J_+)}{\sqrt{[N]_{q^2}}} \to a_q ,$$

$$\frac{\Delta_{N-1}(t)}{q^N} \to r_q^{-1} , \quad \frac{\Delta_{N-1}(t^{-1})}{q^{-N}} \to r_q .$$

Thus, in this case, if we put $v = J_-, w = J_+$, the generalized creation and annihilation operators $A(v), A(w)$, respectively, are unitarily equivalent to a_q^*, a_q, respectively, and their bracket - to r_q^{-2}.

Acknowledgements

I would like to thank Prof. Marek Bożejko for many discussions on this and related subjects.

REFERENCES

[AFL] Accardi L., Frigerio A., Lewis J.T., "Quantum stochastic processes", Publ. RIMS Kyoto Univ. **18** (1982), 97-133.

[BSp1] Bożejko M., Speicher R.,"An example of a generalized Brownian Motion", Commun. Math. Phys. **137**, 519-531 (1991).

[BSp2] Bożejko M., Speicher R.,"An example of a generalized Brownian Motion II", in *Quantum Probability and Related Topics VII"*, Proceedings, New Delhi 1990, World Scientific, Singapore 1993, 67-77.

[CuH] Cushen D.D., Hudson R.L.,"A quantum mechanical central limit theorem", J. Appl. Probability **8**, 454-469 (1971).

[Dri] Drinfel'd V. G.,"Quantum groups", Proc. Int. Congress of Math., Berkeley, California, 1986, 799-820.

[FQu] Fannes M., Quaegebeur J., "Central limits of product mappings between CAR-algebras". Publ. RIMS Kyoto **19**, 469-491 (1983).

[GVV] Goderis D., Verbeure A., Vets P, "Non-commutative central limits", Probab. Th. Rel. Fields **82**, 527-544 (1989).

[GvW] Giri N., von Waldenfels W.,"An algebraic version of the central limit theorem", Z. Wahr. Verw. Gebiete **42**, 129-134 (1978).

[Hud] Hudson R.L.,"A quantum mechanical central limit theorem for anticommuting observables", J. Appl. Probability **10**, 502-509 (1973).

[HuP] Hudson R.L., Parthasarathy K.R.,"Quantum Ito's formula and stochastic evolution", Commun.Math.Phys. **93**, 301-323 (1984).

[Jim] Jimbo M.,"A q-difference analogue of $U(g)$ and the Yang-Baxter equation", Lett. Math. Phys. **10**, 63-69 (1985).

[Küm] Kümmerer B., Markov dialations and non-commutative Poisson processes, preprint.

[LiP] Lindsay, J.M., Parthasarathy K.R., "Cohomology of power sets with applications in quantum probability", Commun Math.Phys. **124**, 337-364 (1989).

[LPo] Lenczewski R., Podgorski K.,"A q-analog of the quantum central limit theorem for $SU_q(2)$", J. Math. Phys. **33**, 2768- 2778 (1992).

[Len1] Lenczewski R., "A q-analog of the quantum central limit theorem for $SU_q(2)$, q complex", J. Math. Phys.**34**, 480-489 (1993).

[Len2] Lenczewski R.,"On sums of q-independent $SU_q(2)$ quantum variables", Commun. Math. Phys. **154**, 127-134 (1993).

[Len3] Lenczewski R.,"Addition of independent variables in quantum groups", Rev. Math. Phys., Vol.6, No.1, 135-147 (1994).

[Len4] Lenczewski R.,"On certain states on free *-algebras and their GNS representation", preprint (1994).

[Qua] Quaegebeur J., "A non-commutative central limit theorem for CCR algebras", J. Funct. Anal. **57**, 1-20 (1984).

[Sch1] Schürmann M.,"A central limit theorem for coalgebras", in *Probability measures on groups VIII*, Proceedings, Oberwolfach 1985, Ed. Heyer H., Lect. Notes in Math. **1210**, 153-157, Springer 1986.

[Sch2] Schürmann M.,"Quantum q-white noise and a q-central limit theorem", Commun. Math. Phys. **140**, 589-615 (1991).

[Sch3] Schürmann M., *White Noise on Bialgebras*, Lecture Notes in Math. **1544**, Springer -Verlag, Berlin, 1993.

[Spe1] Speicher R.,"A new example of "Independence" and "White Noise"", Probab. Th. Rel. Fields **84**, 141-159 (1990).

[Spe2] Speicher R., "A non-commutative central limit theorem", Math. Z. **209**, 55-66 (1992).

[SpB] Speicher R., Bożejko M.,"Interpolations between Bosonic and Fermionic relations given by Generalized Brownian Motions", preprint.

[Voi1] Voiculescu D., "Symmetries of some reduced free product C^*-algebras", in *Operator Algebras and their Connections with Topology and Ergodic Theory*, 556-588, Lecture Notes in Mathematics **1132**, Springer Verlag, Berlin, 1985.

[Voi2] Voiculescu D., "Addition of certain non-commuting random variables", J. Funct. Anal. **66**, 323-346 (1986).

[vWa] von Waldenfels W.,"An algebraic central limit theorem in the anticommuting case", Z. Wahr. Verw. Gebiete **42**, 135-140 (1979).

NON-EUCLIDEAN CRYSTALLOGRAPHY

Miguel Lorente[1] and Peter Kramer[2]

[1]Departamento de Fisica, Universidad de Oviedo,
E-33007 Oviedo, Spain
[2]Institut für Theoretische Physik der Universität
D 72076 Tübingen, Germany

ABSTRACT

Following standard methods of crystallography, we investigate the classification of
Bravais lattices with respect to vector spaces with a non-Euclidean metric, i.e. we
describe sets of transformations of non-compact type that keep an isometry of the
Bravais lattice, and apply the method to two and three dimensional spaces.

INTRODUCTION

In the last Symposium in Science VI, one of the authors [10] explored the technique to
represent classical groups on a cubic lattice. Meanwhile, the other author [8] presented
the realization of mathematical models for quasicrystals, using the embedding of
crystallographic objects in hypercubic lattices of higher dimension, as in the case of
icosahedral tilings.

Here we describe joint work addressed to the problem of finding some lattices -
not necessarily hypercubic - which are invariant under Euclidean or Non-Euclidean
transformations. In the case of Euclidean crystallography the idea is very well known,
compare [7, 1, 13]. Given a lattice, not necessarily cubic, find all transformations,
including translations, that keep the lattice fixed. If the set of transformations is
maximal, the lattice is called a Bravais lattice, and the maximal set of point transformations is called its holohedry. In the case of non-Euclidean crystallography, we
define some lattice invariant under transformations from a non-compact group plus
translations. The holohedries may be infinite in this case, but we can define Bravais
lattices as before.

The motivation for this paper is twofold. The first, more mathematical, consists
in finding discrete subgroups of the classical, especially the non-compact groups, that
leave invariant some indefinite quadratic form on the lattice. The result can be
used to investigate the wave equation on a lattice invariant under discrete Lorentz
transformations.

The second motivation is more philosophical. Recent literature has been occupied
with realistic models of discrete space and time, the foundations of which are based on

relational theories of space and time [11]. According to these theories, the space and time structure is identified with the set of relations among fundamental entities. In the first motivation the space-time lattice is considered as an artificial model. In the second motivation the discreteness of space-time is considered as some fundamental reality [11].

The general outline of the paper is as follows: In section 1 we present the basic definitions and state the general problem: Find the set of all point transformations (holohedry) that leave some (Bravais) lattice invariant. In section 2, we classify all the holohedries in two dimensional crystallography, using the properties of eigenvalues of a two dimensional real matrix. In section 3, we apply the method to two dimensional rotations and corresponding Bravais lattices. In section 4, we analyze proper and improper Lorentz transformations that keep a (not necessarily square) lattice invariant. In section 5, we explore the Lorentz transformations that keep a certain type of three-dimensional lattice invariant. Here we use the homomorphism between the groups $SL(2,R)$ and $SO(2,1)$. Finally in section 6 we describe a scheme for a wave equation in a Bravais lattice that is invariant under crystallographic Lorentz transformations in two dimensions.

1 BASIC NOTIONS

Consider a *vector space* V in n dimensions with basis $B = (e_1, \ldots, e_n)$. A change of basis is $B \to B' = BS$ with $S \in GL(n, R)$, with the new basis $B \to B' = (e'_1, \ldots, e'_n)$. To get metrical notions we define an M-isometry with respect to some non-degenerate matrix M by

$$GL_M(n, R) = \{L | L^t M L = M\} \in GL(n, R). \tag{1}$$

The elements of $GL_M(n, R)$ form a subgroup of $GL(n, R)$. We consider a symmetric bilinear form $(,)$ such that

$$M = g(B)) = (e_i, e_j), \quad i, j = 1, \ldots, n, \tag{2}$$

and call $g(B)$ the metric tensor. Under the transformation S, the metric tensor becomes

$$g(B) \to g(B') = S^t g(B) S, \tag{3}$$

and the M-isometry group becomes a conjugate subgroup of $GL(n, R)$, namely

$$GL_M(n, R) \to GL'_M(n, R) = S^{-1} GL_M(n, R) S, \tag{4}$$

which determines an equivalence class of isometries.

With a basis B we associate a lattice $\Lambda(B)$,

$$\Lambda(B) = \{\sum_{i=1}^{n} z_i e_i, \ (z_1, \ldots, z_n) \in Z^n\}, \tag{5}$$

which is a set of vectors in V. For the new basis B' we have the lattice

$$\Lambda(B') = \{\sum_{i=1}^{n} z_i e'_i, \ (z_1, \ldots, z_n) \in Z^n\}, \tag{6}$$

If T is a unimodular transformation and $B" = B'T$, then $\Lambda(B") = \Lambda(B')$ if and only if $T \in GL(n, Z)$. To a given lattice there corresponds the whole set $\{B'T | T \in$

$GL(n, Z)\}$ of its possible bases, and thus an arithmetic class $\{g(B')\}$ of metric tensors, and also an arithmetic class of point groups. Two metric tensors are *arithmetically equivalent* if $g(B_1) = T^t g(B_2) T$ and two point groups are *arithmetically equivalent* if $L_2 = T^{-1} L_1 T$ for some $T \in SL(n, Z)$.

Let us turn now to crystallographic transformations with M-isometry. Consider triplets

$$(\tilde{L}, L, S), \tilde{L} \in GL(n, Z), L \in GL_M(n, R), S \in GL(n, R) \; : \tilde{L} = S^{-1} L S. \tag{7}$$

Since the elements of $GL_M(n, R)$ have $det(L) = \pm 1$, the unimodularity is preserved under this transformation, namely $det(\tilde{L}) = \pm 1$. By definition $L^t g(B) L = g(B)$, therefore

$$\tilde{L}^t g(B') \tilde{L} = g(B'), \tag{8}$$

which means that $\tilde{L}$ belongs to the M-isometry with respect to the new basis.

Eq.(7) can be written as $LS = S\tilde{L}$. If we take a basis formed by the columns of B, the image of any basis vector under L can be written as an integral linear combination of basis vectors, and hence the lattice $\Lambda(S)$ is transformed into itself. Now for fixed $S \in GL(n, R)$, collect all the pairs $(\tilde{L}, L)$ which fulfill Eq.(7). They form a group which is a subgroup of $GL(n, Z)$ and leaves the lattice invariant. Take for the basis of a lattice the columns of S. Then from Eq.(7) it follows that

$$\Lambda(LS) = \Lambda(S\tilde{L}) = \Lambda(S). \tag{9}$$

The largest group $H \in GL_M(n, R)$ that leaves this lattice invariant and satisfies Eq.(7) is called the *holohedry group* corresponding to the lattice generated by S, and the lattice $\Lambda(S)$ is called the *Bravais lattice* corresponding to the basis generated by the columns of S.

Given a Bravais lattice, we can consider other Bravais lattices that are equivalent and therefore belong to the same *Bravais class*. The matrix S that generates the Bravais lattice can be modified by the following method:

$$(i) \; : \; S \to S" = L^0 S, \; L^0 \in GL_M(n, R), \tag{10}$$
$$L \to L" = L^0 L (L^0)^{-1},$$
$$\tilde{L} \to \tilde{L} = (S")^{-1} L" S".$$

The lattice $\Lambda(S)$ will be "rotated" by the transformation $S" = L^0 S$, namely,

$$\Lambda(S") = L^0 \Lambda(S), \tag{11}$$

but it will be an M-equivalent lattice. All the "rotated" lattices belong to the same class, with the equivalent sets $(\tilde{L}, L, S) \leftrightarrow (\tilde{L}, L", S")$.

$$(ii) \; : \; S \to S" = ST, \; T \in GL(n, Z), \tag{12}$$
$$L \to L,$$
$$\tilde{L} \to \tilde{L}" = T^{-1} \tilde{L} T = (S")^{-1} L S".$$

The matrices $\tilde{L}$ and $\tilde{L}"$ are arithmetically equivalent and the Bravais lattice $\Lambda(S)$ is transformed into itself because $\Lambda(S") = \Lambda(ST) = \Lambda(S)$. We have the equivalent triples $(\tilde{L}, L, S) \leftrightarrow (\tilde{L}", L, S")$.

Finally we have the scaling

$$(iii) \quad : \quad S \to S" = \lambda S, \lambda \neq 0, \lambda \in R, \tag{13}$$
$$L \to L,$$
$$\tilde{L} = S^{-1}LS \to \tilde{L} = (S")^{-1}LS",$$

where the last two equalities arise since λI commutes with all the matrices. The lattice is transformed by the scaling $\Lambda(S") = \lambda\Lambda(S)$, but the defining equation for the Bravais lattice is unchanged. The equivalent triples become $(\tilde{L}, L, S) \leftrightarrow (\tilde{L}, L, S")$.

Lattices $\Lambda(S)$ which are transformed by some operation of type $(i), (ii)$ or (iii) belong to the same *Bravais class*. If they cannot be related by a combination of these three transformations they belong to different Bravais classes.

2 BRAVAIS LATTICES IN TWO-DIMENSIONAL SPACE

Let $S \in GL(2, R)$ be a 2×2 matrix generating a basis formed by the columns of S, with $B = (e_1, e_2)$ an orthonormal basis, and let $L \in GL_M(2, R)$ be an M-isometry satisfying

$$S^{-1}LS = \tilde{L} \in SL(2, Z). \tag{14}$$

In order to classify all the M-isometries of $L \in GL_M(2, R)$ we can calculate the eigenvalues of $\tilde{L}$ since L and $\tilde{L}$ satisfy the same characteristic equation. Upon putting $a = Tr(\tilde{L}), \Delta = det(\tilde{L})$, the eigenvalues of $\tilde{L}$ are the solutions of

$$\lambda^2 - a\lambda + \Delta = 0, \tag{15}$$
$$\lambda = \frac{1}{2}a \pm ((\frac{a}{2})^2 - \Delta)^{\frac{1}{2}}.$$

We can distinguish the following cases and standard forms:

1. $\Delta = 1$.

1.1, $|\frac{a}{2}| > 1$: Two real eigenvalues (λ, λ^{-1}) and a standard form

$$L = \begin{pmatrix} \cosh\beta & \sinh\beta \\ \sinh\beta & \cosh\beta \end{pmatrix} \in SO(1, 1). \tag{16}$$

Since $a = 2\cosh\beta$ must be an integer number we have $a = \pm 3, \pm 4, \pm 5, \ldots$.

1.2, $|\frac{a}{2}| = 1$: One real eigenvalue (degenerate case) with the standard Jordan form

$$L = \begin{pmatrix} 1 & 0 \\ 1 & 1 \end{pmatrix}, \quad a = 2. \tag{17}$$

1.3, $|\frac{a}{2}| < 1$: Two complex conjugate eigenvalues $(\lambda, \overline{\lambda})$ and a standard form

$$L = \begin{pmatrix} \cos\phi & \sin\phi \\ -\sin\phi & \cos\phi \end{pmatrix} \in SO(2). \tag{18}$$

Since $a = 2\cos\phi$ must be an integer we have $a = 2, 1, 0, -1, -2$ corresponding to $\phi = 2\pi, \frac{1}{3}\pi, \frac{1}{2}\pi, \frac{2}{3}\pi, \pi$.

2. $\Delta = -1$.

2.1, Improper rotations:

$$L = \begin{pmatrix} \cos\phi & \sin\phi \\ -\sin\phi & \cos\phi \end{pmatrix} \begin{pmatrix} 1 & 0 \\ 0 & -1 \end{pmatrix} \in O(2) \tag{19}$$

with $\lambda = \pm 1, 2\cos\phi = 2, 1, 0, -1, -2$ corresponding to rotations $\times$ reflections on the x-axis.

2.2, Improper Lorentz transformations:

$$L = \begin{pmatrix} \cosh\beta & \sinh\beta \\ \sinh\beta & \cosh\beta \end{pmatrix} \begin{pmatrix} 1 & 0 \\ 0 & -1 \end{pmatrix} \in O(1,1) \tag{20}$$

with $\lambda = \pm 1, 2\cosh\beta = \pm 3, \pm 4, \pm 5, \ldots$.

2.3, Rotations $\times$ permutation of coordinates:

$$L = \begin{pmatrix} \cos\phi & \sin\phi \\ -\sin\phi & \cos\phi \end{pmatrix} \begin{pmatrix} 0 & 1 \\ 1 & 0 \end{pmatrix} \in O(2) \tag{21}$$

with $\lambda = \pm 1, 2\sin\phi = 2, 1, 0, -1, -2$.

2.4, Lorentz transformations $\times$ permutation of coordinates:

$$L = \begin{pmatrix} \cosh\beta & \sinh\beta \\ \sinh\beta & \cosh\beta \end{pmatrix} \begin{pmatrix} 0 & 1 \\ 1 & 0 \end{pmatrix} \tag{22}$$

The M-isometry of this group corresponds to transformations leaving invariant the quartic expression $(x^2 - y^2)^2$. Since $Tr(L) = Tr(\tilde{L}) = $ integer, we have $2\cosh\beta = 0, \pm 1, \pm 2, \pm 3 \ldots$. A particular example for $2\cosh\beta = 1$ gives the matrix

$$\begin{pmatrix} 1 & 1 \\ 1 & 0 \end{pmatrix} \tag{23}$$

that appears in the Fibonacci substitution, compare [9]. In general for

$$S = \begin{pmatrix} 1 & -\sinh\beta \\ 0 & \cosh\beta \end{pmatrix} \tag{24}$$

one finds

$$\tilde{L} = S^{-1}LS = \begin{pmatrix} a & 1 \\ 1 & 0 \end{pmatrix}, \tag{25}$$

with $a = 2\sinh\beta = $ integer. The last case has been called by Janner and Ascher [5] a "negautomorph" of the Lorentz group because

$$L^t g(B')L = -g(B'). \tag{26}$$

3 TWO-DIMENSIONAL CRYSTALLOGRAPHIC ROTATIONS

Let B be an orthonormal basis (e_1, e_2) and $g(B)$ the metric tensor. The isometries for this tensor obey

$$L^t g(B)L = g(B). \tag{27}$$

If $g(B) = I$ then $L \in O(2)$. We want to find a triple $(\tilde{L}, L, S)$ such that

$$S^{-1}LS = \tilde{L}, \ S \in GL(2, R), \ \tilde{L} \in GL(2, Z). \tag{28}$$

The matrix S applied to B gives a new basis B' and a new metric tensor

$$g(B') = S^t g(B) S = S^t S = \begin{pmatrix} a & \frac{1}{2}b \\ \frac{1}{2}b & c \end{pmatrix}. \tag{29}$$

From the columns of S we form the new basis and the lattice $\Lambda(B')$. The largest point group for this lattice, with elements which satisfy Eq.(28), is the holohedry of $\Lambda(B')$. We use the freedom in the choice of B' and take

$$S = \begin{pmatrix} \sqrt{a} & \frac{1}{2}\frac{b}{\sqrt{a}} \\ 0 & \frac{1}{2}\frac{\sqrt{d}}{\sqrt{a}} \end{pmatrix}, \tag{30}$$

$$B' = \left\{ \begin{array}{rcl} e'_1 &=& \sqrt{a}\, e_1, \\ e'_2 &=& \frac{1}{2}\frac{b}{\sqrt{a}} e_1 + \frac{1}{2}\frac{\sqrt{d}}{\sqrt{a}} e_2 \end{array} \right\}.$$

Now for proper rotations we find the conditions

$$\tilde{L} = S^{-1} L S \tag{31}$$

$$= \frac{2}{\sqrt{d}} \begin{pmatrix} \frac{1}{2}\frac{\sqrt{d}}{\sqrt{a}} & -\frac{1}{2}\frac{b}{\sqrt{a}} \\ 0 & \sqrt{a} \end{pmatrix} \begin{pmatrix} \cos\phi & \sin\phi \\ -\sin\phi & \cos\phi \end{pmatrix} \begin{pmatrix} \sqrt{a} & \frac{1}{2}\frac{b}{\sqrt{a}} \\ 0 & \frac{1}{2}\frac{\sqrt{d}}{\sqrt{a}} \end{pmatrix}$$

$$= \begin{pmatrix} \frac{1}{2}(n + bu), & cu \\ -au, & \frac{1}{2}(n - bu) \end{pmatrix},$$

with $n = 2\cos\phi$, $u := \pm\sqrt{\frac{4-n^2}{d}}, d = 4ac - b^2$. If $a, b, c \in Z$ then $\tilde{L} \in GL(2, Z)$ provided that

$$det(\tilde{L}) = \frac{1}{4}(n^2 + du^2) = 1. \tag{32}$$

The solutions of this diophantine equation are

$$\begin{array}{ccccc}
n & u & d & (a, b, c) & \phi \\
2 & 0 & any & any & 0 \\
-2 & 0 & any & any & \pi \\
1 & \pm 1 & 3 & (1, 1, 1) & \frac{1}{3}\pi \\
-1 & \pm 1 & 3 & (1, 1, 1) & \frac{2}{3}\pi \\
0 & \pm 1 & 4 & (1, 0, 1) & \frac{1}{2}\pi
\end{array} \tag{33}$$

For improper rotations we have

$$L = \begin{pmatrix} \cos\phi & \sin\phi \\ -\sin\phi & \cos\phi \end{pmatrix} \begin{pmatrix} 1 & 0 \\ 0 & -1 \end{pmatrix} = \begin{pmatrix} \cos\phi & \sin\phi \\ \sin\phi & -\cos\phi \end{pmatrix} \tag{34}$$

and we require

$$\tilde{L} = S^{-1} L S = \begin{pmatrix} \frac{1}{2}(n - bu), & \frac{1}{2a}(bn + du) - cu \\ au, & -\frac{1}{2}(n - bu) \end{pmatrix} \in GL(2, Z) \tag{35}$$

with $n = 2\cos\phi$, $u := \pm\sqrt{\frac{4-n^2}{d}}, d = 4ac - b^2$. From the integrality of $\tilde{L}$ it follows that $n = 0, \pm 1, \pm 2$ and $du^2 = $ integer. Since

$$det(\tilde{L}) = -\frac{1}{4}(n^2 + du^2) = -1, \tag{36}$$

we have to solve the diophantine equations $n^2 + du^2 = 4$. The solutions are the same
as for proper rotations. Therefore the holohedries of the lattice $\Lambda(B')$ are the five
types of proper rotations and the reflections with respect to some principal axes. The
Bravais lattices can be constructed with the metric tensor

$$g(B') = \begin{pmatrix} (e_1' \cdot e_1') & (e_1' \cdot e_2') \\ (e_2' \cdot e_1') & (e_2' \cdot e_2') \end{pmatrix} \tag{37}$$

corresponding to the triples (a, b, c) and angles ϕ. In the following Table we give in
four columns the Bravais lattice, the metric tensor, the rotation angle and the symbol
taken from the International Tables for Crystallography [7].

$$\begin{pmatrix} 1 & 0 \\ 0 & 1 \end{pmatrix} \quad \phi = \tfrac{\pi}{2} \quad [tp]$$

$$\begin{pmatrix} 1 & \tfrac{1}{2} \\ \tfrac{1}{2} & 1 \end{pmatrix} \quad \phi = \tfrac{\pi}{3} \quad [kp]$$

$$\begin{pmatrix} a & b \\ b & a \end{pmatrix} \quad \phi = \pi \quad [oc]$$

$$\begin{pmatrix} a & \tfrac{b}{2} \\ \tfrac{b}{2} & c \end{pmatrix} \quad \phi = \pi \quad [kp]$$

$$\begin{pmatrix} a & 0 \\ 0 & c \end{pmatrix} \quad \phi = \pi \quad [op]$$

4 CRYSTALLOGRAPHIC LORENTZ TRANSFORMATIONS IN 2D

Given a set $(\tilde{L}, L, S)$ with $L \in O(1,1), S \in GL(2, R), \tilde{L} \in GL(2, Z)$ we wish to find
an $\tilde{L}$ that satisfies

$$\tilde{L} = S^{-1} L S \tag{38}$$

For an orthonormal basis $B = (e_1, e_2)$, L satisfies

$$L^t g(B) L = g(B), \quad g(B) = \begin{pmatrix} 1 & 0 \\ 0 & -1 \end{pmatrix} \tag{39}$$

For a new basis $B' = BS$, the metric tensor becomes

$$g(B') = S^t g(B) S = \begin{pmatrix} a & \tfrac{1}{2}b \\ \tfrac{1}{2}b & c \end{pmatrix}. \tag{40}$$

From the columns of S we form the new basis $B' = (e'_1, e'_2)$. Given the lattice $\Lambda(B')$, the largest point group whose elements $L, \tilde{L}$ satisfy Eq.(38) is the holohedry of Λ, and the corresponding lattice is a Bravais lattice.

For S we choose

$$S = \begin{pmatrix} \sqrt{a} & \frac{1}{2}\frac{b}{\sqrt{a}} \\ 0 & \frac{1}{2}\frac{\sqrt{d}}{\sqrt{a}} \end{pmatrix}, \quad d = b^2 - 4ac. \tag{41}$$

For proper Lorentz transformations we take

$$L = \begin{pmatrix} \cosh\beta & \sinh\beta \\ \sinh\beta & \cosh\beta \end{pmatrix} \tag{42}$$

such that

$$\tilde{L} = S^{-1}LS = \begin{pmatrix} \frac{1}{2}(n - bu), & -cu \\ au, & \frac{1}{2}(n + bu) \end{pmatrix} \tag{43}$$

with $n = 2\cosh\beta$, $u = \pm\sqrt{\frac{n^2-4}{d}}$, $det(\tilde{L}) = \frac{1}{4}(n^2 - du^2) = 1$.

A metric tensor $g(B')$ is called *integral* when a, b, c are integral numbers, and *primitive* if a, b, c are relative prime, $g.c.d.(a, b, c) = 1$. Janner and Ascher [5] proved that the general form of a proper crystallographic Lorentz transformation is given by Eq.(43) with $(a, b, c) \in Z^3$ and n, d, u integral solutions of the diophantine Pell's equation

$$n^2 - du^2 = 4, \quad d := b^2 - 4ac. \tag{44}$$

It can also be proven that each Bravais class of lattices that have holohedries of infinite order contains a primitive lattice with $a > 0$, and at most two primitive lattices.

Janner and Ascher [5] characterize properties of lattices occurring in Bravais classes of crystallographic groups. First they introduce the notion of the *mirror lattice* $\Lambda' := X\Lambda$, $X \in GL(2, Z)/SL(2, Z)$ and of the *inverse lattice* $\overline{\Lambda} := \{-g(B)\}_+$ and classify the Bravais lattices into five types in terms of these notions. They call the metric tensors $g(B') = (a, b, c)$ *ambiguous* if a divides b, *rectangular* if b/a is even, and *rhombic* if b/a is odd. They also describe the possible lattice types for finite and for infinite holohedries.

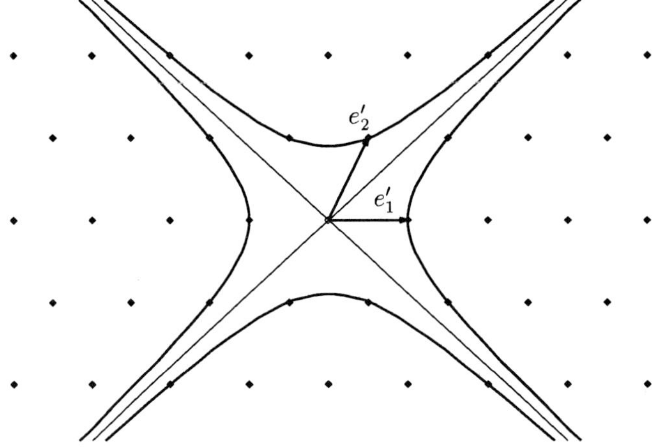

Fig.1. The Bravais lattice $n = 3, (a, b, c) = (1, 1, -1)$ with the basis (e'_1, e'_2).

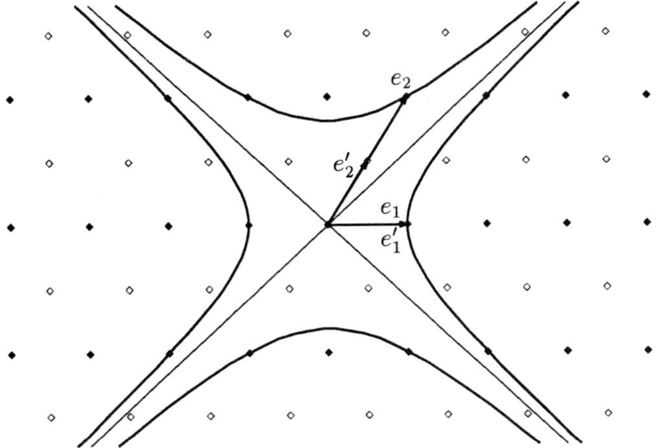

Fig.2. The Bravais lattices $n = 4, (a, b, c) = (1, 2, -2)$ with the basis (e_1, e_2) and $n = 4, (a, b, c) = (2, 2, -1)$ with the basis (e'_1, e'_2). The second lattice (open squares) is a centering of the first one (full squares).

They determine the classes of primitive relativistic lattices, using the correspondence between quadratic forms and lattices: The proper arithmetic classes of primitive indefinite quadratic forms are in one-to-one correspondence with the Bravais classes of metric lattices. Since there are infinitely many equivalent forms, one can choose a reduced form to characterize a Bravais class. The conditions for a form $[a, b, c]$ to be reduced are

$$0 < \sqrt{d} - b < 2|a| < \sqrt{d} + b,$$
$$0 < \sqrt{d} - b < 2|c| < \sqrt{d} + b, \tag{45}$$

The number of equivalent reduced forms is finite and they correspond to the primitive lattices. Each Bravais class of lattices that has holohedries of infinite order contains at most two primitive lattices with $a > 0$. In the Appendix of [5] the description of all Bravais classes for $n \leq 25$ are given. The geometrical interpretation of these Bravais classes can be drawn with the correspondence between the quadratic forms $[a, b, c]$ and the metric tensor $g(B') = (a, b, c)$. In Figs.1,2 we illustrate the following cases:

$$n = 3 \qquad (1, 1, -1) \tag{46}$$
$$\left\{ \begin{array}{ll} e'_1 = & (1, 0) \\ e'_2 = & (\tfrac{1}{2}, \tfrac{1}{2}\sqrt{5}) \end{array} \right\}$$
$$n = 4 \qquad (1, 2, -2)$$
$$\left\{ \begin{array}{ll} e'_1 = & (1, 0) \\ e'_2 = & (1, \sqrt{3}) \end{array} \right\} \tag{47}$$
$$(2, 2, -1)$$
$$\left\{ \begin{array}{ll} e'_1 = & (\tfrac{1}{\sqrt{2}}, 0) \\ e'_2 = & (\tfrac{1}{\sqrt{2}}, \tfrac{\sqrt{3}}{2}) \end{array} \right\}.$$

5 CRYSTALLOGRAPHIC LORENTZ TRANSFORMATIONS
IN $3D$

The program outlined in section 2 can be enlarged to dimension 3. The real subgroups of $SL(3,R)$ are $SO(3)$ and $SO(2,1)$. The eigenvalues of $SO(3)$ matrices are $(\lambda_1, \lambda_2, \lambda_3) = (1, \lambda_2, \overline{\lambda}_2)$ with $|\lambda_2| = 1$. For $SO(2,1)$ the eigenvalues are $(1, \lambda_2, \lambda_3)$, $\lambda_2 \lambda_3 = 1$, where the last pair consists of two different or equal real, or of two complex conjugate numbers.

For the crystallographic rotations the Bravais classes are well known [7]. For the crystallographic Lorentz transformations the construction of Bravais lattices can be done as in the 2-dimensional case. We develop a method for the Bravais classes when the metric tensor $g(B) = diag(1, 1, -1)$ and the basis is orthonormal. The method is based on the homomorphism $SL(2,R) \sim SO(2,1)$.

First of all we introduce the Cayley parametrization [10] for $SO(2,1)$. Given the Cayley transform for the general element of $SO(2,1)$,

$$L = \frac{1-X}{1+X}, \quad L^t g(B) L = g(B), \quad X^t g(B) + g(B) X = 0. \tag{48}$$

we construct

$$X = \frac{1}{m} \begin{pmatrix} 0 & p & r \\ -p & 0 & t \\ r & t & 0 \end{pmatrix}, \tag{49}$$

hence

$$L = \Delta^{-1} \begin{pmatrix} m^2 - p^2 + r^2 - t^2, & 2mp + 2rt, & -2mr - 2pt, \\ -2mp + 2rt, & m^2 - p^2 - r^2 + t^2, & -2mt + 2pr, \\ -2mr + 2pt, & -2mt - 2pr, & m^2 + p^2 + r^2 + t^2 \end{pmatrix}, \tag{50}$$

with $\Delta = m^2 + p^2 - r^2 - t^2$. If we define the general element of $SL(2,R)$ by

$$M = \begin{pmatrix} a & b \\ c & d \end{pmatrix}, \quad ad - bc = 1 \tag{51}$$

and identify

$$a = m - t \quad , \quad b = -p - r, \tag{52}$$
$$c = p - r \quad , \quad d = m + t$$

we obtain the desired homomorphism. To M and $-M$ corresponds the same L.

Now suppose $M \in SL(2,Z)$, $a+b+c+d$ =even. Since $ad-bc = m^2 + p^2 - r^2 - t^2 = 1$ we have $L \in SL(3,Z)$.

The converse is also true. If $L \in SL(3,Z)$ in the Cayley transform we solve for X,

$$X = \frac{1-L}{1+L} = \frac{2}{1+L} - 1 = \frac{1}{m} \begin{pmatrix} 0 & p & r \\ -p & 0 & t \\ r & t & 0 \end{pmatrix}. \tag{53}$$

We can identify the parameters p, r, t in terms of the complementary minors of $\frac{1-L}{1+L}$ and m with the determinant of $(1 + L)$. Because L is integral, the complementary minors of $(1 + L)$ and $det(1 + L)$ are always integers, and therefore a, b, c, d will also be integers.

The only exception to the last proposition is when we solve for m, p, r, t in eq.(53) in terms of integral elements of L, but $m^2 + p^2 - r^2 - t^2 = 2$. In this case from the homomorphism between $SL(2, R)$ and $SO(2, 1)$ we find

$$a = \tfrac{1}{\sqrt{2}}(m - t), \quad b = \tfrac{1}{\sqrt{2}}(-p - r),$$
$$c = \tfrac{1}{\sqrt{2}}(p - r), \quad d = \tfrac{1}{\sqrt{2}}(m + t), \tag{54}$$

with $ad - bc = \tfrac{1}{2}(m^2 + p^2 - r^2 - t^2) = 1$, but a, b, c, d are not integral numbers.

When $m^2 + p^2 - r^2 - t^2 = N, N \neq 2$, in Eq.(53) we get an integral matrix for $SL(2, Z)$ for the following reasons:

Given m, p, r, t from Eq.(53) we want to recover the matrix L from the Cayley parametrization Eq.(50). We can write this expression as a new integral matrix $L + 1$,

$$L + 1 = \frac{2}{m^2 + p^2 - r^2 - t^2} \begin{pmatrix} m^2 - t^2, & mp + rt, & -mr - pt \\ -mp + rt, & m^2 - r^2, & -mt + pr \\ -mr + pt, & -mt - pr, & m^2 + p^2 \end{pmatrix} \tag{55}$$

If $m^2 + p^2 - r^2 - t^2 = N \neq 2$, N should be simultaneously a divisor of the matrix elements of $A = (a_{ij})$, and of the linear combinations $a_{ij} \pm a_{ji}, i \neq j$, namely

$$mp, \; mr, \; mt, \; pr, \; pt, \; rt, \; m^2 - t^2, \; m^2 - r^2, \; m^2 + p^2 \tag{56}$$

Now suppose we decompose N into prime factors, $n = N_1 N_2 \ldots N_i N_k$. Take N_i. It should divide at least some of the integers , say m, that appears in the products mp, mr, mt. Because m does not appear in pr, pt, rt, N_i should divide also one of the remaining parameters, say p, and consequently some of the parameters of the product rt, say r. But if N_i divides m, p, r, it should divide also t, because N_i is a divisor of $m^2 - t^2$. Finally, if N_i divides m, p, r, t simultaneously, from the condition $m^2 + p^2 - r^2 - t^2 = N$ it follows that N_i should be contained twice in N, and we can simplify N_i^2 in both sides of $m^2 + p^2 - r^2 - t^2 = N$.

We can apply the same argument to all prime factors of N, and after simplification we obtain $m'^2 + p'^2 - r'^2 - t'^2 = 1$, where $m' = m/N, p' = p/N, r' = r/N, t' = t/N$. From the correspondence between the Cayley parameters for $SO(2, 1)$ and the matrix elements of $SL(2, R)$,

$$a = m' - t', \; b = -p' - r', \; c = p' - r', \; d = m' + t', \tag{57}$$

with $ad - bc = \tfrac{1}{2}(m'^2 + p'^2 - r'^2 - t'^2) = 1$ we obtain an integral matrix $M \in SL(2, Z)$.

We address ourselves to the problem of finding all the integral elements of $SL(2, Z)$. We identify the elements $M, -M$ because they correspond to the same L. According to Fricke and Klein [3] and to Kurosh [4], any integral unimodular 2×2 matrix can always be written as the free product of a cyclic group of order 2 and a cyclic group of order 3, namely

$$M = t^\alpha u^{\pm 1} \ldots t u^{\pm 1} t^\beta, \; \alpha, \beta = 0, 1, \tag{58}$$

where

$$t = \begin{pmatrix} 0 & -1 \\ 1 & 0 \end{pmatrix}, \; u = \begin{pmatrix} 0 & -1 \\ 1 & 1 \end{pmatrix}, \; t^2 = 1, u^3 = 1, \tag{59}$$

with the condition that we identify t and $-t$, u and $-u$. The converse is also true: if we write down all the possible words with t and u as in Eq.(58), we will get all the unimodular integral 2×2 matrices.

The equivalent classes of the elements of $SL(2, Z)$ can be obtained with the help of cyclic words. We bend the word into a circle, that is, we write down its end in front of its beginning, and carry out the cancellations. In this way we obtain the cyclic word. Now two words are conjugate if they have the same cyclic word. For example, take

$$M = tututut = \begin{pmatrix} 3 & -1 \\ 1 & 0 \end{pmatrix} \tag{60}$$

to find the conjugates belonging to the same class:

$$ututu = \begin{pmatrix} 0 & -1 \\ 1 & 3 \end{pmatrix} \quad , \quad tutuu = \begin{pmatrix} 2 & 1 \\ 1 & 1 \end{pmatrix},$$

$$utuut = \begin{pmatrix} 1 & -1 \\ 1 & 2 \end{pmatrix} \quad , \quad tuutu = \begin{pmatrix} 1 & 1 \\ 1 & 2 \end{pmatrix}, \tag{61}$$

$$uutut = \begin{pmatrix} 2 & -1 \\ -1 & 1 \end{pmatrix} \quad .$$

Nevertheless these elements do not lead to integral matrices of the Lorentz group in 3 dimensions because $a + b + c + d =$odd. Another example that satisfies the last condition is

$$M = tutut = \begin{pmatrix} 2 & -1 \\ 1 & 0 \end{pmatrix} \tag{62}$$

with arithmetically equivalent elements

$$utu = \begin{pmatrix} 0 & 1 \\ -1 & -2 \end{pmatrix} \quad , \quad tuu = \begin{pmatrix} 1 & 0 \\ 1 & 1 \end{pmatrix} \tag{63}$$

$$uut = \begin{pmatrix} 1 & -1 \\ 0 & 1 \end{pmatrix}$$

Only the first and second matrix satisfy the condition $a + b + c + d =$even and hence lead to integral Lorentz transformations that are conjugate in $SL(3, Z)$.

6 KLEIN-GORDON EQUATION ON A LATTICE

Let B be an orthonormal basis (e_1, e_2) and the metric tensor $g(B) = diag(1, -1)$. We introduce the method of finite differences [12] for the Klein-Gordon field on the lattice. A scheme for the wave equation consistent with the continuum case can be constructed:

$$(\frac{1}{\tau^2}\Delta_n\tilde{\Delta}_j\nabla_n\tilde{\nabla}_j - \frac{1}{\epsilon^2}\Delta_j\tilde{\Delta}_n\nabla_j\tilde{\nabla}_n + M^2\tilde{\Delta}_n\tilde{\Delta}_j\tilde{\nabla}_n\tilde{\nabla}_j)\Phi_j^n = 0, \tag{64}$$

where the field is defined in the grid points

$$\Phi_j^n = \Phi(j\epsilon, n\tau), \tag{65}$$

with ϵ, τ being the fundamental space and time intervals, and (j, n) integer numbers,

$$\Delta_j\Phi_j^n = \Phi_{j+1}^n - \Phi_j^n \quad , \quad \nabla_j\Phi_j^n = \Phi_j^n - \Phi_{j-1}^n, \tag{66}$$

$$\tilde{\Delta}_j\Phi_j^n = \frac{1}{2}(\Phi_{j+1}^n + \Phi_j^n) \quad , \quad \tilde{\nabla}_j\Phi_j^n = \frac{1}{2}(\Phi_j^n + \Phi_{j-1}^n)$$

and similarly for the time index.

It can be proved by direct substitution that the following functions of discrete variables are (plane wave) solutions of Eq.(64):

$$f_j^n(\kappa,\omega) = \left(\frac{1 + \frac{1}{2}i\epsilon\kappa}{1 - \frac{1}{2}i\epsilon\kappa}\right)^j \left(\frac{1 - \frac{1}{2}i\tau\omega}{1 + \frac{1}{2}i\tau\omega}\right)^n, \tag{67}$$

provided that the dispersion relation $\omega^2 - \kappa^2 = M^2$ is satisfied, where ω is the angular frequency and κ the wave number. Imposing boundary conditions on the solutions, $f_0^n = f_N^n$, $f_j^0 = f_j^N$, N fixed, we get

$$\kappa_m = \frac{2}{\epsilon}\tan\frac{\pi m}{N}, \quad \omega_m = \frac{2}{\tau}\tan\frac{\pi m}{N}, \quad m = 0, 1, \ldots N - 1, \tag{68}$$

from which we get the discrete mass spectrum [12]

$$M^2 = \omega_m^2(1 - v_g^2), \tag{69}$$

with v_g being the group velocity of the particle.

Now we turn to a Bravais lattice $[a, b, c]$ with basis

$$B' = \left\{ \begin{array}{rcl} e_1' &=& \sqrt{a}\,e_1, \\ e_2' &=& \frac{1}{2}\frac{b}{\sqrt{a}}e_1 + \frac{1}{2}\frac{\sqrt{d}}{\sqrt{a}}e_2 \end{array} \right\}, \tag{70}$$

and a metric tensor

$$g(B') = \left(\begin{array}{cc} a & \frac{b}{2} \\ \frac{b}{2} & c \end{array} \right) := g^{AB}. \tag{71}$$

The length of the basis vectors become $|e_1'| = \sqrt{a}$, $|e_2'| = \sqrt{c}$. The wave equation now reads in tensorial notation

$$(g^{AB}\delta_A\delta_B + M^2\tilde{\delta}_A\tilde{\delta}_B)\Phi = 0, \quad A, B = j, n, \tag{72}$$

where $\delta_j := \Delta_j\tilde{\Delta}_n$, $\delta_n := \Delta_n\tilde{\Delta}_j$, $\tilde{\delta}_j = \tilde{\delta}_n := \tilde{\Delta}_j\tilde{\Delta}_n$. Explicitly this equation is written

$$(a\frac{1}{\tau'^2}\Delta_n\tilde{\Delta}_j\nabla_n\tilde{\nabla}_j + b\frac{1}{\tau'}\Delta_n\tilde{\Delta}_j\frac{1}{\epsilon'}\nabla_j\tilde{\nabla}_n + c\frac{1}{\epsilon'^2}\Delta_j\tilde{\Delta}_n\nabla_j\tilde{\nabla}_n + M^2\tilde{\Delta}_n\tilde{\Delta}_j\tilde{\nabla}_n\tilde{\nabla}_j)\Phi_j^n = 0, \tag{73}$$

with the new fundamental space and time intervals $\tau' = \sqrt{a}$, $\epsilon' = \sqrt{c}$. As before, a plane wave solution is

$$f_j^n(\kappa,\omega) = \left(\frac{1 + \frac{1}{2}i\epsilon'\kappa}{1 - \frac{1}{2}i\epsilon'\kappa}\right)^j \left(\frac{1 - \frac{1}{2}i\tau'\omega}{1 + \frac{1}{2}i\tau'\omega}\right)^n, \tag{74}$$

provided that the dispersion equation

$$a\kappa^2 + b\kappa\omega + c\omega^2 = M^2 \tag{75}$$

is satisfied. Obviously this expression is invariant under crystallographic Lorentz transformations corresponding to the Bravais lattice $[a, b, c]$ as defined in section 4.

As mentioned in the introduction, the method of finite differences can be used to solve non-perturbatively gauge field theories on the lattice, or can be interpreted in terms of realistic models for describing field equations on a discrete space-time.

APPENDIX: A BRAVAIS CLASS FOR $3D$ LORENTZ TRANSFORMATIONS

There are 4 arithmetic classes of shortest words in $SL(2, Z)$ with $a + b + c + d =$ even : $(tutut, utu), (tutu, utut), (tu^{-1}tu^{-1}t, u^{-1}tu^{-1}), (tu^{-1}tu^{-1}, u^{-1}tu^{-1}t)$. We take

$$A = tutut = \begin{pmatrix} 2 & -1 \\ 1 & 0 \end{pmatrix} \tag{76}$$

which from section 5 corresponds to the proper Lorentz transformation

$$L = \begin{pmatrix} -1 & 2 & 2 \\ -2 & 1 & 2 \\ -2 & 2 & 3 \end{pmatrix} \in SO(2, 1) \tag{77}$$

with $m = 1, p = 1, r = 0, t = 1$, $m^2 + p^2 - r^2 - t^2 = 1$. We also construct the proper rotation around the z-axis:

$$R = \begin{pmatrix} 0 & 1 & 0 \\ -1 & 0 & 0 \\ 0 & 0 & 1 \end{pmatrix} \in SO(2, 1) \tag{78}$$

with $m = 1, p = 1, r = t = 0$, $m^2 + p^2 - r^2 - t^2 = 2$.

In the defining equation equation $S^{-1}LS = \tilde{L}$, we take $S = Id$. Accordingly

$$B = (e_1, e_2, e_3), \ e_1 = (1, 0, 0), \ e_2 = (0, 1, 0), \ e_3 = (0, 0, 1) \tag{79}$$

The Bravais class is given by the Bravais lattice

$$\Lambda = [z_1 e_1 + z_2 e_2 + z_3 e_3, \ (z_1, z_2, z_3) \in Z] \tag{80}$$

and the holohedry is $R^k \times L^n$, $(k = 0, 1, 2, 3; \ n = 0, \pm 1, \pm 2, \pm 3, \ldots)$. In the following Figure we give the time-like $(x^2 + y^2 - z^2 < 0)$ lowest vectors.

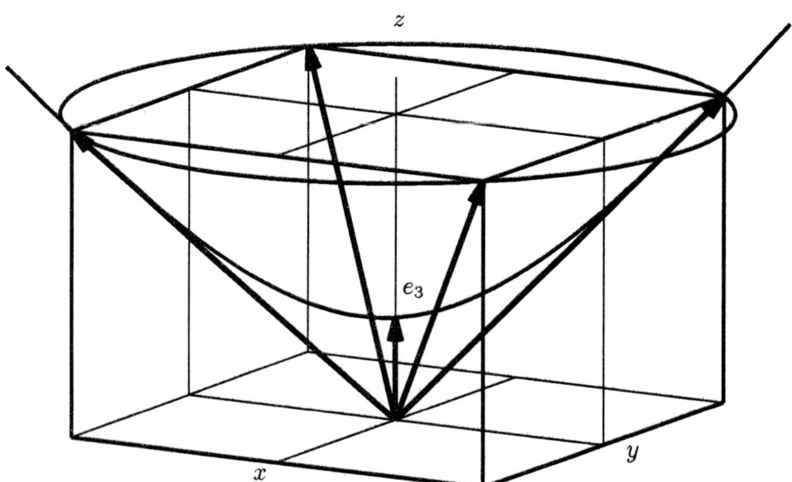

Fig. 3. The lowest vectors $v = R^k L e_3$, $(k = 0, 1, 2, 3)$ in the $SO(2, 1)$ Bravais lattice.

REFERENCES

[1] II Brown, R Bülow, J Neubüser, II Wondratschek, H Zassenhaus 1978, *Crystallographic Groups of Four-Dimensional Space*, Wiley, New York

[2] M Baake, D Joseph and P Kramer 1992, Phys Lett **A168** 199-208

[3] F Klein and R Fricke 1890-92, *Vorlesungen über die Theorie der Elliptischen Modulfunktionen*, vol.1, pp. 218-219, 452-455,

[4] A G Kurosh 1960, *The Theory of Groups*, vol. 2, pp. 261-264 Chelsea, New York

[5] A Janner, E Ascher 1969, Physica **45** 33-66

[6] A Janner 1991, Acta Cryst **A47** 577-590

[7] T Hahn Ed. 1983, *International Tables for Crystallography*, Reidel Publishing, Dordrecht

[8] P Kramer 1993, *Free Groups, their Automorphism Groups, Applications*, in: Symmetries in Science VI, 415-430 Ed. B. Gruber, Plenum, New York

[9] P Kramer 1993, J Phys **A26** 213-228

[10] M Lorente 1993, *Representations of Classical Groups on the Lattice and Its Application to the Field Theory on Discrete Space-Time*, in: Symmetries in Science VI, 437-454, Ed. B Gruber, Plenum, New York

[11] M Lorente 1994, *Quantum processes and the foundation of relational theories of space and time.* in: Encuentros Relativistas Espanoles **93**, Ed. J Diaz Alonso, M Lorente, Editorial Lumiere, Paris

[12] M Lorente 1993, J Group Theory in Physics **1**, 105-121

[13] R L E Schwarzenberger 1980, *N-Dimensional Crystallography*, Pitman, San Francisco

DIFFERENT BASES OF κ-DEFORMED POINCARÉ ALGEBRA

J. Lukierski *

Institute of Theoretical Physics
University of Wrocław
Pl. Maksa Borna 9
PL-50-204 Wrocław
Poland

1 INTRODUCTION

The contraction of $\mathcal{U}_q\left(O(3,2)\right)$ ($|q| = 1$ [1,2] or q real [3]) provided first quantum deformations $\mathcal{U}_\kappa\left(\mathcal{P}_4\right)$ of $D = 4$ Poincaré algebra $\mathcal{P}_4 \equiv \left(M_{\mu\nu}, P_\mu\right)$ with κ describing the mass-like deformation parameter [1]. These so-called κ-deformations are considered in the class of noncommutative and noncocommutative Hopf algebras [4-6], with modified classical coalgebra sector. It should be stressed that the choice of ten generators obtained in [3] is not unique: one can distinguish at least two other bases, with quite interesting properties:

i) The bicrossproduct basis, obtained in [7]. In such a basis the quantum algebra $\mathcal{U}_\kappa\left(\mathcal{P}_4\right)$ can be written in the form [2]

$$\mathcal{U}_\kappa\left(\mathcal{P}_4\right) = \mathcal{U}(O(3,1)) \bowtie T_4^\kappa \tag{1.1}$$

where
$-\mathcal{U}(O(3,1))$ describes the Hopf algebra generated by classical Lorentz generators, with commutative coproducts,
$- T_4^\kappa$ describes the κ-deformed Hopf algebra of fourmomenta, with commuting generators in algebra sector and κ-deformed coalgebra relations.

ii) The classical Poincaré algebra basis, obtained in [11][3]. In such a framework

---

*supported by KBN grant 2P 302 087 06.

[1]Further we shall consider only the deformation $\mathcal{U}_\kappa\left(\mathcal{P}_4\right)$ obtained in [3], with standard reality structure.

[2]For the bicrossproduct formulation and the notations see [8-10].

[3]First choice of κ-Poincaré algebra with classical Poincaré generators was given in [12], but it leads to the difficulties with the reality conditions. Special case of the formulae given in [11] was found by Ruegg [13].

the algebra is a standard Lie algebra, but the coproducts are very complicated noncocommutative expressions.

The aim of this lecture is to present these three distinguished bases. It appears that to each of these bases one can introduce the dual ones, defining particular versions of κ-deformed quantum Poincaré groups. At present it is known [14] only the quantum Poincaré group dual to the κ-deformed Poincaré algebra in bicrossproduct basis. It appears that such a quantum Poincaré group coincides with the one obtained in [15] by quantization of the Lie-Poisson brackets on Poincaré group.

2 STANDARD BASIS

Firstly κ-deformed Poincaré algebra with standard reality structure was given in [3]. It is described by the following Hopf algebra structure, with real ten generators:

i) Algebra structure:

$$[M_i, M_j] = i\epsilon_{ijk}M_k \quad , \qquad [P_\mu, P_\nu] = 0 \,, \tag{2.1a}$$

$$[L_i, M_j] = i\epsilon_{ijk}L_k \,, \tag{2.1b}$$

$$[M_i, P_j] = i\epsilon_{ijk}P_k \quad , \qquad [M_i, P_0] = 0 \,, \tag{2.1c}$$

$$[L_i, P_j] = i\kappa\delta_{ij}\sinh\frac{P_0}{\kappa} \quad , \qquad [L_i, P_0] = iP_i \,, \tag{2.1d}$$

$$[L_i, L_j] = -i\epsilon_{ijk}\left(M_k\cosh\frac{P_0}{\kappa} - \frac{1}{4\kappa^2}P_k(\vec{P}\vec{M}) \right) , \tag{2.1e}$$

where κ is a mass-like deformation parameter, $\vec{M} = (M_1, M_2, M_3)$ and $\vec{P} = (P_1, P_2, P_3)$, $(\vec{P}\vec{M} = P_1M_1 + P_2M_2 + P_3M_3)$.

ii) Coalgebra:

$$\Delta(M_i) = M_i \otimes 1 + 1 \otimes M_i \,, \tag{2.2a}$$

$$\Delta(L_i) \ = L_i \otimes e^{\frac{P_0}{2\kappa}} + e^{-\frac{P_0}{2\kappa}} \otimes L_i +$$
$$+ \tfrac{1}{2\kappa}\epsilon_{ijk}(P_j \otimes M_k e^{\frac{P_0}{2\kappa}} + e^{-\frac{P_0}{2\kappa}}M_j \otimes P_k) \,, \tag{2.2b}$$

$$\Delta(P_i) \ = P_i \otimes e^{\frac{P_0}{2\kappa}} + e^{-\frac{P_0}{2\kappa}} \otimes P_i \tag{2.2c}$$

$$\Delta(P_0) \ = P_0 \otimes 1 + 1 \otimes P_0 \,.$$

The counits ϵ of all generators are zero.

iii) Antipodes:

$$S(M_i) = -M_i, \quad S(P_\mu) = -P_\mu, \quad S(L_i) = -L_i + \frac{3}{2}\frac{i}{\kappa}P_i. \tag{2.3}$$

The deformed two Casimirs for the κ-Poincaré algebra are the following:

1) Mass square operator:

$$C_1 = \vec{P}^2 + 2\kappa^2\left(1 - \cosh\frac{P_0}{\kappa}\right) = \vec{P}^2 - \left(2\kappa\sinh\frac{P_0}{2\kappa}\right)^2. \tag{2.4}$$

The eigenvalues $C_1 = -M_0^2$ determine the κ-relativistic rest mass M_0.

2) Relativistic spin square operator:

$$C_2 = \left(\cosh\frac{P_0}{\kappa} - \frac{\vec{P}^2}{4\kappa^2}\right)W_0^2 - \vec{W}^2, \tag{2.5}$$

where the κ-deformed Pauli-Lubanski four-vector is given by the formulae

$$W_i = \kappa M_i \sinh\frac{P_0}{\kappa} + \epsilon_{ijk}P_j L_k, \quad W_0 = \vec{P}\vec{M}. \tag{2.6}$$

The simplest spinless realization of the κ-Poincaré for which $\vec{P}\vec{M} = 0$ and

$$P_\mu f(p) = p_\mu f(p)$$

is

$$M_i = -\epsilon_{ijk}p_j\frac{\partial}{\partial p_k}, \tag{2.7a}$$

$$L_i = p_i\frac{\partial}{\partial p_0} - [p_0]_\kappa\frac{\partial}{\partial p_i}, \tag{2.7b}$$

where

$$[p_0]_\kappa = \kappa\sinh\left(\frac{p_0}{\kappa}\right). \tag{2.8}$$

From the Casimir (2.4) ($C_1 \equiv -M_0^2$) we obtain the following κ – deformation of the mass-shell condition

$$\vec{P}^2 - \left(\left[\frac{P_0}{2}\right]\right)^2 = -M^2 \tag{2.9}$$

which implies that the following dispersion relation for the momentum dependence of the energy

$$E = \omega_\kappa(\vec{P}^2) = 2\kappa\,\mathrm{arc}\sinh\frac{\omega}{2\kappa} \tag{2.10}$$

where $\omega = \sqrt{\vec{P}^2 + M^2}$. The formula for the momentum dependence of the velocity has been considered at previous Bregenz Symposium [16] and looks as follows:

$$v_i = \frac{\partial\omega_\kappa(\vec{P}^2)}{\partial P_i} = \frac{1}{\left(1 + \frac{\omega^2}{4\kappa^2}\right)^{1/2}}\frac{P_i}{\omega}. \tag{2.11}$$

Asymptotically one obtains (see also [17])

$$v^2 \xrightarrow[|\vec{P}|^2 \to \infty]{} \frac{4\kappa^2}{\vec{P}^2} \tag{2.12}$$

in contrast with the Einstein's result that

$$v^2 \xrightarrow[|\vec{P}|^2 \to \infty]{} 1 \,. \tag{2.13}$$

It is interesting to consider different bases of κ-Poincaré algebras and discuss the modifications of the formula (2.11)

3 BICROSSPRODUCT BASIS

The classical Poincaré algebra is an example of a semidirect product $\tilde{g} \oplus \tilde{t}$ of simple Lie algebra ($\tilde{g}$ = Lorentz sector) and Abelian algebra ($\tilde{t}$ = fourmomentum sector). The general deformation theory of such nonsemisimple Lie algebras is provided by the formalism of bicrossproducts of Hopf algebras (see [8–10]). The first example of such a type of deformation is the bicrossproduct deformation of $D = 3$ Euclidean algebra $O(3) \oplus T_3$ [8,9]; recently also the κ-deformation of $D = 4$ Poincaré algebra has been put in the bicrossproduct form [7].

In order to obtain the bicrossproduct basis for κ-Poincaré algebra one introduces the following redefinitions of the standard generators from the previous section (the space rotations generators M_i and energy operator P_0 are not changed):

$$N_i^{(\pm)} = \frac{1}{2}\{L_i, e^{\mp \frac{P_0}{2\kappa}}\} \mp \frac{\epsilon_{ijk}}{4\kappa}\{M_j, P_k\}e^{\mp \frac{P_0}{2\kappa}} \,, \tag{3.1a}$$

$$\mathcal{P}_j^{(\pm)} = P_j e^{\mp \frac{P_0}{2\kappa}} \,, \tag{3.1b}$$

where in comparison with [7] we introduced also the second choice, obtained by the change of the sign of κ-parameter (see also [18]). After the change (3.1a)-(3.1b) we obtain the following two Hopf algebras $\mathcal{U}_{\pm\kappa}(\mathcal{P}_4)$ in bicrossproduct basis:

i) Classical Lorentz subalgebra ($\dot{M}_{\mu\nu} \equiv (M_i, N_i)$)

$$[M_i, M_j] = i\epsilon_{ijk}M_k \,,$$

$$[M_i, N_j] = i\epsilon_{ijk}N_k \,, \tag{3.2}$$

$$[N_i, N_j] = -i\epsilon_{ijk}M_k \,.$$

ii) Cross-relations between classical Lorentz and fourmomentum generators:

$$[M_i, \mathcal{P}_j^{(\pm)}] = i\epsilon_{ijk}\mathcal{P}_k^{(\pm)} \,, \qquad [M_i, P_0] = 0 \,, \tag{3.3a}$$

$$[N_i^{(\pm)}, \mathcal{P}_j^{(\pm)}] = \pm i\delta_{ij}\left[\frac{\kappa}{2}(1 - e^{\mp \frac{2P_0}{\kappa}}) + \frac{1}{2\kappa}(\vec{\mathcal{P}}^{(\pm)})^2\right] \mp \frac{i}{\kappa}\mathcal{P}_i^{(\pm)}\mathcal{P}_j^{(\pm)} \,, \tag{3.3b}$$

$$[N_i^{(\pm)}, P_0] = i\mathcal{P}_i^{(\pm)} \,. \tag{3.3c}$$

It should be added that the fourmomentum generators $\mathcal{P}_\mu^{(\pm)} = (\mathcal{P}_i^{(\pm)}, P_0)$ commute.

iii) The coproduct formulae look as follows:

$$\Delta(M_i) = M_i \otimes 1 + 1 \otimes M_i \,, \tag{3.4a}$$

$$\Delta(N_i^{(+)}) = N_i^{(+)} \otimes 1 + e^{-\frac{P_0}{\kappa}} \otimes N_i^{(+)} + \frac{\epsilon_{ijk}}{\kappa}\mathcal{P}_j^{(+)} \otimes M_k \,, \tag{3.4b}$$

$$\Delta(N_i^{(-)}) = N_i^{(-)} \otimes e^{\frac{P_0}{\kappa}} + 1 \otimes N_i^{(-)} - \frac{\epsilon_{ijk}}{\kappa}M_k \otimes \mathcal{P}_j^{(-)} \,, \tag{3.4c}$$

$$\Delta(\mathcal{P}_i^{(+)}) = \mathcal{P}_i^{(+)} \otimes 1 + e^{-\frac{P_0}{\kappa}} \otimes \mathcal{P}_i^{(+)} \,, \tag{3.4d}$$

$$\Delta(\mathcal{P}_i^{(-)}) = \mathcal{P}_i^{(-)} \otimes e^{\frac{P_0}{\kappa}} + 1 \otimes \mathcal{P}_i^{(-)} \,, \tag{3.4e}$$

$$\Delta(P_0) = P_0 \otimes 1 + 1 \otimes P_0 \,, \tag{3.4f}$$

and we obtain the following antipodes

$$S(M_i) = -M_i \qquad S(P_0) = -P_0 \,,$$

$$S(\mathcal{P}_i^{(\pm)}) = -e^{\pm\frac{P_0}{\kappa}}\mathcal{P}_i^{(\pm)} \,, \tag{3.5}$$

$$S(N_i^{(\pm)}) = -e^{\pm\frac{P_0}{\kappa}}N_i^{(\pm)} \pm \frac{1}{\kappa}\epsilon_{ijk}e^{\frac{P_0}{\kappa}}\mathcal{P}_j^{(\pm)}M_k \mp \frac{i}{\kappa}\mathcal{P}_i^{(\pm)}e^{\pm\frac{P_0}{\kappa}} \,,$$

In such a way we obtained two κ-deformations of $D = 4$ Poincaré algebra, differing by the sign of the deformation parameter κ. The mass shell condition (2.9) in bicrossproduct basis takes the form:

$$\left(\mathcal{P}_j^{\pm}\right)^2 e^{\pm\frac{P_0}{\kappa}} - \left(\left[\frac{P_0}{2}\right]\right)^2 = -M^2 \tag{3.6}$$

with the following modification of the energy dispersion relation $\omega(\vec{P}) = \sqrt{\vec{P}^2 + M^2}$

i) $\vec{\mathcal{P}} = \vec{\mathcal{P}}^-$

$$\omega_\kappa^\pm(\vec{P}) = \kappa \ln\left[\left(1 + \frac{M^2}{2\kappa^2}\right) \pm \frac{M^2}{2\kappa^2}\sqrt{1 + \frac{4\kappa^2}{M^4}(\vec{P}^2 + M^2)}\right] \,. \tag{3.7}$$

We should choose the sign "+" in order to obtain the consistent solution for large $|\vec{p}|$ (for negative sign the function $e^{\frac{P_0}{\kappa}}$ takes negative values). One obtains the following formulae for the velocity

$$V_i^+(\vec{P}) = \frac{\partial\omega_\kappa^+(\vec{P})}{\partial P_i} = \frac{2\kappa}{M^2}\frac{p_i}{\sqrt{1 + \frac{4\kappa^2}{M^4}(\vec{P}^2 + M^2)}}e^{\frac{-\omega_\kappa^+(\vec{p})}{\kappa}} \,, \tag{3.8}$$

providing the limit

$$V_i^+(\vec{p}) \xrightarrow[|\vec{p}|^2 \to \infty]{} \frac{p_i}{|\vec{p}|}\frac{\kappa}{|p_i|} \,, \tag{3.9}$$

or

$$(\vec{V}^+)^2 \longrightarrow \frac{\kappa^2}{|p_i|^2} \,. \tag{3.10}$$

Comparing (3.10) with (2.12) we see that one obtains the asymptotic formulae differing only by numerical factor.

ii) $\vec{\mathcal{P}} = \vec{\mathcal{P}}^+$

In such a case one obtains the formula for the energy

$$\omega^{\pm}(\vec{\mathcal{P}}^2) = \kappa \ln \left[\frac{M^2 + 2\kappa^2}{2(\kappa^2 - \vec{\mathcal{P}}^2)} \left(1 \mp \sqrt{1 + \frac{4\kappa^2(\vec{\mathcal{P}}^2 - \kappa^2)}{(M^2 + 2\kappa^2)^2}} \right) \right] . \tag{3.11}$$

It is easy to check that only for the sign "+" one obtains the consistent formula for energy ω, which however for large values of $|\vec{p}|$ goes to minus infinity

$$\omega^+(\vec{p}^2) \xrightarrow[|\vec{p}|^2 \to \infty]{} -\kappa \ln \frac{|\vec{p}|}{\kappa} . \tag{3.12}$$

One obtains the formula for velocity $\vec{V}$ which for large $\vec{p}$ is oriented in opposite direction of the threemomenta

$$V_i^+ \xrightarrow[|\vec{p}| \to \infty]{} -\frac{\kappa}{|\vec{p}|} \frac{p_i}{|\vec{p}|} . \tag{3.13}$$

4 CLASSICAL POINCARÉ ALGEBRA BASIS

Let us derive the generators of the bicrossproduct basis of κ-Poincaré algebra satisfying the relations (3.3a)-(3.3c) (for simplicity we consider only the sign "+") as functions of classical Poincaré generators $(M_i^{(0)}, N_i^{(0)}, P_i^{(0)}, P_0^{(0)})$. Because in the bicrossproduct basis (see Sect. 3) the Lorentz algebra is classical we put

$$
\begin{aligned}
M_i &= M_i^{(0)}, & N_i &= N_i^{(0)} , \\
P_0 &= f(P_0^{(0)}, M_0^2) , & P_i &= g(P_0^{(0)}, M_0^2) P_i^{(0)}
\end{aligned}
\tag{4.1}
$$

where

$$M_0^2 = (P^{(0)})^2 - (\vec{P}^{(0)})^2 \tag{4.2}$$

is the mass Casimir of the classical Poincaré algebra. One obtains from the substitution of the relations (4.1) in the relations (2.2a)-(2.2c) that $\left(f' \equiv \frac{\partial f}{\partial P_0^{(0)}} \right)$

$$g' = -\frac{1}{\kappa} g^2 , \qquad f' = g . \tag{4.3}$$

The general solutions of equation (4.3) are the following

$$g = \frac{\kappa}{P_0^{(0)} + C} , \qquad f = \kappa \ln \frac{P_0^{(0)} + C}{A} , \tag{4.4}$$

where from the relations

$$\lim_{\kappa \to \infty} P_i = P_i^{(0)} , \qquad \lim_{\kappa \to \infty} P_0 = P_0^{(0)} , \tag{4.5}$$

one gets

$$\lim_{\kappa \to \infty} \frac{A}{\kappa} = \lim_{\kappa \to \infty} \frac{C}{\kappa} = 1 . \tag{4.6}$$

Besides one obtains the relation

$$g P_0^{(0)} = \frac{\kappa}{2}(1 - e^{-2\frac{P_0}{\kappa}}) + \frac{1}{2\kappa} P_i^2 \,, \tag{4.7}$$

which implies that

$$C^2 - A^2 = M_0^2 \,. \tag{4.8}$$

We see therefore that one obtains the family of solutions depending on one arbitrary function $A(\kappa; M_0^2)$, satisfying the condition (4.6).

The κ-deformed Casimir in Majid-Ruegg basis takes the form (see (3.6) and take the sign "+")

$$\left(2\kappa \sinh \frac{P_0}{2\kappa}\right)^2 - e^{\frac{P_0}{\kappa}} P_i^2 = M^2 \,. \tag{4.9}$$

Substituting the formulae (4.1) in the relation (3.6) one gets the following formula for the Casimir

$$M^2 = 2\kappa^2 \left(\frac{C}{A} - 1\right) \,. \tag{4.10}$$

Using (4.8) one obtains the following relations between two Casimirs (4.2) and (3.6)

$$M_0^2 = A^2 \left[\left(\frac{M^2}{2\kappa^2} + 1\right)^2 - 1\right] \,. \tag{4.11}$$

The general inverse formulae look as follows:

$$P_0^{(0)} = A\left(e^{\frac{P_0}{\kappa}} - \frac{C}{A}\right) = A\left(e^{\frac{P_0}{\kappa}} - 1 - \frac{M^2}{2\kappa^2}\right) \,, \tag{4.12a}$$

$$P_i^{(0)} = \frac{A}{\kappa} e^{\frac{P_0}{\kappa}} P_i \,. \tag{4.12b}$$

We shall consider the following two particular solutions:

a)

$$A = \kappa - \frac{M_0^2}{4\kappa} \,, \qquad C = \kappa + \frac{M_0^2}{4\kappa} \,. \tag{4.13}$$

We get

$$P_0 = \kappa \ln\left(\frac{P_0^{(0)} + \kappa + \frac{M_0^2}{4\kappa}}{\kappa - \frac{M_0^2}{4\kappa}}\right) \,,$$
$$P_i = \frac{\kappa}{P_0^{(0)} + \kappa + \frac{M_0^2}{4\kappa}} P_j^{(0)} \tag{4.14}$$

and the inverse formulae

$$P_0^{(0)} = \kappa\left(e^{\frac{P_0}{\kappa}} - 1\right) - \frac{M_0^2}{4\kappa}\left(e^{\frac{P_0}{\kappa}} + 1\right) \,,$$
$$P_i^{(0)} = \left(1 - \frac{M_0^2}{4\kappa^2}\right) P_i e^{\frac{P_0}{\kappa}} \tag{4.15}$$

where in (4.15) one should use the formulae (4.11) giving for the choice (4.13) the result

$$M_0^2 = M^2 \left(1 + \frac{M^2}{4\kappa^2}\right)^{-1} \,, \tag{4.16}$$

b)

$$A = \kappa\,, \qquad C = \left(M_0^2 + \kappa^2\right)^{\frac{1}{2}}\,. \tag{4.17}$$

One obtains

$$
\begin{aligned}
P_0 &= \kappa \ln\left[P_0^{(0)} + \left(M_0^2 + \kappa^2\right)^{\frac{1}{2}}\right] - \kappa \ln \kappa\,, \\
P_i &= \frac{\kappa}{P_0^{(0)} + (M_0^2 + \kappa^2)^{\frac{1}{2}}} P_i^{(0)}
\end{aligned}
\tag{4.18}
$$

and

$$
\begin{aligned}
P_0^{(0)} &= \kappa \sinh \frac{P_0}{\kappa} + \frac{1}{2\kappa} e^{\frac{P_0}{\kappa}} \vec{P}^2\,, \\
P_i^{(0)} &= e^{\frac{P_0}{\kappa}} P_i\,.
\end{aligned}
\tag{4.19}
$$

The formula (4.11) takes the form

$$M_0^2 = M^2 \left(1 + \frac{M^2}{4\kappa^2}\right)\,. \tag{4.20}$$

It should be mentioned that the formulae (4.19)-(4.20) have been firstly obtained in [19], and it has been observed by Ruegg [13] that the formulae (4.19) with (P_i, P_0) in bicrossproduct basis provide classical Poincaré algebra.

In the basis given by (4.12a)-(4.12b) (see also (4.15) and (4.19)) all the quantum deformation is contained in the coproducts (coalgebra sector). The relations are quite complicated and the coproduct for the energy operator $P_0^{(0)}$ does not lead to the additivity of the energy.

5   FINAL REMARKS

In order to apply the quantum deformations of Poincaré algebra to dynamical theories one should firstly understand well the κ-deformation of relativistic kinematics. At present, when several bases of κ-Poincaré algebra are known, it is not clear which basis of $\mathcal{U}_\kappa(\mathcal{P}_4)$ should be considered physical. There is therefore an important question to be answered: which basis describes <u>physical</u> generators of fourmomenta and relativistic angular momenta when κ is finite? Until 1994 there were studied the physical consequences of standard basis (see Sect. 2); but quite recently it was realized that other bases of κ-Poincaré (see Sect. 3 and 4) have very interesting properties. In Sect. 3 we discussed some physical consequences (the velocity formula) for bicrossproduct basis and did show that the one possibility out of four which might be acceptable gives the same asymptotic behaviour for velocity as the standard basis (modulo numeric factor).

In order to describe κ-deformed kinematics one should understand the physical meaning of noncocommutative coproducts of the momenta generators. Because the Minkowski space is described by Hopf algebra dual to the Hopf subalgebra of fourmomenta, one obtains therefore noncommutative Minkowski coordinates [15,7,17]. One can further argue that the physical content of the κ-deformation of relativistic symmetries requires the consideration of the dual pairs of κ-deformed Poincaré algebra and κ-deformed Poincaré group. It is therefore such a framework of noncommutative geometry which will provide in a future the possible physical interpretation of κ-deformed relativistic framework.

ACKNOWLEDGEMENT

The author would like to thank Prof. Bruno Gruber and the organizers of "Science VI" for their warm hospitality in Bregenz during the Symposium.

REFERENCES

[1] J. Lukierski, A. Nowicki, H. Ruegg and V. Tolstoy, Phys. Lett. **B264**, 331 (1991),

[2] S. Giller, J. Kunz, P. Kosiński, M. Majewski and P. Maślanka, Phys. Lett. **B286**, 57 (1992),

[3] J. Lukierski, A. Nowicki, H. Ruegg, Phys. Lett. **B293**, 334 (1992),

[4] V.G. Drinfeld, *Quantum groups*, Proc. Intern. Congress of Mathematical Physics, Berkeley, USA, 1986, p.198,

[5] S. Woronowicz, Comm. Math. Phys. **111**, 613 (1987); and **112**, 125 (1989),

[6] L. Faddeev, N.Reshetikhin and L.Takhtajan, Algebra i Analiz **1**, 178 (1989),

[7] S. Majid, and H. Ruegg, Phys. Lett. **B334**, 348 (1994),

[8] J. Lukierski, A. Nowicki and H. Ruegg, Phys. Lett. **B271**, 321 (1991),

[9] S. Majid, J. Algebra **130**, 17 (1990),

[10] S. Majid, ,,Foundations of Quantum group Theory", Chapter 6; Cambridge Univ. Press, in press.

[11] P. Kosiński, J. Lukierski, P. Maślanka and J. Sobczyk, Wrocław Univ. preprint, November 1994,

[12] P. Maślanka, Journ. Math. Phys. **34**, 6025 (1993),

[13] H. Ruegg, private communication, June 1994,

[14] P. Kosiński, P. Maślanka, Łódź Univ. preprint, November 1994; hep-th 9411033,

[15] S. Zakrzewski, Journ. Phys. **A27**, 2075 (1994),

[16] J. Lukierski, A. Nowicki and H. Ruegg, Proceeding of "Symmetries in Science VI", Bregenz 1992, ed. B. Gruber (Plenum Press, 1993),

[17] J. Lukierski, H. Ruegg and W.J. Zakrzewski, submitted to Annals of Physics,

[18] P. Kosiński, J. Lukierski, P. Maślanka and J. Sobczyk, SISSA preprint, September 1994; submitted to Jourm. of Phys. A,

[19] H. Ruegg and V.N. Tolstoy, Lett. Math. Phys. **32**, 85 (1994).

q-NONLINEARITY, DEFORMATIONS
AND PLANCK DISTRIBUTION

V. I. Man'ko★†, G. Marmo★, and F. Zaccaria★

★ Dipartimento di Scienze Fisiche,
   Universitá di Napoli "Federico II"
   Istituto Nazionale di Fisica Nucleare, Sezione di Napoli
   Mostra d'Oltremare, Pad.19 - 80125 Napoli, Italy

† Lebedev Physics Institute
   53 Leninsky Prospekt, 117924 Moscow, Russia

INTRODUCTION

In this contribution we will review a new approach to some nonlinear dynamical systems which is related to the q-deformation (or other types of deformations) of linear classical and quantum systems considered in [1]. The main idea of this approach is to replace constants like frequency or mass, etc., which are parameters of the linear systems with constants of the motion of the system. This procedure produces from the initial linear system a nonlinear one and as it was demonstrated in [1], [2] the q-oscillator of [3], [4] may be considered as a physical system with a specific nonlinearity which was called q-nonlinearity. In the example of the q-oscillator the constant parameter which was replaced by the constant of the motiom was the frequency, which became dependent on the amplitude of the vibration.

But this approach may be used for many dynamical systems. So, the constant masses in Klein-Gordon and Dirac equations or the constant signal velocity in the wave equation may be replaced by dynamical variables but these variables are chosen to be constants of the motion of the dynamical system under consideration. Conceptually we address the problem (which is not new) whether the physical constant parameters like light velocity, Planck constant or gravitational constant are constant parameters in reality or they may change, for example, depending on initial conditions. The same question may be put about such characteristics of elementary particles as the electric charge or mass, say, of electron or fine structure constant which play the role of constant parameters in the dynamical equations of the theory. The example of the q-nonlinearity of [1] showed that a linear equation or system of equations may be considered as a limit of nonlinear system characterized by a nonlinearity parameter which is not important for small intensities but starts to play important role for high field intensities or for large energy densities.

Taking into account such phenomenon naturally yields the deformation of the linear equations of the motion which are transformed due to the deformation into the nonlinear ones but of specific type. The corresponding nonlinear system of equations in these cases is simply a "reparametrized" initial linear system of equations in which constant parameters are replaced by the constants of the motion of the nonlinear system. This specific procedure of nonlinearization of the initial linear system of equations selects a large class of integrable dynamical systems which are simple and nevertheless could describe the mentioned above physical phenomena. Also such dynamical systems are appropriate to treat the q-deformations and other aspects of quantum qroups [5], [6], [7] and quantum geometry [8] as influence of different types of nonlinearities in the corresponding physical processes.

Of course, the formulated approach may be extended in the sense that one could deform not only initial linear system to make it the nonlinear one using method of replacing constant parameters by constants of the motion. In principle, one could deform a simple initial nonlinear system containing some constant parameters and to transform it into another nonlinear system by replacing the constants with the integrals of the motion of the nonlinear system. So, it is possible to extend the class of integrable nonlinear systems starting from simple integrable nonlinear systems and "reparametrizing" them, i. e. replacing the constant parameters by the constants of the motion. Such approach may be useful, for instance, in q-deforming general relativity or Yang-Mills type gauge theories. Up to our knowledge such classes of both linear and nonlinear deformed integrable dynamical systems have not been studied till now. It should be noted that the method of obtaining the integrable nonlinear systems starting from the linear ones is close in spirit to the generalized reduction procedure of considering the integrable dynamical systems [9], [10].

The goal of our work is to present a general scheme for the described procedure of deforming the linear systems with finite number of degrees of freedom (next Section). In subsequent sections we also will give some examples of the deformation including infinite number of degrees of freedom (wave equation). Then the physical consequencies of the q-nonlinearity as blue shift effect [2], deformation of Planck distribution formula [2], [11], and change of a charge form-factor [12] will be discussed. In the last two sections deformed Klein-Gordon equation and Maxwell equation in the vacuum will be considered.

SPECIAL NONLINEAR SYSTEMS OBTAINED FROM LINEAR ONES

We consider a carrier space R^n. With any matrix A we associate the (linear) vector field

$$X_A = x^i A^i_j \frac{\partial}{\partial x^j}.$$

This association is actually a Lie algebra isomorphism between the set of all matrices and the linear vectorfields, since

$$aA + bB \to aX_A + bX_B, \quad [A, B] \to X_{[A,B]},$$

with A, B matrices and a, b real numbers.

We study the dynamics described by a linear vectorfield X_A and make first some considerations about its infinitesimal symmetries, to which correspond those vectorfields which commute with X_A. In particular mutually commuting symmetries are obtained by means of powers of A and the above association. In facts, using the associative product of matrices, we can consider $A^k \doteq A \cdot A \cdot .....A$ (k times). If we associate vector fields with anyone of these powers we get

$$X_k = (A^k)_n^m x_m \frac{\partial}{\partial x_n} \tag{1}$$

and

$$[X_k, \ X_j] = 0. \tag{2}$$

Furthermore

$$L_{X_A} X^k = [X_A, X^k] = 0$$

and we have so obtained mutually commuting symmetries for our initial dynamical system X_A. They span a linear s-dimensional real space, where s is the degree of the minimal polynomial associated with A.

For matrices A^k such that $\det(1 \pm A^k) \neq 0$, we can define invertible transformations

$$\Phi_k = \frac{1 - A_k}{1 + A_k} \tag{3}$$

which are finite symmetries for X_A, i. e.

$$\Phi_k^{-1} A \Phi_k = A, \tag{4}$$

or consider $e^{sA^k} = \varphi_k$, which are also finite symmetries for X_A.

We notice that for a generic matrix A there will be n independent such matrices and they are a basis of all the infinitesimal symmetries for X_A. When A is not generic, there are other infinitesimal symmetries which are not obtained from powers of A and moreover the algebra of symmetries needs not be Abelian. Again we can use either the Cayley map or the exponentiation to get finite symmetries out of the infinitesimal ones.

As a general remark, we see that if $\mathcal{F}_A \in \mathcal{F}(R^n)$ is the subring of the constants of the motion for X_A, and $X_1, X_2, \ldots, X_S$ is a basis of linear infinitesimal symmetries for X_A, we get nonlinear infinitesimal symmetries like $X = f^1 X_1 + f^2 X_2 + \cdots + f^S X_S$ for any choice of $f^1, f^2, \ldots, f^S \in \mathcal{F}_A$. Furthermore, if X_A is "reparametrized", i. e. replaced with $f X_A$, $f \in \mathcal{F}_A$, the previous symmetries are broken with the exception of those which preserve f, i. e. $L_{X_j} f = 0$ (j= 1,..,S).

We will make now the dynamics nonlinear by replacing the entries $A_j^i \in R$ with constants of the motion for X_A. With this procedure, infinitesimal symmetries are no more linear and they do not close on a finite dimensional Lie algebra.

To make contact with the quantum situation we shall consider the previous procedure in the framework of Hamiltonian dynamics. If we consider the carrier space R^{2n}, with the symplectic matrix

$$J = \begin{pmatrix} 0 & 1 \\ -1 & 0 \end{pmatrix} \tag{5}$$

the dynamics represented by X_A will be Hamiltonian if ${}^t A \, J + J \, A = 0$ (A is now a 2n-dimensional matrix). It follows that $J A$ is a symmetric matrix and we can define as function $f_A = {}^t x J A x$. This function turns out to be the Hamiltonian function giving rise to X_A. In constructing infinitesimal symmetries for X_A by taking powers of A, not all powers will give rise to matrices in the symplectic Lie algebra and we have to restrict to odd powers, i. e. A^1, A^3, ..., A^{2k+1}, ..., A^{2n+1}, indeed

$$^t(A^{2k+1})J + JA^{2k+1} = 0 \tag{6}$$

and we find functions

$$f_k(x) = {}^t x (J A^{2k+1}) x. \tag{7}$$

From $[A^{2k+1}, A^{2j+1}] = 0$ and the Lie algebra isomorphism

$$\{f_B, \ f_C\} = f_{[B,C]}$$

where $f_B(x) = {}^t x B x$ and $\{,\}$ the Poisson bracket given by (5), we get $\{f_k, \ f_j\} = 0$. Therefore odd powers of A give rise to constants of the motion which are in involution. For a generic matrix A, X_A turns out to be a completely integrable system in the Liouville sense.

The even powers of A give rise to symmetries which are not canonical.

We can turn symmetries into nonlinear ones by using the procedure we have already illustrated previously.

Therefore on R^{2n} we have the dynamics $X_A = x^i A^i_j \partial / \partial x^j$, the symplectic structure $\omega = \sum_i dx_i \wedge dx_{i+n}$ and canonical symmetries X_1, X_2, ..., X_n associated with odd powers of A (and functions f_1, f_2, ..., f_n constants of the motion), while those Y_1, Y_2, ..., Y_n associated with even powers of A are noncanonical. By exponentiating the noncanonical infinitesimal symmetries we get invertible transformations from R^{2n} into R^{2n} which take us from one Hamiltonian description for X_A to a different Hamiltonian description.

As a final comment from linear algebra we recall that any matrix A can be decomposed into the sum

$$A = S + N \tag{8}$$

where S is semisimple and N is nilpotent, moreover $[S, N] = 0$. From here it follows that the flow associated with A is related to

$$e^{tA} = e^{tS} e^{tN}. \tag{9}$$

This formula shows that if we want (non trivial) bounded motions associated with one Hamiltonian vector field X_A, we have to impose $N = 0$ and S should have purely imaginary (non zero) eigenvalues.

Therefore a generic linear Hamiltonian system with bounded motions must be a collection of noninteracting harmonic oscillators.

To be more specific we turn to consider a two–dimensional harmonic oscillator.

THE HARMONIC OSCILLATOR

On R^4 with coordinates $(\tilde{x}, \tilde{v})$ we consider the equations of motion

$$\frac{d\tilde{x}_k}{dt} = \tilde{v}_k, \qquad \frac{d\tilde{v}_k}{dt} = -\omega_k^2 \tilde{x}_k, \qquad k = 1, 2. \tag{10}$$

We introduce new coordinates

$$q_k = \tilde{x}_k, \qquad p_k = \frac{\tilde{v}_k}{\omega_k}, \qquad k = 1, 2 \tag{11}$$

and get

$$\frac{dq_k}{dt} = \omega_k p_k, \qquad \frac{dp_k}{dt} = -\omega_k q_k. \tag{12}$$

The matrix associated with the dynamical vectorfield X_A is

$$A = \begin{pmatrix} 0 & \omega_1 & 0 & 0 \\ -\omega_1 & 0 & 0 & 0 \\ 0 & 0 & 0 & \omega_2 \\ 0 & 0 & -\omega_2 & 0 \end{pmatrix}. \tag{13}$$

We find the commuting symmetries

$$A^0 = \begin{pmatrix} 1 & 0 & 0 & 0 \\ 0 & 1 & 0 & 0 \\ 0 & 0 & 1 & 0 \\ 0 & 0 & 0 & 1 \end{pmatrix}, \tag{14}$$

$$A^2 = \begin{pmatrix} -\omega_1^2 & 0 & 0 & 0 \\ 0 & -\omega_1^2 & 0 & 0 \\ 0 & 0 & -\omega_2^2 & 0 \\ 0 & 0 & 0 & -\omega_2^2 \end{pmatrix}, \tag{15}$$

and

$$A^3 = \begin{pmatrix} 0 & -\omega_1^3 & 0 & 0 \\ \omega_1^3 & 0 & 0 & 0 \\ 0 & 0 & 0 & -\omega_2^3 \\ 0 & 0 & \omega_2^3 & 0 \end{pmatrix}. \tag{16}$$

We introduce the functions

$$h_1 = p_1^2 + q_1^2, \qquad h_1 = p_2^2 + q_2^2 \tag{17}$$

and consider the nonlinear transformation

$$\begin{aligned} x_k &= f_k(h_k)p_k, \\ y_k &= f_k(h_k)q_k. \end{aligned} \tag{18}$$

which is a symmetry for the dynamics but is noncanonical. In facts

$$\sum_k dx_k \wedge dy_k = \sum_k d(f_k p_k) \wedge d(f_k q_k) = \tilde{\omega} \tag{19}$$

is different from $\omega = \sum_k dp_k \wedge dq_k$. The latter gives rise to the standard Hamiltonian description of our harmonic oscillator, i.e. with the Hamiltonian

$$H = \sum_k \frac{\omega_k}{2}(p_k^2 + q_k^2).$$

The picture we are presented with is now the following:
X_A has Hamiltonian descriptions given by (ω, H) or $(\tilde{\omega}, \tilde{H})$ with $\tilde{H}$ obtained from H through the inverse transformation of (18), i. e.

$$i_{X_A}\omega = -dH, \qquad i_{X_A}\tilde{\omega} = -d\tilde{H}. \tag{20}$$

What happens if we associate a vector field X with $\tilde{H}$ by using ω? Then X should satisfy the equation

$$i_X\omega = -d\tilde{H}, \tag{21}$$

and it turns out that X is a reparametrization of X_A by a constant of the motion.

At this point we can consider the coordinate system where ω has the standard form (Heisenberg coordinates) and $\tilde{H}$ is a deformation of the quadratic expression or use a coordinate system where $\tilde{H}$ has the standard quadratic expression but ω is "deformed" (via the nonlinear noncanonical transformation). As a symplectic form gives rise to Poisson brackets ("commutation relations"), this second choice can be interpreted via a deformation of the commutation relations and is the starting point for "q-deformed oscillators."

For our further needs we can repeat the analysis by introducing complex coordinates in R^4

$$\alpha_k = x_k + iy_k, \qquad \alpha_k^* = x_k - iy_k. \tag{22}$$

The complex structure in R^4 is given by the matrix

$$J = \begin{pmatrix} 0 & 1 & 0 & 0 \\ -1 & 0 & 0 & 0 \\ 0 & 0 & 0 & 1 \\ 0 & 0 & -1 & 0 \end{pmatrix}$$

satisfying $J^2 = -1$. This matrix defines a complex structure commuting with the dynamical vectorfield

$$\Gamma = i(\alpha_k\frac{\partial}{\partial\alpha_k} - \alpha_k^*\frac{\partial}{\partial\alpha_k^*}). \tag{23}$$

In the (p, q) coordinates we have new complex coordinates

$$\xi_k = p_k + iq_k, \qquad \xi_k = p_k - iq_k, \tag{24}$$

and, setting

$$n_k = \alpha_k\alpha_k^*, \tag{25}$$

$$\xi_k = f_k(n_k)\alpha_k, \qquad \xi_k^* = f_k(n_k)\alpha_k^*. \tag{26}$$

This transformation is not "analytic" (we notice that analyticity depends on the complex structure and here we have two alternative complex structures compatible with the dynamics). In these complex coordinates, if we take the new point, steming from qantum mechanics, which takes the Hamiltonian as a primitive concept for the dynamics, we are naturally led to consider the following two Hamiltonians

$$H_1 = \frac{1}{2}\sum_k n_k \tag{27}$$

in the α-coordinates and

$$H_2 = \frac{1}{2} \sum_k \xi_k \xi_k^* \tag{28}$$

in the ξ-coordinates. To compare, we express them in the same variables to find

$$H_1 = \frac{1}{2} \sum_k n_k \tag{29}$$

and

$$H_2 = \frac{1}{2} \sum_k f_k^2(n_k) n_k. \tag{30}$$

We now use the same bracket for them, say

$$\{\alpha_k, \alpha_k^*\} = i. \tag{31}$$

and obtain two different dynamical systems, where the one associated with H_2 is even not necessarily isotropic. The evolution goes from periodic orbit to orbit whose closure is a 2-dimensional torus, i.e. the associated systems are completely different. Of course, there is no contradiction. Indeed to have the same dynamics we should use different Poisson Brackets, namely, the one we get transforming Eq. (31).

It is now clear that under quantization these complex coordinates go into creation and annihilation operators. Therefore for the corresponding commutators we can repeat what we have said for the Poisson Bracket.

As a final remark, we notice that the notion of coherent states in the way it is normally introduced is a kinematical notion. Coherent states are discussed in quantum mechanics from the point of view of the states the closest as possible to classical ones. According to this physical property the coherent states of the electromagnetic field harmonic oscillator were introduced in [13]. The coherent states (classical Gaussian packets) for a charge moving in magnetic field were introduced in [14], [15]. Coherent states of spin were introduced in [16].

We close this Section with few remarks concerning the structures connected with coherent states. We start with a Poisson bracket compatible with a complex structure and consider a canonical chart to be anyone such that

$$\{\chi_k, \chi_j^*\} = i\delta_{kj}. \tag{32}$$

Then we associate creation and annihilation operators with these variables and define the associated coherent states. It is clear therefore that the ingredients needed for coherent states are:

    1) a symplectic structure

    2) a compatible complex structure

and together they imply the existence of an Hermitian structure. When there are various structures compatible with a given dynamical evolution, we find various coherent states (i.e.associated with various symplectic structures) preserved by this evolution. The discussed complex structure has been used to construct Jordan-Schwinger map for some functional groups in [17].

CLASSICAL q-OSCILLATOR AS "REPARAMETRIZATION"

In previous section we described the general scheme of deformations. Now we give in detail the example of q-oscillator. Also, the q-oscillator mathematical properties and its modifications have been used to associate them with some physical phenomena [18]–[26]. The q-oscillators are interesting from the point of view of mathematical structure [27]–[32].

Let us start from considering the one-dimensional linear harmonic oscillator with frequency ω and corresponding nonlinear q-oscillator following the approach of Ref. [1]. The linear second order equation for the coordinate of this oscillator x is

$$\ddot{x} + \omega^2 x = 0. \tag{33}$$

while the form of the first order equations for the position and momentum of the harmonic oscillator is

$$\begin{aligned} \dot{p} &= -\omega^2 x, \\ \dot{x} &= p. \end{aligned} \tag{34}$$

The variables x and p are real ones. If we introduce complex variables

$$\begin{aligned} \alpha &= \frac{1}{\sqrt{2}}(\sqrt{\omega}x + \frac{i}{\sqrt{\omega}}p), \\ \alpha^* &= \frac{1}{\sqrt{2}}(\sqrt{\omega}x - \frac{i}{\sqrt{\omega}}p) \end{aligned} \tag{35}$$

the harmonic oscillator is described by a first order differential equation

$$\dot{\alpha} = -i\omega\alpha \qquad \dot{\alpha}^* = i\omega\alpha^*. \tag{36}$$

Both systems (34) and (35) may be rewritten in matrix form of Ref. [1]

$$\underline{\dot{X}} = A\underline{X}. \tag{37}$$

The real two-vector $\underline{X} \in \mathbf{R}^2$ and the matrix A for the system (34) are

$$\underline{X} = \begin{pmatrix} p \\ x \end{pmatrix}, \qquad A = \begin{pmatrix} 0 & -\omega^2 \\ 1 & 0 \end{pmatrix}. \tag{38}$$

The complex vector $\underline{X}$ and matrix A for the system (35) are

$$\underline{X} = \begin{pmatrix} \alpha \\ \alpha^* \end{pmatrix}, \qquad A = -i\omega \begin{pmatrix} 1 & 0 \\ 0 & -1 \end{pmatrix}. \tag{39}$$

According to our procedure let us replace the constant matrix elements A_i^j of matrix A by functions of the constants of the motion. The system obtained is nonlinear and it may be integrated by exponentiation as easily as the initial linear system. We illustrate this procedure again on the harmonic oscillator equations of motion (39) since in this case the matrix A is diagonal one. Thus the constant frequency ω should be replaced by the constant of the motion $\tilde{\omega}(\alpha, \alpha^*)$ and we have new system for one and the same variable α

$$
\begin{aligned}
\dot{\alpha} &= -i\tilde{\omega}(\alpha, \alpha^*)\alpha, \\
\dot{\alpha}^* &= i\tilde{\omega}(\alpha, \alpha^*)\alpha^*.
\end{aligned} \tag{40}
$$

If one takes the constant of the motion $\tilde{\omega}(\alpha, \alpha^*)$ to be an integral of the motion without explicit dependence on time we have

$$
\frac{d}{dt}\tilde{\omega}(\alpha, \alpha^*) = \frac{\partial \tilde{\omega}}{\partial \alpha}\dot{\alpha} + \frac{\partial \tilde{\omega}}{\partial \alpha^*}\dot{\alpha}^* = 0 \tag{41}
$$

which after using the equation of motion (40) may be integrated and the solution for this equation is

$$
\tilde{\omega}(\alpha, \alpha^*) = \Omega(\alpha\alpha^*), \tag{42}
$$

i. e. the frequency can be any function Ω of the modulus of the amplitude and is a constant of the motion. We will take this function in the form

$$
\Omega = \omega f_q(\alpha\alpha^*). \tag{43}
$$

The function $f_q(\alpha\alpha^*)$ will specify the q-deformation of the harmonic oscillator and we will take it in the form

$$
f_q(z) = \frac{\lambda}{\sinh \lambda} \cosh \lambda z, \qquad q = e^\lambda, \tag{44}
$$

where λ is the parameter of q-nonlinearity of the vibrations. In case $\lambda \to 0$ the function $f_q(z) \to 1$. The relation to classical q-oscillator (and its quantum counter part) consists in the existence the nonlinear noncanonical transform in the phase space of the oscillator which in the variables (α, α^*) look [1] as

$$
\alpha_q = \sqrt{\frac{\sinh \lambda\alpha\alpha^*}{\alpha\alpha^* \sinh \lambda}}\,\alpha, \qquad \alpha_q^* = \sqrt{\frac{\sinh \lambda\alpha\alpha^*}{\alpha\alpha^* \sinh \lambda}}\,\alpha^*. \tag{45}
$$

The Poisson Brackets of the variables α_q reproduce the structure of the annihilation and creation operators and the commutation relations of the q-oscillator [3] (see below). The dynamics of the physical variables α due to the Hamiltonian

$$
H = \omega\alpha_q^*\alpha_q, \tag{46}
$$

which is taken in form to be the same as the Hamiltonian of the harmonic oscillator

$$
H = \omega\alpha^*\alpha, \tag{47}
$$

is the nonlinear vibration corresponding to Eq. (40). The linear dynamics of the variable α corresponding to the Hamiltonian (47) is given by the solution to the Eq. (36). It is necessary to point out that in the suggested approach the physical meaning of the variables α, related to physical position x and physical momentum p in the sense of the measurement procedure by the formula (35), is preserved for nonlinearized dynamics. It means that the dynamics of the coordinate x and momentum p of the nonlinear q-oscillator is described by the system of equations

$$
\begin{aligned}
\dot{x} &= f_q(\alpha\alpha^*)p, \\
\dot{p} &= -\omega^2 f_q(\alpha\alpha^*)x
\end{aligned}
\tag{48}
$$

where

$$
\alpha\alpha^* = \frac{1}{2}(\omega x^2 + \frac{1}{\omega}p^2).
\tag{49}
$$

For the deformed equations of motion of q-oscillator as the formula (48) shows the momentum is a function of the velocity and position. This function may be obtained as the solution to the functional equation

$$
p(x,\ \dot{x}) = \frac{\sinh\lambda}{\lambda}\frac{\dot{x}}{\cosh[\frac{\lambda}{2}\{\omega x^2 + \frac{1}{\omega}p^2(x,\ \dot{x})\}]}
\tag{50}
$$

considered as the implicit equation for the given momentum as function of the position x and velocity $\dot{x}$. The solution to q-oscillator equation of motion

$$
\dot{\alpha} = -i\omega\alpha\frac{\lambda}{\sinh\lambda}\cosh\lambda\alpha\alpha^*
\tag{51}
$$

is

$$
\alpha(t) = \alpha_0\exp[-i\omega t\frac{\lambda}{\sinh\lambda}\cosh\lambda\alpha_0\alpha_0^*]
\tag{52}
$$

where

$$
\alpha_0 = \alpha(t=0)
$$

is the initial complex amplitude of the nonlinear q-oscillator. The solution to the equation of motion for the coordinate x of the nonlinear q-oscillator

$$
\ddot{x} + \omega^2\frac{\lambda^2}{\sinh^2\lambda}\cosh^2\{\frac{\lambda}{2\omega}[x^2\omega^2 + p^2(x,\ \dot{x})]\} = 0
\tag{53}
$$

where the function $p(x,\ \dot{x})$ is given explicitly by solving the relation (50), may be written in the form

$$
\begin{aligned}
x(t) &= \frac{x_0}{2}\{\exp[\frac{i\lambda\omega t}{\sinh\lambda}\cosh\{\frac{\lambda}{2\omega}[x_0^2\omega^2 + p^2(x_0,\ \dot{x}_0)]\}] \\
&+ \exp[-\frac{i\lambda\omega t}{\sinh\lambda}\cosh\{\frac{\lambda}{2\omega}[x_0^2\omega^2 + p^2(x_0,\ \dot{x}_0)]\}]\} \\
&+ \frac{\dot{x}_0\sinh\lambda}{2i\lambda\omega}\cosh^{-1}\{\frac{\lambda}{2\omega}[x_0^2\omega^2 + p^2(x_0,\ \dot{x}_0)]\} \\
&\times \{\exp[\frac{i\lambda\omega t}{\sinh\lambda}\cosh\{\frac{\lambda}{2\omega}[x_0^2\omega^2 + p^2(x_0,\ \dot{x}_0)]\}] \\
&- \exp[-\frac{i\lambda\omega t}{\sinh\lambda}\cosh\{\frac{\lambda}{2\omega}[x_0^2\omega^2 + p^2(x_0,\ \dot{x}_0)]\}]\}.
\end{aligned}
\tag{54}
$$

Here $x_0 = x(t=0)$ and $\dot{x}_0 = \dot{x}(t=0)$ are the initial position and velocity of the nonlinear q-oscillator. In the limit $\lambda \to 0$ we have the usual solution for the linear harmonic oscillator.

One can find for small nonlinearity $\lambda \ll 1$ the approximate expression for the momentum solving the equation (50) by iteration method. We have

$$p = \dot{x}[1 + \frac{\lambda^2}{6} - \frac{\lambda^2}{8}(\omega x^2 + \frac{\dot{x}^2}{\omega})^2]. \tag{55}$$

This formula may be interpreated as the negative shift of the mass of the oscillator by the factor depending quadratically on the energy of the oscillations.

TWO–DIMENSIONAL CLASSICAL q-OSCILLATOR

The dynamics of generically deformed two–dimensional oscillator is described in previous sections. In this Section we will give the formula for specific q-deformation of this oscillator [1]. If one considers two degrees of freedom, the amplitudes of the first harmonic oscillator α_+ and of the second oscillator α_- may be q-deformed by two different methods. One is to have frequency of the first oscillator to be dependent only on the energy of this oscillator $n_+ = |\alpha_+|^2$ and the frequency of the second oscillator to be dependent only on its energy $n_- = |\alpha_-|^2$. Another method of deformation is to make the frequencies of both oscillators to be dependent on the full energy of vibrations

$$n = n_+ + n_-.$$

We now will describe the q-deformed variables $\alpha_{q\pm}$ in the case of q-deformation when the frequency of vibrations depends on full energy. Then we have the following nonzero Poisson Brackets

$$\{\alpha_{q\pm}, \alpha_{q\mp}^*\} = i\alpha_{\pm}\alpha_{\mp}^* \frac{(\lambda n)\cosh n\lambda - \sinh n\lambda}{n^2 \sinh \lambda}, \tag{56}$$

and

$$\{\alpha_{q\pm}, \alpha_{q\pm}^*\} = \frac{i}{n \sinh \lambda}[(1 - \frac{n_\pm}{n})\sinh n\lambda + \lambda n_\pm \cosh n\lambda]. \tag{57}$$

Here

$$\alpha_{q\pm} = \alpha_\pm F(n) \tag{58}$$

where

$$F(n) = \sqrt{\frac{\sinh \lambda n}{n \sinh \lambda}}. \tag{59}$$

The introduced deformation of two harmonic oscillators implies the interaction of these oscillators which exists due to q-nonlinearity. Thus we have not only selfinteraction of the different modes but the mutual influence of the motion of one oscillator onto the other.

q-DEFORMED WAVE EQUATION

Now we consider the dynamics of a system with infinite number of degrees of freedom.

The example we consider first is the one-dimensional wave equation for the real scalar field, with the constant parameter being the wave velocity. This constant parameter we take to be unity. That is

$$(\frac{\partial^2}{\partial t^2} - \frac{\partial^2}{\partial x^2})\varphi(x,t) = 0. \tag{60}$$

In order to clarify the deformation procedure we represent this equation as a system of equations for decoupled oscillators. To do this we rewrite this equation in momentum representation

$$\ddot{\varphi}(k,t) + k^2\varphi(k,t) = 0 \tag{61}$$

where the complex Fourier amplitude

$$\varphi(k,t) = \frac{1}{2\pi} \int \varphi(x,t)\exp(-ikx)\,dx, \tag{62}$$

plays the role of new coordinate. Since $\varphi(x,\,t) = \varphi^*(x,\,t)$ we have $\varphi(k,\,t) = \varphi^*(-k,\,t)$. Eq. (61) describes a two–dimensional oscillator with equal frequencies for both modes labelled by k and $-k$. Writing the Eq. (61) in the form

$$\begin{aligned} \dot{\varphi}(k,\,t) &= \pi(k,\,t), \\ \dot{\pi}(k,\,t) &= -k^2\varphi(k,\,t) \end{aligned} \tag{63}$$

we have the equations of the form (38) in which frequency $\omega = k^2$, $p \to \pi(k,\,t)$ and $x \to \varphi(k,\,t)$. Due to this we can deform this linear system taking the integral of the motion

$$\mu = \int dk \frac{k^2|\varphi|^2(k,t) + |F_k|^2}{2|k|}\,, \tag{64}$$

in which the function F_k is the complex momentum of k–field mode solution to the infinite system of equations

$$\dot{\varphi}(k,t) = F_k \ f_q(\int dk' \frac{k'^2|\varphi|^2(k',t) + |F_{k'}|^2}{2|k'|}). \tag{65}$$

The function f_q is given by the Eq. (44). Then the parameter μ plays the role of the initial number of vibrations corresponding to given Cauchy initial conditions.

Thus we made the deformation by method used in the previous Section for two–dimensional oscillator introducing the interaction of modes through the parameter μ. The deformed wave equation may be written as

$$\ddot{\varphi}(x,t) = - \int k^2 f_q^2(\mu)\varphi(k,t)e^{ikx}dk. \tag{66}$$

Otherwise we can have it in the form for which only scalar field in time–space coordinate $\varphi(x,t)$ is present. For this we replace in the integrand the Fourier components $\varphi(k,t)$ by its expression in terms of $\varphi(x,t)$ as well as in the integral of motion μ which, in principle, may depend on the momentum vector k through the dependence on the Fourier components of the initial field $\varphi(x,0)$ and $\dot{\varphi}(x,0)$. Thus we have the deformed wave equation

$$\ddot{\varphi}(x,t) = -\frac{1}{2\pi} \int\int k^2 \exp[ik(x - x')]\varphi(x',t)f_q^2\{\mu[\varphi(x,0),\dot{\varphi}(x,0)]\}\,dk\,dx', \tag{67}$$

We have pointed out here that the integral of motion μ is a functional of the field $\varphi(x,t)$ of a special form. Being an integral of the motion it depends only on the initial values of the field and of its velocity, i.e.

$$\mu = \mu[\varphi(x,0), \dot{\varphi}(x,0)].$$

Since the parameter μ is chosen as the common integral of motion for all field oscillators it does not depend on wave vector k, and the function $f_q^2(\mu)$ can be considered as a commom factor. Hence the wave equation may be rewritten in the form

$$\ddot{\varphi}(x,t) = f_q^2(\mu)\frac{\partial^2}{\partial x^2}\varphi(x,t). \tag{68}$$

We have the differential–functional equation which looks like usual wave equation with the wave velocity $f_q(\mu)$ which is constant of the motion. Thus this procedure of deformation yields us an equation for which the velocity of wave propagation depends on the initial configuration of the field and its time derivative. This equation may be appropriate to describe a field behaviour in a media with strong nonlinear response of its properties to the presence of the quanta of the field.

Now we will obtain the solutions to nonlinear deformed wave equation (68).

We write first the solution to the nonlinear mode–vibration equation (61) which is

$$\begin{aligned}
\varphi(k,t) &= \frac{1}{2}\varphi(k,0)\{\exp[i|k|f_q(\mu)]t + \exp[-i|k|f_q(\mu)]t\} \\
&+ \frac{1}{2i}\dot{\varphi}(k,0)\{\exp[i|k|f_q(\mu)]t - \exp[-i|k|f_q(\mu)]t\}\frac{1}{i|k|f_q(\mu)},
\end{aligned} \tag{69}$$

and

$$\begin{aligned}
\varphi(k,0) &= \frac{1}{2\pi}\int \varphi(x,0)\exp(-ikx)\,dx, \\
\dot{\varphi}(k,0) &= \frac{1}{2\pi}\int \dot{\varphi}(x,0)\exp(-ikx)\,dx.
\end{aligned} \tag{70}$$

It is the multimode generalization of the solution (54) of the one–mode nonlinear oscillator. Thus given initial condition $\varphi(x,0)$, $\dot{\varphi}(x,0)$ implies that $\varphi(k,0)$ and $\dot{\varphi}(k,0)$, and μ are fixed, too. In terms of these values we have the k-th mode solution $\varphi(k,t)$ and the solution to nonlinear q-deformed wave equation (68) are given by Eqs. (62), (69).

It is now easy to prove that the q-deformed wave equation (68) has the soliton–like solution

$$\varphi_\pm(x,\,t) = \Phi(x \pm f_q(\mu)t) \tag{71}$$

where Φ is a function fixed by the Cauchy conditions. In facts the discussed q-deformation implies the existence of nonlinear interaction among the modes.

The generalization to the case of three space coordinates may be performed following the same reparametrization procedure. The q-deformed massless Klein-Gordon equation is considered by this method in [33]. From the point of view of deformed Poincare' symmetry group the relativistic equations are discussed in [34], [35].

QUANTUM q-OSCILLATOR

To make more clear the relation of the suggested approach to classical equations of motion with standard quantum q-oscillator formalism of [3] we review the description of the oscillator given in [1]. Let us introduce the usual creation and annihilation oscillator operators a and $a^\dagger$ obeying bosonic commutation relations

$$[a, a^\dagger] = 1. \tag{72}$$

Below we assume the classical dynamical variables to which a and $a^\dagger$ correspond to oscillate with a frequency $\omega = 1$. It is known that the operators a, $a^\dagger$, 1 form the Lie algebra of Heisenberg-Weyl group. So, the linear harmonic oscillator may be connected with the generators of pure Heisenberg-Weyl Lie group. In view of the commutation relation (72) the usual scheme for generating the states of the harmonic oscillator is based on the properties of the Hermitean number operator $\hat{n} = a^\dagger a$

$$[a, \hat{n}] = a, \quad [a^\dagger, \hat{n}] = -a^\dagger. \tag{73}$$

Thus constructing the vacuum state $|0\rangle$ obeying the equation

$$a|0\rangle = 0, \tag{74}$$

and the excited states

$$|n\rangle = \frac{a^{\dagger\,n}}{\sqrt{n!}}|0\rangle \tag{75}$$

which are eigenstates of the number operator $\hat{n}$

$$\hat{n}|n\rangle = n|n\rangle, n \in Z^+ \tag{76}$$

the matrix representation of the operators a and $a^\dagger$ in the basis (75) have the known expressions

$$a = \begin{pmatrix} 0 & \sqrt{1} & 0 & \cdots \\ 0 & 0 & \sqrt{2} & 0 \\ 0 & 0 & 0 & \sqrt{3} \\ \cdots & \cdots & \cdots & \cdots \end{pmatrix},$$

$$a^\dagger = \begin{pmatrix} 0 & 0 & 0 & \cdots \\ \sqrt{1} & 0 & 0 & \cdots \\ 0 & \sqrt{2} & 0 & \cdots \\ \cdots & \cdots & \cdots & \cdots \end{pmatrix} \tag{77}$$

while the number operator $\hat{n}$ is described by the matrix

$$\hat{n} = \begin{pmatrix} 0 & 0 & 0 & \cdots \\ 0 & 1 & 0 & \cdots \\ 0 & 0 & 2 & \cdots \\ \cdot & \cdot & \cdot & \cdots \end{pmatrix}. \tag{78}$$

The Hamiltonian for such a system is defined as

$$H = \frac{a^\dagger a + a a^\dagger}{2}. \tag{79}$$

The q-oscillators may be introduced by generalizing the matrices (77) and (78) with the help of the q–integer numbers n_q,

$$n_q = \frac{\sinh n\lambda}{\sinh \lambda}, \quad q = e^\lambda. \tag{80}$$

Here λ and q are dimensionless c-numbers, which appear at this purely mathematical level. When $\lambda = 0$, $q = 1$ and the q-integer n_q coincides with n. Then, replacing the integers in (77) and (78) by q-integers we obtain matrices which define the annihilation and creation operators of the quantum q-oscillator,

$$a_q = \begin{pmatrix} 0 & \sqrt{1_q} & 0 & \cdots \\ 0 & 0 & \sqrt{2_q} & \cdots \\ 0 & 0 & 0 & \sqrt{3_q} \\ \cdots & \cdots & \cdots & \cdots \end{pmatrix},$$

$$a_q^\dagger = \begin{pmatrix} 0 & 0 & 0 & \cdots \\ \sqrt{1_q} & 0 & 0 & \cdots \\ 0 & \sqrt{2_q} & 0 & \cdots \\ \cdots & \cdots & \cdots & \cdots \end{pmatrix},$$

$$\hat{n}_q = \begin{pmatrix} 0 & 0 & 0 & \cdots \\ 0 & 1_q & 0 & \cdots \\ 0 & 0 & 2_q & \cdots \\ \cdots & \cdots & \cdots & \cdots \end{pmatrix}, \tag{81}$$

since the action of $\hat{n}_q$ on eigenstates $|n>$ is given by

$$\hat{n}_q|n> = \frac{\sinh n\lambda}{\sinh \lambda}|n> . \tag{82}$$

The above matrices obey the commutation relation

$$[a_q, \hat{n}] = a_q, \quad [a_q^\dagger, \hat{n}] = -a_q^\dagger \tag{83}$$

but the commutation relations of the operators a_q and $a_q^\dagger$ do not coincide with the boson commutation relations. Eq. (72) is replaced by

$$[a_q, a_q^\dagger] = F(\hat{n}) \tag{84}$$

where the function $F(\hat{n})$ has the form

$$F(\hat{n}) = \frac{\sinh \lambda(\hat{n} + 1) - \sinh \lambda\hat{n}}{\sinh \lambda}. \tag{85}$$

For $\lambda = 0$ (84) reduces to (72). In addition to the above commutation relation there exists the reordering relation

$$a_q a_q^\dagger - q a_q^\dagger a_q = q^{-\hat{n}} \tag{86}$$

which usually is taken as the definition of q-oscillators.

It is worthy noting that the operators a_q and $a_q^\dagger$ can be expressed in terms of the operators a and $a^\dagger$ (see, for example, [1])

$$a_q = a f(\hat{n}), \quad a_q^\dagger = f(\hat{n}) a^\dagger \tag{87}$$

where

$$f(\hat{n}) = \sqrt{\frac{\hat{n}_q}{\hat{n}}}. \tag{88}$$

The comparison of formulae (45) and (87) shows the complete analogy of the quantum q-oscillator and the classical q-oscillator discussed in previous sections. We have also

$$\hat{n}_q = a_q^\dagger a_q \tag{89}$$

and

$$[a_q, \hat{n}_q] = F(\hat{n}) a_q, \qquad [a_q^\dagger, \hat{n}_q] = -a_q^\dagger F(\hat{n}). \tag{90}$$

In the Schrödinger representation the evolution operator of the harmonic oscillator

$$U(t) = \exp\left[-i\omega \frac{(a^\dagger a + a a^\dagger) t}{2}\right] \tag{91}$$

gives the possibility to find out explicitly linear integrals of motion which depend on time

$$\begin{aligned}
A(t) &= U(t) a U^{-1}(t) = e^{i\omega t} a, \\
A^\dagger(t) &= U(t) a^\dagger U^{-1}(t) = e^{-i\omega t} a^\dagger.
\end{aligned} \tag{92}$$

The matrices of the integrals of motion (92) in Fock basis may be obtained from the equations

$$\begin{aligned}
A(t)|n\rangle &= e^{i\omega t} \sqrt{n} |n-1\rangle, \\
A^\dagger(t)|n\rangle &= e^{-i\omega t} \sqrt{n+1} |n+1\rangle.
\end{aligned} \tag{93}$$

Let us now introduce the Hamiltonian

$$\hat{H} = \omega \frac{a_q a_q^\dagger + a_q^\dagger a_q}{2} \tag{94}$$

for which the evolution operator takes the form

$$U_q(t) = \exp\left[-i\omega t \frac{(a_q a_q^\dagger + a_q^\dagger a_q)}{2}\right]. \tag{95}$$

We have for the integrals of motion

$$A_q(t) = U_q(t) a_q U_q^{-1}(t), \qquad A_q^\dagger(t) = U_q(t) a_q^\dagger U_q^{-1}(t) \tag{96}$$

the following explicit matrix expressions [1]

$$A_q(t) = \begin{pmatrix} 0 & \sqrt{1_q}\,e^{i(1_q-0_q)\omega t} & 0 & \cdots \\ 0 & 0 & \sqrt{2_q}\,e^{i(2_q-1_q)\omega t} & \cdots \\ \cdots & \cdots & \cdots & \cdots \end{pmatrix},$$

$$A_q^\dagger(t) = \begin{pmatrix} 0 & 0 & 0 & \cdots \\ \sqrt{1_q}\,e^{-i(1_q-0_q)\omega t} & 0 & 0 & \cdots \\ 0 & \sqrt{2_q}\,e^{-i(2_q-1_q)\omega t} & 0 & \cdots \\ \cdots & \cdots & \cdots & \cdots \end{pmatrix}. \tag{97}$$

These operators are the generalizations of the linear integrals of motion (92) to the case of nonlinear Hamiltonian. This result is a generalization for q-oscillators of such integrals of motion for usual and parametric quantum oscillator which have been found in [36] and discussed for constructing coherent states in [37] and [38].

ANALOGY OF CLASSICAL AND QUANTUM DEFORMATIONS

By using the example of the oscillator we clarify now the analogy which exists in the suggested approach to deformation of classical systems and quantum ones. Now we recall what has been done in the previous Section for q-deformed quantum oscillator to make clear the connection with the procedure of deforming the classical oscillator of first sections.

Equations of motion for the harmonic oscillator amplitude a we rewrite for another operator

$$A = h(a^\dagger a)a$$

where h is a real function, and its hermite conjugate $A^\dagger = a^\dagger\, h(a^\dagger a)$ in the same form

$$\dot{A} = -i\omega A, \qquad \dot{A}^\dagger = i\omega A^\dagger.$$

We have in our Hilbert space the vacuum state $|0>$ which satisfies

$$a\,|0> = 0, \qquad A\,|0> = 0.$$

We can construct two basis in this vector space. One is the standard basis

$$|n> = \frac{a^{\dagger n}}{\sqrt{n!}}\,|0>$$

which is orthonormal in standard scalar product

$$< n|m > = \delta_{nm}.$$

Another basis is constructed using the operator $A^\dagger$

$$|\tilde{n}> = \frac{(A^\dagger)^n}{\sqrt{n!}}\,|0> .$$

and define a new scalar product in the same vector space which is given by

$$< \tilde{n} | \widetilde{m} > = \delta_{nm}.$$

The adjoint with respect to this new scalar product needs not coincide with the old one. We can define the operators

$$b^* | \tilde{n} > = \sqrt{n+1} | \widetilde{n+1} > \qquad b | \tilde{n} > = \sqrt{n} | \widetilde{n-1} >$$

where $*$ means the adjoint in the new scalar product. These operators satisfy the commutation relation $[b, b^*] = 1$. Taking the Hamiltonian $H = \omega b^* b$ we have for the operators b, b^* the equation of motion of the harmonic oscillator. Thus for one and the same vector space we have introduced two Hilbert space structures. As for the dynamics we have, like in the classical case, two different descriptions.

Very much as we did for the classical case we can use the new Hamiltonian and the old commutator relations to get a "deformed" dynamics. As for the partition function, similarly to the classical case, we can use the trace defined via the two different scalar products to get the same result, i. e. the partition function depends only on the dynamics and not on the particular Hamiltonian description we use. Either we change the Hamiltonian and for the old scalar product we obtain new dynamics. Or changing Hamiltonian and simultaneously the scalar product we obtain the same dynamics. For partition function in such case we have the same value that was for nondeformed oscillator.

DEFORMED PLANCK DISTRIBUTION

In this Section we will discuss what physical consequences may be found if the considered q-nonlinearity influences the vibrations of the real field mode oscillators like, for example, electormagnetic fields ones or the oscillations of the nuclei in polyatomic molecules. First of all this nonlinearity changes the specific heat behaviour. To show this we have to find the partition function for a single q-oscillator corresponding to the Hamiltonian $H = \hat{n}_q$

$$Z(T) = \sum_{n=0}^{\infty} \exp(-\beta n_q) \tag{98}$$

where the variable β is the function of the temperature T^{-1}. The evaluation of the quantum partition function of the q-oscillator yields for the specific heat that it decreases for $T \to \infty$ as

$$C \propto \frac{1}{\ln T}. \tag{99}$$

Thus the behaviour of the specific heat of the q-oscillator found in [1] for $\lambda \ll 1$ is different from the behaviour of the usual oscillator in the high temperature limit. This property may serve for an experimental check of the existence of vibrational nonlinearity of the q-oscillator fields.

q-Deformed Bose distribution can be obtained by the same method starting from the Hamiltonian $H = \frac{1}{2}\{a_q^\dagger, a_q\}_+$ and one obtains [1]

$$< n > = \bar{n}_0 - \beta \frac{\lambda^2}{6} \left[\frac{1}{2}((n^2)_0 - (\bar{n})_0^2) + \frac{3}{2}((n^3)_0 - \bar{n}_0(n^2)_0) + (n^4)_0 - \bar{n}_0(n^3)_0) \right] \tag{100}$$

in which $\bar{n}_0$ is the usual Bose distribution function and

$$(\bar{n^k})_0 = 2\sinh\frac{\beta}{2}\sum_{n=0}^{\infty} n^k e^{-\beta(n+1/2)}. \tag{101}$$

Calculating the partition function for small q-nonlinearity parameter we have also the following q-deformed Planck distribution formula

$$<n> = \frac{1}{e^{\hbar\omega/kT}-1} - \lambda^2\frac{\hbar\omega}{kT}\frac{e^{3\hbar\omega/kT}+4e^{2\hbar\omega/kT}+e^{\hbar\omega/kT}}{(e^{\hbar\omega/kT}-1)^4}. \tag{102}$$

It means that q-nonlinearity deforms the black body radiation formula [1].

One can write down the high and low temperature approximations for the deformed Planck distribution formula [11]. For small temperature the behaviour of the deformed Planck distribution differs from the usual one

$$<n> -\bar{n}_0 = -\lambda^2\frac{\hbar\omega}{kT}e^{-\hbar\omega/kT}. \tag{103}$$

For the high temperature the nonlinear correction to the usual Planck distribution also depends on temperature

$$<n> -\bar{n}_0 = -6\lambda^2(\frac{\hbar\omega}{kT})^{-3}. \tag{104}$$

As it was seen, the discussed q-nonlinearity produces a correction to Planck distribution formula and also this may be subjected to an experimental test.

As it was suggested in [2] the q-nonlinearity of the field vibrations produces blue shift effect which is the efeect of the frequency increase with the field intensity. For small nonlinearity parameter λ and for large quantity of photons n in a given mode the relative shift of the light frequency is

$$\frac{\delta\omega}{\omega} = \frac{\lambda^2}{2}(n-\frac{1}{3}).$$

This phenomenon of possible existence of the q-nonlinearity may be essential for the models of the early stage of the Universe.

Another possible phenomenon related to the q-nonlinearity was considered in [12] where it was shown that if one deforms the electrostatics equation using the method of deformed creation and annihilation operators the formfactor of a point charge appears due to q-nonlinearity.

NONLINEAR KLEIN-GORDON EQUATION

To demonstrate how the q-nonlinearity may appear in Klein-Gordon equation we start from the consideration of usual Klein-Gordon equation with mass equal to zero ($c=1$)

$$(\frac{\partial^2}{\partial t^2} - \Delta)\varphi(\mathbf{x},t) = 0. \tag{105}$$

Let us take the plane wave solutions of the equation, i.e., we represent the field $\varphi(\mathbf{x},t)$ in the form

$$\varphi(\mathbf{x},t) = \int \varphi(\mathbf{k},t)\exp(i\mathbf{kx})\,d\mathbf{k}, \tag{106}$$

where Fourier amplitude

$$\varphi(\mathbf{k}, t) = \frac{1}{(2\pi)^3} \int \varphi(\mathbf{x}, t) \exp(-i\mathbf{k}\mathbf{x}) \, d\mathbf{x}, \tag{107}$$

plays the role of new coordinate. It satisfies the integral equation

$$\int \ddot{\varphi}(\mathbf{k}, t) \exp(i\mathbf{k}\mathbf{x}) \, d\mathbf{k} = \int (-k^2)\varphi(\mathbf{k}, t) \exp(i\mathbf{k}\mathbf{x}) \, d\mathbf{k}. \tag{108}$$

This integral equation is equivalent to the differential one

$$\ddot{\varphi}(\mathbf{k}, t) + k^2 \varphi(\mathbf{k}, t) = 0, \tag{109}$$

which is an infinite system of decoupled oscillators with frequencies $\omega^2 = k^2$. According to suggested procedure we replace this equation by

$$\ddot{\varphi}(\mathbf{k}, t) + k^2 f_q(\mu)\varphi(\mathbf{k}, t) = 0. \tag{110}$$

Here the new frequency of $\mathbf{k}$–th mode appeared

$$\omega^2 = k^2 f_q(\mu), \tag{111}$$

where the choice of parameter μ determines the deformation. According to our ideology it is possible to take it to be an integral of motion of the Klein-Gordon equation. There exists a common integral of motion which is the full number of the scalar field quanta, and we take it to be equal μ. As well as for the wave equation case the parameter μ behaves as constant for any choice of the initial conditions $\varphi(\mathbf{x}, t = 0)$, $\dot{\varphi}(\mathbf{x}, t = 0)$. Due to this the solution to the nonlinear mode-vibration equation is the following

$$\begin{aligned}
\varphi(\mathbf{k}, t) &= \frac{1}{2}\varphi(\mathbf{k}, 0)\{\exp[i|k|f_q(\mu)]t + \exp[-i|k|f_q(\mu)]t\} \\
&+ \frac{1}{2i}\dot{\varphi}(\mathbf{k}, 0)\{\exp[i|k|f_q(\mu)]t - \exp[-i|k|f_q(\mu)]t\}\frac{1}{i|k|f_q(\mu)},
\end{aligned} \tag{112}$$

Here

$$\begin{aligned}
\varphi(\mathbf{k}, 0) &= \frac{1}{(2\pi)^3} \int_{-\infty}^{\infty} \varphi(\mathbf{x}, 0) \exp(-i\mathbf{k}\mathbf{x}) \, d\mathbf{x}, \\
\dot{\varphi}(\mathbf{k}, 0) &= \frac{1}{(2\pi)^3} \int_{-\infty}^{\infty} \dot{\varphi}(\mathbf{x}, 0) \exp(-i\mathbf{k}\mathbf{x}) \, d\mathbf{x}.
\end{aligned} \tag{113}$$

Thus the given initial condition $\varphi(\mathbf{x}, 0)$, $\dot{\varphi}(\mathbf{x}, 0)$ determine also $\varphi(\mathbf{k}, 0)$, $\dot{\varphi}(\mathbf{k}, 0)$, and μ. In terms of these values we have the $\mathbf{k}$–th mode solution $\varphi(\mathbf{k}, t)$ and the solution to nonlinear q-deformed Klein-Gordon equation

$$(\frac{\partial^2}{\partial t^2} - f_q^2(\mu)\Delta)\varphi(\mathbf{x}, t) = 0. \tag{114}$$

The constant of motion $f_q(\mu)$ plays the role of signal velocity.

q-DEFORMED ELECTRODYNAMICS

We will consider the system of usual linear Maxwell equations in vacuum for the field $\mathbf{E}(\mathbf{x}, t)$, $\mathbf{H}(\mathbf{x}, t)$ expressed in terms of the potentials $\mathbf{A}(\mathbf{x}, t)$, $\varphi(\mathbf{x}, t)$

$$
\begin{aligned}
\mathbf{E} &= -\frac{1}{c}\frac{\partial \mathbf{A}}{\partial t} - \frac{\partial \varphi}{\partial \mathbf{x}}, \\
\mathbf{H} &= rot\ \mathbf{A},
\end{aligned}
\tag{115}
$$

in the form ($c = 1$)

$$
(\frac{\partial^2}{\partial t^2} - \Delta)\varphi(\mathbf{x}, t) = 0,
\tag{116}
$$

$$
\begin{aligned}
(\frac{\partial^2}{\partial t^2} - \Delta)\mathbf{A}(\mathbf{x}, t) &= 0, \\
\frac{\partial \varphi}{\partial t} + div\ \mathbf{A} &= 0.
\end{aligned}
\tag{117}
$$

The suggested approach for scalar field of previous Section may be applied if one reexpresses the fields $\varphi(\mathbf{x}, t)$, $\mathbf{A}(\mathbf{x}, t)$ in Fourier basis

$$
\begin{aligned}
\varphi(\mathbf{x}, t) &= \int \varphi(\mathbf{k}, t) \exp(i\mathbf{k}\mathbf{x})\ d\mathbf{k}, \\
\mathbf{A}(\mathbf{x}, t) &= \int \mathbf{A}(\mathbf{k}, t) \exp(i\mathbf{k}\mathbf{x})\ d\mathbf{k}.
\end{aligned}
\tag{118}
$$

Then we have

$$
\begin{aligned}
\ddot{\varphi}(\mathbf{k}, t) + k^2\varphi(\mathbf{k}, t) &= 0, \\
\ddot{\mathbf{A}}(\mathbf{k}, t) + k^2\mathbf{A}(\mathbf{k}, t) &= 0
\end{aligned}
\tag{119}
$$

together with gauge constraint

$$
\dot{\varphi}(\mathbf{k}, t) + i\mathbf{k}\mathbf{A}(\mathbf{k}, t) = 0.
\tag{120}
$$

The approach is based on the taking into account that the equations describing the linear forced oscillators of the electromagnetic field are deformed due to dependence of the frequency of the oscillations on the energy of the electromagnetic vibrations. As in the case of the scalar field we will introduce the total "number of quanta" of the defiormed electromagnetic field μ which is integral of motion. Then in this case we have the deformed nonlinear equations of vibrations

$$
\begin{aligned}
\ddot{\varphi}(\mathbf{k}, t) + f_q^2(\mu)k^2\varphi(\mathbf{k}, t) &= 0, \\
\ddot{\mathbf{A}}(\mathbf{k}, t) + f_q^2(\mu)k^2\mathbf{A}(\mathbf{k}, t) &= 0,
\end{aligned}
\tag{121}
$$

The sense of the function $f_q(\mu)$ is the signal velocity. Due to this we have to reparametrize the gauge condition to become

$$
f_q^{-1}(\mu)\varphi(\mathbf{k}, t) + i\mathbf{k}\mathbf{A}(\mathbf{k}, t) = 0.
\tag{122}
$$

After this we could reconstruct Maxwell equations in space–time analogously to the wave equation case. It should be noted that parameter μ in the Klein-Gordon and Maxwell equations is given by Eq. (64). The limit of electrostatics in the suggested type of q–deformation coincides with the usual electrostatics described by the Laplace equation (compare with [12]). It would be interesting to take into account interaction of the deformed electromagnetic field with the sources, but it will be taken up elsewhere.

References

[1] V. I. Man'ko, G. Marmo, S. Solimeno, and F. Zaccaria, Int. J. Mod. Phys. A **8**, 3577 (1993).

[2] V. I. Man'ko, G. Marmo, S. Solimeno, and F. Zaccaria, Phys. Lett. A **176**, 173 (1993).

[3] L. C. Biedenharn, J. Phys. A **22**, L873 (1989).

[4] A. J. Macfarlane, J. Phys. A **22**, 4581 (1989).

[5] V. G. Drinfeld, in: *Quantum Groups, Proc. Int. Conf. of Math.* (MSRI, Berkeley, CA., 1986) p. 798.

[6] M. Jimbo, Int. J. Mod. Phys. **A4**, 3759 (1989).

[7] 3. S. Woronowicz, Comm. Math. Phys. **111**, 615 (1987).

[8] Y. Manin, Comm. Math. Phys. **123**, 163 (1989).

[9] V. I. Man'ko and G. Marmo, Mod. Phys. Lett. A **7**, 3411 (1992).

[10] J. Grabowski, G. Landi, G. Marmo, and G. Vilasi, Fortschr. Phys. **42**, 393 (1994).

[11] V. I. Man'ko, G. Marmo, S. Solimeno, and F. Zaccaria, "Q-nonlinearity of Electromagnetic Field and Deformed Planck Distribution," in: *Technical Digests of EQEC'93-EQUAP'93, Firenze, September 10-13, 1993*, eds. P. De Natale, R. Meucci, and S. Pelli, vol. **2** (1993).

[12] V. I. Man'ko, G. Marmo, and F. Zaccaria, Phys. Lett. A **191**, 13 (1994).

[13] R. J. Glauber, Phys. Rev. Lett. **10**, 84 (1963).

[14] I. A. Malkin and V. I. Man'ko, Sov. Phys. JETP **28**, 527 (1969).

[15] A. Feldman and A. H. Kahn, Phys. Rev. B **1**, 4584 (1970).

[16] J. M. Radcliffe, J. Phys. A **4**, 313 (1971).

[17] V. I. Man'ko, G. Marmo, P. Vitale, and F. Zaccaria, Int. J. Mod. Phys. A (to appear, 1994).

[18] K. N. Ilinski and V. M. Uzdin, Phys. Lett. A **174**, 179 (1993).

[19] S.V.Shabanov, *Quantum and Classical Mechanics and q-deformed Systems*, Preprint BUTP 92/24 (1992).

[20] E. Celeghini, M. Rasetti, G. Vitiello, Phys. Rev. Lett. **66**, 2056 (1991).

[21] G. Su and M. Ge, Phys. Lett. A **173**, 17 (1993).

[22] P. P. Kulish and E. V. Damashinsky, J. Phys. A **23**, L415 (1990).

[23] R. N. Alvarez, D. Bonatsos, and Yu. F. Smirnov, Phys. Rev. A **250**, 1088 (1994).

[24] P. Shanta, S. Chaturvedi, V. Srinivasan, and R. Jagannathan, J. Phys. A **27**, 6433 (1994).

[25] G. Brodimas, A Jannusis, and R. Mignani, J. Phys. A **25**, L323 (1992).

[26] M. Chaichian, D. Ellinas, P. Kulish, Phys. Rev. Lett. **65**, 980 (1990).

[27] D. B. Fairlie and C. K. Zachos, Phys. Lett. B **256**, 43 (1991).

[28] D. Bonatsos and C. Daskaloyannis, Phys. Lett. B **307**, 100 (1993).

[29] C. Daskaloyannis, J. Phys. A **24**, L789 (1991); **25**, 2261 (1992).

[30] C. Quesne, Phys. Lett. A **193**, 245 (1994).

[31] A. P. Polychronakos, Mod. Phys. Lett. A **5**,2335 (1990).

[32] R. Floreanini, V. P. Spiridonov and L. Vinet, in: *Group Theoretical Methods in Physics, Proc. Inter. Colloquium, Moscow, 4-9 June, 1990*, eds. V. V. Dodonov and V. I. Man'ko, Lecture Notes in Physics **382** (Springer Verlag, 1991).

[33] V. I. Man'ko, G. Marmo, and F. Zaccaria, Phys. Lett. A (to be submitted).

[34] J. Lukierski, A. Nowicki, and H. Ruegg, Phys. Lett. B **293**, 344 (1992).

[35] M. Pillin, J. Math. Phys. **35**, 2804 (1994).

[36] I. A. Malkin and V. I. Man'ko, Phys. Lett. A **32**, 243 (1970).

[37] I. A. Malkin and V. I. Man'ko *Dynamical Symmetries and Coherent States of Quantum Systems* (Nauka Publishers, Moscow, 1979) [in Russian].

[38] V. V. Dodonov and V. I. Man'ko, *Invariants and Evolution of Nonstationary Quantum Systems, Proc. of Lebedev Physics Institute* **183**, ed. M. A. Markov (Nova Science, Commack, N. Y., 1989).

HIDDEN SYMMETRIES, SYMMETRY-BREAKING
AND EMERGENCE OF COMPLEXITY

Koichiro Matsuno

Department of BioEngineering
Nagaoka University of Technology
Nagaoka 940-21, Japan

INTRODUCTION

Symmetry presumes the presence of a certain set of representations to be acted upon and transformed among themselves. The phase-space point in classical mechanics and the wavefunction in quantum mechanics are two of the most quoted representations or, more precisely, state-representations in physics. However, representation pointing to a physical object already assumes by itself an implicit operation of getting it while referring to the object in the first place. Any representation of a physical object is intrinsically dynamic in the act of associating the object to its representation. The present implicit dynamics underlying the representation now makes a sharp contrast to the dynamics in terms of representations alone. At issue is how the transformation dynamics of a physical object to its representation would influence the dynamics of representations themselves, especially with regard to its dynamic symmetry. This interference would become most acute when it is focused upon how to conceive of the physical origin of symmetry-breaking acting upon the dynamic symmetry that those representations are supposed to maintain.

Symmetry-breaking associated with any representations would become extremely crucial when one faces emergence of a new representation or splitting of a representation into its constituent sub-representations. Representations can neither integrate nor disintegrate dynamically by themselves. For instance, the wavefunction of a hydrogen nucleus, being apart from the physical object thus represented, does presume by itself neither dynamic integration from nor disintegration into the wavefunctions of the constituent quark particles. The wavefunction is in itself a consequence of the dynamics of transformation from the physical object to its representation, though what sort of transformation has actually been taken is left unidentified.

Accordingly, the wavefunction as a representation-in-itself is consistent in itself and nonlocal in having accomplished the transformation dynamics globally. The relationship between the wavefunction of a hydrogen nucleus and those of the constituent quark particles is claimed to be descriptively consistent and is not dynamic anymore. Descriptive consistency between any representation and its constituent sub-representations is by definition nonlocal, extending over to the whole description undertaken. The dynamics in terms of only representations-in- themselves like that tried in the form of Schroedinger's equation of motion for the wavefunction is necessarily consistent internally and nonlocal.

However, descriptive consistency between local and global representations does not imply a similar physical consistency between the corresponding local and global objects. Even if a hydrogen nucleus can be represented as an internally consistent organization of three quarks, a mere aggregation of the three quarks cannot necessarily form a hydrogen nucleus physically. Descriptive consistency between the local and the global forcibly assumes a successful completion of the dynamics transforming physical objects into their representations in spite of the fact that it remains silent about how the transformation dynamics could have successfully been completed. Whether the physical consistency between the corresponding local and global objects could be established at all remains uncertain. If the likelihood and possibility of such a physical consistency becomes a matter of concern, the scheme forcing the complete association of a physical object to its representation cannot stand because it would dismiss the whole issue by requiring the consistency from the start. Instead, we require an alternative scheme such that it does not assume the completion of the transformation from a physical object into its representation.

A representation-in-itself as with the case of the wavefunction is already completed and does not fulfill the requirement for the incompleteness of the transformation. Compared to the representation-in-itself as a product of the act of representing, what is sought after is the on-going act of representing an object, or the representation-in-an-object. Unique to the representation-in-an-object is that it constantly turns out to be an object to be further represented in a successive manner. Moreover, the act of representing an object is no more than a form of interaction actualized between two parties influencing each other. It is interaction that makes one party constantly be involved in representing in its own way how the other party would behave. That is to say, interaction embodied in the one party's act of representing the other is simply a form of measurement proceeding internally (Matsuno, 1985).

Internal measurement thus underlies a representation-in-an-object so long as the descriptive completion of interaction is not forcibly imposed. Although interaction as a representation-in-itself would dismiss internal measurement altogether, the present dismissal is totally descriptive in its origin, and by no means physical.

It is one thing to have the consistency between the local and the global once the description of whatever object has been attempted, but quite another to ascertain whether there could be such a physical object that could guarantee the descriptive consistency before the description is actually undertaken. Interaction as a representation-in-itself would survive only when the underlying interaction as a representation-in-an-object could be dismissed physically even if in an approximate sense. Otherwise, interaction as a representation-in-an-object has to be focused. In contrast to interaction-in-itself that presumes its internal consistency in an atemporal manner, interaction as a representation-in-an-object claims its internal consistency only to the extent of what has been completed and thus described. Interaction as a representation-in-an-object constantly maintains those contributions that remain yet to be represented because of the on-going and incomplete act of representing an object involved there.

The difference between interaction as a representation-in-itself and that as a representation-as-an-object can be intensified if one examines it from the perspectives of symmetry and symmetry-breaking. Interaction as a representation-in-itself is necessarily destined to maintain a certain dynamic symmetry simply because it presumes an atemporal internal consistency to be described. In contrast, interaction as a representation-in-an-object takes dynamic symmetry-breaking for granted because of the dichotomy between what has been represented and what remains to be represented. One thus observes that internal measurement underlying interaction as a representation-in-an-object assumes the major role for the occurrence of dynamic symmetry-breaking. Symmetries hidden in the interaction as a representation-in-itself can be broken by evoking the underlying interaction as a representation-in-an-object.

COMPLEXITY AND EMERGENCE

Recent upheaval of the attention toward complexity and emergence urges us to take a new look at what the underlying dynamics is all about especially with regard to whether symmetries could be preserved or broken, because both complexity and emergence are the attributes of dynamic process par excellence. Biological evolution as a mode of dynamics seems full of complexity and emergent phenomena (Salthe, 1993). This quick perusal of dynamics leads us to ask ourselves what should be required of the basic framework of dynamics that can cope with complexity and emergence. At issue is the foundation of dynamics, no matter what it may refer to (Matsuno, 1989).

Underlying any description of dynamics is the activity of identifying the object to be described (Matsuno, 1985, 1993). However, descriptive identification unique to the subject who intends to describe whatever object raises its own problem. The activity of forming and identifying an object for description goes beyond what the description can give in the end (Pattee, 1993; Rosen, 1991). Object-oriented activities are more than what the resulting object tells us about. Once the capacity of descriptive identification is taken to be an object to be described, an infinite regression would become inevitable such as asking the capacity of identifying the capacity of identifying the object.

In contrast, descriptive realization simply as a consequence of describing the object thus identified can stand on its own if there are no further object-oriented activities to intervene in the description. If the separation between descriptive identification and realization is completed with no remnant of object-oriented activities within the resulting description, one can see the description as a structured organization in which every participating component identifies itself and its place by establishing and referring to the firm and invariant relationship with all the others existing there (Farre, 1994). Each participant's identifying itself and others is by no means an actual identification in the empirical sense, but is already structured in the organization.

A greatest advantage of the description as a structured organization is that one can get rid of any interference from object-oriented activities whose description may end up with an infinite regression (Kauffman, 1993). Object-oriented activities cannot be descriptively structuralized because the implicit dichotomy between the subject and the object constantly prohibits us from identifying who the subject is. Nonetheless, descriptive identification as a mode of object-oriented activity cannot be dismissed simply because it would cause a formidable malaise in description. Any attempt to describe dynamics is first required to answer the question of whether the dynamic realization to be described as such may remain free from any interference of object-oriented activities such as dynamic identification proceeding internally.

A clue to answering the question comes from scrutinizing the time-honored champion of structured dynamics of all, namely, mechanics. Galilean-Newtonian mechanics is structuralized in time because of the assumed presence of the global time entailed by absolute time flowing uniformly and homogeneously in absolute stationary space. The structure of time latent in Galilean-Newtonian mechanics is within the uniformity of absolute time. The global assignment of absolute time yields a global structuralization of time, and the structuralization by itself is atemporal. What is peculiar to classical mechanics is a form of atemporal structuralization of time that serves as synchronizing all the dynamic movements in absolute space to the progression of the common and absolute time.

Identifying the synchronization is not an instance of actual identification, but is already structuralized in the framework of mechanics in an atemporal manner.

The similar atemporal structuralization of time is also latent in quantum mechanics, in which the only difference from classical mechanics is about the nature of the descriptive object to be identified, that is to say, the wavefunction in quantum mechanics whereas the phase-space point in classical mechanics. Special relativity of Einstein also maintains a global structuralization of time in the form of the Lorentz invariance. Even general relativity is atemporally structuralized in the sense of holding a common covariant transformation available to all the conceivable local manifolds in four-dimensional space-time space.

Atemporal structuralization of time can certainly bring about   global synchronization as a structuralized form of dynamic identification. Nonetheless, this does not imply that all the possible forms of dynamic identification could equally be structuralized atemporally. Global synchronization of absolute time in classical mechanics, for instance, refers only to the space and time structure of the underlying absolute space-time space and has nothing to do with material components that could be present there. Whether atemporal structuralization of dynamic identification on material grounds could be feasible has to be examined in the light of material constraints to be observed by all means.

A most significant constraint of material origin is that nothing propagates faster than the speed of light (Matsuno, 1982). Finiteness of light velocity suggests that dynamic identification of material origin, of whatever sort it may be, is local and object-oriented, that is to say, detection of material origin is intrinsically local both in space and in time. Although there has been an attempt to let dynamic identification of local character be subject to mechanistic atemporal structuralization of time with the help of definite boundary conditions applied externally, this scheme does not squarely face the problem of whether dynamic identification of material origin could be dispensed with. Rather, it begs the further questions. Imposing boundary conditions in itself raises the question of how they could be identified as such and then realized accordingly. Both dynamic identification and realization are indispensable to imposing boundary conditions. Structured dynamics of mechanics is singularly exceptional in letting dynamic identification of boundary conditions be structuralized atemporally, whereas the actual identification remains unstructured as being object-oriented.

Impossibility in identifying what will be detected before it has actually been detected or in structuralizing dynamic identification atemporally lets the participating dynamic detection be necessarily local both in space and in time. The resulting dynamics of local detection is inevitable in accommodating in itself object-oriented activities that cannot be structuralized. The problem of complexity and emergence within the scheme of the dynamics of local detection will have to be addressed accordingly.

Interestingly enough, however, the similar problem of complexity and emergence can also be dealt with by employing a form of structured dynamics supplemented by external disturbances or contingencies (Ehresmann and Vanbremeersch, 1987; Baas, 1994). The universal objective observer claiming an instantaneous bird's-eye view on the global scale, if available, can regard the consequence of the preceding structured dynamics perturbed by external disturbances as the emergence of a new structured dynamics. Crucial to the theoretical artifact of introducing the universal objective observer is that the capacity of identification on the part of the observer is neither structuralized in the dynamics nor taken to be object-oriented while exhibiting its observational activity. However, once the material basis of the capacity of observing is focused, the conception of the universal objective observer would turn out to be immaterial because of the absence of any material means claiming instantaneous detection or communication on the global scale. Examining material grounds of complexity and emergent phenomena is to place the dynamics of local detection at its inner-most core.

DYNAMICS OF LOCAL DETECTION

Given a set of available degrees of freedom in motion, the underlying dynamics comes to face the problem of how each degree of freedom determines its value in time. Noting that dynamic identification of material origin is necessarily local, one is led to observe that the identification remains indefinite about what will be identified, whereas definite about what has already been realized. The temporal asymmetry between the possible yet to come and the actual already established now lets the law of motion be of a one-to-many temporal mapping as contrasting the indefiniteness in the future to the definiteness in the past (Matsuno, 1989).

One of the attempts toward addressing ourselves to the law of motion of a one-to-many mapping has been Boltzmann's statistical mechanics, though to a limited extent. The postulate of stosszahl ansatz or molecular chaos of Boltzmann stating that each interacting particle loses its past memory after a few or at most several collisions with others in the immediate past points to the limitedness of dynamic identification on the part of interacting particles. Boltzmann's dynamics of local detection lets each interacting particle be an agent doing detection locally as utilizing those collisions with others only in the immediate past (Kreuzer, 1981). Recognizing the activity of agents doing local detection makes Boltzmann's dynamics describable in terms of microstates and the probability distribution defined over the latter, instead of referring directly to each constituent degree of freedom and its value identification.

Identifying the microstate corresponding to how each particle behaves between successive collisions with others is local both in space and in time, and is at the same time structuralized in Boltzmann's statistical mechanics. If one emphasizes the structuralized aspect, Boltzmann's dynamics of local detection will determine the probability distribution of microstates. On the other hand, however, if agents doing local detection are emphasized, Boltzmann's dynamics will fix the probability distribution of those events of finding each microstate occupied. It is an agent doing local detection conceived within Boltzmann's dynamics that relates the probability distribution of microstates to that of identifying these microstates occupied.

The difference between the presence of microstates and the event of identifying the latter disappears within the scheme of Boltzmann's dynamics of local detection. Disappearance of the difference will become more evident when one compares both Boltzmann's and Shannon's entropy.

Boltzmann's entropy has been defined in the manner being independent of who and what the observer is insofar as the probability distribution of microstates remains unaltered, whereas Shannon's depends entirely upon who the observer is. Nonetheless, Boltzmann's dynamics of local detection makes it possible to equate the probability distribution of microstates to that of those events of identifying each microstate occupied. When a particular observer is singled out that is of course of local character, there certainly exists an identical relationship between Boltzmann's and Shannon's entropy in the eyes of the observer. If an arbitrary interacting particle is singled out as a local observer, the resulting Boltzmann's entropy of the whole system would decrease while the observer is identifying the microstate which it occupies (Schneider, 1991). The decrease of the Boltzmann's entropy is due to the change in the probability distribution of microstates as a consequence of singling out one particle as an observer.

Once the observer is identified, Shannon's entropy to the observer becomes obvious. Any decrease in Boltzmann's entropy due to observation is identical to the similar decrease in Shannon's entropy to the observer. Decrease in Shannon's entropy by the amount of, say, h (bits) identically induces decrease in Boltzmann's entropy by s=khln2, in which k is Boltzmann's constant measured in units of erg/Kelvin.

Decrease in Boltzmann's entropy due to local detection in the light of the second law of thermodynamics implies more than just the operation of local detection. It also entails heat generation so as to fulfill the second law of thermodynamics as an effect. If the local detection causing the decrease of Boltzmann's entropy by the amount of s is accompanied by the generation of heat q in the immediate neighborhood, the second law yields

$$-s + q/T > 0,$$

in which T is the temperature measured in Kelvin. The second law suggests that any decrease in Boltzmann's entropy due to observation proceeding internally has to be compensated by heat generation through the mediating dissipative process.

Moreover, if the presence of a thermal equilibrium is guaranteed as demonstrated in Boltzmann's statistical mechanics, there would be neither net decrease in Boltzmann's entropy nor net energy dissipation even though local detection is kept going on everywhere. The energy source driving the local detection that could decrease Boltzmann's entropy internally comes entirely from the heat energy dissipated through preceding local detections of similar nature having taken place in the neighborhood. Although the minimum energy to be dissipated for the local detection gaining one bit of Shannon's information turns out to be kTln2 (erg/bit), this doesn't imply either net information generation nor net energy dissipation, but is rather a coupled form of both fluctuations and dissipation occurring around a thermal equilibrium.

We have noted that the dynamics of local detection including even Boltzmann's as its special case holds the law of motion of a one-to-many temporal mapping with regard to its constituent degrees of freedom. In contrast, thermodynamics is unique in maintaining the law of motion of a many-to-one mapping type guaranteeing approach toward a thermal equilibrium at the asymptotic time limit. The many-to-one temporal mapping inherent in thermodynamics is, however, of imposed character by the external observer who limits the available dynamic degrees of freedom only to the thermodynamic variables. Reduction of an enormous number of degrees of freedom with the original dynamics of local detection to a set of extremely limited ones of the thermodynamic variables necessarily yields a many-to-one temporal mapping with an irreversible characteristic letting each different input transform into a unique common output with the elapse of time. The factors other than those taken up as thermodynamic variables are regarded simply as random forces acting upon the variables and causing fluctuations in the latter. The second law of thermodynamics conceived within the present scheme of reducing the number of degrees of freedom is just synonymous with the irreversibility of the underlying many-to-one temporal mapping.

At issue now is whether a one-to-many temporal mapping of the dynamics of local detection could effectively be transformed into a form of a many-to-one mapping supplemented by random fluctuations. If the present transformation is practically justifiable, attainment of a thermal equilibrium would be guaranteed with no net energy dissipation and no net information generation. Whether a one-to-many mapping of local detection could be substituted by a theoretical artifact of a many-to-one mapping supplemented by random fluctuations depends exclusively upon the nature of agents doing local detection. Boltzmann's interacting particle losing its past memory after a few, or at most several, collisions with others in the immediate past is certainly one prototype of agents doing local detection, but is not exclusively exhaustive. The nature of agents doing local detection needs its further scrutiny.

NET INFORMATION GENERATION

The one-to-many temporal mapping upholding the dynamics of local detection is unquestionably informational in contrasting the multitude of possibilities perceived in the future to the fixed actuality in the recorded past. Even interacting particles in Boltzmann's dynamics of local detection are informational as limiting the extent of being possible when they realize their own movement. On the other hand, how they generate information or how specific the actualized movement could be compared to the possible ones depends upon how the movement would be represented. If the Boltzmann's dynamics of a one-to-many mapping is represented by the thermodynamic scheme of a many-to-one mapping supplemented by random fluctuations, there would be no net information generation because every excited movement is destined to approach the common goal of thermal equilibrium, as exhibiting that individual specific movements altogether would not lead to any specific movement as a collection.

The thermodynamic framework of a many-to-one temporal mapping supplemented by random fluctuations as a means of representation serves as a theoretical artifact, while the dynamics of local detection to be represented is of material origin. If each individual degree of freedom loses its past memory after a few, or at most several, collisions with others in the immediate past as Boltzmann perceived, can one be led to conceive of the thermodynamics whose constituent microscopic degrees of freedom remain unaltered as given. Boltzmann's dynamics of local detection addresses itself to both what constituent individual degrees of freedom are and how they maintain their individuality while interacting with each other. Nonetheless, thermodynamics as a many-to-one mapping supplemented by random fluctuations is at best methodological in taking constituent individual degrees of freedom as given. Methodologically, there is allowed no net information generation within the framework of thermodynamics at thermal equilibrium. It is the dynamics of local detection that is responsible for constructing and altering each individual degree of freedom. A possible enhancement of specificity, that is to say, information generation that is of course of emergent character, may be sought within constructing and rearranging individual degrees of freedom to be actualized in the dynamics of local detection.

Thermodynamics at thermal equilibrium would neither enhance its own specificity by itself nor increase the complexity accompanying emergent phenomena insofar as its constituent degrees of freedom remain fixed. The invariant nature of the constituent degrees of freedom provides the thermodynamics with a methodological reference against which whether or not there would be any information generation or enhancement of specificity can be identified. Crucial to the presence of such a methodological reference for identifying whether there could be any emergent phenomena is the nature of each constituent degree of freedom for the thermodynamics especially with regard to whether or not it could remain immutable.

If immutable, the resulting thermodynamics could hold no capacity for generating emergent phenomena. This is because whatever spontaneous emergences of local character, even if excitable as fluctuations, would come to be mutually cancelled with each other in the long run due to the presumed persistence of a thermal equilibrium. Accordingly, emergent properties are to reside within the mutability of each degree of freedom that constitutes a thermodynamics as a consequence. It is the dynamics of local detection, instead of thermodynamics itself, which specifies how individual degrees of freedom could be mutable.

The present perspective on the mutability of individual degrees of freedom contrasts with the time-honored practice of letting the distinguishability or indistinguishability of individual degrees of freedom be left unchanged. If each degree of freedom is taken to be indivisible or irreducible, this would satisfy a necessary condition that these individual degrees of freedom may remain immutable. In addition, it would also be possible to construct an organized aggregation from those degrees of freedom while observing the immutability of each individual. However, this scheme is not free from its own internal inconsistency. If individuality is restricted only to those constituent degrees of freedom, it would not be feasible to assign a new individuality to the resulting aggregation.

On the other hand, if a newly organized aggregation acquires its own individuality, the former monopoly of individuality limited only to the constituent degrees of freedom would break down. In this regard, Boltzmann's dynamics of local detection is exceptional in the monopoly of individuality by the constituent degrees of freedom as dismissing other possibilities of individuation. Elimination of further individuation within Boltzmann's dynamics is evident in the postulate of molecular chaos. Nonetheless, this does not exclude the possibility that the dynamics of local detection may enhance its own individuation depending upon the nature of local detection proceeding there. Prerequisite to further possibility for individuation within the dynamic framework of local detection is the plasticity of each individual degree of freedom such that it may be dissociated into its sub-individuals or may participate in forming a new individual degree of freedom while being associated with others.

The degree of freedom in the dynamics of local detection cannot be absolute in holding its individuality, but rather relative in changing its nature in time. A new association of some of existing degrees of freedom is certainly an emergent phenomenon. Likewise, dissociation of a degree of freedom into its sub-individual ones is also of emergent character. The dynamics of local detection, if it is framed in the form of a thermodynamics after the change in the individuality, turns out to have an emergent characteristic compared to that framed before the change. In this sense, the problem of enhancement of complexity and emergent phenomena is to reduce to how the participating individual degrees of freedom could be mutable through their association and dissociation. Enhancement of specificity or net information generation can be sought within the mutability of constituent individual degrees of freedom.

EMERGENCE AS A FREEZING OF DEGREES OF FREEDOM

It is of course methodologically legitimate to conceive of the dynamics of local detection as referring to its thermodynamic framework. However, there would inevitably exist some residues that could not be incorporated in an arbitrary form of thermodynamics with fixed constituent degrees of freedom. It is this residual part which can make the constituent degrees of freedom mutable. One major mode of making degrees of freedom in motion mutable is their association or freezing. In fact, a nucleon such as a proton and neutron as an emergent material aggregate from quarks can be seen as a frozen aggregate of quark-degrees of freedom as much as an atomic nucleus as a frozen aggregate of nucleon-degrees of freedom. Each degree of freedom in motion necessarily remains indefinite in what it implies.

When a test degree of freedom happens to be frozen into another with the consequence of forming a new combined degree of freedom, the association process can be parameterized in terms of the time required for completing the association after getting the initial impetus for the association. The initial impetus can be detected internally between those degrees of freedom that will eventually be associated. The initiating local detection is of course due to thermal fluctuations among those constituent degrees of freedom whose thermodynamics serves as the driving factor for the fluctuations. However, it does not follow that the resulting thermodynamic fluctuations could cover all the excitations being possible in the original dynamics of local detection. It is one thing to construct a thermodynamics from a given set of individual degrees of freedom interacting with each other, but quite another to see whether the resulting thermodynamics would faithfully reproduce the original dynamics of local detection. If the time required for completing the association of the test degree of freedom with others happens to become far greater than the thermal relaxation time of the resulting thermodynamics, such an association of degrees of freedom would go beyond the consequent thermodynamic fluctuations.

One quantitative figure to represent the association process of the test degree of freedom is the probability that the association has not yet been completed since having the initiating local detection. In particular, since the association is dichotomous between having been completed and not yet and since the probability of the association not yet being completed after having the initiating impetus is sequentially cumulative in time, the association process can be parameterized in terms of bits per unit time per degree of freedom. If the probability that the test degree of freedom does not yet complete the association with the other over a unit time after having the initiating local detection is just one half, the association process can be parameterized in terms of the rate of information generation as much as one bit per unit time per degree of freedom. The rate of information generation with regard to the association process measures how fast the association of the test degree of freedom with the other could be completed once the associative interaction has got started as having the initiating local detection due to thermal fluctuations. Consequently, the association process accompanied by information generation turns out to be a dynamic mode of holding thermal fluctuations over the period much longer than the thermal relaxation time.

What is of emergent character in freezing those degrees of freedom through their association is the spontaneous holding of thermal fluctuations locally even though initiated by themselves. The source of enhancing the holding time of thermal fluctuations can be found within the original dynamics of local detection.

One of the constraints which any local detection has to fulfill is the observation of various conservation laws such as energy flow continuity. Global realization of energy flow continuity in a system does not imply a similar global identification of the continuity. To the contrary, that detection has to be local necessarily induces the lasting process of equilibration for energy flow continuity everywhere at every moment because detecting the effect of realizing energy flow continuity in the neighborhood constantly serves as a cause for the action for the sake of recovering the continuity there (Matsuno, 1989). If the process of equilibration for energy flow continuity fails to survive, the empirically incontrovertible conservation of energy would come to be violated.

Equilibration for energy flow continuity constantly survives simply due to the unbeaten contrast between global realization and local detection. The indefinite continuation of equilibration that prohibits itself from reaching a stationary equilibrium provides the dynamics of local detection with the capacity of generating those fluctuations having ever-increasing relaxation times as being subject to no definite limit. Approaching the dynamics of local detection as utilizing the framework of thermodynamics comes to approve that there may exist a possibility of freezing degrees of freedom through their association even though initiated by thermal fluctuations of theirs.

An advantage of employing the framework of thermodynamics to cope with the dynamics of local detection is that the property of being emergent can be seen as an occurrence of freezing of degrees of freedom. The onset of freezing is triggered by thermal fluctuations and the process of association is seen as an instance of holding the thermal energy from the ambient. This sequential process of the triggering then followed by the holding simply points to an energy flow into those aggregates of degrees of freedom just to be frozen while holding the thermal energy in a limited region spatio-temporally.

Spontaneous appearance of a frozen aggregate of degrees of freedom is information generative during the formative period due to the dichotomous nature of its event, and is at the same time boundary formative for the crossing of energy flow (Farre, 1994). The generation of information comes to be associated with the formation of a spatial boundary across which energy can flow. The thermodynamic perspective on the dynamics of local detection enables us to see occurrence of an emergent phenomenon as a consequence of generating information through forming a spatial boundary across which energy can flow.

The energy to be fed into the emergent frozen aggregate of degrees of freedom is from thermal fluctuations. Although forming such an emergent aggregate may look a work done by thermal fluctuations, this does not violate the second law of thermodynamics in any sense. The spontaneous emergence of frozen aggregates of degrees of freedom from what looks like thermal fluctuations is due to the occurrence of fluctuations that could survive over an indefinitely long period of time in the original dynamics of local detection.

The dynamics of local detection is always more than what thermodynamics of any sort could imply, even if it is extended into a nonequilibrium situation. Although a nonequilibrium thermodynamics can bring about a certain emergent characteristic compared to the equilibrium counterpart, the source of emergent properties is strictly of external origin. Once one decides to ask the source of novelties exclusively within external contingencies, anything can go. This attitude however is not a legitimate one, because any emergent property assumes the presence of a transformation from being implicit to explicit. In other words, thermodynamics does not provide itself with a sound ontological basis upon which emergent phenomena could satisfactorily be explicated. This is because thermodynamics does not supply itself with the capacity of determining where to put itself whether near at the thermal equilibrium or far from it. Both thermodynamics and something else that is responsible for letting it be placed arbitrarily relative to thermal equilibrium are necessary in order to cope with emergent properties. Precisely for this reason, emergent characteristic latent in the dynamics of local detection can be made explicated by consulting thermodynamics as a methodological reference. What is most significant to thermodynamics methodologically is the postulate of thermal equilibrium as a frame of reference that gives no emergent phenomena nor net information generation whatever individual degrees of freedom may be taken as its constituent members.

Freezing of degrees of freedom in the dynamics of local detection is certainly of emergent character and information generative because of the dichotomous nature of the underlying events. The event description carries with it the capacity of coping with emergent phenomena. In contrast, if one limits oneself only to the state description that has been taken for granted in the practice of physics, the subject matter of information generation and emergent phenomena would be destined to slip out of that descriptive enterprise even if the notion of state is relaxed as in the case of microstates in Boltzmann's statistical mechanics.

State dynamics is as a matter of principle of non-emergent character because the state as a global realization at every moment is defined as being definitely identifiable. By definition, there can be no emergent property within the state that can be definitely identifiable. That the state description of dynamics does not face squarely with the problem of emergent phenomena is due to an obvious observation that nothing of emergent character is allowed to exist there. Dynamic realization and identification both on the global scale that underlie the unique determination of the state dismiss the problem of emergent phenomena altogether on the methodological ground.

Emergent phenomena can be focused when dynamic identification is taken to be local both in space and in time. What now come to the fore is dichotomous or, more generally, discrete events. Any event referring to what happens over a limited time interval lacks its uniqueness of occurrence unless forcibly stipulated otherwise. Events are discrete in two different manners. One is in picking up one of the many happenings that could be potentially possible, and the other is in identifying, even though locally, the time point at which a particular happening completes itself. Both types of discreteness lack uniqueness of their occurrences. However, since the completed events again become identifiable in the external record, the representation of the consequence of the local dynamics has to be unambiguous.

Securing the unambiguous representation can be accomplished by focusing on countable quantities when the events are referred to. If the event refers to the one that is uncountable, such as the event that the weather will be fine tomorrow, it cannot be dichotomous even after the event because the distinction between the event and no such event remains fuzzy. The number of degrees of freedom and its change through their association certainly satisfies the condition for serving as an attribute to represent dichotomous or discrete events. Those emergent events of discrete nature occur in time domain, and what makes the progression of time punctuated discontinuously is the discreteness latent in the countability of degrees of freedom in motion.

Local identification of degrees of freedom with regard to their countability contrasts with the global identifiability of the state that consists of those degrees of freedom. Identification of the state such as conceived in state dynamics requires the value of each constituent degree of freedom to be specified, while the dynamics of local detection being capable of emergent phenomena restricts itself only to specifying the number of degrees of freedom. The value identification of each degree of freedom constituting the state however does necessitate global identification, thus dismissing the likelihood of emergent phenomena altogether. In contrast, the dynamics of local detection supplies itself with the capacity of being emergent as leaving the value of each degree of freedom indefinite, though fixed in the external record. Unique to the dynamics of local detection is its methodological competence in coping with emergent phenomena, whereas there are no such phenomena within the scheme of state dynamics allowing realization and identification both on the global scale.

BIRD'S-EYE AND WORM'S-EYE VIEWS

An irony about emergent phenomena is that they can be globally represented even though their genesis owes processes being local both in space and in time. The nexus connecting local detection in the making to global identification in the record should be a quantitative figure, because the numerical figure once identified remains as it is whether in a local context or in a global counterpart. The number of degrees of freedom and its change through their dissociation certainly satisfy the condition for becoming an objective figure whether locally or globally.

In particular, if all the dynamic variables are uniquely identifiable in a global manner, there would be no room of emergent phenomena. The one-to-one temporal mapping guaranteeing the unique development of the dynamics renders the underlying temporal transformation to be a computational process that can be logically reversible. Computational process of the one-to-one temporal mapping can also be made dissipationless by maintaining the whole system undisturbed externally without invoking the second law of thermodynamics. Dissipation is simply irrelevant and foreign to the one-to-one temporal mapping. That is to say, being dissipationless is as much physical and real as the one-to-one mapping is.

In contrast, the one-to-many temporal mapping underlying the dynamics of local detection makes the corresponding computational process logically irreversible because of the absence of bilaterality of the mapping. The one-to-many mapping constantly making a choice out of possible many alternatives is not deductively logical in its operation, but synthetically contingent. That the one-to-many mapping is computationally irreversible is not a matter of approximation, but an incontrovertible physical reality. The computational irreversibility, when perceived from a thermodynamic perspective, can be seen accomplished by a physically dissipative process as implied in the second law of thermodynamics. The present dissipative process, however, cannot be identified as such externally if there exists a thermal equilibrium. The computational irreversibility will constantly be offset, though incoherently, by those fluctuations originating in the dissipation due to the preceding process of similar nature.

Insofar as one sticks to a one-to-one temporal mapping, there is neither physical nor logical irreversibility. Only when the thermodynamic framework forcing us to frame whatever dynamics in terms of a many-to-one temporal mapping supplemented by random fluctuations is taken, the computational irreversibility latent in the one-to-many temporal mapping can gain its thermodynamic implication, especially with regard to the physical irreversibility embodied in the second law of thermodynamics.

Nonetheless, there would arise a confusion if one accepts a one-to-one temporal mapping and then applies to it a many-to-one mapping to get a thermodynamics. The capacity of global identification taken for granted for confirming the one-to-one mapping would be invalidated in the resulting thermodynamics that is methodologically incompetent in relating macroscopic variables such as an entropy to the underlying microscopic dynamics whether it may be of a one-to-one mapping or of a one-to-many. In other words, insofar as thermodynamics near thermal equilibrium is guaranteed, there is no net dissipation even if dissipations are locally excitable. For the energy from the preceding dissipations somewhere constantly serves as an impetus for exciting the coming local dissipations elsewhere within the framework of equilibrium thermodynamics.

There is no direct relationship between the reversibility of the one-to-one temporal mapping and no net dissipation in equilibrium thermodynamics. It is thus inappropriate to ask the one-to-one temporal mapping the cause of the lack of both net dissipation and net irreversibility in equilibrium thermodynamics as much as to seek in thermodynamics the remnant of the reversibility of the original mapping.

Neither thermodynamics nor the one-to-one temporal mapping can address itself to the problem of how dissipation or irreversibility could be confirmed physically or ontologically. This observation may suggest to us a choice between whether or not the problem of dissipation or irreversibility would be basically ill-stated as dismissing emergent phenomena simply as an illusion. In order to face this problem squarely, one requires a firm ontological ground other than a one-to-one temporal mapping on a microscopic scale or a many-to-one mapping on a thermodynamic scale. As a matter of fact, the one-to-many mapping underlying the dynamics of local detection that we have observed can cope with dissipation and irreversibility, and accordingly with emergent phenomena. Its methodological competence is rooted in the ontological ground admitting that nothing propagates faster than the speed of light.

The difference between the one-to-one temporal mapping as embodied in classical and quantum mechanics and the one-to-many mapping with the dynamics of local detection is deep. The capacity of global identification latent in the one-to-one mapping makes it possible to specify the global state of the dynamics at every moment, thus eliminating the likelihood of emergent phenomena. In contrast, the one-to-many mapping for the dynamics of local detection makes emergent phenomena as being quite natural. The difference is found within where to place the activity of identification.

If global identification is the case like in the bird's eye, the state on the global scale can be claimed without taking the actual procedure of observation in the empirical sense. On the other hand, if local detection is taken seriously as it should be, the source of being emergent can be sought within the contrast between what has been detected and will be detected. Furthermore, the problem of how to represent the emergent phenomena will require to be focused in its turn, since the representation completed again points to the global situation. The representation is global in the record. What can be in the worm's eye has to be translated in the manner that the bird's eye can grasp.

The problem of emergent phenomena is intrinsically convoluted. If the global representation of a dynamics is taken before it is examined whether the actual dynamic process of identification could be accomplished globally in an instantaneous manner, there would be no emergent phenomena unless exerted upon externally. If the local process of identification is taken before the global representation of the dynamics is tried, on the other hand, it would be required for the representation to coordinate various local perspectives in the manner such that emergent phenomena may be brought about on the global scale as the effect. The number of degrees of freedom and its change through their association just fulfill the condition for connecting emergent phenomena in the making in the worm's eye to the similar phenomena in the product in the bird's eye.

LINGUISTIC CONDITIONS FOR PHYSICAL SYMMETRY-BREAKING

Once the dynamics of local detection is taken as referring to the fundamental material process, the breaking of temporal symmetry is to become the rule and not the exception. What does matter is only how significant the symmetry-breaking would be.

If the extent of the symmetry breaking can be neglected for some reasons, the symmetry-preserving dynamics could be saved. The presence of such a dynamic symmetry can make the association of any physical object with the corresponding representation-in-itself completed because the process of associating the object with its representation is deemed to be atemporal. The representations-in-themselves thus obtained are closed of themselves without allowing any further interferences from the underlying objects. The symmetry property latent in the physical objects can completely be transferred into that found in the representations-in-themselves.

Symmetries latent in the representations-in-themselves can be reduced to their irreducible units as much as an arbitrary formal language can be reduced to its irreducible predicates. Conversely, a possible symmetry property conceived within in the representations-in-themselves can be constructed from their irreducible units as much as any theorem being possible in a formal language can be constructed from its axioms in terms of the irreducible predicates.

In contrast, the dynamics of local detection, once taken seriously, dismisses the possibility of complete association of a physical object with its representation. What is possible instead is a representation-in-an-object or, figuratively a representation only half done. The dynamics certainly admits in itself the process of representing others through the interaction between arbitrary parties. But, the process of representing is always in its own making and never completed because of the locality of detection. Any representation-in-an-object constantly makes itself a further object to be represented subsequently. The process of representing is to be regressed infinitely. As a matter of fact, representations admitting an infinite regression are realizable only in natural languages in view of the fact that any formal language prohibits itself from being entrapped by the trouble of infinite regression while permitting a set of irreducible fundamental predicates.

Representation-in-an-object conceived within any dynamics of local detection is dynamic in the sense that once the process of detection or representation is got started, it cannot stop because the representations can never be completed. The situation is in parallel to the case of a natural language that admits a sequence of infinite regression because of the absence of irreducible basics there. Once a question asking the meaning of a predicate appearing in a natural language is raised, there is no means to prohibit any one from questioning the question.

The contrast between symmetry and symmetry-breaking in material processes is intriguing. If one commits oneself to the conceptual scheme that a complete representation of any physical object is always available at any time, the underlying dynamics cannot be anything other than that preserving the dynamic symmetry. Ironically enough, however, the origin of such dynamic symmetry is methodological rather than being intrinsically physical. The symmetry property latent in any formal language preserving its own set of irreducible fundamental predicates invariant can also be seen as a physical symmetry.

The present perspective provides two totally incompatible views on symmetry and symmetry-breaking. If a formal language based upon a set of irreducible fundamental predicates is prior to a natural language, the preservation of symmetry would be the rule and the occurrence of symmetry-breaking might be trivial appendages upon the former. The symmetry hidden in a formal language could be broken only sporadically as responding to disturbances originating elsewhere. On the other hand, if natural languages are the rule and the articulation of a formal language is taken to be at most an artifact conceived out of the former, symmetry-breaking will be the rule.

The physics of symmetry-preservation and symmetry-breaking has to be founded upon material grounds, to be sure. Nevertheless, unless it is described linguistically, we cannot even talk about the physics. This convoluted relationship between a physics and a means of describing it necessarily imposes a certain restriction upon what has been described. It is of course one thing to think much of both a formal language and the consequent preservation of symmetry, but quite another to examine the role of the formal language compared to natural languages as its mentor. The dynamics in the scheme of a formal language is actually static and thus symmetry-preserving. In contrast, the dynamics conceived in the scheme of natural languages is really dynamic and accordingly symmetry-breaking, although what the physics written in a natural language looks like still remains to be seen.

REFERENCES

Baas, N. A., 1994. Emergence, hierarchies and hyperstructures, to appear.

Ehresmann, A. C. and Vanbremeersch, J.-P., 1987. Hierarchical evolutionary systems: a mathematical model for complex systems. Bull. Math. Biol. 49, 13-50.

Farre, G. L., 1994. Wittgenstein's philosophy of science, to appear.

Kauffman, S. A., 1993. The Origin of Orders: Self-Organization and Selection in Evolution. Oxford University Press, New York.

Kreuzer, H. J., 1981. Nonequilibrium Thermodynamics and its Statistical Foundations. Clarendon Press, Oxford.

Matsuno, K., 1982. A theoretical construct of protobiological synthesis: from amino acids to functional protocells. Int. J. Quant. Chem. QBS 9, 181-193.

Matsuno, K., 1985. How can quantum mechanics of material evolution be possible?: symmetry and symmetry-breaking in protobiological evolution. BioSystems 17, 179-192.

Matsuno, K., 1989. Protobiology: Physical Basis of Biology, CRC Press, Boca Raton Florida.

Matsuno, K., 1993. Being free from ceteris paribus: a vehicle of founding physics on biology rather than the other way around. Appl. Math. Comp. 56, 261-279.

Pattee, H. H., 1993. The limit of formal models of measurement, control and cognition. Appl. Math. Comp. 56, 111-130.

Rosen, R., 1991. Life Itself: A Comprehensive Inquiry into The Nature, Origin and Fabrication of Life. Columbia University Press, New York.

Salthe, S. N., 1993. Development and Evolution: Complexity and Change in Biology, MIT Press, Cambridge Mass.

Schneider, T. D., 1991. Theory of molecular machine. II. energy dissipation from molecular machines. J. Theor. Biol. 148, 125-137.

MEYER SETS AND
THE FINITE GENERATION OF QUASICRYSTALS

R. V. Moody[†]

Department of Mathematics
University of Alberta
Edmonton, Alberta, T6G 2G1 Canada

ABSTRACT: A monk saw a turtle walking in the garden of Ta-sui's
monastery and asked his teacher " All beings cover their bones with
flesh and skin. Why does this being cover its flesh and skin with
bones?" Ta-sui, the master, took off one of his sandals and covered
the turtle with it. ——The Iron Flute

1. INTRODUCTION

The basic paradigm upon which, until recently, the crystallography of this century has been founded can be summed up in the following scheme of supposedly equivalent concepts:

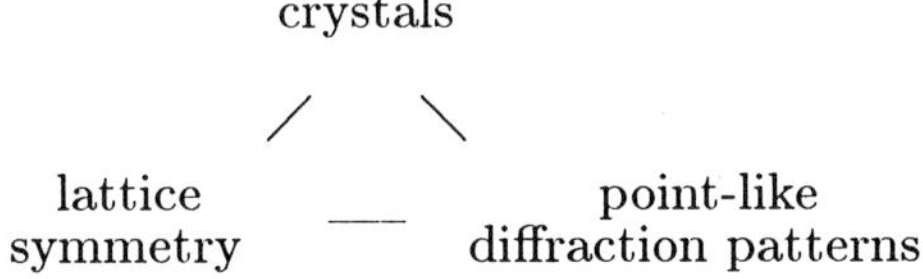

Crystals can be recognized by the appearance of their diffraction patterns which are composed of discrete bright spots. These signify the existence of underlying lattice symmetries (translational symmetry in three independent directions) and a primary task is to determine the fundamental cell and the atomic distribution within this cell.

But lattice symmetry puts severe constraints on the possible rotational symmetries. Among the most famous of these "crystallographic restrictions" is that 5-fold symmetry is incompatible with $\mathbb{Z}^2$-lattice symmetry and icosahedral symmetry is incompatible with $\mathbb{Z}^3$-lattice symmetry. It is interesting to observe that these same restrictions appear in the theory of compact Lie groups. Indeed, let K be such a group

---

[†] Work supported in part by a grant from the Natural Sciences and Engineering Research Council of Canada and the Aspen Center for Physics.

and T a maximal torus. The universal covering group of T is $\mathbb{R}^k$, where $k = \operatorname{rank} K$ and we have the exact sequence

$$0 \longrightarrow L \longrightarrow \mathbb{R}^k \longrightarrow T \longrightarrow 1, \tag{1.1}$$

where L is a lattice. The Weyl group W is the finite group $N(T)/T$, where $N(T)$ is the normalizer of T in K. The point is that W is a finite reflection group (Coxeter group) and it acts on the lattice L as a point-group. But now the crystallographic restriction comes into the picture and forbids the appearance of the non-crystallographic Coxeter groups. The indecomposable Coxeter groups of this non-crystallographic type are the dihedral groups $I_2^{(n)}$, $n \geq 7$, of rank 2 and the three groups H_2, H_3, H_4, whose Coxeter diagrams are

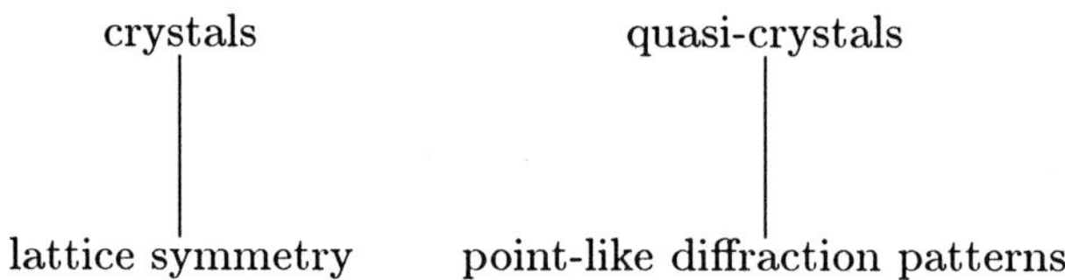

$$\tag{1.2}$$

All other indecomposable Coxeter groups do appear in the theory of compact Lie groups.

The announcement by Schechtman et al. [Sc] a decade ago and the extensive literature that refining and clarifying their work show that point-like diffraction patterns can occur in the solid state that possess non-crystallographic symmetries (in particular icosahedral (H_3) symmetries) and hence are not the reciprocal lattices of any crystal lattice. For an up-to-date references on this whole subject, see [Ja, Se].

In view of these revelations a new paradigm has emerged:

crystals   quasi-crystals

| |

lattice symmetry   point-like diffraction patterns

So quasicrystals are generalizations of crystals. This new paradigm is useful phenomenologically but it poses a serious mathematical problem. We do not know how to characterize point-sets that have point-like diffraction patterns.

Here we should pause to clarify exactly what we mean. If Λ is a set of points in $\mathbb{R}^k$ then its "diffraction pattern" is based on its Fourier transform

$$f(\lambda) = \sum_{x \in \Lambda} e^{2\pi i \lambda x}, \tag{1.3}$$

which has to be understood in the sense of distributions (actually what is obsserved is the square modulus of f). What appears to happen with physical quasicrystals (and this agrees well with the mathematical models) is that f is a sum of weighted delta functions

$$f = \sum_{\mu \in \Lambda'} a_\mu \delta_\mu \tag{1.4}$$

where $\Lambda' \subset \mathbb{R}^k$ is a countable set and δ_μ is the Dirac delta function centred at μ. The set Λ' is not actually necessarily discrete, but if

$$\Lambda'_c := \{\mu \in \Lambda' \mid |a_\mu| > c\} \tag{1.5}$$

then for each $c > 0$, Λ'_c is discrete. As the resolution of the experiments increases (c decreases) the number of delta peaks visible (Λ'_c) increases.

There are several ways known in which to construct mathematical models of quasicrystals, each bringing its own insights and each having its own limitations. One of the most useful approaches is the cut and project method. The main objections to this method have been its use of higher (than 3) dimensional spaces and/or its use of totally discontinuous mappings. After reviewing this method and its strengths and weaknesses, we introduce a very beautiful theory of Y. Meyer.

This theory, that predates quaiscrystals by some 15 years, has not been utilized very much either by the mathematics or physics communities. We will see (and Y. Meyer himself has recently made this point [Me2]) that Meyer's theory entirely encompasses the cut and project method; in fact it is essentially equivalent to it. However it also presents us with two quite different alternative veiwpoints. Both of these viewpoints are worthy of far more thought than has so far been given to them. One of them in particular, that is based on the idea of uniform approximation of group characters by continuous group characters, leads us to a new way of generating quasicrystals. In the last part of this paper we introduce this idea which was formulated [MP2]. When it is applicable, it offers a method of generating (growing) a quasicrystal by the use of a simple building block and a finite set of continuous guiding conditions (**coherent phase conditions**). The virtue of this is that it avoids all mention of unphysical dimensions and/or discontinuous processes.

2. THE CUT AND PROJECT METHOD

In this paper crystals and quasicrystals refer to point-sets Λ in Euclidean space $\mathbb{R}^k$. A subset Λ of $\mathbb{R}^k$ satisfies the **Delaunay property** (or is an (r, R)-set [Se]) if

D1 (no clustering): there is an $r > 0$ so that $|x - y| \geq r$ for all $x, y \in \Lambda$; and

D2 (no holes): there is an $R > 0$ so that every ball of radius R in $\mathbb{R}^k$ contains at least one point of Λ.

(D1) represents the required separation of atoms in a physical solid and (D2) their idealized infinite and (crudely) uniform distribution in $\mathbb{R}^k$.

The idea behind the cut and project method is that although a given finite point-symmetry G may be impossible in a lattice of some fixed rank k, it is always possible if the rank is allowed to be sufficiently large. This is simply because any finite group G can be embedded in $\mathrm{GL}_n(\mathbb{Z})$ for some n. We begin then with a lattice L of rank $k+m$ in $\mathbb{R}^{k+m}$ for some m (which is supposed to have the required symmetry G) and form two orthogonal projections (called the parallel and orthogonal projections)

$$p_{\parallel} \cdot \mathbb{R}^{k+m} \longrightarrow \mathbb{R}^k, \quad p_{\perp} : \mathbb{R}^{k+m} \longrightarrow \mathbb{R}^m. \tag{2.1}$$

What is required here is that

$$\mathrm{Ker}(p_{\parallel}) \cap L = (0) \quad \text{and} \quad p_{\perp}(L) \text{ is dense in } \mathbb{R}^m \tag{2.2}$$

and that

$$p_{\parallel} \quad \text{and} \quad p_{\perp} \quad \text{are } G\text{-invariant.} \tag{2.3}$$

The details of how this might be accomplished is not the point here (see [LP]). The idea is to form our quasicrystal using $p_{\parallel}(L)$. In fact $p_{\parallel}(L)$ necessarily has accumulation points (violates (D1)). To avoid this a window Ω is taken in orthogonal space $p_{\perp}(\mathbb{R}^{k+m}) = \mathbb{R}^m$. We require that Ω contains a non-empty open subset of $\mathbb{R}^m$ and is bounded. Then

$$\Lambda = \Lambda(\Omega) := \{p_{\parallel}(x) \mid x \in L,\ p_{\perp}(x) \in \Omega\} \tag{2.4}$$

is a point-set in $\mathbb{R}^k$ satisfying (D1) and (D2). If Ω is G-invariant, so also is Λ. This is then the basic cut and project method for quasicrystals, initiated in [KN]. An extension of (2.4) that we will find useful is to replace Λ by a a subset of a finite number of its translates

$$\Sigma = \Lambda + F, \quad F \text{ some finite set.} \tag{2.5}$$

An enormous variety of quaiscrystals Λ can be constructed in this way. Whole continua of quasicrystals can be obtained by varying Ω continuously, and as long as Ω is reasonable (namely that the Fourier transform of its characteristic function χ_{Ω} is reasonable) then the Fourier transform (1.3) of Λ satisfies (1.4).

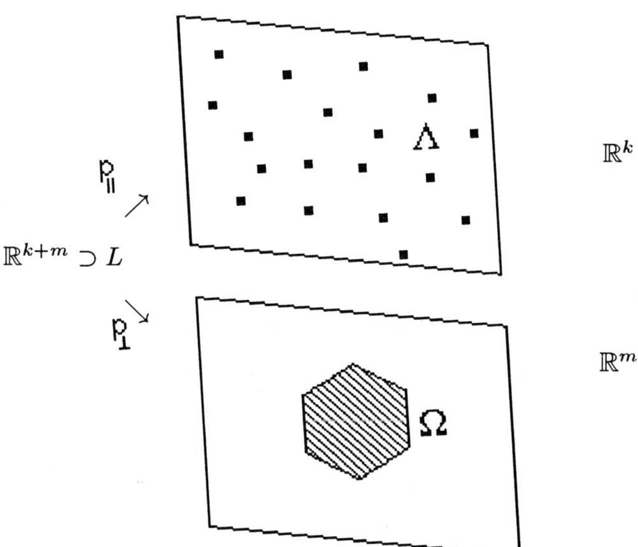

FIGURE 1. The quasicrystal Λ constructed as the projection $p_{\parallel}$ of those points of L whose orthogonal projection lies in the window Ω

The introduction of higher rank lattices is physically reasonable since the experimental evidence supports the fact that the points of the quasicrystal (or its diffraction pattern) can only be indexed (coordinatized over $\mathbb{Z}$) with more than 3 coordinates (typically 6) for 3D-quasicrystals [Ja,3.3]. However the method has been seriously criticised for its introduction of non-physical dimensional spaces ($\mathbb{R}^{k+m}$ for quasicrystals in $\mathbb{R}^k$).

Since $p_\parallel$ is 1–1 on L by (2.2) we can form the composite mapping

$$p := p_\perp \cdot (p_\parallel)^{-1} : \quad p_\parallel(L) \longrightarrow \mathbb{R}^m$$

so Λ can equally be described

$$\Lambda = \{x \in p_\parallel(L) \mid p(x) \in \Omega\}, \tag{2.6}$$

which seemingly avoids using $\mathbb{R}^{k+m}$. This looks particularly attractive when $k = m$ (which in fact is often the case in the modelling of icosahedral quasicrystals) since then Λ can be defined using only the physical space, $\mathbb{R}^k$. The problem now, however, is that p is hopelessly discontinuous. Any extension of p to a function on $\mathbb{R}^k$ will, in general, be discontinuous everywhere. This is directly related to the way in which L has to be embedded in $\mathbb{R}^{k+m}$ in order to get (aperiodic) quasicrystals Λ instead of (periodic) crystals. Such a discontinuous map is just as unphysical as the extra dimensions.

Of itself, the cut and project method gives no hint as to any growth mechanisms for the quasicrystals it produces. This is a question that we address in §5.

3. SOME EXAMPLES

In this section we introduce two families of quasicrystals that we will later use to exemplify the growth models that we have in mind. Both examples are closely related to the H-family of symmetries. At the end of this section we place these examples into the context of finite root systems, which may help to clarify how they appear more naturally.

EXAMPLE 3.1.   We are going to produce a quasicrystal in $\mathbb{R}^2$ with H_2-symmetry. It is mathematically efficient to identify $\mathbb{R}^2$ with the complex numbers $\mathbb{C}$. For $x, w \in \mathbb{C}$ the standard dot product in $\mathbb{R}^2$ reads

$$z \cdot w = \mathrm{Re}(z\bar{w}). \tag{3.1}$$

Set $\zeta := e^{2\pi i/5}$ and let

$$\mathbb{Z}[\zeta] = \sum_{j=0}^{4} \mathbb{Z}\zeta^j \subset \mathbb{C}. \tag{3.2}$$

Then $\mathbb{Z}[\zeta]$ is a $\mathbb{Z}$-module of rank 4 with the relation $1 + \zeta + \cdots + \zeta^4 = 0$. Although we do not need this, $\mathbb{Z}[\zeta]$ is the ring of integers of the cyclotomic field $\mathbb{Q}(\zeta)$.

There is a unique ring automorphism (Galois automorphism)

$$* : \mathbb{Z}[\zeta] \longrightarrow \mathbb{Z}[\zeta] \tag{3.3}$$

defined by

$$\zeta \longmapsto \zeta^2.$$

Figure 3.1 illustrates the effect of this automorphism on the vertices of the regular decagon

$$D_{10} := \{\pm\zeta^j \mid j = 0,\ldots,4\}. \tag{3.4}$$

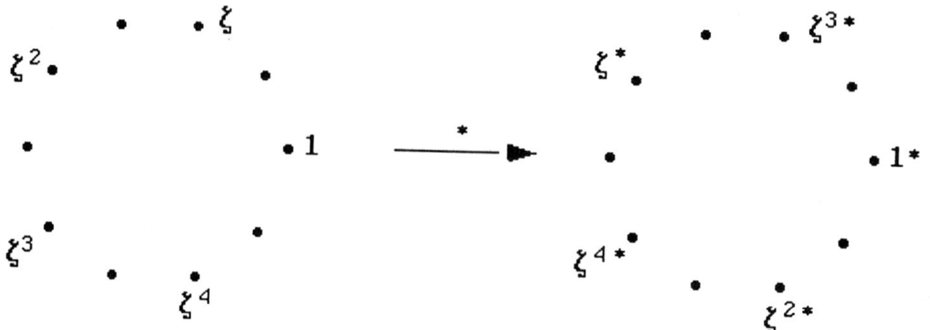

FIGURE 3.1

Now $\mathbb{Z}[\zeta] \simeq \mathbb{Z}^4$ as a $\mathbb{Z}$-module, but as a subset of $\mathbb{R}^2 = \mathbb{C}$ it is dense. This is not surprising because $\mathbb{Z}[\zeta]$ is invariant under the rotation $z \longmapsto z\zeta$ of order 5 and we know this is incompatible with discrete groups in $\mathbb{R}^2$. To make $\mathbb{Z}[\zeta]$ discrete we embed it in $\mathbb{R}^4$: define

$$L := \{(x, x^*) \mid x \in \mathbb{Z}[\zeta]\} \subset \mathbb{C} \times \mathbb{C} \simeq \mathbb{R}^4. \tag{3.5}$$

Then L is a lattice in $\mathbb{R}^4$ and $L \simeq \mathbb{Z}[\zeta]$ under the map $(x, x^*) \longmapsto x$.

This leads us to the following picture

$$
\begin{array}{ccc}
\mathbb{Z}[\zeta] \subset \mathbb{R}^2 & \qquad & x^* \\
\nearrow & & \nearrow \\
\mathbb{R}^4 \supset L & & (x, x^*) \\
\searrow & & \searrow \\
\mathbb{Z}[\zeta] \subset \mathbb{R}^2 & & x
\end{array}
$$

which corresponds exactly with (2.1) and (2.2). Then for $\Omega \subset \mathbb{R}^2$, Ω bounded with non-empty interior, we obtain a quasicrystal

$$\Lambda(\Omega) := \{x \in \mathbb{Z}[\zeta] \mid x^* \in \Omega\} \tag{3.6}$$

(see(2.4)). $\Lambda(\Omega)$ carries decagonal (H_2)-symmetry if Ω does. For the purposes of this paper we will restrict ourselves to the case

$$\Omega = \{z \in \mathbb{C} \mid |w \cdot \zeta^j| \le 1, j = 0, \dots, 4\}, \tag{3.7}$$

which is the regular decagon of Figure 3.2

EXAMPLE 3.2.   We define
$$\tau = (1 + \sqrt{5})/2 \tag{3.8}$$
and form the ring
$$\mathbb{Z}[\tau] = \mathbb{Z} + \mathbb{Z}\tau \subset \mathbb{R}.$$
The map
$$\tau \longmapsto \tau' := (1 - \sqrt{5})/2$$

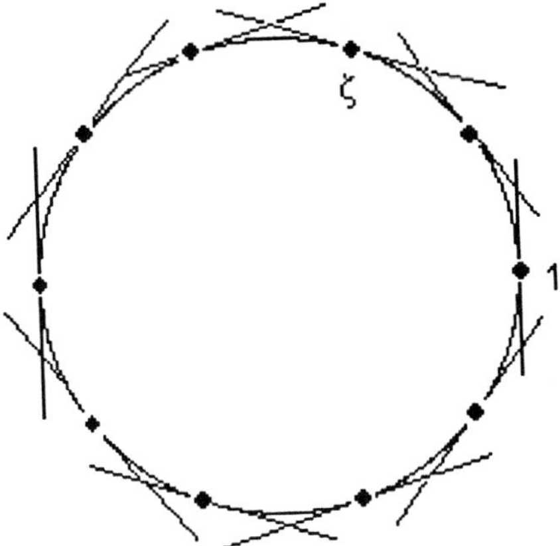

FIGURE 3.2. The decagon dual to the regular decagon D_{10}.

determines a unique (Galois) automorphism($'$) of $\mathbb{Z}[\tau]$ and, just as in the previous example,

$$L := \{(x, x') \mid x \in \mathbb{Z}[\tau]\} \subset \mathbb{R}^2$$

is a lattice. We then obtain one-dimensional quasicrystals

$$\Lambda(\Omega) := \{x \in \mathbb{Z}[\tau] \mid x' \in \Omega\} \subset \mathbb{R}$$

is the usual way. In this paper we restrict ourselves to the special class of these

$$\Lambda_a := \{x \in \mathbb{Z}[\tau] \mid x' \in [-a, a]\}, \quad a > 0. \tag{3.9}$$

Λ_1 has the simple description

$$\pm \big\{ \{1\} \cup \{ \lceil b\tau^{-1} \rceil + b\tau \mid b \in \mathbb{Z} \} \big\}. \tag{3.10}$$

Around 0 it looks like

The Examples 3.1 and 3.2 may give an incorrect impression that the quasicrystals that we are discussing are dependent on some additional algebraic structure since both $\mathbb{Z}[\tau]$ and $\mathbb{Z}[\zeta]$ are rings. This is not the case, nor is there any need to restrict to one and two dimensional spaces. The examples here are chosen for their simple descriptions and because it is easier to draw pictures in low dimensional spaces.

We close this section with a few remarks about the special features that are associated with the non-crystallographic Coxeter groups H_2, H_3, and H_4. Although, as we have pointed out, these groups do not appear in the theory of Lie groups, Lie theorists have often found that the combinatorial aspects of the subject can be extended to include them. The root systems are one such example. Root systems arise from the study of the eigenspaces of a Cartan subalgebra of a simple Lie algebra $\mathfrak{g}$ acting on $\mathfrak{g}$ via the adjoint representation, and as such do not appear in association with the non-crystallographic groups (see §1). V. Deodhar has developed a version of the theory of root systems that includes all the finite Coxeter groups. Briefly a finite set Δ of vectors of the same length in $\mathbb{R}^k$ is a root system if for each $\alpha \in \Delta$ the reflection

$$r_\alpha \colon x \longmapsto x - \left(\frac{2x \cdot \alpha}{\alpha \cdot \alpha} \right) \alpha \tag{3.11}$$

maps Δ into itself. The group

$$W := \langle r_\alpha | \alpha \in \Delta \rangle \tag{3.12}$$

is then a Coxeter group. For H_2, H_3, H_4 the root systems are

Δ_2: the 10 vertices of a regular decagon in $\mathbb{R}^2$.

Δ_3: the 30 vertices of a regular iscosadodecahedron in $\mathbb{R}^3$.

Δ_4: the 120 vertices of a regular 600-cell in $\mathbb{R}^4$.

For more on this see [De, Hu, CMP].

The $\mathbb{Z}$-span of Δ_k, which we denote by M_k, should be regarded as a fundamental object in the quasicrystallography of these groups. M_k is dense in $\mathbb{R}^k$ and as a $\mathbb{Z}$-module is of rank $2k$. This is quite different from crystallographic root systems where the $\mathbb{Z}$-span is a lattice of rank k equal to the dimension of the ambient space $\mathbb{R}^k$. It is a simple consequence of the five-fold symmetries inherent in H_k that $\tau M_k = M_k$.

Forming the ring $\mathbb{Z}[\tau]$, some order is restored: M_k is a free $\mathbb{Z}[\tau]$-module of rank k:

$$M_k \simeq \mathbb{Z}[\tau] \oplus \cdots \oplus \mathbb{Z}[\tau] \quad (k \text{ summands}) \tag{3.13}$$

Thus the decagon (3.4) is in fact Δ_2 and $\mathbb{Z}[\zeta] \simeq M_2$. Example 3.2 should be thought of as a member of the same family, but being one-dimensional it does not exhibit any five-fold symmetry. Any H_k-module of rank $2k$ in $\mathbb{R}^k$ is necessarily very closely tied to M_k (see [CMP] for more on this). For more on the extraordinary connections between these root systems and the root systems of types A_4, D_6, E_8 see [ES] and [MP1].

4. MEYER'S THEORY

This section is based on [Me1]. Meyer introduces the concept of harmonious sets in the context of locally compact abelian groups. We use a slightly stronger concept and restrict ourselves to the case of euclidean space $\mathbb{R}^k$, coining the term Meyer set in the process. For more on the Fourier analysis side of this theory see [Me2].

(I) A subset $\Lambda \subset \mathbb{R}^k$ is a **Meyer set** if and only if

(i) Λ is Delaunay;

(ii) there is a finite set F such that $\Lambda - \Lambda \subset \Lambda - F$.

If $F = \{0\}$ then (ii) says that Λ is a subgroup of $\mathbb{R}^k$. Then, together with (i), we see that Λ is a lattice. So Meyer sets are a generalization of lattices and (ii) is a weakening of the subgroup condition. Unlike groups, the structure of a Meyer set is not destroyed by translation. For example, $F = \{a\}$ then $\Lambda - a$ is a group and so Λ is a coset of a group. For card $F \geq 2$, Λ is no longer necessarily particularly group-like.

For example, to use the notation of §3, let

$$\Lambda_1 := \{x \in \mathbb{Z}[\tau] \mid x' \in [-1, 1]\}. \tag{4.1}$$

Then Λ is a Delaunay set and we can see

$$\Lambda - \Lambda \subset \Lambda - \{0, 1, -1\}.$$

Indeed, for $x, y \in \Lambda$, $(x - y)' \in [-2, 2]$. If $(x - y)' \in [-1, 1]$ then $x - y \in \Lambda$. If $(x - y)' \in [1, 2]$ then $x - y - 1 \in \Lambda$ and $x - y = (x - y - 1) + 1 \in \Lambda + 1$. The remaining case is similar.

Meyer goes on to give two further very different characterizations of these sets.

Let $L \subset \mathbb{R}^{k+m}$ be a lattice and let $p_{\parallel}, p_{\perp}$ be orthogonal projections (2.1) satisfying (2.2). Let Ω be a bounded subset of $\mathbb{R}^m$ that contains a non-empty open subset of $\mathbb{R}^n$ and form $\Lambda(\Omega)$ as in (2.4). Let F be any finite subset of $\mathbb{R}^k$. Then

(II)

    (i) $\Lambda(\Omega) + F$ is a Meyer set;

    (ii) any subset Σ of $\Lambda(\Omega) + F$ that is also a Delaunay set is a Meyer set;

    (iii) any Meyer set can be constructed in the form of (ii).

This is the second characterization of Meyer sets: the subsets of $\mathbb{R}^k$ generated by the cut and project method are Meyer sets and visa-versa. It follows that the Meyer sets are the same thing as the quasicrystals (2.5) determined by the cut and project method.

The third characterization involves duality theory. Suppose $\Lambda \subset \mathbb{R}^k$ is any subset. We let $[\Lambda]$ denote the subgroup of $\mathbb{R}^k$ generated by Λ. A **character** (or **algebraic** character) is a homomorphism

$$\chi : [\Lambda] \longrightarrow U(1) := \{z \in \mathbb{C} \mid |z| = 1\}. \tag{4.2}$$

Thus for all $x, y \in \Lambda$, $\chi(x + y) = \chi(x)\chi(y)$. We also call χ a **character on** Λ (even though Λ is not itself a group in general).

It is well known that the set of continuous characters $\mathbb{R}^k \longrightarrow U(1)$ are all of the form

$$\chi_\mu : x \longmapsto e^{2\pi i \mu \cdot x} \tag{4.3}$$

for some $\mu \in \mathbb{R}^k$, where $\cdot$ is the usual scalar product in $\mathbb{R}^k$. The set of all these characters form the dual group $\widehat{\mathbb{R}^k}$ which itself is clearly isomorphic to $\mathbb{R}^k$.

Returning to our character $\chi : [\Lambda] \longrightarrow U(1)$, we can ask whether or not χ is the restriction to $[\Lambda]$ of a continuous character χ_μ of $\mathbb{R}^k$. The answer is emphatically no, even if Λ is a Delaunay set. For example, for $\Lambda_1 \subset \mathbb{R}$ given by (4.1), $[\Lambda_1] = \mathbb{Z} + \mathbb{Z}\tau$ and the mapping

$$\chi : x + \tau y \longmapsto e^{2\pi i (3x + 4y)/5} \tag{4.4}$$

is a character which takes only five values: the fifth roots of 1. However, since it takes on all five values infinitely often in every neighbourhood of $\mathbb{R}$, it is impossible to extend χ to a continuous mapping on all of $\mathbb{R}$.

Let $\Lambda \subset \mathbb{R}^k$ be a Delaunay set. Let χ be a character on Λ and let $\varepsilon > 0$ be arbitrary. $\chi_\mu \in \widehat{\mathbb{R}^k}$ is an ε-uniform approximation of χ on Λ if,

$$\text{for all } x \in \Lambda, \quad |\chi(x) - \chi_\mu(x)| < \varepsilon. \tag{4.5}$$

Note that (4.5) is not required to hold on $[\Lambda]$! Also note that if $\varepsilon \geq 2$ the concept becomes trivial: every character is a 2-uniform approximation of χ.

Now we have the third characterization:

(III)

A Delaunay set Λ is a Meyer set if and only if for all characters $\chi \colon [\Lambda] \longrightarrow U(1)$ and for all $\varepsilon > 0$ there exists an ε-uniform approximation $\chi_\mu \in \widehat{\mathbb{R}^k}$ of χ on Λ (μ depends on Λ, χ, and ε).

One might think that if χ is actually the restriction to $[\Lambda]$ of a continuous characer (for instance, the trivial character $\chi(x) = 1$ for all $x \in [\Lambda]$) then the ε-approximation would necessarily be χ itself. This is far from the case. Indeed for $\varepsilon > 0$ define:

$$\Lambda^\varepsilon := \left\{ \chi_\mu \in \widehat{\mathbb{R}^k} \;\middle|\; \chi_\mu \;\; \begin{array}{l} \text{is an } \varepsilon\text{-\textbf{uniform approximation}} \\ \text{of} \\ \text{the trivial character on } [\Lambda] \end{array} \right\}. \tag{4.6}$$

Thus $\chi_\mu \in \Lambda^\varepsilon \Leftrightarrow |e^{2\pi i \mu \cdot x} - 1| < \varepsilon$ for all $x \in \Lambda$. Then Λ^ε is itself a Meyer set in $\widehat{\mathbb{R}^k} \simeq \mathbb{R}^k$.

The illustration of this in the case of Λ of (4.1) is very revealing of how the characterizations (II) and (III) are linked. Let Λ_1 be defined by (4.1). For all $\mu, w \in \mathbb{Z}[\tau]$

$$\mu \cdot w + \mu' \cdot w' \in \mathbb{Z} \tag{4.7}$$

(here $\cdot$ is the scalar product ($=$ multiplication) in $\mathbb{R}$) so

$$e^{2\pi i \mu \cdot w} = e^{-2\pi i \mu' \cdot w'}. \tag{4.8}$$

Now $w \in \Lambda_1 \iff w' \in [-1, 1]$ and so if μ' is small, $|e^{-2\pi i \mu' \cdot w'} - 1|$ is uniformly small on Λ_1. But then by (4.8) $|e^{2\pi i \mu \cdot w} - 1|$ is small too, so χ_μ is an ε_a - uniform approximation. For example, suppose $a < \frac{1}{3}$ and let

$$\mu \in \Lambda_a := \{\mu \in \mathbb{Z}[\tau] \mid \mu' \in [-a, a]\}. \tag{4.9}$$

Then

$$|e^{-2\pi i \mu' \cdot w'}| < |e^{2\pi i a} - 1| =: \varepsilon_a \tag{4.10}$$

on Λ_1 so χ_μ is an ε_α-uniform approximation on Λ_1 and $\Lambda_a \subset \Lambda^{\varepsilon_a}$.

We call Λ^ε the ε-**dual** of Λ. It is interesting to see what the ε-dual means in the case that Λ is actually a lattice. If $\chi_\mu \in \Lambda^\varepsilon$ then for all $x \in \Lambda$, $|e^{2\pi i \mu \cdot x} - 1| < \varepsilon$. Since Λ is a group it follows that also $|e^{2\pi i \mu \cdot n x} - 1| < \varepsilon$, for all $n \in \mathbb{Z}$.

Now if $\varepsilon < \sqrt{3}$ this is possible only if $e^{2\pi i \mu \cdot x} = 1$, for all $x \in \Lambda$ and that simply says that $\mu \cdot x \in \mathbb{Z}$ for all $x \in \Lambda$. Thus if $\varepsilon < \sqrt{3}$ and Λ is a group

$$\Lambda^\varepsilon = \Lambda^0 := \{\mu \in \mathbb{R}^k \mid \mu \cdot x \in \mathbb{Z}\}, \tag{4.11}$$

388

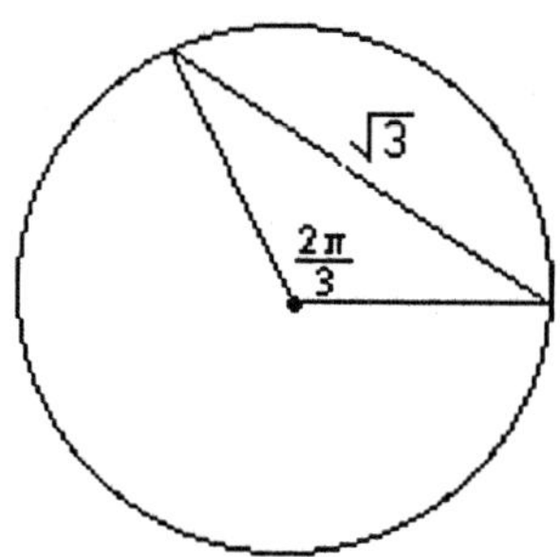

Λ^0 being the standard $\mathbb{Z}$-dual (reciprocal lattice) of Λ. Thus the sets Λ^ε are a generalization of the dual of a lattice.

It is also interesting to note that $\Lambda^{\varepsilon\varepsilon} \supset \Lambda$ and $(\Lambda^{\varepsilon\varepsilon})^{\varepsilon\varepsilon} = \Lambda^{\varepsilon\varepsilon}$, so $\Lambda \longrightarrow \Lambda^{\varepsilon\varepsilon}$ is a type of "closure operation" on the set of all Meyer sets in $\mathbb{R}^k$ (of course this depends on ε).

The proofs of the equivalence of the three Meyer conditions are quite non-trivial and can be found in his [Me1]

5. GENERATING QUASICRYSTALS

This section reports the main result of our work with J. Patera [MP2]. See also [KMP].

If Λ is a lattice then Λ^0 is also a lattice and Λ can be completely determined by a finite number of elements of Λ^0, namely a basis of Λ^0. It seems too optimistic to expect that Λ can be completely determined by a finite number of elements of Λ^ε (for some fixed ε, but we are now going to see that in fact can be the case.

The objective then is to give a set of rules that allow one to construct (generate) a given quasicrystal Λ. In fact the method we propose is not always applicable but it does work in a great variety of cases, as we will see. The key points of this approach are that it avoids all montion of orthogonal (window)-spaces and discontinuous processes, and that it generates the quasicrystal in an organized way outward from a starting seed set of points. One way to think of this is as a dynamical process, and indeed this is physically appealing. Whether or not this generating process can be viewed as the beginnings of a truly physical model is something that is not clear at the moment.

The ingredients are listed below:

1. a **starting seed**:   a finite set S of points of Λ which serve as the points from which further growth will procede;
2. an **ideal local configuration**(i.l.c.):   a set B of vectors which are to serve as the basic generators for the growth;
3. the **generation or growth process**:   if z is an existing (already created) point of Λ then the set of points

$$x = z + b, \qquad b \in B$$

are new **potential** points of Λ.

4. the **selection process**: a decision process (based on a finite set of elements from Λ^ε for some ε) which selects from the potential points those that will be accepted as points of Λ. The rules of the selection process are also called **coherent phase conditions** for reasons that will become clearer.

The physical intuition is this: starting with the seed atomic sites, each created site z considers itself as the centre of a potentially "perfect" atomic configuration $z + B$. The selection rules decree that some of these new sites have to be declared unacceptable (in the interests of continuing the structure indefinitely as a Delaunay set). Each site will also come with a finite set of phases that will be required to sufficiently closely match some globally imposed continuous phase conditions.

We illustrate the method with a pair of examples.

EXAMPLE 5.1. Consider Λ_1 of Example 3.2. Then it is quite easy to check that if $x \in \Lambda_1$ then one of $x - 1$ or $x - \tau$ lies in Λ_1 and similarly one of $x + 1$ or $x + \tau$ lies in Λ_1. This shows that

$$S := \{0, \pm 1\}$$
$$B := \{\pm 1, \pm \tau\}$$

will serve as a suitable seed and i.l.c.. Now fix any $a < \frac{1}{3}$ and choose $\mu \in \Lambda_a$. Set

$$\varepsilon := |e^{2\pi i \mu'} - 1|. \tag{5.1}$$

We claim that the condition

$$|e^{2\pi i \mu x} - 1| \le \varepsilon \tag{5.2}$$

will serve as the (single) phase coherence condition that will generate Λ_1.

To see this first observe that for $x \in \Lambda_1$, $x' \in [-1, 1]$ and so

$$|e^{2\pi i \mu' x'} - 1| \le |e^{2\pi i \mu'} - 1| = \varepsilon.$$

But from (4.8) (with x replacing w) we see that then

$$|e^{2\pi i \mu x} - 1| = |e^{2\pi i \mu' x'} - 1| \le \varepsilon \tag{5.3}$$

and so (5.2) holds for all $x \in \Lambda_1$.

Now let us see that condition (5.2) and the generation process together conspire to keep all points accepted in step 4 inside Λ_1. For suppose that $z \in \Lambda_1$, and $b \in B$. We want to show that if $x := z + b$ satisfies (5.2) then $x \in \Lambda_1$, i.e. $x' \in [-1, 1]$. By assumption $z' \in [-1, 1]$ and using (5.3)

$$|e^{2\pi i \mu \cdot z'} - 1| \le \varepsilon. \tag{5.4}$$

Figure 5.1 shows the graph of $|e^{2\pi i \mu' \cdot y} - 1|$ as a function of y.

Since $x' = z' + b' \in [-1, 1] + b' \subset [-2, 2]$ and (5.2) holds, $x' \in [-1, 1]$ also (see the caption to Fig. 5.1). Thus one element of Λ^ε, namely $\chi_\mu : x \longmapsto e^{2\pi i \mu x}$ together with S and B determines Λ_1 completely. We may say that (5.2) "locally defines" the acceptance window $[-1, 1]$.

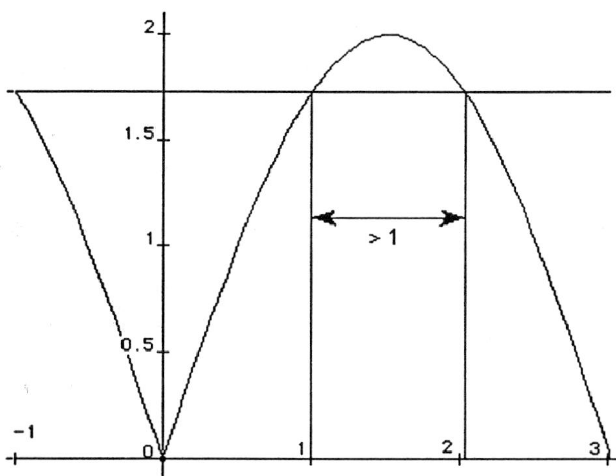

FIGURE 5.1.   Graph of $f(y) := |e^{2\pi i \mu' y} - 1|$. Note that the period $\frac{1}{|\mu'|}$ satisfies $\frac{1}{|\mu'|} - 2 > 1$ since $|\mu'| < a < \frac{1}{3}$. Condition (5.2) together with the growth mechanism traps points in the region [-1,1]

EXAMPLE 5.2.   (see [MP2]) We consider the model in Example 3.1 given by (3.5) and (3.6). We use the following ingredients:

1. starting seed: $\{0\}$;
2. the i.l.c.: $B = \{\pm 1, \pm\zeta, \pm\zeta^2, \pm\zeta^3, \pm\zeta^4\}$;
3. growth process: as above.

    Define five conditions $(C_j), j = 0, \ldots, 4$ by

$$|e^{2\pi i 2\tau^6 \zeta^j \cdot x} - 1| < 0.686; \qquad\qquad (C_j)$$

4. selection process: accept x produced in step 3. if and only if x satisfies all five conditions $(C_0), \ldots, (C_4)$.

Figure 5.2 illustrates the growth process which works radially outwards.

The five continuous characters

$$x \longmapsto e^{2\pi i 2\tau^6 \zeta^j \cdot x}$$

are all in $\Lambda^{0.686}$. These five characters do not on their own determine Λ in the strongest sense: namely $\Lambda \subset [B] = \mathbb{Z}[\zeta]$ but the conditions $x \in \mathbb{Z}[\zeta]$ and x satisfies $(C_j), j = 0, \ldots, 4$, are not sufficient to force $x \in \Lambda$. But combined with the growth process 3, they are enough: if $z \in \Lambda, b \in B$ then $x := z + b \in \Lambda$ if and only if (C_j), $j = 0, \ldots, 4$ hold. This occurs for the same sort of reason as in Example 5.1. The conditions (C_j) "locally define" the acceptance window of Λ.

What we have achieved here is summarized in the scheme:

$$\text{finite subset of } \Lambda^\varepsilon + \text{ growth mechanism } \implies \Lambda. \qquad (5.5)$$

A finite subset of Λ^ε can appear not only separately and overtly, as in Examples 5.1 and 5.2 but also in numerous other forms In [MP2] we use the same ingredients as in

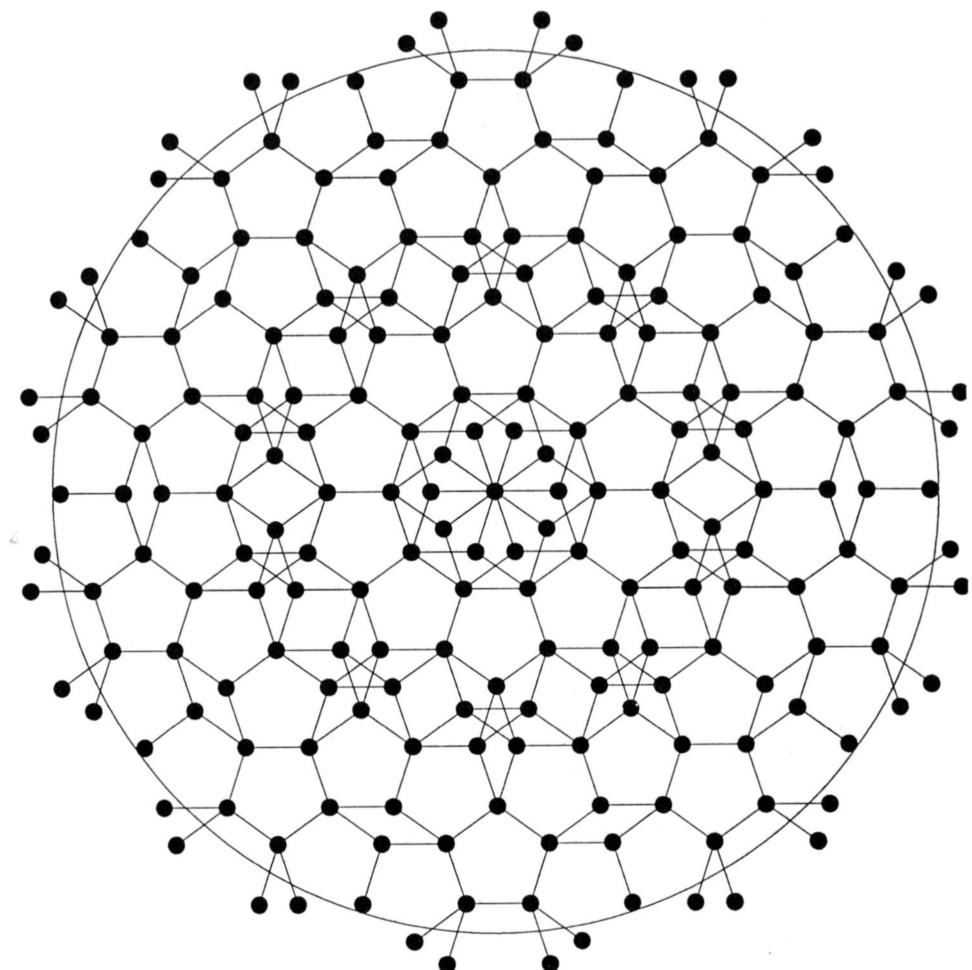

FIGURE 5.2. Illustration of the growth process of Example 5.2. The growth procedes radially outwards. The line segments indicate which vectors of B are actually used. This quasicrystal is technically speaking singular in the sense that the origin is a distinguished point of overall 5-fold symmetry. This sort of phenomenon is not intrinsic to the growth process and can be avoided by minor alteration of the conditions (C_j).

Example 5.2 but replace the five conditions (C_j) by an averaged version of them:

$$f(x) = \frac{1}{11}\left\{1 + \sum_{j=0}^{4} e^{2\pi i \zeta^j \cdot x} + \sum_{j=0}^{4} e^{-2\pi i \zeta^j \cdot x}\right\} \tag{5.6}$$

and use the single coherent phase condition

$$f(2\tau^4 x) \geq 0.3823 \tag{5.7}$$

Now the acceptance window becomes the region bounded by a countour of the associated function $f(2\tau^{-4}x)$. The variety of quasicrystals that can be generated by this process is extremely large. It is a challenging problem to shape the coherent phase conditions into physically meaningful forms.

What are the requirements on a quasicrystal $\Lambda = \Lambda(\Omega) \subset \mathbb{R}^k$ that it can be constructed in the way outlined here? Basically they are two. The first requirement is that a seed S and a set B can be found to serve as the i.l.c. for it. The quasicrystal $\Lambda(\Omega)$ is said to be **amenable** to the finite subset $B \subset \mathbb{R}^k$ if there exists and $R > 0$ such that for all $x \in \Lambda$ with $|x| > R$ there is a $b \in B$ such that $x - b \in \Lambda$ and $|x - b| < |x|$.

Under these conditions

$$S := \{x \in \Lambda \mid |x| \le R\}$$

will serve as the seed and B as the i.l.c.. In [CMP] we give straightforward conditions for determining amenability for the quasicrystals of the H_2, H_3, H_4 families. In general, it is not hard to find such amenable sets (though optimizing them, i.e. minimizing B and S is not easy at all).

The second requirement is finding coherent phase conditions. This is relatively straightforward if Ω is a convex polyhedral domain whose boundary (to be included or not) is given by a set of equations

$$y \cdot w_j \le r_j \qquad j = 1, \dots, n \tag{5.8}$$

in $\mathbb{R}^k$, where the $w_j \in p_\perp(L)$, $r_j \in \mathbb{R}$. (Since $p_\perp(L)$ is assumed dense in $\mathbb{R}^k$ the special form of the w_j in (5.8) is, for all practical purposes, not a constraint.)

For non-polyhedral windows the situation is far from clear (at least to the author) at the present time. The question is basically what sort of regions can be defined by a finite number of inequalities of the form

$$\sum_{j=1}^{n} a_j e^{2\pi i w_j \cdot y} \le r, \tag{5.9}$$

where $a_j, r \in \mathbb{R}$, $w_j \in p_\perp(L)$ are given.

Whatever the ultimate usefulness of this approach to the generation of quasicyrstals, the main point of this paper should be to emphasize the value of Meyer's theory to the theoretical side of quasicrystals and to point out that the consequences of his equivalent conditions (I) and (III) are largely unexplored.

Thanks to M. Senechal for her helpful comments in preparing this paper. Figure 5.2 was computed and drawn by the software *simpLie* that has been developed by D. Rand, J. Patera, and the author.

References

[CMP]   Chen, L., Moody, R. V., and Patera, J., *Non-crystallographic root systems and quasicrystals* (to appear).

[De]   Deodhar, V., *On the root system of a Coxeter group*, Comm. in Algebra **10** (1982), 611.

[ES]   Elser, V. and Sloane, N. J. A., *A highly symmetric four-dimensional quasicrystal*, J. Phys. A: Math. Gen. **20** (1987), 6161.

[Hu]   Humphreys, J.E., *Relection groups and Coxeter groups*, Cambridge Univ. Press, 1990.

[Ja]   Janot, C., *Quasicrystals : A Primer*, Monographs on the Physics and Chemistry of Materials, vol. 48, Clarendon Press, Oxford, 1992.

[KMP] Kramer, P., Papadopolis, Z., and Moody, R. V., *A growth mechanism for the $T^{(2F)}$-tiling* (to appear).

[KN] Kramer, P. and Nori, R., *On periodic and no-periodic space fillings obtained by projections*, Acta Cryst. **A40** (1984), 580.

[LP] Lunnon, W. F. and Pleasants, P., *Quasicrystallographic tilings*, J. Math. pures et appl. **66** (1987), 217.

[Me1] Meyer, Y., *Algebraic numbers and harmonic analysis*, North-Holland Publ., Amsterdam, 1972.

[Me2] ______ , *Quasicrystals, diophantine approximation, and algebraic numbers* (1994), Beyond Quasicrystals, Les Houches,1994.

[MP1] Moody, R. V. and Patera, J., *Quasicrystals and icosians*, J.Phy.A:Math.GEn. **26** (1993), 2829.

[MP2] Moody, R. V. and Patera, J., *Local dynamical generation of quasicrystals* (to appear).

[Sc] Schechtman, D., Blech, I., and Gratias, D., *Metallic phase with long-range orientational order and no translational symmetry*, Phys. Rev. Lett. **53** (1984), 1951.

[Se] Senechal, M., *Quasicrystals and Geometry*, Cambrdige Univ. Press (to appear1995).

A TOPOLOGICAL STUDY OF
INDUCED REPRESENTATION

Kazuhiko Odaka

Department of Mathematics and Physics
National Defence Academy
Yokosuka 239 Japan

1 INTRODUCTION

In 1939 E. Wigner[1,2] proposed the induced representation technique when he was obtaining the unitary representation of the Poincaré group. This technique is very useful for elementary particle physics. All elementary particles can be corresponded to the induced representations of the Poincaré group. We have not found any exception.

The Poincaré group P is composed of two parts:

$$P = T \otimes L. \tag{1.1}$$

One part T is the parallel translation in space-time and another part L is the Lorentz transformation. The symbol $\otimes$ means the semi-direct product. We take $\hat{k}_\mu$ to be the generators of the translation, that is, momentum and energy. The summation of their square

$$\hat{k}_0^2 - \sum_{i=1}^{3} \hat{k}_i^2 = \hat{\kappa}^2 \tag{1.2}$$

is invariant under the translation and the Lorentz transformation. $\hat{\kappa}^2$ is the Casimir operator of the Poincaré group and so the representations are classified by the eigenvalues of $\hat{\kappa}$ whose physical meaning is the mass of the particle. These classes are as follows:

$$[M_\pm], \quad [0_\pm], \quad [L], \quad [T]$$

where $\pm$ is $\text{sgn}(k_0)$ and $+$ and $-$ indicate the particle and the anti-particle, respectively. The first classes correspond to the usal massive particle and the second ones are the massless particle. We should notice that these classes do not include the zero energy mode, which belongs to the third class $[L]$. The last case is the tachyon.

A vector space is introduced to provide the representation in the concrete. We take the momentum diagonal state $\mid k, \xi >$ as a vector for the representation. Here,

Symmetries in Science VII, Edited by
B. Gruber, Plenum Press, New York, 1995

ξ means the spin degree. The inner product $< k, \xi \mid k', \xi' >$ must be invariant under the Lorentz transformation.

For the massive case, the Lorentz invariant inner product is given by

$$< k, \xi \mid k', \xi' > = \delta_{\xi \xi'} \omega_k \delta^3(\vec{k} - \vec{k'}) \tag{1.3}$$

where $\omega_k = \sqrt{\vec{k}^2 + m^2}$. The state vectors are well-defined everywhere on the momentum space, which is equivalent to the 3-dimensional Euclidean space. This space is topologically trivial.

For the massless case, the Lorentz invariant inner product is

$$< k, \xi \mid k', \xi' > = \delta_{\xi \xi'} k_0 \delta^3(\vec{k} - \vec{k'}) \tag{1.4}$$

where $k_0 = \mid \vec{k} \mid$. On the occasion of defining the states, one point is subtracted from the momentum space. Therefore, the state vectors are on the $R^3 - \{0\}$ space, which is homotopically equivalent to S^2. It is a topologically non-trivial space.

There exist many problems in the quantum field theory of the massless particle[3], for example, gauge anomaly[4]. Most problems are related with the zero energy mode, which is subtracted from the region of definition of the state vectors. In order to settle these problems, we must study the induced representaion technique on the topologically non-trivial space and its application to quantum theories.

In 1968, furthermore, E. Mackey[5,6] generalized Wigner's technique to other groups and he study quantm mechanics on a homogeneous space (G/H) by using the technique. In his research he discoursed that there exist many inequivalent quantizations for quntum mechanics on a topologically non-trivial configulation space and that they can be classified according to the representaions of H. This situation does not appear in the usual approach[7]. We introduce a position operator on the homogeneous space and the diagonal state vector of this operator is taken to be one for the representation. But, such vectors is not always defined over the homogeneous space, since it is not guarenteed that vecor fields do not vanish anywhere. We need to study quantum mechanics of the case that the state vector can not be defined over all.

In this note, we will study the problem how Wiger's argument about the induced representaion technique is changed for the topologically non-trivial case. Our talk was in the following order; § 2. Wigner's argument and toplogy, § 3. Euclidean group and gauge structrue and § 4. Quantum mechanics on a sphere.

2 WIGNER'S ARGUMENT AND TOPOLOGY

Let us review Wigner's argument[1,2] about the induced representation technique. The group, with which we are concerned, is

$$M = V \otimes G \tag{2.1}$$

where V and G are an abelian group and a transformation group, respectively. The group law on $M = V \otimes G$ is

$$(a_2, \Lambda_2) \cdot (a_1, \Lambda_1) = (a_2 + \Lambda_2 * a_1, \Lambda_2 \Lambda_1). \tag{2.2}$$

The action of G on a is remained on a homogeous space $(G/H \equiv X)$ such as

$$\Lambda * a \in X. \tag{2.3}$$

In order to obtain the representation of G on the space X, we take the following state of the position operator $\hat{x}$ on X,

$$\hat{x} \mid x, \xi > = x \mid x, \xi > . \tag{2.4}$$

These state vectors should be single-valued. If they are degenerate, a new degree ξ is introduced to remove their degeneracy. We consider the representaion matrix $G(\Lambda)$ on the state vector $\mid x, \xi >$. The representaion matrix $G(\Lambda)$ is divided into two parts as

$$G(\Lambda) = Q(\Lambda, \hat{x})P(\Lambda) \tag{2.5}$$

where $P(\Lambda)$ and $Q(\Lambda, \hat{x})$ are diagonal for ξ and $\hat{x}$, respectively. They act on the state vector in the following manner,

$$P(\Lambda) \mid x, \xi > = \mid \Lambda * x, \xi >, \quad Q(\Lambda, \hat{x}) \mid x, \xi' > = \mid x, \xi > Q^{\xi\xi'}(\Lambda, x). \tag{2.6}$$

Their product rules

$$P(\Lambda')P(\Lambda) = P(\Lambda'\Lambda), \quad Q(\Lambda', \hat{x})Q(\Lambda, \Lambda'^{-1}\hat{x}) = Q(\Lambda'\Lambda, \hat{x}) \tag{2.7}$$

are derived from $G(\Lambda')G(\Lambda) = G(\Lambda'\Lambda)$, eq.(2.5) and eq.(2.6). $P(\Lambda)$ satisfies the properties of the representation matrix but $Q(\Lambda, \hat{x})$ does not do so.

If $\Lambda'^{-1}\hat{x} = \hat{x}$, $Q(\Lambda', \hat{x})Q(\Lambda, \hat{x}) = Q(\Lambda'\Lambda, \hat{x})$ and then $Q(\Lambda', \hat{x})$ is the representaion matrix. Then, Wigner define a little group by

$$\lambda l = l, \tag{2.8}$$

where l is some fixed point on X. It is obvious that the little is the subgroup H. Next, he show that any element of G can be reduced to the representation of the little group and the translation on X by taking a suitable unitary transformation

$$G(\Lambda) \simeq U(\hat{x})G(\Lambda)U^{-1}(\hat{x}) = U(\hat{x})Q(\Lambda, \hat{x})U^{-1}(\Lambda^{-1}\hat{x})P(\Lambda), \tag{2.9}$$

where $\simeq$ means the unitary equivalence. This unitary matrix $U(\hat{x})$ is determined through the following procedure. The boost transformation is introduced such as

$$\alpha_x l = \hat{x} \quad (\alpha_x \in G). \tag{2.10}$$

The element λ_x, given by $\alpha_x^{-1}\Lambda\alpha_{\Lambda^{-1}x}$, satisfies $\lambda_x l = l$ and then any element Λ is expressed by

$$\Lambda = \alpha_x \lambda_x \alpha_{\Lambda^{-1}x}^{-1}. \tag{2.11}$$

Substituting this form into $Q(\Lambda, \hat{x})$, we have

$$Q(\Lambda, \hat{x}) = Q(\alpha_x \lambda_x \alpha_{\Lambda^{-1}x}^{-1}, \hat{x}) = Q(\alpha_x, \hat{x})Q(\lambda_x, l)Q(\alpha_{\Lambda^{-1}x}^{-1}, l). \tag{2.12}$$

Let the unitary matrix $U(\hat{x})$ in eq.(2.9) be $Q(\alpha_x, \hat{x})$. It is easy to see that

$$G(\Lambda) \simeq Q(\lambda_x, l)P(\Lambda). \tag{2.13}$$

Here, $Q(\lambda_x, l)$ is just a representation matrix of the subgroup H and it is called Wigner rotation. We arrive at the unitary equivalent representation which is divided into the translation and some representaion of the subgroup H.

The boost operator α_x moves a vector at l to one at $\hat{x}$, and then a vector is assigned smoothly to each point of X by α_x. It is called a vector field over X. However, if a d-dimensional manifold M is topologically non-trivial, it is not necessarily possible to define d vector fields which are linearly independent everywhere. When the d-dimensional homogeous space X does not admit d linearly independent vector fields over X, the boost operator α_x is not well defined over X and then we can not determine the unitary matrix $U(\hat{x})$. Moreover, the invariant inner product of the $\hat{x}$-diagonal states includes d δ-functions, but these δ-functions are not defined at a point where d vector fields are not linearly independent. Therefore the state vecors are not defined over all.

Wigner's technique, reviewed in this section, must be amended for the manifold which is not parallelisable. In the next section, we take S^d as a homogeous space and show how Wigner's argument is modified.

3 EUCLIDEAN GROUP AND GAUGE STRUCTURE

We take the Euclidean group which is composed of the translation and the rotation group $SO(d+1)$ in the d+1 dimensional Euclidean space. When the induced representation is obtained, we do not use the momentum operator but the position operator $\hat{x}$ which is a vector in the d+1 dimensinal Euclidean space. The summation of the square of its components $\hat{x}_i$; $r^2 = \sum_i^{d+1} \hat{x}_i^2$ is invariant under the rotation. r is fixed to be 1. The boost transformation is defined by

$$\alpha_x l = \hat{x}, \qquad \alpha_x \in SO(d+1), \tag{3.1}$$

where l indicates some fixed point on S^d.

This operator also translates the vector fields at l to ones at $\hat{x}$. However, it is well known that the sphere S^d, except for $d = 1, 3$ and 7, does not admit d vector fields which are linearly independent everywhere on S^d. Therefore, the boost operators are not well defined over all. Moreover, the state vectors, which are the eigenstates of the opsition operator, are not well defined over S^d, since the definition of their inner product includes the δ-functions which can not defined at the point where a vector field vanishes.

Wigner's argument should be madified as follows. Firstly, we take two charts on S^d, which are named N and S, respectively. We introduce two fixed points (l^N and l^S), the boost operators (α_x^N and α_x^S) and the state vectors. The induced representation technique, shown in §2, is applied to the group $SO(d+1)$ on each chrat. The unitary matrix $U(\hat{x})$ in eq(2.9) is given by $Q(\alpha_x^N, \hat{x})$ and $Q(\alpha_x^S, \hat{x})$ on each chart.

Secondly, these representations on each chart must be unitary-equivalently connected at the same point in the overlap region such that

$$Q(\alpha_x^N, \hat{x})G(\Lambda)Q(\alpha_x^N, \hat{x})^{-1} \simeq Q(\alpha_x^S, \hat{x})G(\Lambda)Q(\alpha_x^S, \hat{x})^{-1}. \tag{3.2}$$

From the uniraty equivalence $Q(\alpha_x^N, \hat{x})$ and $Q(\alpha_x^S, \hat{x})$ are required to satisfy the condition that $R(\hat{x})$, defined by $Q(\alpha_x^S, \hat{x})^{-1}Q(\alpha_x^N, \hat{x})$, must be a single-valued unitary matrix which is assigned smoothly to each point of the overlap region. $R(\hat{x})$ is rewritten by $Q(\alpha_x^{S-1}\alpha_x^N, l^S)$ and it is obvious that it is not a representation matrix. Then it is difficult to confirm directly that the above condition is satisfied by $R(\hat{x})$.

We comment on the gauge structue in the induced representation technique[8]. The relation of the unitary equivalence at some point is

$$R(\hat{x})G^N(\Lambda, \hat{x})R(\hat{x})^{-1} = G^S(\Lambda, \hat{x}), \tag{3.3}$$

where $G^i(\Lambda, \hat{x})$ $(i = S, N)$ is the representation matrix divided into the translation and the representation of $SO(d)$. Let us consider the infinitesimal rotation $(\Lambda = 1 + \omega)$ and the displacement δx on S^d which is given by $x + \delta x = (1 + \omega)x$. From eq.(3.3), the unitary equivalence of the infinitesimal rotation is expressed by

$$\lim_{\delta x_\mu \to 0} \frac{R(\hat{x})G^N(1 + \omega, \hat{x})R(\hat{x})^{-1} - 1}{\delta x_\mu} = \lim_{\delta x_\mu \to 0} \frac{G^S(1 + \omega, \hat{x}) - 1}{\delta x_\mu} \qquad (3.4).$$

Noting that

$$\frac{1}{\delta x_\mu} = \frac{1}{\omega_{\mu\nu} x_\nu} = \frac{x_\nu}{x^2} \frac{1}{\omega_{\mu\nu}}, \qquad (3.5)$$

we have

$$R(\hat{x})\partial_\mu R(\hat{x})^{-1} + R(\hat{x})A_\mu^N(\hat{x})R(\hat{x})^{-1} = A_\mu^S(\hat{x}), \qquad (3.6)$$

where

$$R(\hat{x})\partial_\mu R(\hat{x})^{-1} \equiv \frac{x_\nu}{x^2} \lim_{\omega_{\mu\nu} \to 0} \frac{R(\hat{x})R(\hat{x} + \omega\hat{x})^{-1} - 1}{\omega_{\mu\nu}}, \qquad (3.7a)$$

$$A_\mu^i(\hat{x}) \equiv \frac{x_\nu}{x^2} \lim_{\omega_{\mu\nu} \to 0} \frac{Q^i(1 + \omega, \hat{x}) - 1}{\omega_{\mu\nu}} \qquad (i = N, S). \qquad (3.7b)$$

∂_μ in eq.(3.7a) means the differential on the sphere. $A_\mu^i(\hat{x})$ can be regarded as gauge potentials on each chart, since eq.(3.6) expresses the gauge transformation.

Now, $R(\hat{x})$ is classified by the homotopy group $\pi_{d-1}(U(n))$, since $R(\hat{x})$ is a map $R : S^{d-1} \to U(n)$ where $U(n)$ is a unitary group. Then the winding (wrapping) number can be defined and if $R(\hat{x})$ satisfy the above condition, the number must be a integer. Let us examine this point for S^2 and S^4 whose homotopy groups are $\pi_1(U(1)) = Z$ and $\pi_3(U(n)) = Z$, respectively. The results for the general cases are published elsewhere[9]. The winding (wrapping) numbers are given by

$$k = \frac{1}{2\pi} \int_{S^1} R^{-1} dR, \qquad (3.8a)$$

$$k = \frac{1}{24\pi^2} \int_{S^3} tr\left(R^{-1}dR \wedge R^{-1}dR \wedge R^{-1}dR\right). \qquad (3.8b)$$

Fortunately, we can calculate these numbers without the explicit form of $R(\hat{x})$ such that

$$1 + R(\hat{x})^{-1} dR(\hat{x}) = Q(\alpha_x^{S-1}\alpha_x^N, l^S)^{-1} Q(\alpha_{x+dx}^{S-1}\alpha_{x+dx}^N, l^S)$$

$$= Q(\alpha_x^{N-1}\alpha_x^S \alpha_{x+dx}^{S-1}\alpha_{x+dx}^N, \alpha_x^{N-1}\alpha_x^S l^S). \qquad (3.9)$$

Defining χ by $\alpha_x^S \alpha_{x+dx}^{S-1}$, we have

$$1 + R(\hat{x})^{-1} dR(\hat{x}) = Q(\alpha_x^{-1}\chi\alpha_{\chi^{-1}x}, l^N). \qquad (3.10)$$

Then, we can show that $1 + R(\hat{x})^{-1} dR(\hat{x})$ is just the Wigner rotation.

We use the stereographic projection:

$$(S) \quad (\hat{x}_i, \hat{x}_{d+1}) = \frac{1}{1 + \hat{y}^2}(2\hat{y}_i, \hat{y}^2 - 1), \qquad (3.11a)$$

$$(N) \quad (\hat{x}_i, \hat{x}_{d+1}) = \frac{1}{1 + \hat{z}^2}(2\hat{z}_i, 1 - \hat{z}^2), \qquad (3.11b)$$

where $\hat{y} = \frac{\hat{z}}{\hat{z}^2}$ in the overlap region, and the winding (wrapping) number is written down explicitly through the following calculation. Firstly, taking the spinor representation of the group $SO(d+1)$, we express the boost trsnsformation as

$$\alpha_x^S = \frac{1}{\sqrt{1+\hat{y}^2}}(1 + \gamma_{d+1} \sum_{i=1}^{d} \gamma_i \hat{y}_i), \qquad (3.12a)$$

$$\alpha_x^N = \frac{1}{\sqrt{1+\hat{z}^2}}(1 - \gamma_{d+1} \sum_{i=1}^{d} \gamma_i \hat{z}_i), \qquad (3.12b)$$

where γ_j is the Hermitian matrix satisfying the Clifford algebra of order $d+1$. χ is given by

$$\chi = 1 + i \sum_{i=1}^{d+1} \frac{d\hat{y}_i}{1+\hat{y}^2} \sigma_{i\,d+1}, \qquad (3.13)$$

where $\sigma_{ij} = \frac{1}{2i}[\gamma_i, \gamma_j]$. Then, after somewhat lengthy calculation we get

$$\alpha_x^{N-1} \chi \alpha_{\chi^{-1}x} = 1 - i \sum_{j,i=1}^{d} \frac{\hat{z}_i d\hat{z}_j}{\hat{z}^2+1} \sigma_{ji}. \qquad (3.14)$$

Next, replacing $\frac{\sigma_{ij}}{2}$ with the generator (S_{ij}) of some representation of the group $SO(d)$, we arrive at

$$Q(\alpha_x^{N-1} \alpha_x^S \alpha_{x+dx}^{S-1} \alpha_{x+dx}^N, l^N) = 1 - i \sum_{j,i=1}^{d} \frac{\hat{z}_i d\hat{z}_j}{\hat{z}^2+1} 2S_{ji}. \qquad (3.15)$$

Taking account of the metric, we express the winding (wrapping) numbers on the unit sphere such as

$$k = \frac{1}{\pi} \int_{S^1} \hat{z}_i d\hat{z}_j S_{ij}, \qquad (3.16a)$$

$$k = \frac{1}{3\pi^2} \int_{S^3} tr\Big(\hat{z}_{i_1} d\hat{z}_{j_1} S_{i_1 j_1} \wedge \hat{z}_{i_2} d\hat{z}_{j_2} S_{i_2 j_2} \wedge \hat{z}_{i_3} d\hat{z}_{j_3} S_{i_3 j_3}\Big). \qquad (3.16b)$$

By the way, it is noteworthy that the winding (wrapping) can be written in terms of the gauge fields defined in eq.(3.7b) such that

$$k = -\frac{1}{4\pi} \int_{S^2} F, \qquad (3.17a)$$

$$k = -\frac{1}{8\pi^2} \int_{S^4} tr\Big(F \wedge F\Big), \qquad (3.17b)$$

where F is the field tensor of A.

4 QUANTUM MECHANICS ON A SPHERE

We apply the induced representation technique to quantum mechanics on a d-dimensional sphere $(\simeq SO(d+1)/SO(d))$[6,8] and formulate its path integral for the transition amplitude by using the semi-classical approximation[10].

A state vector $| \psi(t) >$ satisfies the Schrödinger equation

$$i\frac{d}{dt} | \psi(t) >= \hat{H} | \psi(t) > \tag{4.1}$$

whose formal solution is $| \psi(t) >= e^{-i\hat{H}t} | \psi(0) >$. The quadratic Casimir operator of $SO(d+1)$ is taken as the Hamiltonian $\hat{H}$, because it is invariant under the translation on X and is simple.

We take two charts on X and the following program is executed on each chart. The coordinate operator $\hat{x}$ is diagonalised and the state vectors are spanned by

$$| x, \xi >= G(\Lambda) | l, \zeta > \tag{4.2}$$

where $G(\Lambda)$ is the representation matrix of $SO(d+1)$, l is some fixed point and $x = \Lambda * l$.

The transition amplitude $\Im(T, 0)$ for a particle to start at x_0 at $t = 0$ and end up at x_f at $t = T$, is given by

$$\Im(T, 0) = < x_f, \xi_f | e^{-i\hat{H}T} | x_0, \xi_0 > . \tag{4.3}$$

We divide it as

$$< x_f, \xi_f | \prod_{n=1}^{N} e^{-i\hat{H}n\tau} | x_0, \xi_0 >, \tag{4.4}$$

where $\tau = T/N$. Inserting the identity

$$\sum_{\xi_n} \int d\mu(x_n) | x_n, \xi_n >< x_n, \xi_n | = 1 \tag{4.5}$$

where $d\mu(x)$ is the measure satisfying $d\mu(x) = d\mu(\Lambda * x)$, we have

$$\Im(T, 0) = \sum_{\xi_1} \int d\mu(x_1) \cdots \sum_{\xi_{N-1}} \int d\mu(x_{N-1}) < x_f, \xi_f | e^{-i\hat{H}\tau} | x_{N-1}, \xi_{N-1} >$$

$$\cdots < x_1, \xi_1 | e^{-i\hat{H}\tau} | x_0, \xi_0 > . \tag{4.6}$$

Each transition amplitude

$$K(x_{n+1}, \xi_{n+1}; x_n, \xi_n) = < x_{n+1}, \xi_{n+1} | e^{-i\hat{H}\tau} | x_n, \xi_n > \tag{4.7}$$

at infinitesimal interval τ is estimated by using the semi-classical approximation. Noting that $| x_n, \xi_n >= G(\Lambda_n) | l, \zeta >$ where $\Lambda_n * l = x_n$, we find

$$K(x_{n+1}, \xi_{n+1}; x_n, \xi_n) = < x_n, \zeta_{n+1} | G^{-1}(\omega)e^{-i\hat{H}\tau} | x_n, \xi_n > . \tag{4.8}$$

where $\omega \equiv \Lambda_{n+1}\Lambda_n^{-1}$. Since the Hamiltonian is invariant under the action of G, this is rewritten by

$$K(x_{n+1}, \xi_{n+1}; x_n, \xi_n) = < x_n, \xi_{n+1} | e^{-i\hat{H}\tau} G^{-1}(\omega) | x_n, \zeta_n > \tag{4.9}$$

According to the argument in §2, we have

$$K(x_{n+1}, \xi_{n+1}; x_n, \xi_n) \ = \ < x_n, \xi_{n+1} \mid e^{-i\hat{H}\tau} P(\omega)^{-1} \mid x_n, \zeta_n > Q^{\xi_n \zeta_n}(\omega, x_n)^{-1}. \tag{4.10}$$

All possible representations of the group $SO(d+1)$, which are laveled by S, are put between $< x_n, \zeta_{n+1} \mid$ and $e^{-i\hat{H}\tau} P(\omega)^{-1}$. The amplitude is now given by

$$K(x_{n+1}, \xi_{n+1}; x_n, \xi_n) \ = \ Q^{\xi_n \zeta_n}(\omega, x_n)^{-1} \sum_S e^{-iH(S)\tau} D^S_{\xi_{n+1} \zeta_n}(\omega), \tag{4.11}$$

where

$$D^S_{\xi_{n+1} \zeta_n}(\omega) \ \equiv \ < x_n, \xi_{n+1} \mid S >< S \mid P(\omega)^{-1} \mid x_n, \zeta_n > . \tag{4.12}$$

Since the Hamiltonian is quadratic Casimir operator of the group $SO(d+1)$, it is easy to calculate its value for the representations. $D^S_{\xi_{n+1} \zeta_n}(\omega)$ can be expressed by the Gegenbauer polynomial and $Q^{\xi_n \zeta_n}(\omega, x_n)$ can reduced to the representaion matrix of the little group H by applying Wigner's technique. To carry out the summation of S in eq.(4.11) corresponds to integrating of momentum in the usual path integral formalism based on the canonical quantization. It is very hard to carry out the summation without any approximation.

Here, we take quantum mechanics on S^2 as the simplest example and show its results. The investigation on the other cases are in progress. Now, G and H are $SO(3)$ and $SO(2)(\simeq U(1))$, respectively. The Hamiltonian $\hat{H}$ is given by $\frac{1}{2}\vec{L}^2$ where $\vec{L}$ are the generators of $SO(3)$ and their reprentations are specified by the eigenvalues (j, m) of $\vec{L}^2$ and L_z. Then the transition amplitude is

$$K(x_{n+1}; x_n, s) \ = \ Q^s(\omega, x_n)^{-1} \sum_{j=|s|}^{\infty} \frac{2l+1}{4\pi} e^{(-i\frac{\tau j(j+1)}{2})} d^l_{ss}(\omega), \tag{4.13}$$

where s is an integer and $d^l_{ss}(\omega)$ is a well-known representation function. ω is specified by the inner product $x_{n+1} \cdot x_n = cos\Theta$ and we use the semi-classical approximation; $\tau \ll 1$, $\omega^2 \simeq O(\tau)$. Then after somewhat lengthy calculations we are let to

$$K(x_{n+1}; x_n, s) \ \simeq \ \frac{1}{2\pi i \tau} Q^s(\omega, x_n)^{-1} e^{(\frac{4i}{\tau}(1-cos\frac{\Theta}{2}))} e^{(-i\tau(\frac{s^2}{2}-\frac{1}{4}))}. \tag{4.14}$$

$Q^s(\omega, x_n)$, by using the semi-classical approximationis, is written in terms of the gauge field discussed in § 3 such that

$$Q^n(\omega, x_n) \ \simeq \ 1 + iA, \tag{4.15}$$

where

$$A \ = \ \frac{-s dx_1 x_2}{1 + x_3} + \frac{s dx_2 x_1}{1 + x_3}. \tag{4.16}$$

For general cases, if we use the semi-classical approximation, $Q^{\xi_n \zeta_n}(\omega, x_n)$ can be also written in terms of the gauge fields. For example, the instanton gauge field appears in S^4. This is pointed out from the different approach[11]. However, it is emphasized that the appearance of gauege fields is not exact but based on the semi-classical approximation.

When we take the limit $(N \to \infty)$, the transition amplitude (4.6) is given symbolically by

$$\Im(T,0) \; = \; \int D\mu(x(t)) \; e^{i(S_{eff}+S_{top})}, \tag{4.17}$$

where

$$S_{eff} \; = \; \int_0^T dt \Big(\frac{1}{2} \sum_{i=1}^{3} (\frac{dx_i}{dt})^2 + \frac{1}{4} - \frac{s^2}{2} \Big), \tag{4.18a}$$

$$S_{top} \; = \; s \int_0^T dt \Big(\frac{-dx_1}{dt} \frac{x_2}{1+x_3} + \frac{dx_2}{dt} \frac{x_1}{1+x_3} \Big). \tag{4.18b}$$

S_{eff} is the classical action of free particle on S^2 except for $(\frac{1}{4} - \frac{s^2}{2})$.

We can also put the similar calculation in practice on another chart. The different result appears only in S_{top} such that

$$S_{top} \; = \; s \int_0^T dt \Big(\frac{dx_1}{dt} \frac{x_2}{1-x_3} - \frac{dx_2}{dt} \frac{x_1}{1-x_3} \Big). \tag{4.19}$$

Quantum mechanics on S^2 is equal to the quantum dynamics in the background field of s magnetic monopoles within the semi-classical approximation.

ACKNOWLEDGEMENTS

We would like to thank W. Thirring for his warm hospitality at Wiener University. We gratefully acknowledge numerous conversations with Y. Ohnuki and S. Kamefuchi.

REFERENCES

1.   E. Wigner, Ann. Math. 40(1939)149.
2.   Y. Ohnuki, Unitary Representation of the Poincaé Group and Relativistic Wave Equations (World Scientific, Singapore 1988).
3.   M. Flato, D. Sternheimer and C. Fronsdal, Commun. Math. Phys. 90(1983)563.
4.   T. Itoh and K. Odaka, Fortschr. Phys. 39(1991)557.
5.   G.W. Mackey, Induced Representaions of Groups and Quantum Mechanics (Benjamin, New York 1969).
6.   C.J. Isham, in Relativisty, Group and Topology II (ed. B.S. de Witt and R. Stora, North-Holland, Amsterdam 1984), and references cited therein.
7.   P.A.M. Dirac, Lectures on Quantum Mechanics (Yeshiva, New York 1964).
8.   N.P. Landsman and N. Linden, Nucl. Phys. 34(1991)121, Y. Ohnuki and S. Kitakado, J. Math. Phys. 34(1993)2827.
9.   K. Odaka, in preparation.
10.   M.S. Marinov and M.V. Terentyev, Fortschr. Phys. 27(1979)511, and references cited therein.
11.   D. McMullan and I. Tsutsui, PLY-MS-93-04(1993) (To be published in Phy. Lett.).

HAMILTONIAN STRUCTURE OF MULTICOMPONENT KdV EQUATIONS

Ömer Oğuz

Physics Department
and
Center for Turkish-Balkan Physics
Research and Applications
Boğaziçi University
80815, Bebek, Istanbul, Turkey

INTRODUCTION

Recently a lot of attention has been given to generalizations of KdV equations more than one KdV field (see refs.[1-7] and references therein). These integrable coupled equations can play an important role in both physics and mathematics. It is known that there exists a relation between the Poisson bracket structure of integrable systems and the conformal algebras (see refs.[8-13] and references therein). Furtheremore, a connection between integrable systems and conformal field theory has been established after quantizing the integrable systems (see e.g. [14, 15]). Very recently it has also been shown that there is a relation between a matrix KdV hierarchy and a non-linear and non-local Poisson bracket algebra, or the so-called V-algebra [16].

It has also been discussed how bi-Hamiltonian structure of a multicomponent KdV equation can be constructed starting from a Lagrangian for an N-component field in 1+1 dimensions [17]. In this paper we investigate the problem in detail and introduce a multicomponent generalization of modified KdV equations. In the second section we obtain the first Hamiltonian structure of multicomponent KdV equations via a canonical Hamiltonian formalism. This formalism can be relevant for the quantization of the system. The second Hamiltonian structure is obtained by introducing another Hamiltonian as a conserved quantity with the corresponding Hamiltonian operator. As is well known a linear operator, J is called a Hamiltonian operator if it satisfies the condition of skew-symmetry

$$\int G \wedge JF dx = - \int F \wedge JG dx \tag{1}$$

Symmetries in Science VII, Edited by
B. Gruber, Plenum Press, New York, 1995

and the Jacobi identity

$$\int \widetilde{J\Theta} \wedge E(I)dx = 0 \qquad (2)$$

where E denotes the variational derivative and I is a cosymplectic functional two-vector

$$I = \frac{1}{2}\tilde{\Theta} \wedge J\Theta$$

where $\tilde{\Theta}$ is the transpose of Θ. In general, the tangent vectors $\Theta = \{\theta_i\}$ form a basis of univectors. In our case the second Hamiltonian operator satisfies the Jacobi identity (2) with a set of constraints on the parameters introduced in the Lagrangian. We then consider N=1 and N=2 as particular cases. A generalization of the Miura transformation to a multicomponent form is given in the third section. A multicomponent modified KdV equation is derived by using the multicomponent Miura transformation from the multicomponent KdV equations. The Lax hierarchy of the multicomponent KdV equations and the construction of the recursion operator for the system under consideration are also discussed. The recursion relations of the Hamiltonian operators give rise to infinitely many conserved quantities. It is well known that the Poisson bracket algebra of the KdV equation is related to the Virasoro algebra [8,9]. In this paper we consider a generalization of the KdV equation. Therefore we expect that such a generalization must yield a possible generalization of the Virasoro algebra. The last part of the paper is devoted to a discussion of this generalized conformal algebra.

BI-HAMILTONIAN STRUCTURE

The starting point for our consideration is the Lagrangian density written in terms of an N-component field, $\phi_i(x,t)$,

$$L = \frac{1}{2}\delta_{ij}\phi_{i,x}\phi_{j,t} + \frac{1}{2}b_{ij}\phi_{i,xx}\phi_{j,xx} - \frac{1}{3}c_{ijk}\phi_{i,x}\phi_{j,x}\phi_{k,x} \qquad (3)$$

where $i,j = 1,\dots,N$ and b_{ij} and c_{ijk} are completely symmetric constant coefficients. In eq.(3) and everywhere below, summation over repeated indices is assumed. The Lagrangian (3) is invariant under the transformations

$$\phi'_i = T_{ij}\phi_j, \qquad (4)$$

$$c'_{ijk} = c_{lmn}\tilde{T}_{nk}\tilde{T}_{mj}\tilde{T}_{li}, \qquad (5)$$

$$b'_{ij} = b_{mn}\tilde{T}_{nj}\tilde{T}_{mi} \qquad (6)$$

where T is an arbitrary orthogonal matrix with constant entries.

We obtain equations of motion from eq.(3)

$$\phi_{i,xt} = b_{ij}\phi_{j,xxxx} + c_{ijk}(\phi_{j,x}\phi_{k,x})_x. \qquad (7)$$

The velocity potentials, u_i, are defined by

$$u_i(x,t) = \phi_{i,x}(x,t). \qquad (8)$$

One can also relate u_i fields to a fundamental field of a conformal field theory via second quantization as pointed out in [14, 15]. Using the fields u_i in eq.(7) the multicomponent KdV equation

$$u_{i,t} = b_{ij}u_{j,xxx} + c_{ijk}(u_ju_k)_x \qquad (9)$$

is obtained. An integrable multicomponent KdV equation,

$$u^i_{,t} = u^i_{,xxx} + a^i_{jk} u^j u^k_{,x} \tag{10}$$

has been considered by Svinolupov [7] where the constant parameters a^i_{jk}'s are symmetric with respect to the subindices (i.e. $a^i_{jk} = a^i_{kj}$) and are the structure constants of a Jordan algebra. The integrability criteria of the multicomponent KdV equation (10) are associated with the presence of higher symmetries and the corresponding recursion operator [7].

In general constructing a recursion operator is a complicated problem which involves guesswork. On the other hand, the recursion operators can be easily constructed in bi-Hamiltonian systems [18]. In this work we consider a different approach for the integrability criteria of the multicomponent KdV equation (9) which are based on the bi-Hamiltonian formalism.

Let us now consider the Hamiltonian formulation of our system. With this aim we introduce the canonical momentum

$$\Pi_i = \frac{\partial L}{\partial(\phi_{i,t})} = \frac{1}{2}\phi_{i,x} \tag{11}$$

which is conjugate to ϕ_i. The canonical Poisson bracket is

$$[\phi_i(x), \Pi_j(x')] = \frac{1}{2}\delta_{ij}\delta(x - x'). \tag{12}$$

On the other hand, one can easily calculate that

$$[u_i(x), \Pi_j(x')] = \frac{1}{2}\delta_{ij}\delta'(x - x') \tag{13}$$

or in terms of vector potentials

$$[u_i(x), u_j(x')] = \delta_{ij}\delta'(x - x') \tag{14}$$

where prime on the delta function denotes a partial derivative with respect to x . Furthermore, it is easy to realize directly from eq.(14) that the first Hamiltonian operator is

$$J^{(0)}_{ij} = \delta_{ij} D_x \tag{15}$$

where $D_x = \frac{\partial}{\partial x}$. This operator is skew-symmetric and satisfies the Jacobi identity. The corresponding Hamiltonian is

$$H_1 = \int (\delta_{ij}\Pi_i\phi_{j,t} - L)dx \tag{16}$$

or

$$H_1 = \int (-\frac{1}{2}b_{ij}\phi_{i,xx}\phi_{j,xx} + \frac{1}{3}c_{ijk}\phi_{i,x}\phi_{j,x}\phi_{k,x})dx. \tag{17}$$

In view of eq.(8), the Hamiltonian becomes

$$H_1 = \int (-\frac{1}{2}b_{ij}u_{i,x}u_{j,x} + \frac{1}{3}c_{ijk}u_i u_j u_k)dx. \tag{18}$$

Thus the first Hamiltonian structure of multicomponent KdV equations (9) is obtained. One can now obtain eqs.(9) from

$$u_{i,t} = [u_i, H_1] = J^{(0)}_{ij} E_j(H_1) \tag{19}$$

where the Euler derivative E_j is

$$E_j = \frac{\delta}{\delta u_j} = \sum_{m=0}^{n} (-1)^m D_x^m \frac{\partial}{\partial u_{i,nx}}.$$

It is also possible to show that eq.(9) admits

$$H_0 = \int (\frac{1}{2}\delta_{ij} u_i u_j) dx \tag{20}$$

as a conserved quantity in addition to H_1 . Then the required second Hamiltonian operator is

$$J_{ij}^{(1)} = b_{ij} D_x^3 + m_{ij} D_x + D_x m_{ij} \tag{21}$$

where $m_{ij} = \frac{2}{3} c_{ijk} u_k$. The skew symmetry of the operator (21) is manifest. In order to verify that $J_{ij}^{(1)}$ is Hamiltonian we must verify that it satisfies the Jacobi identities. Following the procedure of ref.[18], we introduce an arbitrary basis of tangent vectors $\Theta = \{\theta_i\}$ which are then conveniently manipulated according to the rules of exterior calculus. In our case the Jacobi identity (2) can explicitly be expressed as

$$\int J_{ij}^1 \theta_i \wedge E_j(I) dx = 0 \tag{22}$$

where the functional two vector I is

$$I = \frac{1}{2}\theta_i \wedge J_{ij}^{(1)} \theta_j \tag{23}$$

For the Hamiltonian operator (21) we obtain

$$I = \frac{1}{2} b_{ij}\theta_i \wedge \theta_{j,xxx} + \frac{2}{3} c_{ijk} u_k \theta_i \wedge \theta j, x$$

and the Euler derivative of the bi-vector I is

$$E_i(I) = \frac{2}{3} c_{ijk}\theta_j \wedge \theta_{k,x}.$$

When b_{ij} and c_{ijk} satisfy the following constraints

$$b_{ij} c_{klj} - b_{lj} c_{kij} = 0 \tag{24}$$

$$c_{ijm} c_{klm} - c_{kjm} c_{ilm} = 0 \tag{25}$$

we can easily derive that

$$J_{ij}^{(1)} \theta_i \wedge E_j(I) = \frac{2}{3} b_{ij} c_{klj}(\theta_{i,xx} \wedge \theta_k \wedge \theta_{l,x})_x + \frac{4}{9} c_{ijm} c_{klm}(u_j \theta_i \wedge \theta_k \wedge \theta_{l,x})_x$$

is a total derivative so that the Jacobi identities (22) are satisfied.

We now consider some particular cases .

i) N=1, $u_1 = u$, $c_{111} = a$, $b_{11} = \alpha$

The Hamiltonians are

$$H_0 = \int \frac{1}{2} u^2 dx, \tag{26}$$

$$H_1 = \int (-\frac{1}{2}\alpha u_x^2 + \frac{1}{3} a u^3) dx. \tag{27}$$

The equation of motion is the usual KdV equation

$$u_t = \alpha u_{xxx} + 2auu_x \qquad (28)$$

and eqs.(24,25) are automatically satisfied.

ii) N=2, $u_1 = u$, $u_2 = v$, $b_{11} = \alpha$, $b_{22} = \beta$, $b_{12} = \gamma$, $c_{111} = a$, $c_{222} = b$, $c_{112} = c$, $c_{122} = d$,
With the Hamiltonians

$$H_0 = \int \frac{1}{2}(u^2 + v^2)dx, \qquad (29)$$

$$H_1 = \int [-\frac{1}{2}(\alpha u_x^2 + \beta v_x^2 + 2\gamma u_x v_x) + \frac{1}{3}(au^3 + bv^3 + 3cu^2 v + 3duv^2)]dx \qquad (30)$$

the coupled KdV equations are

$$u_t = \alpha u_{xxx} + \gamma v_{xxx} + 2auu_x + 2c(uv)_x + 2dvv_x, \qquad (31)$$

$$v_t = \beta v_{xxx} + \gamma u_{xxx} + 2bvv_x + 2d(uv)_x + 2cuu_x. \qquad (32)$$

In this case, eqs.(24,25) become

$$ad + bc = d^2 + c^2, \qquad (33)$$

$$c(\alpha - \beta) - \gamma(d - a) = 0, \qquad (34)$$

$$d(\alpha - \beta) - \gamma(b - c) = 0. \qquad (35)$$

As a special case we take $\gamma = c = b = 0$. Then a=d and $\alpha = \beta$ from eqs.(33-35). Eqs.(31,32) become

$$u_t = \alpha u_{xxx} + 2auu_x + 2avv_x, \qquad (36)$$

$$v_t = \alpha v_{xxx} + 2a(uv)_x \qquad (37)$$

which are Ito's equations with a dispersion term [2]. A different generalization of Ito's equations with dispersion is given in ref.[3].

As an another example, let $\gamma = 0$, $c = d = \frac{1}{2}$. Then eqs.(33-35) become

$$a + b = 1,$$

$$\alpha = \beta$$

and eqs.(31,32) are now

$$u_t = \alpha u_{xxx} + 2auu_x + (uv)_x + vv_x, \qquad (38)$$

$$v_t = \alpha v_{xxx} + 2bvv_x + (uv)_x + uu_x \qquad (39)$$

which are the KdV equations given in ref.[6].

THE MIURA TRANSFORMATION

It is well known that there exists the so-called modified KdV equation related to the KdV equation. This equation can be obtained from the KdV equation by a Miura transformation. The Miura transformation also plays an important role in a systematically construction of the conserved quantities of the KdV equation. The modified KdV equation is integrable and its conserved quantities give rise to the conserved quantities of the KdV equations via the Miura transformation [19].

Let us now introduce a multicomponent generalization of the Miura transformation as

$$u_i = -c_{ijk}w_j w_k + \sqrt{3}w_{i,x} \tag{40}$$

where u_i is the solution of the multicomponent KdV equation with $b_{ij} = \delta_{ij}$,

$$u_i = u_{i,xxx} + 2c_{ijk}u_j u_{k,x}. \tag{41}$$

Substituting the transformation (40) into the equation (41) and using the constraints on c_{ijk}'s given in equation (25) we obtain

$$(-c_{ijk}w_j + \sqrt{3}\delta_{ik}D_x)(w_{i,t} - w_{k,xxx} - 2c_{klm}c_{npm}w_l w_n w_{p,x}) = 0. \tag{42}$$

Thus, the multicomponent modified KdV equations are

$$w_{i,t} = w_{i,xxx} + 2c_{ijk}c_{mnk}w_j w_m w_{n,x}. \tag{43}$$

As can easily be seen, any solution of multicomponent modified KdV equations (43) gives a solution of the multicomponent KdV equations (41) through the Miura transformations (40). A method based on a Lie algebra valued soliton connection is disscussed and a different generalization of modified KdV equations is introduced as an integro-differential equations in [20].

THE RECURSION OPERATOR

A family of nonlinear partial differential equations generated by a recursion operator of increasing order and nonlinearity are known as the Lax hierarchy of KdV equations. A recursion operator in a bi-Hamiltonian system is defined as [18]

$$R_{ij} = J_{ik}^{(1)}(J^{-1})_{kj}^{(0)} \tag{44}$$

and for the system under consideration it is given by

$$R_{ij} = b_{ij}D_x^2 + \frac{4}{3}c_{ijk}u_k + \frac{2}{3}c_{ijk}u_k D_x^{-1}. \tag{45}$$

The recursion operator produces an infinite sequence of generalized symmetries when applied successively to the initial equation. It also yields the recursion relations

$$J_{ij}^{(1)}E_j(H_{k-1}) = J_{ij}^{(0)}E_j(H_k) \tag{46}$$

and gives rise to infinitely many conserved Hamiltonians H_k with $k = 1, 2, 3, \ldots$. One can easily obtain the next conserved quantity in this sequence by taking $k = 2$ in the eq.(46). Then the third Hamiltonian is given by

$$H_2 = \int (\frac{1}{2}b_{im}b_{mj}u_{i,xx}u_{j,xx} + \frac{10}{36}c_{ijm}c_{klm}u_i u_j u_k u_l - \frac{5}{3}b_{im}c_{jkm}u_i u_{j,x}u_{k,x})dx. \tag{47}$$

The next evolution equation in this sequence is

$$u_{i,t} = J_{ij}^{(0)}E_j(H_2) \tag{48}$$

or

$$u_{i,t} = b_{im}b_{mj}u_{j,5x} + \frac{10}{3}c_{ijm}c_{klm}u_j u_k u_{l,x} + \frac{20}{3}b_{km}c_{ijm}u_{j,x}u_{k,xx} + \frac{10}{3}b_{km}c_{ijm}u_j u_{k,xxx}. \tag{49}$$

The higher order symmetries and higher order multicomponent KdV equations in this hierarchy can be obtained by continuing with $k = 3, 4, \ldots$.

A GENERALIZED POISSON BRACKET ALGEBRA

Hamiltonian operators associated with integrable non-linear evolution equations are related to the extended conformal algebras. In our case this relation provides us with a consistent generalization of the Virasoro algebra. To this end we write down the Poisson bracket between the velocity potentials as

$$[u_i(x), u_j(x')] = J_{ij}^{(1)}\delta(x - x').\tag{50}$$

Then we expand u_i in a Fourier series

$$u_i = \sum_{n=-\infty}^{\infty} e^{inx} L_i(n) + \frac{e_i}{2}\tag{51}$$

where e_i's are constants. The corresponding linear functional is

$$L_i(n) = \frac{1}{(2\pi)} \int_0^{2\pi} e^{inx} u_i(x)dx - \frac{e_i}{2}\delta(n)\tag{52}$$

where the Kronecker delta $\delta(n) = \delta_{n,0}$. Multiplying the right hand side of eq.(50) by 2π we obtain the Poisson brackets of the $L_i(n)$'s as

$$[L_i(m), L_j(n)] = i(m - n)E_{ijk}L_k(m + n) + F_{ij}(m)\delta(m + n)\tag{53}$$

where

$$E_{ijk} = \frac{2}{3}c_{ijk},$$

$$F_{ij}(m) = im(m^2 b_{ij} - \frac{2}{3}c_{ijk}e_k).$$

One can also obtain the same extended Virasoro algebra with the aid of a non-linear functional $L_i(n)$ as pointed out in reference [21]. As is expected, the above algebra contains the Virasoro subalgebra. For $N = 1$ the algebra reduces to the Virasoro algebra

$$[L_m, L_n] = i(m - n)L_{m+n} + imb(m^2 - 1)\delta_{m+n,0}\tag{54}$$

where $L_1(n) \equiv L_n$ is the usual variable of the Virasoro algebra. Here we have taken $c_{111} = \frac{3}{2}b$, $b_{11} = b$ and $e_1 = 1$.

Furthermore, if b_{ij} and c_{ijk} can be simultaneously diagonalized by the O(N) transformation given in equations (5) and (6) in a consistent way along with constraints in equations (24) and (25), then the algebra given by eq.(53) reduces to a product of 1-dimensional Virasoro algebras. This also means that the multicomponent KdV equations (9) decouple. However, for N=2, due to fact that we have only a single unknown parameter in O(2) the simultaneous diagonalization of b_{ij} and c_{ijk} clearly can not be achieved. For $N > 2$ the situation becomes more complicated. Thus, the simultaneous diagonalization of b_{ij} and c_{ijk} for an arbitrary N poses an interesting problem for future study.

CONCLUSION

In this paper we have introduced the Hamiltonian formalism for a class of the integrable multicomponent KdV equations, starting from a Lagrangian and expressing the dynamics in terms of velocity potentials. The bi-Hamiltonian structure of the system enabled us to obtain an infinite number of conservative quantities. Some special cases of the generalized KdV equations we have studied reduce to earlier known generalizations of the KdV equations. The Lax hierarcy of yhe multicomponent KdV equations is investigated. It is also given a generalization of modified KdV equation through a multicomponent Miura tranformation. The Poisson bracket structure of the multicomponent KdV equations led us to a nontrivial generalization of the Virasoro algebra. We expect to relate this Poisson bracket algebra to the operator algebra of conformal field theory after quantizing the multicomponent KdV system within the Lagrangian approach.

ACKNOWLEDGEMENTS

I would like to thank M.Arık, M.Gürses, R.Güven, F.Neyzi and Y.Nutku for very useful discussions and comments. This work was in part supported by the Scientific and Technical Research Council of Turkey-TUBITAK and Bogazici University Foundation.

REFERENCES

[1] Hirota R. and Satsuma J.:Phys. Lett. **85**A (1981) 407.

[2] Ito M.:Phys. Lett. **91**A (1982) 335.

[3] Kupershmidt B.A.:J.Phys. A **18** (1985) L571.

[4] Antonowicz M. and Fordy A.P.:Physica D **28** (1987) 345.

[5] Athorne C. and Fordy A.P.:J.Phys. A **20** (1987) 1377.

[6] Nutku Y. and Oğuz Ö.: Il Nuo. Cim. **105**B (1990) 1381.

[7] Svinolupov S.I.:Theor.Math. **87** (1991) 611.

[8] Gervais J.L. and Neveu A.:Nucl. Phys. B **209** (1982) 125.

[9] Gervais J.L.:Phys. Lett. **160**B (1985) 277.

[10] Bilal A. and Gervais J.L.:Phys. Lett. **211**B (1988) 95.

[11] Goddard P. and Olive D.I.: Intern. Journ. Mod. Phys. A **1** (1986) 303.

[12] Bakas I.:Comm. Math. Phys. **123** (1989) 627.

[13] Mathieu P.:Phys. Lett. **208**B (1988) 101.

[14] Schiff J.:"The KdV action and deformed minimal models" Institute for Advanced Study Preprint, IASSNS-HEP-92/28 (1992).

[15] Schiff J.:"Actions for integrable systems and deformed conformal theories" Institute for Advanced Study Preprint, IASSNS-HEP-92/76 (1992).

[16] Bilal A.: "Multicomponent KdV Hierarchy, V-Algebra and Non-Abelian Toda Theory". Princeton University Preprint PUPT-1446 (1994) and HEP-TH 9401167.

[17] Oğuz Ö.: "A canonical Lagrangian formalism of multicomponent KdV equations" Bogazici University preprint (1994).

[18] Olver P.J.: "Applications of Lie groups to differential equations" Graduate texts in Mathematics Vol.107 Springer Verlag (1986).

[19] For example see in A. Das "Integrable Models" World Scientific Lecture Notes in Physics, Vol.30, (1989).

[20] Gürses M., Oğuz Ö. and Salihoğlu Ş.:Int.Jour.Mod.Phys.5, (1990), 1801.

[21] Fokas A.S. and Gel'fand I.M. :J.Math.Phys. 35 (1994) 3117.

IRREDUCIBLE REPRESENTATIONS OF FUNDAMENTAL ALGEBRA FOR QUANTUM MECHANICS ON S^D AND GAUGE STRUCTURES

Yoshio Ohnuki

Nagoya Women's University
1302 Takamiya Tempaku Nagoya 468, Japan

ABSTRACT

All possible irreducible representations of fundamental algebra for a particle moving on S^D are determined by applying the induced representation technique developed by Wigner. It is shown that the theory is automatically equipped with a monopole-like gauge potential. Some topological properties of the gauge potential are also discussed. Examining a relation of our theory with Dirac's formulation for a constrained system we determine the irreducible representation of the Dirac algebra for a particle constrained on S^D.

1. INTRODUCTION

It is known that the self-adjoint operators x_j and p_j $(j = 1, 2, \cdots, n)$, which satisfy the canonical commutation relations

$$[x_j, x_k] = [p_j, p_k] = 0,$$

$$[x_j, p_k] = i\, \delta_{jk}, \tag{1.1}$$

take continuous eigenvalues belonging to $(-\infty, \infty)$. Thus the canonical commutation relations (1.1) are powerless to describe a particle constrained to move on a finite domain, say, S^D. To avoid this situation we rewrite (1.1) into the form such that

$$[x_j, x_k] = 0,$$

$$U(a)U(b) = U(a+b), \quad U(a)^\dagger x_j\, U(a) = x_j - a_j, \tag{1.2}$$

where $U(a)$ is a unitary operator that is a function of arbitrary real parameters a_j. Starting with (1.1) and writing

$$U(a) = \exp[i \sum_j p_j\, a_j] \tag{1.3}$$

we easily obtain (1.2). Conversely, if assuming (1.2) we find from the left relation of the second line of (1.2) that the unitary operator $U(a)$ is expressed as (1.3) with self-adjoint operators p_j, thereby immediately arriving at (1.1) by virtue of the last relation of (1.2). Accordingly we see that set of the position and displacement operators in R^n leads us to the canonical quantization (1.1) for a system of n degrees of freedom. We will apply this point of view to general cases.

Let M be a smooth manifold on which a particle is constrained to move. Then we introduce two kinds of operators; the one is the position operator, whose eigenvalue can uniquely describe a position of the particle on M, and the other is the displacement operator which can move any point on M to an arbitrary position on it. The algebra among these operators will be called the fundamental algebra[1] for quantum mechanics on M. In this connection it is noted that the manifold M under consideration is a configuration space but not a phase space and the fundamental algebra defined thereon is completely independent of the dynamics.

The simplest example of fundamental algebra is seen for $M = S^1$. It was given[2] by

$$[G, W] = W, \tag{1.4}$$

where G and W are assumed to be self-adjoint and unitary operators respectively. It was shown[2] that the irreducible representation of the algebra (1.4) is uniquely specified by a real parameter α that satisfies the condition $0 \leq \alpha < 1$, and in W-diagonal representation the operators W and G are expressed as

$$W = e^{i\theta}, \qquad G = \frac{1}{i} \frac{\partial}{\partial \theta} + \alpha \qquad (0 \leq \alpha < 1) \tag{1.5}$$

together with the inner product

$$<\varphi \,|\, \chi> = \int_0^{2\pi} d\theta \; \varphi^*(\theta) \chi(\theta) \qquad \text{for } \varphi, \chi \in \mathfrak{h}_\alpha, \tag{1.6}$$

where $\mathfrak{h}_\alpha$ stands for the irreducible representation space corresponding to the parameter α. Thus the operator W, whose eigenvalue has the period 2π with respect to θ, can be regarded as a position operator on S^1.   On the other hand, using (1.5) we obtain for an arbitrary real parameter ε

$$e^{i\varepsilon G} W e^{-i\varepsilon G} = e^{i(\theta + \varepsilon)},$$

which shows that the operator G plays a role of the generator of the displacement of a point on S^1. For these reasons we can regard the relation (1.4) as the fundamental algebra on S^1. We note that the representation space of the algebra (1.4) is not uniquely given unlike the case of canonical commutation relations. In fact, as seen in the above, there exists an infinite number of inequivalent representations corresponding to α in $[\,0, 1\,)$.

It is not a difficult task to generalize the above argument to the case of S^D. To this end, we will rewrite the unitary operator W to $W = (x + iy)/r$ and embed S^1 into R^2, where x and y are self-adjoint operators and the constant r (>0) denotes the radius of S^1. Then from the unitarity of W we have

$$x^2 + y^2 - r^2 = 0, \tag{1.7}$$

$$[x, y] = 0, \tag{1.8}$$

and on the other hand the algebra (1.4) can equivalently be written as

$$[G, x] = i\,y, \qquad\qquad [G, y] = -i\,x. \qquad\qquad (1.9)$$

We may generalize Eqs. (1.7) ~ (1.9). Namely, we introduce a self-adjoint position operator $x = (x_1, x_2, \cdots, x_{D+1})$ that satisfies the following constraint:

$$x^2 - r^2 = 0, \qquad\qquad (1.10)$$

where

$$x^2 \equiv \sum_{\alpha=1}^{D+1} x_\alpha^2, \qquad\qquad (1.11)$$

and assume the algebra

$$[x_\alpha, x_\beta] = 0,$$

$$[x_\lambda, G_{\alpha\beta}] = i\,(x_\alpha \delta_{\lambda\beta} - x_\beta \delta_{\lambda\alpha}),$$

$$[G_{\alpha\beta}, G_{\lambda\mu}] = i\,(\delta_{\alpha\lambda} G_{\beta\mu} - \delta_{\alpha\mu} G_{\beta\lambda} + \delta_{\beta\mu} G_{\alpha\lambda} - \delta_{\beta\lambda} G_{\alpha\mu}), \quad (1.12)$$

$$(\alpha, \beta = 1, 2, \cdots, D+1)$$

where the operators $G_{\alpha\beta}\ (= -G_{\beta\alpha})$ are the self-adjoint generators of SO($D+1$). We can move any point P on S^D to an arbitrary point P' on it by repeated use of unitary transformations $U = \exp[i\,\varepsilon\,G]$. Thus we may call Eq.(1.12) the fundamental algebra on S^D.

In the next section we will determine all possible irreducible representations of (1.12) by applying the induced representation technique developed by Wigner[3] and show that there exists an infinite number of inequivalent representations[1,4,5] for (1.12). In doing this a special caution will be used for $D = 1$, since the configuration space S^1 in this case is a multiply connected space. We will derive (1.5) and (1.6) in a different way from that done in Ref. 2.

In section 3 it is shown that the irreducible representation of (1.12) automatically contains a gauge structure in itself, topological properties of which will also be examined in some details.

We will discuss, in section 4, a relation of our theory with the Dirac formalism for a particle constrained to move on S^D and using the result in section 2 we will derive all irreducible representations of Dirac's algebra in this case and explicitly give an analytic form of the canonical momentum p_α in the x-diagonal representation.

The final section is devoted to some additional remarks.

2. IRREDUCIBLE REPRESENTATIONS OF FUNDAMENTAL ALGEBRA

We note that the algebra (1.12) is isomorphic to the $(D+1)$-dimensional Euclidean group E($D+1$), in which the momentum p_α is used in place of our x_α. Thus we can apply the induced representation technique to obtain irreducible representations.

Let us denote, in the x-diagonal representation, a wave function $\varphi(x)$ that belongs to an irreducible representation of (1.12) as

$$\varphi(x) = (\varphi_1(x), \varphi_2(x), \cdots, \varphi_s(x)), \tag{2.1}$$

where the indices ξ ($= 1, 2, \cdots, s$) of φ_ξ have been introduced to denote those degrees of freedom other than x which uniquely specify a state vector in the irreducible representation space. We write the inner product invariant under E($D+1$) as

$$<\varphi | \chi> = N \sum_{\xi=1}^{s} \int dx^{D+1} \delta(x^2 - r^2) \, \varphi_\xi^*(x) \chi_\xi(x) \tag{2.2}$$

with a suitable normalization constant N (> 0).

Corresponding to W in (1.4) we introduce the unitary operator $T(a)$ defined as

$$T(a) = \exp\left[i \sum_{\alpha=1}^{D+1} a_\alpha x_\alpha\right] \tag{2.3}$$

with real parameter $a = (a_1, a_2, \cdots, a_{D+1})$. This plays a role of the position operator in the irreducible representation. Moreover corresponding to the orthogonal transformation $x'_\alpha = \Sigma_\beta R_{\alpha\beta} x_\beta$ under SO($D+1$) we introduce the unitary operators $D(R)$ which also acts on the state vector in the irreducible representation space. They are seen to obey, in the neighborhood of the identity transformation of SO($D+1$), the relations such that

$$T(a)T(b) = T(a+b), \tag{2.4}$$

$$D(R)D(R') = D(RR'), \tag{2.5}$$

$$D(R)T(a) = T(Ra)D(R), \tag{2.6}$$

where $(Ra)_a = \Sigma_\beta R_{\alpha\beta} a_\beta$. Needless to say, $D(R)$ describes the transformation

$$\varphi'_\xi(x) = \sum_{\xi'=1}^{s} D(R)_{\xi\xi'} \, \varphi_{\xi'}(x)$$

$$= \sum_{\xi'=1}^{s} \left(1 + \frac{i}{2} \sum_{\alpha,\beta=1}^{D+1} \varepsilon_{\alpha\beta} G_{\alpha\beta}\right)_{\xi\xi'} \varphi_{\xi'}(x) \tag{2.7}$$

for the infinitesimal rotation $R_{\alpha\beta} = \delta_{\alpha\beta} + \varepsilon_{\alpha\beta}$, where $\varepsilon_{\alpha\beta}$ ($= -\varepsilon_{\beta\alpha}$) is a small angle of rotation in the α-β plane.

Now, following Wigner[3] we write $D(R)$ as

$$D(R)_{\xi\xi'} = Q(R, x)_{\xi\xi'} P(R), \tag{2.8}$$

where $P(R)$ is a unitary operator defined by

$$P(R)\varphi_\xi(x) = \varphi_\xi(R^{-1}x). \tag{2.9}$$

Since by definition the operator $P(R)$ satisfies the relation of the same form as (2.6), i.e.,

$$P(R)T(a) = T(Ra)P(R), \tag{2.10}$$

we find that the matrix $Q(R,x)(=D(R)P^{\dagger}(R))$ is commutable with $T(a)$. Hence we have

$$[Q(R,x)_{\xi\xi'},\, x_{\alpha}]=0,\qquad(2.11)$$

which implies that the matrix element $Q(R,x)_{\xi\xi'}$ is diagonal with respect to the position x. The operator $Q(R,x)$ is called the Wigner rotation. Here it is noted that $Q(R,x)$ and the wave function (2.1) are single valued functions of x. This property is automatically guaranteed for $D\geq 2$, because in this case S^D is a simply connected space so that the possibility of defining a multivalued function on it is excluded from the outset.

Taking account of the relation $P(R)P(R')=P(RR')$ we obtain from (2.8) and (2.9)

$$D(R)D(R')=Q(R,x)P(R)Q(R',x)P(R')$$

$$=Q(R,x)Q(R',R^{-1}x)P(R)P(R')$$

$$=Q(R,x)Q(R',R^{-1}x)P(RR').$$

which leads us to

$$Q(RR',x)=Q(R,x)Q(R',R^{-1}x).\qquad(2.12)$$

Then if $R'=R^{-1}$, we find that the above relation turns out to be

$$Q^{\dagger}(R,x)=Q(R^{-1},R^{-1}x),\qquad(2.13)$$

which holds true at least in the neighborhood of the identity transformation.

Let $U(x)$ be an $(s\times s)$-unitary matrix, which is assumed to be a single-valued function of x. Then, the unitary equivalence of the two representations $\{D(R),T(a)\}$ and $\{U^{\dagger}(x)D(R)U(x),\,T(a)\}$ is enable us to use the Wigner rotation

$$Q'(R,x)\equiv U^{\dagger}(x)Q(R,x)U(R^{-1}x)\qquad(2.14)$$

in place of the original $Q(R,x)$, since

$$U^{\dagger}(x)Q(R,x)P(R)U(x)=U^{\dagger}(x)Q(R,x)U(R^{-1}x)P(R).$$

With these preparations we will determine the form of the Wigner rotation. For this purpose we introduce a notion of the little group, which will be denoted as H_l. Following Wigner[3] we define it by

$$H_l=\{\lambda\mid\lambda l=l,\,\lambda\in SO(D+1)\},\qquad(2.15)$$

where l is an arbitrary vector satisfying (1.10) and is related to x through a rotation under $SO(D+1)$ in the following way:

$$x=\alpha_x l,\qquad\alpha_x\in SO(D+1).\qquad(2.16)$$

It is not a difficult task to see that

$$\lambda_x\equiv\alpha_x^{-1}R\,\alpha_{R^{-1}x}\qquad(2.17)$$

is an element of the little group, i.e., $\lambda_x\, l = l$. Then inserting $R = \alpha_x\, \lambda_x\, \alpha_{R^{-1}x}^{-1}$ into $Q(R, x)$ and applying Eq.(2.12) we are led to

$$Q(R, x) = Q(\alpha_x, x)\, Q(\lambda_x, l)\, Q(\alpha_{R^{-1}x}^{-1}, l)$$

$$= Q(\alpha_x, x)\, Q(\lambda_x, l)\, Q^{\dagger}(\alpha_{R^{-1}x}, R^{-1}x), \qquad (2.18)$$

where use has been made of $\alpha_x^{-1} x = l$ together with

$$Q^{\dagger}(\alpha_{R^{-1}x}, R^{-1}x) = Q(\alpha_{R^{-1}x}^{-1}, l) \qquad\qquad (2.19)$$

which is obtained from (2.13). Thus we write (2.18) as

$$Q(\lambda_x, l) = Q^{\dagger}(\alpha_x, x)\, Q(R, x)\, Q(\alpha_{R^{-1}x}, R^{-1}x). \qquad (2.20)$$

The unitary matrix $Q(\alpha_x, x)$ is of course a single-valued function of x for $D \geq 2$. Therefore if we use it in place of $U(x)$ in (2.14), then by virtue of (2.20) we can employ $Q(\lambda_x, l)$ as a substitute for the original Wigner rotation $Q(R, x)$.

For $D = 1$, however, we encounter a completely different situation, because S^1 is a multiply connected space. In this case we can by no means exclude the possibility that $Q(\alpha_x, x)$ becomes a multivalued function of x, though the Wigner rotation $Q(R, x)$ has been assumed to be a single-valued function of x. Accordingly under the 2π-rotation of the vector x in R^2 it generally behaves as

$$Q(\alpha_x, x) \quad\underset{2\pi\text{-rotation of } x}{\longrightarrow}\quad e^{-2\pi i\,\alpha}\, Q(\alpha_x, x) \qquad \text{for } D = 1 \qquad (2.21)$$

with a real parameter α belonging to $[0, 1)$. Then with the angle θ defined by

$$x_1 = r\cos\theta, \qquad x_2 = r\cos\theta, \qquad\qquad (2.22)$$

we introduce a single-valued function $\tilde{Q}(\alpha_x, x)$ of x by the relation

$$\tilde{Q}(\alpha_x, x) \equiv e^{i\alpha\theta}\, Q(\alpha_x, x), \qquad\qquad (2.23)$$

which also gives

$$\tilde{Q}(\alpha_{R^{-1}x}, R^{-1}x) = e^{i\alpha(\theta+\varepsilon)}\, Q(\alpha_{R^{-1}x}, R^{-1}x), \qquad (2.24)$$

where ε is an angle of two-dimensional rotation under R. Thus from (2.20) we obtain

$$e^{i\alpha\varepsilon}\, Q(\lambda_x, l) = \tilde{Q}^{\dagger}(\alpha_x, x)\, Q(R, x)\, \tilde{Q}(\alpha_{R^{-1}x}, R^{-1}x). \qquad (2.25)$$

We then find, by virtue of (2.14), that $e^{i\alpha\varepsilon} Q(l_x, l)$ is unitarily equivalent to the original Wigner rotation $Q(R, x)$.

Consequently, without loss of generality we can write the operator $D(R)$ as

$$D(R) = \eta \, Q(\lambda_x, l) P(R) \tag{2.26}$$

with

$$\eta = \begin{cases} e^{i\alpha\varepsilon} & \text{for } D = 1 \\ 1 & \text{for } D \geq 2. \end{cases} \tag{2.27}$$

On the other hand owing to (2.12) we see that there holds the relation

$$Q(\lambda, l) Q(\lambda', l) = Q(\lambda \lambda', l) \qquad \text{for } \lambda, \lambda' \in H_l, \tag{2.28}$$

so that the set of $Q(\lambda, l)$'s forms a unitary representation of the little group H_l. Furthermore, it is evident that if the unitary representation of $\{D(R), T(a)\}$ is irreducible under E($D+1$), then $Q(\lambda_x, l)$ in (2.26) belongs to a unitary irreducible representation of the little group.

In order to obtain an explicit form of $D(R)$ in the irreducible representation we will examine the infinitesimal transformation corresponding to (2.7). Then by definition the operator $P(R)$ is expressed as

$$P(R) = 1 + \frac{1}{2} \sum_{\alpha, \beta = 1}^{D+1} \varepsilon_{\alpha\beta} (x_\alpha \partial_\beta - x_\beta \partial_\alpha), \tag{2.29}$$

while on the other hand we will write $\eta Q(\lambda_x, l)$ as

$$\eta Q(\lambda_x, l) = 1 + \frac{i}{2} \sum_{\alpha, \beta = 1}^{D+1} \varepsilon_{\alpha\beta} f_{\alpha\beta}(x) \tag{2.30}$$

with hermitian matrices $f_{\alpha\beta}(x)$ ($= -f_{\beta\alpha}(x)$; $\alpha, \beta = 1, 2, \cdots, D+1$). Then   the generator $G_{\alpha\beta}$ is given in the form

$$G_{\alpha\beta} = \frac{1}{i} (x_\alpha \partial_\beta - x_\beta \partial_\alpha) + f_{\alpha\beta}(x). \tag{2.31}$$

Thus our next task is to determine the explicit form of the function $f_{\alpha\beta}(x)$ by analyzing irreducible representations of the little group.

To this end we first consider the case of $D = 1$. The little group in this case is SO(1) which consists of a single element representing the identity transformation so that we have $Q(\lambda_x, l) = 1$. Since $\eta = 1 + i \varepsilon \alpha$ for small ε ($= \varepsilon_{12}$) we obtain

$$\begin{cases} G_{12} = \dfrac{1}{i} (x_1 \partial_2 - x_2 \partial_1) + f_{12}, \\[2ex] f_{12} = -f_{21} = \alpha & (0 \leq \alpha < 1). \end{cases} \tag{2.32}$$

Accordingly, if taking account of (2.22) we find that G_{12} obtained here is nothing but the operator G already given by Eq.(1.5).

We now proceed to argue the cases of $D \geq 2$ by examining $Q(\lambda_x, l)$. For this purpose with the aid of Eq.(2.17) we will explicitly calculate λ_x corresponding to the infinitesimal rotation $R_{\alpha\beta} = \delta_{\alpha\beta} + \varepsilon_{\alpha\beta}$. For simplicity we employ the vector l of the following form

$$l = (\underbrace{0, 0, \cdots, 0, r}_{D}) \tag{2.33}$$

and use the spinor representation for the rotations R and α_x, in which they are given by

$$R = 1 + \frac{i}{2} \sum_{\alpha,\beta=1}^{D+1} \varepsilon_{\alpha\beta} \frac{\sigma_{\alpha\beta}}{2} \qquad \text{with} \ \ \sigma_{\alpha\beta} = \frac{[\gamma_\alpha, \gamma_\beta]}{2i}, \tag{2.34}$$

and

$$\alpha_x = \frac{1}{\sqrt{2r}} \left(\sqrt{r + x_{D+1}} + \sqrt{r - x_{D+1}} \frac{(x\gamma)}{|x|} \gamma_{D+1} \right) \tag{2.35}$$

with

$$|x| = \sqrt{\sum_{j=1}^{D} x_j^2}, \qquad (x\gamma) = \sum_{j=1}^{D} x_j \gamma_j,$$

where γ_α ($\alpha = 1, 2, \cdots, D+1$) are irreducible hermitian matrices satisfying the Clifford algebra of order $(D+1)$,

$$\gamma_\alpha \gamma_\beta + \gamma_\beta \gamma_\alpha = 2\delta_{\alpha\beta} \qquad (\alpha, \beta = 1, 2, \cdots, D+1). \tag{2.36}$$

Then after some lengthy calculations we are led to

$$\lambda_x = 1 + \frac{i}{2} \sum_{j,k=1}^{D} \varepsilon_{jk} \frac{\sigma_{jk}}{2} - i \sum_{j,k=1}^{D} \varepsilon_{j\,D+1} \frac{1}{r + x_{D+1}} \frac{\sigma_{jk}}{2}. \tag{2.37}$$

Thus any irreducible representation of $Q(\lambda_x, l)$ is obtainable by replacing $\sigma_{jk}/2$ in (2.37) with the irreducible S_{jk} that satisfies the Lie algebra of SO(D);

$$[S_{jk}, S_{lm}] = i (\delta_{jl} S_{km} - \delta_{jm} S_{kl} + \delta_{km} S_{jl} - \delta_{kl} S_{jm}) \tag{2.38}$$

$$(j, k, l, m = 1, 2, \cdots, D).$$

Namely, we have

$$Q(\lambda_x, l) = 1 + \frac{i}{2} \sum_{j,k=1}^{D} \varepsilon_{jk} S_{jk} - i \sum_{j,k=1}^{D} \varepsilon_{j\,D+1} \frac{1}{r + x_{D+1}} S_{jk}, \tag{2.39}$$

which leads to

$$\begin{cases} f_{jk} = S_{jk}, \\ \\ f_{j\,D+1} = -\frac{1}{r + x_{D+1}} \sum_{k=1}^{D} S_{jk} x_k \qquad (D \geq 2). \end{cases} \tag{2.40}$$

We have so far considered infinitesimal transformations of SO(D+1) to obtain unitary irreducible representations of E(D+1). In order to get to the goal, however, we have to further examine whether or not the global property of the group is correctly reflected in the representations which have been derived using infinitesimal transformations. It is known that the global property in our case requires that for $D \geq 2$ any irreducible representation of SO(D +1) should be realized by a single- or double-valued representation. If $D \geq 3$, we see from (2.38) and (2.39) this property is automatically satisfied. Nevertheless, for $D = 2$ this is not the case. In fact, form (2.38) we obtain $S_{12} = S$, where S is an arbitrary real constant. Then, with the aid of (2.31) and (2.40) we find that the generators of SO(3) are given by

$$
\begin{cases}
G_{23} = \dfrac{1}{i}(x_2\partial_3 - x_3\partial_2) + \dfrac{x_1}{r+x_3}S\,, \\[2mm]
G_{31} = \dfrac{1}{i}(x_3\partial_1 - x_1\partial_3) + \dfrac{x_2}{r+x_3}S\,, \\[2mm]
G_{12} = \dfrac{1}{i}(x_1\partial_2 - x_2\partial_1) + S\,.
\end{cases}
\tag{2.41}
$$

The reader will easily check that they satisfy the Lie algebra of SO(3) for an arbitrary constant S. However, the irreducible representation of SO(3) is either single- or double-valued, so that an arbitrary value of S is not allowable in this case. Namely, in order to attain correct representations of SO(3) the constant S must be restricted to some special values. On the other hand the singe- or double-valuedness of the representation of SO(3) is due essentially to the self-adjointness of the three generators. Imposing this self-adjointness condition on the generators (2.41) and analyzing their properties in the irreducible representation we can actually derive[7] all possible values of S, which are given by

$$
S = 0,\ \pm\frac{1}{2},\ \pm 1,\ \pm\frac{3}{2},\ \pm 2,\ \pm\frac{5}{2},\ \cdots.
\tag{2.42}
$$

As a result we see that any unitary irreducible representations of E(4) can uniquely be specified by the value of S of (2.42)*.

Consequently we have obtained all possible irreducible representations of the algebra (1.12). In this connection it would be worthwhile to remark that for any value of D there always exists an infinite number of inequivalent representations of the fundamental algebra on S^D. Thus the situation is quite different from the usual canonical commutation relations which are known[8] to possess a unique irreducible representation apart from unitary equivalence transformations.

3. GAUGE STRUCTURES

Now let us consider a unitary-equivalent transformation given by the unitary matrix $U(x)$ defined on S^D. Then the fundamental operators submit to the transformation

$$
x_\alpha \rightarrow x_\alpha,
$$

$$
G_{\alpha\beta} \rightarrow G'_{\alpha\beta} = U^\dagger(x)G_{\alpha\beta}U(x).
\tag{3.1}
$$

Hence if we write $G'_{\alpha\beta}$ as

---

* Incidentally, the value of j in the eigenvalue $j(j+1)$ of the operator $G_{12}^2 + G_{23}^2 + G_{31}^2$ is shown[7] to take $|S|, |S|+1, |S|+2, \cdots$ for given S.

$$G'_{\alpha\beta} = \frac{1}{i}(x_\alpha \partial_\beta - x_\beta \partial_\alpha) + f'_{\alpha\beta}(x), \tag{3.2}$$

we find from (2.31) that the transformation (3.1) is reduced to

$$f_{\alpha\beta}(x) \rightarrow f'_{\alpha\beta}(x) = U^\dagger(x) f_{\alpha\beta}(x) U(x) + \frac{1}{i} U^\dagger(x)(x_\alpha \partial_\beta - x_\beta \partial_\alpha) U(x). \tag{3.3}$$

Using this $f_{\alpha\beta}(x)$ we define a quantity $A_\alpha(x)$ by

$$A_\alpha(x) \equiv \frac{1}{r^2} \sum_{\beta=1}^{D+1} f_{\alpha\beta}(x) x_\beta, \tag{3.4}$$

which corresponding to (3.3) undergoes the transformation

$$A_\alpha(x) \rightarrow A'_\alpha(x) = \frac{1}{r^2} \sum_{\beta=1}^{D+1} f'_{\alpha\beta}(x) x_\beta$$

$$= U^\dagger(x) A_\alpha(x) U(x) + i U^\dagger(x) A_\alpha(x) \frac{\partial U(x)}{\partial x_\alpha}. \tag{3.5}$$

Here we have used the relation $\sum_\alpha x_\alpha U^\dagger(x) \partial U(x)/\partial x_\alpha = 0$ since x lies on S^D. It is worthwhile to note that the transformation (3.5) is nothing but the gauge transformation for $A_\alpha(x)$, so that we may regard the quantity $A_\alpha(x)$ as a gauge potential, which is inherent in our theory. In what follows we will argue its properties in respective cases of D.

$D = 1$:

From (2.32) and (3.4) we have

$$A_1(x) = \frac{x_2}{r^2} \alpha, \qquad A_2(x) = -\frac{x_1}{r^2} \alpha, \tag{3.6}$$

which just represent the Aharonov-Bohm gauge potential[9] produced by an extremely thin solenoid with infinite length containing the magnetic flux $\Phi = -2\pi\alpha/e$ in itself. Here it is interesting to remark a relation of our result to the Aharonov-Bohm Hamiltonian

$$H = -\frac{m}{2}\left[\frac{1}{r}\frac{\partial}{\partial r}(r\frac{\partial}{\partial r}) + \frac{1}{r^2}(\frac{1}{i}\frac{d}{d\theta} + \alpha)^2 \right], \tag{3.7}$$

in which the first and second terms in the square bracket in the right hand side correspond to the radial and angular parts of kinetic energy, respectively. Thus we see that the angular momentum $-i\, d/d\theta + \alpha$ in the above Hamiltonian is the same as the generator G in (1.4).

$D = 2$:

In this case, from (2.31), (2.41) and (3.4) the gauge potential is found to be

$$A_1(x) = \frac{x_2}{r(r+x_3)} S, \qquad A_2(x) = -\frac{x_1}{r(r+x_3)} S, \qquad A_3(x) = 0, \tag{3.8}$$

where S is given by (2.42). Clearly this is the gauge potential produced by a magnetic monopole sitting on the origin of S^2 and the magnetic charge g is related to S by the equation

$eg = S$.   Thus the quantization of charge in Dirac's monopole theory just corresponds to the quantization (2.42) of the parameter S which has been derived from the requirement of self-adjointness of the generators.

$D = 2n$ (n = 2, 3, $\cdots$):

Recently McMullan and Tsustui[5] showed that the gauge potential (3.4) for $D = 4$ represents the BPST instanton solution of the Yang-Mills equation in Euclidean 4-dimensional space-time. We will discuss[10] this problem from a general point of view. To this end for the sake of simplicity we assume $r = 1$ and apply the spinor representation for S_{jk} $(j, k = 1, 2, \cdots, 2n ; n \geq 2)$ in (2.38). In this case, as is well known, there are two kinds of irreducible representations, which are respectively given by

(A) $\qquad\qquad\qquad S^A_{ab} = \frac{1}{2} \sigma_{ab} , \qquad\qquad S^A_{a\,2n} = -\frac{1}{2} e_a$ $\qquad\qquad$ (3.9a)

and

(B) $\qquad\qquad\qquad S^B_{ab} = \frac{1}{2} \sigma_{ab} , \qquad\qquad S^B_{a\,2n} = \frac{1}{2} e_a$ $\qquad\qquad$ (3.9b)

$$(a , b = 1, 2, \cdots, 2n - 1),$$

where e_a and σ_{ab} are $(2^{n-1} \times 2^{n-1})$ hermitian matrices defined by

$$\begin{cases} \{ e_a, e_b \} = 2\delta_{ab} \qquad (a , b = 1, 2, \cdots, 2n - 1), \\[2mm] i^{\,n+1} e_1 e_2 \cdots e_{2n-1} = 1 \end{cases}$$

$\qquad\qquad\qquad\qquad\qquad\qquad\qquad\qquad\qquad\qquad\qquad\qquad$ (3.10)

and

$$\sigma_{ab} \equiv \frac{1}{2i} [e_a , e_b]. \qquad\qquad\qquad (3.11)$$

In what follows we will use the representation of type (A) for simplicity. With the aid of (2.40) and (3.4) we find that the gauge potentials $A_\alpha(x)$ $(\alpha = 1, 2, \cdots, 2n - 1)$ are expressed as

$$\begin{cases} A_j (x) = \dfrac{1}{1 + x_{2n+1}} \displaystyle\sum_{k=1}^{2n} S^A_{jk} x_k \\[4mm] A_{2n+1}(x) = 0 \end{cases}$$

$\qquad\qquad\qquad\qquad\qquad\qquad\qquad\qquad\qquad\qquad\qquad\qquad$ (3.12)

$$(j = 1, 2, \cdots, 2n , \qquad \sum_{\alpha=1}^{2n+1} x_\alpha^2 = 1).$$

On the other hand, in $2n$-dimension ($n = 2, 3, \cdots$) a set of topologically non-trivial gauge potentials that satisfy the Yang-Mills equation on S^{2n} was found by Fujii[10] in 1986. He called them generalized BPST configurations, the form of which was given by

$$A_j (\xi) = i \frac{\xi^2}{1 + \xi^2} U^\dagger(\xi) \frac{\partial}{\partial \xi_j} U(\xi), \qquad\qquad (3.13)$$

where

$$\xi^2 \equiv \sum_{j=1}^{2n} \xi_j^2 \qquad (-\infty < \xi_j < \infty), \tag{3.14}$$

$$U(\xi) \equiv \frac{1}{\sqrt{\xi^2}} \left(\xi_{2n} + i \sum_{a=1}^{2n-1} \xi_a\, e_a \right). \tag{3.15}$$

In order to obtain a relation between two types of the gauge potentials (3.12) and (3.13) we utilize the stereographic mapping

$$x_j = \frac{2\xi_j}{1+\xi^2}, \qquad x_{2n+1} = \frac{1-\xi^2}{1+\xi^2}. \tag{3.16}$$

Then after some calculations we arrive at the relation

$$\sum_{k=1}^{2n} A_k(\xi)\, \frac{\partial \xi_k}{\partial x_j} = \frac{1}{2(1+x_{2n+1})} \times \begin{cases} e_a\, x_{2n} + \displaystyle\sum_{b=1}^{2n-1} \sigma_{ab}\, x_b & \text{for } j = a < 2n \\[3ex] \displaystyle\sum_{b=1}^{2n-1} e_b\, x_b & \text{for } j = 2n \end{cases}$$

$$= \frac{1}{1+x_{2n+1}}\, S_{jk}^{A}\, x_k\,, \tag{3.17}$$

where the right hand side is found to be the gauge potential given by (3.12). In this connection it is to be noted that if we use the unitary matrix $U^{\dagger}(\xi)$ in place of $U(\xi)$ in (3.13) and apply the stereographic mapping (3.16) we are then led to a gauge potential that is also a topologically non-trivial solution of the Yang-Mills equation on S^{2n} and has an expression obtained by substituting (3.9b) for S_{jk}^{A} in (3.12). Any way, as a consequence of this procedure we have

$$\sum_{j=1}^{2n} A_j(x)\, dx_j = \sum_{j=1}^{2n} A_j(\xi)\, d\xi_j\,, \tag{3.18}$$

which shows, in the case of $D = 2n$ $(n \geq 2)$, that the gauge potential (3.12) is nothing but the generalized BPST configuration found by Fujii.

$D = 2n-1$ $(n = 2, 3, \cdots)$:

We put $x_{2n} = \xi_{2n} = 0$ and make a substitution $x_{2n+1} \to x_{2n}$ in the above argument. Then Eqs.(3.12) and (3.13) are changed into

$$A_a(x) = \frac{1}{1+x_{2n}} \sum_{b=1}^{2n-1} S_{ab}\, x_b\,, \qquad A_{2n}(x) = 0\,, \tag{3.19}$$

$$(a = 1, 2, \cdots, 2n-1; \quad \sum_{j=1}^{2n} x_j^2 = 1)$$

and

$$A_a(\xi) = i \frac{\xi^2}{1+\xi^2} U^\dagger(\xi) \frac{\partial}{\partial \xi_a} U(\xi), \tag{3.20}$$

respectively, where

$$\xi^2 \equiv \sum_{a=1}^{2n-1} \xi_a^2, \qquad U(\xi) \equiv \frac{i}{\sqrt{\xi^2}} \sum_{a=1}^{2n-1} \xi_a e_a, \tag{3.21}$$

$$(-\infty < \xi_a < \infty).$$

Then if applying the stereographic mapping

$$x_a = \frac{2\xi_a}{1+\xi^2}, \qquad x_{2n} = \frac{1-\xi^2}{1+\xi^2}, \tag{3.22}$$

we obtain

$$\sum_{a=1}^{2n-1} A_a(x)\, dx_a = \sum_{a=1}^{2n-1} A_a(\xi)\, d\xi_a, \tag{3.23}$$

which implies again the equivalence of the gauge potentials $A_a(x)$ and $A_a(\xi)$. They are, however, shown to be topologically trivial though it satisfies the Yang-Mills equation on S^{2n-1}.

Thus we see that the fundamental algebra formulated on S^D is automatically equipped by a special type of gauge potentials that obey the Yang-Mills equation defined on it.

4. RELATION WITH DIRAC FORMALISM

In this section we discuss a relation of our approach with the Dirac formalism[6] for constrained system. Since all constraints in the Dirac formalism are required to be preserved during the time development of the system, the fundamental commutation relations among canonical variables become to depend on the dynamics through the secondary constraints. This is in sharp contrast with our fundamental algebra which has been settled based only upon the structure of configuration space. Thus the Dirac formalism seems more restrictive than ours to describe a particle constrained to move on a manifold.

To make clear the situation let us consider the system constrained on S^D with radius r in the framework of the Dirac formalism, where the Hamiltonian is assumed to be of the form

$$H = \frac{1}{2} \sum_{\alpha=1}^{D+1} p_\alpha^2 + V(x) \tag{4.1}$$

for simplicity. Thus the consistency condition for the primary constraint

$$x^2 - r^2 = 0 \tag{4.2}$$

requires the following relation to hold

$$\sum_{\alpha=1}^{D+1} (p_\alpha x_\alpha + x_\alpha p_\alpha) = 0 \tag{4.3}$$

as the secondary constraint. Then according to the prescription by Dirac we are led to the fundamental commutation relations among canonical variables;

$$\left\{ \begin{array}{l} [x_\alpha, x_\beta] = 0, \\[2ex] [x_\alpha, p_\beta] = i\left(\delta_{\alpha\beta} - \dfrac{1}{r^2} x_\alpha x_\beta\right), \\[2ex] [p_\alpha, p_\beta] = \dfrac{i}{r^2}(p_\alpha x_\beta - p_\beta x_\alpha), \end{array} \right. \tag{4.4}$$

$$(\alpha, \beta = 1, 2, \cdots, D+1).$$

The forms of the secondary constraint (4.3) and the commutation relations (4.4) depends of course on the Hamiltonian chosen at the starting point.

Here it would be worthwhile to show that there exist the self-adjoint operators x_α and p_α that satisfy the conditions (4.2) ~ (4.4). First we will examine their properties assuming that they exist. Using them we define the operator $G_{\alpha\beta}$ by

$$G_{\alpha\beta} \equiv x_\alpha p_\beta - x_\beta p_\alpha . \tag{4.5}$$

Then if use is made of (4.4) we see that this $G_{\alpha\beta}$ satisfies the fundamental algebra (1.12) together with x_α. In other words, the operators $G_{\alpha\beta}$ and x_α form a representation of the Lie algebra of E($D+1$). Thus, among irreducible representations of the fundamental algebra we have to look for a representation $G_{\alpha\beta}$ that satisfy the condition $G_{\alpha\beta} = G_{\alpha\beta}$, which can also be written from (4.5) as

$$G_{\alpha\beta} = x_\alpha p_\beta - x_\beta p_\alpha . \tag{4.6}$$

With the aid of (4.2) and (4.3) we can conversely express the operator p_α as

$$p_\alpha = \frac{1}{2r^2} \sum_{\rho=1}^{D+1} (x_\rho G_{\rho\alpha} + G_{\rho\alpha} x_\rho). \tag{4.7}$$

On the other hand, if starting with the fundamental algebra (1.10) ~ (1.12) and defining p_α by (4.7) we can calculate the left hand sides of the second and third commutation relations in (4.4). Then the result is found to be

$$[x_\alpha, p_\beta] = i\left(\delta_{\alpha\beta} - \frac{1}{r^2} x_\alpha x_\beta\right),$$
$$[p_\alpha, p_\beta] = -\frac{i}{r^2} G_{\alpha\beta} . \tag{4.8}$$

It is remarked that p_α of (4.7) automatically satisfies the constraint (4.3). Hence we see that only if the two the relations (4.6) and (4.7) are assumed, the fundamental algebra (1.10) ~ (1.12) leads us to the Dirac algebra (4.2) ~ (4.4) . Thus eliminating p_α from (4.6) with use of (4.7) we arrive at the condition for $G_{\alpha\beta}$, which is now written as

$$G_{\alpha\beta} = \frac{1}{2r^2} \sum_{\rho=1}^{D+1} (x_\alpha x_\rho G_{\rho\beta} + x_\alpha G_{\rho\beta} x_\rho - x_\beta x_\rho G_{\rho\alpha} - x_\beta G_{\rho\alpha} x_\rho). \qquad (4.9)$$

This is a necessary and sufficient condition for $G_{\alpha\beta}$ that satisfy the fundamental algebra to give an irreducible representation of x_α and p_α obeying the Dirac algebra (4.2) $\sim$ (4.4). In fact, if we have an irreducible generator $G_{\alpha\beta}$ of E(D +1) that satisfies (4.9), then defining p_α by (4.7) and rewriting (4.9) into (4.6) we easily obtain the Dirac algebra as just mentioned above. Conversely, if we have x_α and p_α that satisfy the Dirac algebra, then defining $G_{\alpha\beta}$ by (4.6) that satisfy the fundamental algebra and deriving (4.7) by virtue of (4.2) and (4.3) we can arrive at (4.9).

In order to decide the irreducible representation of the Dirac algebra we insert $G_{\alpha\beta}$ of (2.31) into (4.9). Then it turns out to be

$$f_{\alpha\beta} = \frac{1}{r^2} \sum_{\rho=1}^{D+1} x_\rho (x_\alpha f_{\rho\beta} - x_\beta f_{\rho\alpha}). \qquad (4.10)$$

Now using this equation we first examine the case of $D = 1$, in which $f_{12} = -f_{21} = \alpha$ as shown already. We then find that the condition (4.10) is automatically satisfied for arbitrary α. Thus from (4.7) and the second equation of (1.4) we obtain

$$\begin{cases} p_1 = \dfrac{1}{r}(i \sin\theta \, \dfrac{d}{d\theta} + \dfrac{i}{2} \cos\theta - \alpha \sin\theta), \\[2ex] p_2 = \dfrac{-1}{r}(i \cos\theta \, \dfrac{d}{d\theta} - \dfrac{i}{2} \sin\theta - \alpha \cos\theta), \qquad (D=1). \end{cases} \qquad (4.11)$$

It is interesting to note that the irreducible representation space in this case is the same as that for (1.3) and uniquely specified by the parameter α belonging to $[0, 1)$. Hence we have an infinite number of inequivalent irreducible representations that satisfy the Dirac algebra for $D = 1$. The inner product is the same as (1.5) and with use of it we can show[11] that p_1 and p_2 are both self-adjoint operators, that is, by explicitly solving the eigenvalue problem it is shown that the eigenvalues are all real and the corresponding eigenfunctions as a whole form a complete orthoganal system in $\mathfrak{h}_\alpha$.

For $D \geq 2$, $f_{\alpha\beta}$'s are given in (2.40). Then from (4.10) we have

$$\begin{cases} S_{jk} = \dfrac{1}{r(r + x_{D+1})} \sum_{i=1}^{D} x_i (x_j S_{ik} - x_k S_{ij}), \\[2ex] \sum_{i=1}^{D} x_i S_{ij} = 0, \end{cases} \qquad (4.12)$$

which leads to the solution $S_{jk} = 0$, thereby obtaining

$$f_{\alpha\beta} = 0 \qquad (D \geq 2). \qquad (4.13)$$

Consequently, from (4.7) we have the expression for p_α of the form

$$p_\alpha = \frac{1}{2ir^2} \sum_{\rho=1}^{D+1} \left\{ x_\rho \, (x_\alpha \partial_\beta - x_\beta \partial_\alpha) + (x_\alpha \partial_\beta - x_\beta \partial_\alpha) x_\rho \right\}, \quad (D \geq 2). \qquad (4.14)$$

Eq.(4.13) implies that for $D \geq 2$ the Dirac algebra (4.2) ~ (4.4) can be realized only by the irreducible representation with $Q(\lambda_x, l) = 1$ of the fundamental algebra (1.10) ~ (1.12), where no gauge structure emerges. Accordingly, for $D \geq 2$ the irreducible representation of the Dirac algebra is uniquely determined except for unitarily equivalent representations. The wave function in this case is of the single component and from (2.2) we have the inner product of the following expression:

$$<\varphi \mid \chi> = N \int dx^{D+1} \, \delta(x^2 - r^2) \, \varphi^*(x) \chi(x), \qquad (4.15)$$

with which we can show[11] the self-adjointness of the operators p_α ($\alpha = 1, 2, \cdots, D+1$).

Thus we were able to determine, for the arbitrary dimension D, the irreducible representations of the set of the self-adjoint operators x_α and p_α for the Hamiltonian (4.1) under the constraint $x^2 - r^2 = 0$ by applying the results obtained in Sec.2.

5. CONCLUDING REMARKS

In this paper we have studied a framework of quantum mechanical description of a particle constrained to move on S^D. Unlike the cases often investigated on the phase space our theory has been built on the configuration space and as we have seen in Secs. 2 and 3 we have been able to construct the wave mechanical representation on it. As a result we have seen the emergence of a kind of gauge structure which can be regarded as a characteristic feature of our theory. We have also seen that there exists an infinite number of inequivalent irreducible representations for the fundamental algebra by which properties of operators in our theory are completely determined.

Especially, for $D = 1$ the irreducible representation is specified by the continuous parameter $\alpha \in [0, 1)$. Related to this it is interesting to remark that this parameter is just the same as that which appears in Schulman's path integral formulation in multiply connected space[12]. In fact, we can actually derive out[13] Schulman's path integral representation by applying (1.5). In this sense the algebra (1.4) provides us with a quantum mechanical basis for Schulman's approach.

In the limit of $r \to \infty$ the sphere S^D tends to the D-dimesional flat space. To see the behavior of the operators of our theory in this limit we apply the group contraction method[14] to the fundamental algebra keeping x_j finite. Then taking account of the relations

$$\frac{1}{r} x_{D+1} \to 1, \qquad \frac{1}{r} G_{D+1\,j} \to \frac{1}{i} \partial_j \ (\equiv p_j) \qquad (5.1)$$

in this limit, we see that the algebra (1.12) on S^D reduces to

$$[x_j, x_k] = [p_j, p_k] = 0,$$

$$[x_j, p_k] = i \, \delta_{jk} ,$$

$$[x_j, G_{kl}] = i \, (\delta_{jl} x_k - \delta_{jk} x_l), \qquad [p_j, G_{kl}] = i \, (\delta_{jl} p_k - \delta_{jk} p_l),$$

$$[G_{jk}, G_{lm}] = i \, (\delta_{jl} G_{km} - \delta_{jm} G_{kl} + \delta_{km} G_{jl} - \delta_{kl} G_{jm}), \qquad (5.2)$$

$$(j, k, l, m = 1, 2, \cdots, D)$$

where

$$G_{jk} = \frac{1}{i} \left(x_j\, \partial_k - x_k\, \partial_j \right) + S_{jk}\,. \tag{5.3}$$

It is interesting that in addition to the algebra (1.1) we have the algebra for the total angular momentum including the spin S_{jk} automatically built in the theory.

In the present paper we have been mainly concerned with the representation theory of the fundamental algebra on S^D. However it is not impossible to generalize our approach to construct a fundamental algebra for a particle constrained on a deformed manifold that connects with S^D through a diffeomorphic mapping. In this case we can also find the existence of spin and gauge structures in it which have similar properties to those mentioned in this paper.

Details of this problem will be published[15] elsewhere together with related topics.

REFERENCES

1. Y. Ohnuki and S. Kitakado, Mod. Phys. Lett. **A7** (1992) 2447; J. Math. Phys. **34** (1993) 2827.

2. Sec.2 of the second paper in Ref.1.

3. E. P. Wigner, Ann. Math. **40** (1939) 149.
   See also G.W.Mackey, *Induced Representation of Groups and Quantum Mechanics* (Benjamin, New York, 1969); Y. Ohnuki, *Unitary Representations of the Poincaré Group and Relativistic Wave Equations* (World Scientific, Singapore, 1988).

4. N. P. Landsman and N. Linden, Nucl. Phys. **B365** (1991) 121.

5. D. McMullan and I. Tsutsui, Phys. Lett. **B320** (1994) 287; Ann. Phys. to be published.

6. P. A. M. Dirac, Can. J. Math. 2 (1950) 129; *Lectures on Quantum Mechanics* (Belfer Graduate School of Science, Yeshiva University, New York, 1964).

7. The problem is essentially the same as the eigenvalue problem of the angular momentum of massless particle, which is obtained by substituting its momentum (k_1, k_2, k_3) for (x_1, x_2, x_3) in the expression of (2.41). For a correct treatment of such a type of angular momentum, see Sec.5.4 of the third paper in Ref.5.

8. J. von Neumann, Math. Ann. **104** (1931) 370.

9. Y. Aharonov and D. Bohm, Phys. Rev. **115** (1959).

10. K. Fujii, S. Kitakado and Y. Ohnuki, to be published.

11. Appendix of the third paper of Ref.1.

12. L. S. Schulman, Phys. Rev. **176** (1968) 1558; J. Math. Phys. **12** (1971)305.

13. Y. Ohnuki, M. Suzuki and T. Kashiwa, *The Method of Path Integrals* (in Japanese, Iwanami Shoten Pub., 1992) Sec.3.3.

14. E. Inönü and E. P. Wigner, Nuovo Cim. **9** (1952) 705.

15. Y. Ohnuki and S. Kitakado, to be published.

THE BIRTH OF GAUGE THEORY

L. O'Raifeartaigh

Dublin Institute for Advanced Studies

Introduction

Although future historians of Physics will remember the first part of the twentieth century for the emergence of Special Relativity and Quantum Mechanics they may well judge that the most fundamental physical discoveries of the century were Einsteins theory of gravitation and the gauge-theory of the fundamental forces. These discoveries completely changed our conception of *dynamics* whereas Quantum Mechanics and Special Relativity changed only our ideas about *kinematics*. Furthermore, they showed that Geometry was not only the stage on which physics took place, but was part of the physical drama. The first appreciation of intimate relationship between geometry and physical force came, of course, with the theory of gravitation. In that case the geometry was strictly metrical. The appreciation of the intimate relationship between geometry and the other fundamental forces emerged much more slowly, partly because in those cases the geometry was non-metrical, and non-metrical geometry was itself an innovation. The process took about fifty years altogether, thirty to discover the basic structure of the fundamental forces (a structure which we now know as non-abelian or Yang-Mills gauge theory) and twenty to discover how such a theory should be applied. It is about the first thirty years of this evolution that I wish to speak here.

Strictly speaking, of course, the roots of present-day gauge-theory lie in the early theory of magnetism, when it was realized that magnetic field should be expressed, not as the gradient of a scalar-potential, but as the curl of a vector-potential, $\vec{B} = \vec{\nabla} \times \vec{A}$. In contrast to the scalar-potential, which was thereby defined up to a constant, the vector-potential $\vec{A}$ was thereby defined only up to the gradient of a scalar, and thus was born the concept of gauge-invariance. The idea of the vector-potential $\vec{A}$, and the fact that it was defined only up to the gradient of a scalar, was well-known by the middle of the 19th century (motivating Stokes' theorem for example) but the name vector-potential came much later (though not later than 1892 [1]). Maxwell, for example, called the vector-potential the electromagnetic momentum, presumably because of the analogy between the kinematical momentum $\vec{p}$ and the canonical momentum $\vec{p} + e\vec{A}$. The vector potential $\vec{A}$ was used extensively, of course, in the work of Lorentz and even more extensively after the advent of special relativity, when it could then be extended to the four-vector $\mathbf{A}$ which included the electric scalar potential ϕ. Even then, however, the significance of gauge-invariance does not seem to have been fully appreciated. For example, Mie's 1912 theory of electromagnetism [2] was not gauge-invariant, but was not criticized on those grounds, and Einstein's development of his gravitional theory was delayed because, by failing to take the gauge-freedom (in this context coordinate-freedom) fully into account, he thought for some time that his field equations were underdetermined.

The Aftermath of Einstein's Gravitational Theory

Einstein soon realized the true situation, of course, and proceeded to construct his famous 1916 theory of Gravitation. This theory constituted not only the enormous advance within the context of gravitation for which it is justly celebrated, but provided the starting point for modern gauge theory. Indeed, once the role of Riemannian geometry as the origin of gravity was appreciated the idea that the other forces should have a geometrical origin became quite natural. However, this idea in its simplest form foundered on the fact that four-dimensional Riemannian geometry was restrictive to permit a generalization from gravitation to the other interactions, in particular to electromagnetism and no other geometries were known to the physicists. Accordingly it is not surprising that the first attempt to generalize Einstein's gravitational theory stayed within the context of Riemannian geometry and changed only the dimensions of the Riemannian space. This was the 5-dimensional Kaluza-Klein (KK) theory in which one extra component of the metric tensor was set equal to unity and the other four identified with the electromagnetic potential. The main difficulty with this theory was that one had to make the unwarranted assumption that the dependence of the metric on the fifth dimension was negligible (in the same way that the dependence of a quasi-static Einstein metric on the time dimension is negligible). It might be mentioned that today, in the light of spontaneous symmetry breaking and the consequent possibility of dimensional reduction, the assumption would not appear to be so unwarranted. But because of this difficulty and some other more technical problems the proposal was temporarily abandoned. Nevertheless, as we shall see later, the KK theory was to play an important role in motivating present-day gauge-theory, and in recent times it has been revived in its own right in the context of string theory, for which a higher number of dimensions (ten and twenty-six) is natural.

The major development for modern gauge-theory theory took place, however, not in physics but in pure mathematics, where in 1917 Levi-Civita [5], motivated by some earlier work of Christoffel, introduced the idea of parallel transport. In doing so he replaced the metric by an affine connection and so liberated differential geometry from its traditional metric-dependence. The importance of this development, for both mathematics and physics, can hardly be over-estimated. In mathematics it started the series of developments in both differential geometry and topology which we associate with names like Cartan (who immediately generalized Levi-Civita's symmetric connection to a general one) Whitney, de Rham, Hodge, Chern and Steenrod, which culminated in the theory of fibre bundles [6] about 1950. Although Einstein kept himself abreast of the early mathematical developments through Cartan, who even wrote a review paper [7] at Einstein's specific request, almost nothing of this mathematical progress was known to physicists in general, and nothing at all was known by those physicists who were to invent gauge-theory. Indeed most of them made the connection with fibre-bundles only at a much later date. As we shall see, for some of them the discovery of gauge-theory came from KK theory and for the others it came from an element of Levi-Civita's original theory which had been introduced into physics by Weyl [8] in 1918, almost immediately after Levi-Civita's original paper. What Weyl had proposed in the 1918 paper was that the general coordinate invariance of Einstein's theory should be extended to include invariance with respect to parallel-transport transformations of the form $g_{\mu\nu}(x) \to g_{\mu\nu}(x)\exp(\int dx.A(x))$. where $g_{\mu\nu}$ is the metric tensor. Incidentally, it was in this 1918 paper of Weyl that the word gauge(*Eich*)-transformation first made its appearance and had the everyday meaning of a change of length or a change in calibration. Einstein (in a postscript at the end of Weyl's paper) was quick to point out that Weyl's proposal was not physically acceptable, because the non-integrability of the scale-factor implied that the times of clocks would depend on their history. Nevertheless, the idea was launched, and it was soon realized that it was not the idea itself that was wrong, but only its sphere of application. What was realized, in fact, was that the non-integrable phase-factor should be attached not to the gravitational metric but to the quantum-mechanical wave-function, where it should also pick up a factor i and become a phase-factor. In other words, $g_{\mu\nu}(x) \to g_{\mu\nu}(x)\exp(\int dx.A(x))$ should be replaced by $\psi(x) \to \psi(x)exp(\int dx.A(x))$. Surprisingly, this idea seems to have been first suggested [9] in 1922 before the wave-function was invented but, not so surprisingly, the suggestion came from none other than Schrödinger, who formulated it (rather obscurely) in terms of Bohr quantization rules. Indeed it is claimed in [10] that this 1922 paper of Schrödinger played an important role in his later invention of the wave-function (although he did not refer to it in the 1926 papers). After the invention of the wave-function the idea could be formulated much more clearly, namely, as the statement that in wave-mechanics the change $\vec{p} \to \vec{p} + e\vec{A}$ of classical electrodynamics becomes $\vec{\partial} \to \vec{\partial} + \frac{ie}{h}\vec{A}$ and hence a gauge-transformation of $\vec{A}$ could be viewed as a phase-transformation $\psi(x) \to e^{\frac{ie}{h}\alpha(x)}\psi(x)$ of the wave-function. This observation was made independently by Fock [11] and London [12] in 1927, and an interesting letter from London to Schrödinger linking his observation to the 1922 paper can be read in [13]. The transference of his idea to the wave-function was endorsed enthusiastically by Weyl, and in his now-famous 1929 paper [14] he went even further and pointed out that the standard theory of electromagnetism was not only compatible with the gauge-principle but could actually be derived from it. Unfortunately, Weyl continued to call the transformations $\psi(x) \to e^{\frac{ie}{h}\alpha(x)}\psi(x)$ gauge-transformations, although they no longer had anything to do with length or calibration. But, apart from this minor flaw, the 1929 paper was fundamental in so many ways that it must rank among the most important contributions of the century.

Weyl's 1929 Paper

In this paper [14], which was neglected for many years but has now become famous, Weyl laid the foundation not only for modern gauge-theory but for many other later developments. Among these were

The use of parallel transport in General Relativity
The use of the Vierbein (which he called Achsenkreuz) for gravitation itself and for gravitating spinors
The derivation of Energy, Momentum and Angular Momentum conservation from Euclidean invariance
The introduction of the 2-component Weyl spinor
The first discussion of the possibility of P and T violation

The first mention of a possible relation between the P,T and and charge-conjugation
The use of covariant differentiation for electromagnetism
The derivation of charge conservation from gauge-invariance
The introduction of the non-integrable phase-factor $\exp\left(i\frac{e}{\hbar}\int A_\mu dx^\mu\right)$.
Most important of all, the introduction of the electromagnetic gauge-principle

Not all of these developments were completely new. Some had been introduced by Weyl himself in earlier work, the derivation of conservation laws from symmetry principles was originally due to Noether [15], and the use of the Vierbein for gravitating spinors had been proposed by Wigner [16] shortly before. But in the 1929 paper these and the new concepts were presented in a manner that was consistent and comprehensive. Einstein's comment on Weyl's earlier book 'Raum-Zeit-Materie' [17], namely that it was *a master-symphony, in which each detail had its part to play in the whole, and the total was grandios,* was equally applicable here. Weyl himself laid particular stress on the derivation of conservation laws from symmetry principles (Noether's theorem). Indeed he claimed that it was the derivation of charge conservation from gauge-invariance that convinced him of the correctness of the gauge-principle.

It is amusing to contrast Einstein's enthusiastic welcome for the paper with the reaction of Pauli. The latter, on first seeing a preliminary version in U.S. National Academy Proceedings, made some quite scathing remarks and, on reading the final version, only partially withdrew them . An amusing account of the two letters which he wrote to Weyl can be found in ref [18]. Pauli seems to have been particularly upset by the two-component spinor, on the grounds of parity-violation. But, behind the facade, he really appreciated Weyl's approach, as was shown by subsequent events (see below) and by his Handbuch article on relativity [2].

As is now well-known, Weyl's non-integrable phase-factor plays a crucial role in the Aharonov-Bohm effect [19] and in the Yang-Wu [20] formulation of electrodynamics, and has been generalized to non-abelian theories in the form of the Wilson loop [21]. An interesting comment, due to Yang [22], is that from the non-integrable phase-factor point of view Einstein's objection to Weyl's gravitational theory is just the objection that there is no Aharonov-Bohm effect for clocks in a gravitational field!

In spite of the depth and originality of Weyl's 1929 paper it appears to have had little impact during the following two decades, except for the fact that the concepts of Vierbein and parallel-transport came into common use in General Relativity theory and that it had some influence on the attempts by Einstein and Schrödinger [23] to construct a unified theory of Gravitation and Electromagnetism. It was only toward the end of the forties that the ideas were again taken up, this time by Yang. Today it is well-known that Yang's considerations of Weyl's gauge-principle lead to what is now called Yang-Mills, or non-abelian, gauge theory in 1954. What is perhaps not so well-known is that Yang had already begun to consider the problem when he was a graduate student and that similiar ideas were being considered independently by other people, notably Klein, Pauli, Shaw and Utiyama. What I wish to do here is sketch both the Yang-Mills and the other approaches. I should perhaps hasten to add that am not raising any question of priority since Yang and Mills were the first to formulate the theory in a clear and logical manner and to publish in a regular journal. The other approaches were more speculative (except possibly for Utiyama, whose results were obtained slightly later, though not as late as 1956, as is generally thought). It should also be pointed out, perhaps, that the fact that others were on the same track in no way detracts from the Y-M result. Rather does it emphasize the fundamentality of the Y-M idea by showing how it could emerge from a number of quite different considerations.

Klein's Theory

The very first Yang-Mills-type theory was produced by Oskar Klein in 1938. Up to recently Klein's contribution was almost unknown, partly because it was produced just before the outbreak of World War II, partly because it was produced only as a conference talk [24] and partly because the notation and physical application were rather obscure. Indeed the papewr seems to have been appreciated by only one member, Møller, of the illustrious audience present at the talk, which many of the greats such as Bohr, Gamow, von Neumann, and Wigner, and it was not even quoted, much less developed, in any subsequent paper. What Klein did was to make the natural generalization of the existing Kaluza-Klein theory, which had *one* extra dimension (a circle) to the corresponding theory with *two* extra dimensions (a 2-sphere). Since $S0(3)$ is the symmetry group of the 2-sphere and Klein included the nucleons as a multiplet, it is not surprising that he arrived at an $SU(2)$ gauge-theory. Because of the Kaluza-Klein derivation the field strengths came out automatically and he wrote them in the form

$$A_{rs} = \partial_r A_s - \partial_s A_r + \frac{ie}{hc}(B_r \tilde{B}_s - B_s \tilde{B}_r) \qquad B_{rs} = \delta_r A_s - \delta_s A_r, \qquad (30)(31)$$

and similarly for $\tilde{B}_{rs}$, where δ_r is the EM convariant derivative. He then postulated a Lagrangian and derived the field equations

$$\epsilon_r \partial_r - \frac{ie}{2hc}\epsilon_r \begin{pmatrix} 0 & \tilde{B}_r \\ B_r & 2A_r \end{pmatrix} \begin{pmatrix} n \\ p \end{pmatrix} = \frac{Mc}{h} \begin{pmatrix} n \\ p \end{pmatrix}, \qquad (22)$$

for the nucleons, where ϵ_r are the Dirac matrices, together with the field equations

$$\delta_l B_{kl} - \frac{2ie}{hc}A_{kl}B_l + \frac{\mu^2 c^2}{h^2}B_k = j_l, \qquad (40)$$

and similarly for $\tilde{B}_l$, for the gauge-fields, where j_l is the matter-current. These are essentially the Yang-Mills equations, but in an unconventional notation and with ad hoc masses for the gauge-fields. I have used Klein's equation numbering in order to give some idea of the sequence and the number of intermediate steps.

Klein's problem was how to identify the gauge-fields physically and how to give them masses. As one can see from (40) he solved the mass problem by simply inserting masses for the B-fields in the Lagrangian (in gross violation of gauge-invariance). He then identified the A-field with the electromagnetic field and suggested that the B-fields should mediate the nuclear forces. But it was not clear which of the nuclear forces they should mediate, which is not surprising if we recall that the same problem was still around twenty years later. Recall that in 1938 Yukawa's meson had only recently been proposed, and it was not clear whether it should be scalar or vector! Indeed, the only property of the nuclear forces that had been established beyond doubt was the charge independence of the strong interactions. This property prompted Møller to ask the single question that followed the talk, namely, how the fact that Klein's new mesons were both electrically charged could be reconciled with charge-independence. Astonishingly, Klein responded by pointing out that this difficulty could be overcome by making the following simple change in his array of vector mesons

$$\begin{pmatrix} A_k & \tilde{B}_k \\ B_k & A_k \end{pmatrix} \quad \rightarrow \quad \begin{pmatrix} A_k - C_k & \tilde{B}_k \\ B_k & A_k + C_k \end{pmatrix} \qquad (A)$$

If one takes into account that the array (2) is just diag(1,-1) times the usual present-day array one sees that Klein made the generalization $SU(2) \rightarrow SU(2) \times U(1)$ on the spot! This is just the generalization that Glashow made twenty years later for the *weak* interactions when he generalized Schwinger's 1957 $SU(2)$ model to the standard model. Finally, Klein proceeded to second-quantize the system in the manner used at the time for QED. The full Klein paper, the Møller discussion and a commentary by David Gross can be found in ref [23].

Pauli's Theory

A second Yang-Mills-type theory was produced by Pauli in 1953, as a sequel to a discussion following a talk by Pais at the Lorentz Conference [25] in Leiden in 1953. During the discussion Pauli had said: *I am very much in favour of the general principle to bring empirical conservation laws and invariance properties in connection with the mathematical groups of transformations of the laws of Nature. If, besides the conservation of energy-momentum and of charge, the conservation of the property defined as number of nucleons and charge-independence of the nuclear forces are well established, they have indeed, as Pais tried now to express mathematically, also to be connected with group theoretical proerties of the laws of nature...I would like to ask in this connection whether the transformation group* (isospin) *with constant phases can be amplified in a way analogous to the gauge group for electromagnetic potentials in such a way that the meson-nucleon interaction is connected with the amplified group. The main problem here seems to be the proper incorporation of the coupling constant into the group.* To which Pais replied that *an amplified group is very much on my mind as a means of specifying the coupling in more detail. I have no results so far.* Pauli and Pais were clearly unaware of Klein's contribution, and, stimulated by the discussion, Pauli independently generalized Kaluza-Klein theory from the circle to the 2-sphere. The results, similiar to Klein's, were sent to Pais in letters dated July and December 1953, in which Pauli says he wrote the theory down *just to see how it looks.* He considered his most important result to be the expression for the gauge-field, which, having been induced from the Kaluza-Klein theory, was automatically the correct one. However, Pauli did not attempt to abstract his theory from the Kaluza-Klein context or make any explicit application to the nuclear interactions, possibly because he realized that the classical gauge-fields masses would have to be zero.

Yang-Mills Theory

Both the Klein and Pauli theories were formulated within the context of Kaluza-Klein theory and neither author seems to have envisaged the possibility of constructing a non-abelian gauge theory in its own right. This was left to Yang, who, as mentioned above, had begun to think about this possibility while still a graduate student. He writes [26]: *While a graduate student in Kunming and in Chicago, I had thoroughly studied Pauli's review articles on field theory. I was very much impressed with the idea that charge conservation was related to the invariance of the theory under phase changes, an idea, I later found out, due originally to H. Weyl. I was even more impressed with the fact that gauge invariance determined all the electromagnetic interactions. While in Chicago I tried to generalize this to isotopic spin interactions by the procedure later written up in [27], equations (1) and (2)* (covariant derivative). *Starting from these it was easy to get equation (3)* (transformation law for the gauge-potential) *Then I tried to define the field-strengths $F_{\mu\nu}$ by $F_{\mu\nu} = \partial_\mu B_\nu - \partial_\nu B_\mu$ which was a 'natural' generalization of electromagnetism. This led to a mess, and I had to give up. But the basic idea remained attractive, and I came back to it several times in the next few years, always getting stuck at the same point...As more and more mesons were discovered and all kinds of interactions were being considered, the necessity to have a principle for writing down interactions became more obvious to me. So while at Brookhaven* (in the summer of 1953) *I returned once more to the idea of generalizing gauge invariance. My office mate was R.L.Mills who was about to finish his Ph.D. at Columbia with N.Kroll. We worked on the problem and eventually produced [27]. We also wrote an Abstract for the April 1954 meeting of the AMS in Washington, which became [28]. Different motivations were emphasized in the two papers. The formal aspect of our work did not take long and was essentially finished by February 1954. But we found that we were unable to conclude what the mass of the gauge-particle should be. We toyed with the dimensional argument that, for a pure gauge theory, there is no quantity with the dimension of a mass to start with, and therefore the gauge particle must be massless. But we quickly rejected this line of reasoning.* The first public presentation of their results seems to have been during a seminar given at the Institute in Princeton at the end of February. In this seminar Pauli (who, as mentioned above, had produced a Kaluza-Klein version and was aware of the problem with the gauge-field masses) caused some disruption by refusing to accept Yang's statement that he was not sure whether these masses should be zero. (In the light of our present knowledge concerning spontaneous breakdown and confinement, Pauli's demand for immediate information was quite a tall order!). In any case Yang and Mills soon realized that the mass and renormalization problems for this theory were not going to be solved overnight, and sent the manuscript for publication in June 1954.

Shaw's Theory

One of the most remarkable and least known approaches to non-abelian gauge-theory was that of Ronald Shaw, who in 1954 was studying for his doctorate [29] under Professor Salam at Cambridge. Quite unaware of the work of Yang and Mills, Shaw introduced $SU(2)$ Yang-Mills theory in his thesis, starting from a completely different angle, namely from a preprint of Schwiner in which the usual $U(1)$ group of electromagnetism was replaced by $SO(2)$. According to Shaw, on reading this preprint he saw *in a flash* that the $SO(2)$ formulation could be generalized to $SU(2)$. Even the construction of the kinetic terms does not seem to have given him any difficulty. It may be of interest reproduce the relevant two pages of his thesis (in slightly abbreviated form and with the original equation-numbering):

Now under a general transformation (13) $(\psi' = (1 + \frac{1}{2}ic_i\tau_i)\psi)$ with $\vec{c}$ a function of position, L_0 is no longer invariant but becomes

$$L_0' = L_0 + \frac{1}{2}\partial^\alpha \vec{c}.\bar{\psi}\gamma_\alpha\vec{\tau}\psi. \tag{16}$$

The additional term can be compensated by introducing an interaction

$$L_1 = -\frac{iq}{2}\vec{B}^\alpha.\bar{\psi}\gamma_\alpha\vec{\tau}\psi \qquad \text{where} \qquad q\vec{B}^\alpha = q(\vec{B}^\alpha - \vec{c}\times\vec{B}^\alpha) + \partial^\alpha\vec{c} \tag{17)(18}$$

In order to obtain a theory invariant under general isotopic spin transformations it only remains to construct an invariant free field Lagrangian for the $\vec{B}$-field. It is natural in anology with with electromagnetic case, to define

$$\vec{f}^{\alpha\beta} = \partial^\alpha\vec{b}^\beta - \partial^\beta\vec{b}^\alpha \qquad L_2 = -\frac{1}{4}\vec{f}^{\alpha\beta}.\vec{f}_{\alpha\beta}. \tag{19)(20}$$

However (20) is not invariant under the transformation (18).... A properly invariant Lagrangian can be obtained by defining

$$\vec{F}^{\alpha\beta} = \vec{f}^{\alpha\beta} - q(\vec{B}^\alpha\times\vec{B}^\beta - \vec{B}^\beta\times\vec{B}^\alpha) \qquad L_2 = -\frac{1}{4}\vec{F}^{\alpha\beta}.\vec{F}_{\alpha\beta} \tag{22)(23}$$

L_2 is now invariant under (18) as $\vec{F}^{\alpha\beta}$ transforms solely as a vector in E_3."

This work was done in the Spring of 1954, around the time that Yang was giving his seminar in Princeton, and thus it came a little later than the work of Yang and Mills. But, of course, it was completely independent. The thesis was not submitted until 1955 because it had to be completed by further work on the Poincare and Lorentz groups. (The latter work appeared finally as Part I and the above-quoted extract in Part II, which adds to the confusion in dates!). As Professor Salam was Shaw's thesis adviser it is sometimes maintained that he should have urged Shaw to publish his result at once. But, in Shaw's own words: *Sometime later (still early in 1954?) I showed it to Salam, but I do not think he appreciated it at the time. I do not at all blame him for this–I probably told him about it in a very dismissive way, since the relevant particles, surely m=0, I thought, did not exist. Much later (end of 1954?)*

he told me that Yang and Mills had had the same idea and told me to write mine up (which I did not).

Utiyama's Theory

Utiyama's published contribution [30] is quite well-known. But because it was published only in 1956, quoted Yang and Mills and dealt with all simple compact groups and with Gravitation, it is usually regarded as a generalization of the Yang-Mills paper which was written as a sequel to that paper. This is not the case. As we now know from his published statements, Utiyama derived the YM equations slightly later than Yang and Mills but quite independently. These statements appear in two books which, as far as I can ascertain, are available only in Japanese and Japanese/Russian respectively but two of my colleagues have been kind enough to translate the relevant relevant parts of the two books (the translations agree in detail where they overlap). So I can let Utiyama speak for himself.

Book 1 [31] Brief summary of Chap.10, entitled Memoir of Remorse

1954, January: In the end of January, I (R. Utiyama) received a letter from the Institute for Advanced Study (IAS), Princeton, inviting me to stay at the IAS from September, 1954. I decided to prepare some completed work prior to going to the IAS so that I could publish papers even if I could not work successfuly there. In fact, I had already the idea of the general gauge theory, including the basic structure of the theory, in my mind before I received the letter from the IAS. So what I should do was simply to spell it out in the form of a paper I completed the work at the end of March 1954.

1954, May or June: I gave a talk on the work at a small workshop held at the then Research Institute for Fundamental Physics (now Yukawa Institute for Theoretical Physics.) Kyoto University. I got a negative response.

1954, September: I arrived at Princeton. I met Prof. Minoru Kobayashi (then Professor at Kyoto University) who was my supervisor at Osaka University and had been staying at the IAS from one year earlier as a visiting fellow. When I explained to him my idea of general gauge theory, he told me that recently Prof. Yang announced a theory which is quite similar to mine. Some days later, I found a preprint by Yang in my mailbox. I took a look of it and found many formulae already familiar to me — I immediately realised that he had found the same theory that I had developed. I was shocked too deeply to examine Yang's paper closely and to compare it with my work with care.

1955, March: As my appointment was prolonged for one more year, I felt that I should work on a subject which was more important than those I had engaged so far at the IAS. So I decided to return to the general gauge theory and took a closer look on Yang's paper which had been published in 1954. At this moment I realised for the first time that there was a significant difference between Yang's theory and mine. The difference was that Yang merely found an example of non-Abelian gauge theory whereas I had developed a general idea of gauge theory which could contain gravity as well as electromagnetic theory. Thus I decided to publish my work by translating it into English, and adding an extra section where Yang's theory is discussed as a example of my general gauge theory. I did so partly because his paper had already been published, and partly because I thought that it would not be fair if I did not refer to his paper after I read it, even if I developed the idea of the general gauge theory before knowing his work. My paper was accepted for publication in Phys. Rev. in July 1955 and published early in 1956

Remark: I admit that it was Yang and Mills who first found a new type of (non-Abelian) gauge theory (though not gravity). However, it is very much disappointing for me to see that non-Abelian gauge theory is called by the name 'Yang-Mills theory' only, without a slightest mention to me who developed the general framework of the gauge theory. I very much regret not having submitted my paper to some Japanese journal in March, 1954 when I completed the work.

The most interesting aspect of Utiyama's approach is that it was motivated by gravitational theory and that (partly as a result of this) it was formulated for arbitrary simple Lie groups, not just $U(1)$ and $SU(2)$ or even compact simple groups. In fact, Utiyama had hoped to use this approach to unify the interactions. As we now know (partly thanks to Utiyama himself), strict unification of gravitation with the other interactions along these lines is impossible since gravitation cannot be formulated as a set of local field equations for the gauge-potentials, but only for the metric (or the Vierbein). Nevertheless, Utiyama was the first to suggest that the broader analogy between gravitation and electromagnetism which had so fascinated Weyl, Einstein and Schrödinger [22] could be extended to include the nuclear interactions. As he says himself in the second book [32]:

If I am allowed to refer to my work on the general gauge theory, I would like to stress that it is my paper which first showed clearly that the theory of gravity could fall into the framework of the gauge theory, which would be a first step toward the the grand unified theory in the modern viewpoint. Even more importantly, my paper pointed out that fields carrying a fundamental force — either gravity or electromagnetism — must in fact be those termed as 'connections' in mathematics, which are now called 'gauge fields'. That the notion of connections is indispensable in establishing a theory of interactions was the basic assertion that I wanted to make in my paper.

Summary

As everybody knows the Yang-Mills paper did not immediately herald a new dawn for the nuclear interactions. Indeed the hurdles still to be overcome, the infrared, unitarity and renormalization problems on the theoretical side and the zero-mass, neutral current and gauge-field identifications on the experimental side, were so formidable that for more than a decade the theory languished in a backwater. Only with the advent of the concept of spontaneous symmetry breaking and suitable renormalization techniques did the theory become feasible, and only with the the experimental discovery of neutral currents and the experimental/theoretical discovery of asymptotic freedom did it become completely acceptable. Of course, even when fully accepted gauge theory provides us only with a gemetrical framework for the fundamental interactions but does not yet combine the strong and electroweak interactions in an intrinsic way, much less combine these two with gravitation and there is no shortage of open questions (are there Higgs bosons? what is the origin of the fermion masses? where does CP-violation fit in? and so on). We are still far from the end-station. intrinsic

Acknowledgements: I am indebted to Professor C.N. Yang for a discussion on these matters and for some very relevant material, and to Professor Ronald Shaw for sending me a copy of his thesis and comments. I am indebted to Professor Norbert Straumann for a copies of the Pauli-Pais correspondence and for his own contribution to the Weyl Centenary Volume. Finally, I should like to thank Professor Z. Ajduk for extracts from the Proceedings of the Kasimierz Conferences, Professor Misha Saveliev and Dr. Izumi Tsutsui for obtaining and translating the Utiyama books and Professor A. Wightman for some pertinent historical comments.

References

[1] A. Pais, *Inward Bound* Clarendon Press, Oxford, 1986.

[2] W. Pauli, *Theory of Relativity*, Pergamon Press 1958.

[3] T. Kaluza, Sitzb. Preuss. Akad. Wiss., Berlin Math. Phys. **K1** (1921) 966

[4] O. Klein, Z. Phys. **37** (1926) 895

[5] T. Levi-Civita, Rend. Circ. Mat. Palermo **42**(1917)

[6] C. Ehresmann, Colloque de Topologie, Brussels, 1950 p.29-55, S. Kobayashi and K. Nomizu, Foundations of Differential Geometry, Interscience, I, 1963; II 1969

[7] E. Cartan, Math. Annalen **102** (1930) 114

[8] H. Weyl, S.B. Preuss. Akad. Wiss (1918) 465, Gesammelte Abhandlungen, Springer 1968 II p.1 p.29

[9] E. Schrödinger, Zeit. Phys. **12** (1922) 13

[10] V. Raman and P. Forman, Hist. Studies Phys. Sci. **1** (1969) 291.

[11] V. Fock, Z. Physk. **39** (1927) 226

[12] F. London, Z. Physk. **42** (1927) 375

[13] C. N. Yang in *Schrödinger* ed. C. Kilmister, Cambridge Univ. Press 1987

[14] H. Weyl, Zeitschrift fur Physik **56**(1929) 330.

[15] E. Noether, Nachr. d. Kgl. Ges. d. Wiss. Gottingen (1918)235

[16] E. Wigner, Zeit. f. Phys. **53** (1929) 592

[17] *Space-Time-Matter* Methuen, London 1922

[18] N. Straumann, DMV-Seminar Geschichte der Mathematik H. Weyl's Raum-Zeit-Materie (1992)

[19] M. Peshkin and Y. Tonomura, The Aharonov-Bohm Effect, Springer 1991

[20] T.T. Wu and C.N. Yang, Phys.Rev. **D12**(1975)3845

[21] See for example T-P Cheng and L-F Li, *Gauge Theory of Elementary Particle Physics* Clarendon Press, Oxford 1984.

[22] C.N.Yang, in *Hermann Weyl 1885-1985* ed. K.Chandrasekharan (Springer 1986).

[23] E. Schrödinger, *Space-Time Structure* Cambridge University Press, 1953.

[24] *New Theories in Physics* Proc. XI Warsaw Symposium on Elementary Particle Physics, held at Kasimierz, Poland, May 1988 (eds. Z. Ajduk, S. Pokorski and A. Trautman, World Scientific, 1989)

[25] A. Pais, Physica XIX (1953) 869.

[26] C. N. Yang, *Selected Papers with Commentary* Freeman, 1983

[27] C. N. Yang and R. L. Mills, Phys. Rev. **96** (1954) 191

[28] C. N. Yang and R. L. Mills, Phys. Rev. **95** (1954) 631

[29] R. Shaw, Thesis, Cambridge University, 1955 (unpubl.)

[30] R. Utiyama, Phys. Rev. **101** (1956) 1597. (rec. July 7th 1955)

[31] Book 1: *Butsurigaku wa dokomade susundaka (How far has Physics progressed?)* by Ryoyu Uchiyama (Iwanami Shoten, Japan, 1983).

[32] Book 2: *Ippan gauge ba ron josetsu (Introduction to the General Gauge Field Theory)*, Ryoyu Uchiyama, (Iwanami Shoten, Japan, 1987).

ALGEBRAIC APPROACH TO THE HYPERCOULOMB PROBLEM

E.Santopinto[1], M.Giannini [1], and F. Iachello[2]

[1]Dipartimento di Fisica, Universita' di Genova, Genova , Italy
[2]Center for Theoretical Physics, Sloane Laboratory,Yale University
New Haven, Connecticut 06520-8120

1. INTRODUCTION

The low-lying states of baryons are thought to be mainly composed of three-quarks configurations. Although for light (u,d) quarks a construction of the quark-quark interaction is very difficult, it has been suggested that a good fraction of this interaction is of three-body character [1], and that, in addition, it can be well approximated by a hypercentral potential, i.e. an interaction that is invariant under rotations in the six dimensional space spanned by the intrinsic coordinates of the three-body problem. This suggestion arose from a study of the flux tube configurations of Fig. 1 for heavy-quark baryons. However, in general, the fact that QCD is a non-abelian gauge theory leads to gluon-gluon couplings which, in turn, produce three-body forces, as shown in Fig. 2. One is thus led to the conclusion that three-body forces can be a very important part of the quark interactions and this is particularly so for baryons composed of light quarks where the coupling constant α_s is large.

A study of three-body problems with three-body interactions is rather difficult. Algebraic methods may provide a simplifying way to approach the problem. In this article, we discuss a simple example of a three-body problem with three-body interactions, namely that in which the interaction is the hyper-Coulomb interaction 1/x. This interaction has the advantage of being exactly solvable and provides a starting point from which more elaborate (and realistic) calculations can be done.

2. JACOBI AND HYPERSPHERICAL COORDINATES

The relevant degrees of freedom characterizing the configuration of a three-body system (Fig.3) , after the removal of the center of mass coordinate $\vec{R}$, are described by the two Jacobi coordinates $\vec{\rho}$ and $\vec{\lambda}$,

Symmetries in Science VII, Edited by
B. Gruber, Plenum Press, New York, 1995

$$\vec{\rho} = \frac{1}{\sqrt{2}}(\vec{r}_2 - \vec{r}_1), \tag{1}$$

$$\vec{\lambda} = \frac{1}{\sqrt{6}}(2\vec{r}_3 - \vec{r}_2 - \vec{r}_1), \tag{2}$$

or, equivalently, $\rho, \Omega_\rho (\Omega_\rho = \vartheta_\rho, \varphi_\rho)$ and $\lambda, \Omega_\lambda (\Omega_\lambda = \vartheta_\lambda, \varphi_\lambda)$. Thus one has $v = 3+3$ independent variables.

Instead of the Jacobi coordinates, $\vec{\rho}$ and $\vec{\lambda}$, one can introduce another set of coordinates, the hyperspherical coordinates. These are defined as [2]

$$x = (\rho^2 + \lambda^2)^{1/2} \quad , \quad \xi = \mathrm{arctg}(\rho / \lambda) \quad , \tag{3}$$

where x and ξ are called hyperradius and hyperangle, together with the angles $\Omega_\rho, \Omega_\lambda$.

The hyperspherical coordinates are particularly useful if the three-body interaction is invariant under rotations in the 6-dimensional space spanned by $\vec{\rho}$ and $\vec{\lambda}$.

2.1 THE HYPERSPHERICAL FORMALISM

Using hyperspherical coordinates, the kinetic energy operator of a three-body problem can be written (h=c=1) as

$$-\frac{1}{2m}\left(\nabla_\rho^2 + \nabla_\lambda^2\right) = -\frac{1}{2m}\left(\frac{\partial^2}{\partial x^2} + \frac{5}{x}\frac{\partial}{\partial x} - \frac{\hat{L}^2(\Omega)}{x^2}\right) \quad , \tag{4}$$

where $\hat{L}^2(\Omega)$ $(\Omega = \Omega_\rho, \Omega_\lambda, \xi)$ is the Casimir operator of the rotation group in six dimensions, O(6). The eigenfunctions of $\hat{L}^2(\Omega)$ are the hyperspherical harmonics

$$\hat{L}^2(\Omega_\rho, \Omega_\lambda, \xi) Y_{[\gamma]}(\Omega_\rho, \Omega_\lambda, \xi) = \gamma(\gamma + 4) Y_{[\gamma]}(\Omega_\rho, \Omega_\lambda, \xi) \,, \tag{5}$$

with $\gamma = 2v + l_\rho + l_\lambda$, $v = 0, 1, \dots$. Here l_ρ and l_λ are the angular momenta associated with the $\vec{\rho}$ and $\vec{\lambda}$ variables.

The hyperspherical harmonics form an orthonormal complete basis on the hypersphere Ω. Any three-body state can be expanded as

$$\Psi(\vec{\rho}, \vec{\lambda}) = \sum_\gamma \psi_{[\gamma]}(x) Y_{[\gamma]}(\Omega) \,. \tag{6}$$

If there is hyperrotational invariance, (O(6)-invariance), that is if the potential depends only on the hyperradius x, only one term in Eq. (6) is necessary and so each eigenfunction can be factorized into a hypercentral part $\psi(x)$ and a hyperangular part $Y_{[\gamma]}(\Omega)$. The hyperradial wavefunction $\psi(x)$ is a solution of the equation:

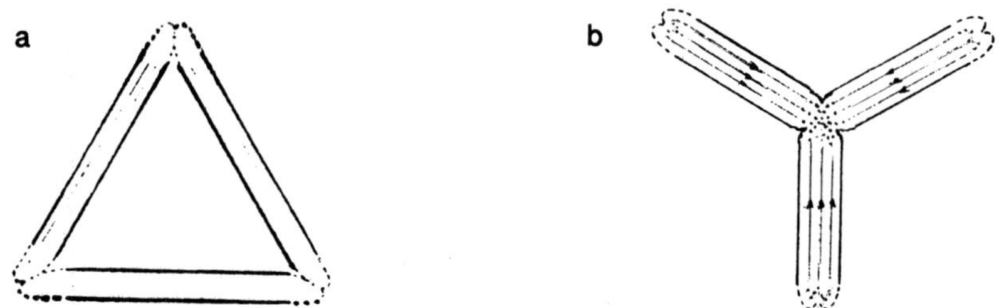

Fig.1 Flux tube configurations in baryons: the flux tubes can connect the three quarks directly, forming a triangle (a) or can assume a Y-shape form (b).

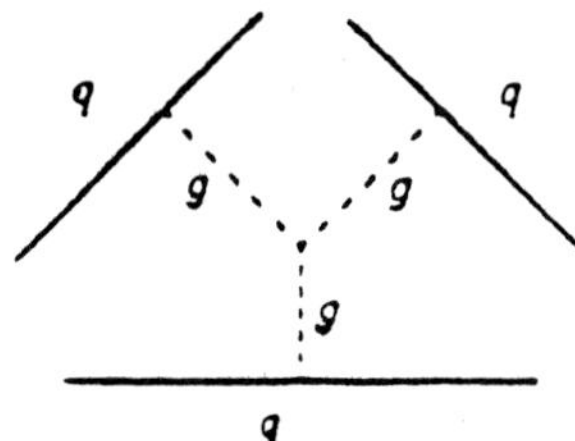

Fig.2 The lowest-order diagram leading to three-body forces among quarks.

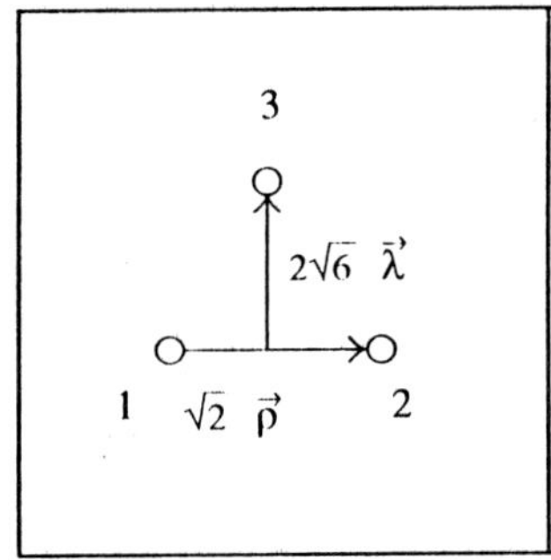

Fig.3 Jacobi coordinates $\vec{\rho}$ and $\vec{\lambda}$ used in the description of three-body problems.

$$\left[\frac{d^2}{dx^2}+\frac{5}{x}\frac{d}{dx}-\frac{\gamma(\gamma+4)}{x^2}\right]\psi(x)=-2m[E-V(x)]\psi(x) \quad . \tag{7}$$

The hyperradial equation (7) can be solved in closed analytic form in two cases : $V(x)=\frac{3}{2}kx^2$ (six-dimensional harmonic oscillator) and $V(x)= -\frac{\tau}{x}$ (six-dimensional Coulomb problem).

3. ALGEBRAIC SOLUTION OF THE HYPERCOULOMB PROBLEM

An algebraic solution of the hyperCoulomb problem can be obtained by generalizing to six dimensions the algebraic solution of the usual Coulomb problem [3]. This generalization has been done both in the context of the six-dimensional Coulomb problem [4] and in that of the many-body problem [5]. Here we briefly review this solution. Introducing the notation,

$$\left(q_i\right)_{i=1,6}=\left(\vec{\rho},\vec{\lambda}\right), \; \left(p_i\right)_{i=1,6}=\left(\vec{p}_\rho,\vec{p}_\lambda\right) \quad ; \qquad i=1,...,6 \quad , \tag{8}$$

one can construct the 15 elements of the algebra o(6) by

$$L_{ij}=q_i p_j - q_j p_i \quad , \qquad i,j=1,...,6 \; , \; i<j \quad . \tag{9}$$

To these, one can add the 6 components of the generalized Runge-Lenz vector,

$$A_i'=\frac{1}{2m}\left(L_{ij}p_j-p_j L_{ji}\right)-\frac{\tau q_i}{x} \quad , \qquad i=1,...,6 \quad . \tag{10}$$

The 21 elements (9) and (10) satisfy the commutation relations:

$$\begin{aligned}
&\left[L_{ij},L_{kl}\right]=i\left(\delta_{ik}L_{jl}+\delta_{jl}L_{ik}-\delta_{il}L_{jk}-\delta_{jk}L_{il}\right) \; , \\
&\left[A_i',A_j'\right]=iL_{ij}\left(-\frac{2H}{m}\right) , \\
&\left[L_{ij},A_k'\right]=i\left(\delta_{ik}A_j'-\delta_{jk}A_i'\right) \quad .
\end{aligned} \tag{11}$$

The 21 elements also commute with the Hamiltonian H

$$\left[L_{ij},H\right]=\left[A_i',H\right]=0 \quad , \tag{12}$$

where $H=\frac{p^2}{2m}-\frac{\tau}{x}$.

One can then replace, for bound states, A_i' by

$$A_i=\sqrt{-\frac{m}{2E}}A_i' \tag{13}$$

in the space where E=<H> (E<0). The resulting commutation relations of L_{ij} and A_i are those of o(7). This is the degeneracy algebra of the hyperCoulomb problem. One can then easily obtain the eigenvalues of the Hamiltonian H by first introducing the quadratic Casimir operator of o(7),

$$C_2(o(7)) = \sum_{i<j}^{6} L_{ij}^2 + \sum_{i}^{6} A_i^2 \; , \tag{14}$$

and then noting that

$$\sum_{i}^{6} A_i'^2 = \frac{2}{m} H \left(\sum_{i<j}^{6} L_{ij}^2 + \frac{25}{4} \right) + \tau^2 \; . \tag{15}$$

(Here $25/4$ is $\left(\dfrac{\nu-1}{2}\right)^2$ and replaces 1 when $\nu=6$.)

Eq. (15) leeds trivially to

$$H = -\frac{\tau^2 m}{2\left[C_2(o(7)) + \dfrac{25}{4} \right]} \; . \tag{16}$$

Denoting by $(\omega,0,0)$ the totally symmetric representations of o(7) and by $\omega(\omega+5)$ the eigenvalues of $C_2(o(7))$, one obtains the eigenvalues

$$E(\omega) = -\frac{\tau^2 m}{2\left[\omega(\omega+5) + \dfrac{25}{4} \right]} \; , \qquad \omega=0, 1, \dots \; . \tag{17}$$

The representation of o(7) that appear are the totally symmetric representations, in view of conditions analogous to $\vec{L} \cdot \vec{A} = 0$ in the three dimensional case. Summarizing the results, one finds, that the hyperCoulomb problem has eigenvalues given by (17). Introducing all quantum numbers that characterize uniquely the states the wave functions can be written as

$$\left| \begin{array}{cccccc} o(7) & \supset & o(6) & \supset & o_\rho(3)\oplus o_\lambda(3) & \supset & o(3) & \supset & o(2) \\ \downarrow & & \downarrow & & \downarrow & & \downarrow & & \downarrow & & \downarrow \\ \omega & & \gamma & & l_\rho & & l_\lambda & & L & & M \end{array} \right\rangle \; , \tag{18}$$

where we have used the notation of the accompanying paper [6]. Here $\gamma = \omega, \omega-1,\dots,0$, the values of l_ρ and l_λ are obtained by partitioning γ as $\gamma = 2\nu + l_\rho + l_\lambda$, $\nu = 0,1,\dots$, and L is obtained by combining l_ρ and l_λ , $|l_\rho + l_\lambda| \geq L \geq |l_\rho - l_\lambda|$.

4. S3-SYMMETRY

When treating non-strange baryons one encounters the problem that the three constituent quarks are identical. The corresponding states must then transform as representations of the permutation group, S_3.

The permutation group S_3 is a six element discrete group which is isomorphic to the dihedral group D_3. In order to characterize the transformation properties of the operators and states under S_3, one can use the labels of either group. A common notation in hadronic physics is that introducing the letters S, M, A for the three representations of S_3

$$
\begin{array}{ccc}
S_3 & & D_3 \\
\square\square\square & \text{S} & A_1 \\
\begin{array}{c}\square\square \\ \square\end{array} & \text{M} & E \\
\begin{array}{c}\square \\ \square \\ \square\end{array} & \text{A} & A_2
\end{array}
\tag{19}
$$

In order to be able to use the results of the previous section in the study of baryon spectra, one must solve the problem of finding the representations of S_3 contained in a given representation of o(7). This problem can be solved in an indirect way by noting that the reduction o(7) $\supset$ o(6) gives [6]

$$\gamma = \omega, \ \omega - 1, \ \dots, 0 \ . \tag{20}$$

Thus the representations of o(6) contained in a given representation of o(7) are easily identified. Since o(6) is also contained in u(6) and the problem of determining the representations of S_3 contained in a given representation of u(6) has been solved up to relatively large values of the quantum numbers[7], one can then determine what representations of S_3 are contained in a given representation of o(7). For the lowest representations the result is given in table I and the corresponding spectrum is shown in Fig. 4 .

Table I . Irreducible representations of S_3 contained in a given representation of o(7).

ω	$\dim[\omega]$	γ	L	S_3	D_3
0	1	0	0	S	A_1
1	7	1	1	M	E
		0	0	S	A_1
2	27	2	$2^2;1;0$	S,M;A;M	$A_1,E;A_2;E$
		1	1	M	E
		0	0	S	A_1

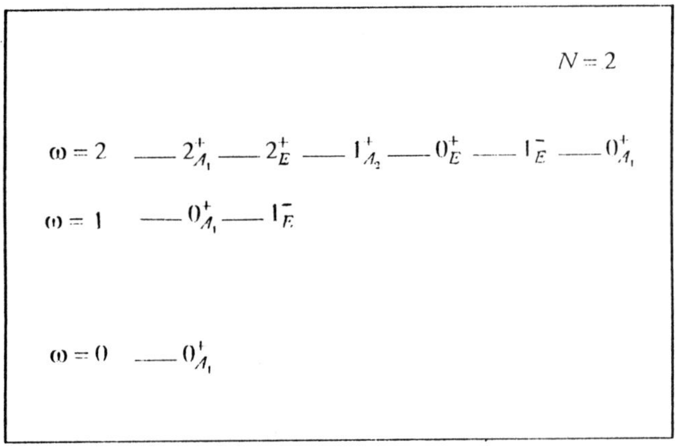

$$N = 2$$

$$\omega = 2 \quad ---\; 2^{+}_{A_1} --- 2^{+}_{E} --- 1^{+}_{A_2} --- 0^{+}_{E} ---- 1^{-}_{E} --- 0^{+}_{A_1}$$

$$\omega = 1 \quad --- 0^{+}_{A_1} --- 1^{-}_{E}$$

$$\omega = 0 \quad --- 0^{+}_{A_1}$$

Fig. 4 Schematic representation of the spectrum of the hyperCoulomb problem. The angular momentum, parity, and D$_3$-representation are indicated next to each level.

The spectrum of Fig.4 may be compared with the experimental spectrum of baryons. An interesting property of Fig.4 is the degeneracy of the L=0 and L=1 states belonging to the representation (1,0,0) of o(7). If one assigns the L=0 state to the Roper resonance N(1440) P$_{11}$ and the L=1 states to N(1520) D$_{13}$, N(1535) S$_{11}$, ... , one observes that indeed this degeneracy appears to occur. One must note that with only two-body forces it is not possible to obtain this degeneracy . (The L=0 state having one more node, must always lie above the L=1 state.) . We thus conclude that the o(7) dynamic symmetry may be relevant to baryon spectroscopy. However, in its Coulomb form, the o(7) symmetry produces a spectrum which converges to zero for large values of ω. This is not the case experimentally. In order to provide a realistic description of the baryon spectra one must therefore modify the hyperCoulomb interaction (at least through the introduction of a confining term.) . The modification must preserve the o(7) symmetry, if one wants to maintain the degeneracies of Fig. 4 .

5. CONCLUSIONS

In this note, we have briefly reviewed the algebraic solution of the hyperCoulomb problem and observed that the baryon spectrum has degeneracies reminiscent of those of o(7). The next step is to exploit this observation by constructing a more realistic model of baryons. In order to do so, we must embed o(7) into a larger algebraic structure. Barut and Kitagawara[5] have suggested to use o(7,2) . This embedding is appropriate for Colomb-like problems. Since we want to investigate more general situations we suggest to embed o(7) onto u(7), as discussed by Bijker and Leviatan[8] and in the accompanying article[6] . Use of u(7) requires a full investigation of its algebraic structure, which remains to be done. We hope to be able to report on this investigation soon.

ACKNOWLEDGMENTS

We are indebted to A.Barut for pointing out to us the earlier references on the six-dimensional Coulomb problem. This work has been supported in part by D.O.E. Grant DE-FG02-91ER40608 .

REFERENCES

[1]P.Hasenfratz, R.R.Horgan, J.Kuti and J.M.Richard, Phys.Lett. **B94**, 401(1980); A.T. Aerts and L.Heller, Phys.Rev. **D23**, 185(1981); Heller L. in "Quarks and Nuclear Forces", Springer Tracts in Modern Physics **100**, ed. D.C. Vries and B. Zeitnitz, Berlin Springer, p. 145-183.

[2] G.Morpurgo, Nuovo Cimento **9**, 461 (1952); J.L.Ballot and M. Fabre de La Ripelle, Ann.Phys.(N.Y.) **127**, 62 (1980); M.Giannini, Nuovo Cimento **A76**, 455 (1983), and references therein; D.Drechsel, M.M.Giannini and L.Tiator, Few-Body Systems, Suppl.2, p.448 (1987).

[3] W.Pauli,Z.Phys. **36**,336 (1926); V.Fock, Z.Phys. **98**, 145 (1935); V.Bargmann, J.Math.Phys. **5**, 1606 (1964).

[4] W.O.Rasmussen and S.Salamo,J.Math.Phys. **20**(6),1064(1979).

[5] A.O.Barut and Y.Kitagawara,J.Phys. **A14**,2581(1981);**A15**,117(1982).

[6] F. Iachello, accompanying paper.

[7] K.C. Bowler, P.J.Corvi, A.J.G. Hey, P.D.Jarvis and R.C.King,Phys.Rev. **D24**,197(1981); A.J.G.Hey and R.L.Kelly, Phys.Rep. **96**,71(1983).

[8] R.Bijker and Leviatan, in "Symmetries in Science VII ", B.Gruber and T. Otsuka eds., Plenum Press, New York (1994), p. 87 ; R.Bijker, F. Iachello and A. Leviatan, Ann. Phys. (N.Y.), in press.

CONFRONTATION OF SUPERSYMMETRY IN ELEMENTARY PARTICLES AND NUCLEAR PHYSICS

Stanisław Szpikowski

Institute of Physics, M. Curie–Słodowska University
20–031 Lublin, Poland

1. INTRODUCTION

At first, I am very pleased to express my sincere gratitude to Professor Bruno Gruber and all of the people whom we owe the organisation of such a nice, not only from scientific point of view, Symposium. My second remark concerns the motivation of the present work. I am involved since several years in supersymmetries in nuclear physics and during seminars I delivered on our supersymmetry results I have got very often questions from elementary particle physicists: what is the difference between supersymmetry in elementary particle and nuclear physics; whether the Poincaré group is involved in nuclear supersymmetries; is there Clifford Algebra applied to fix the content of supermultiplets; and so on. Hence, I have been forced to loook more carefully on the elementary particle supersymmetry from the point of view of similarities and differences with nuclear physics. I have been astonished as a new–comer how beauty and dramatic history of the elementary particle supersymmetry is and how many turning points in a supersymmetry development and application have been met. Hence, I would like to pass to you my astonishment connected with several steps in the historical evolution of elementary particle supersymmetries. Then, in the same way I will present the supersymmetry problem in nuclear physics. I will also illustrate the nuclear supersymmetry by examples of our recent results in light nuclei. In the final part I will compare both supersymmetries from mathematical and physical point of view.

2. ELEMENTARY PARTICLE SUPERSYMMETRY

The beginning of supersymmetry considerations of elementary particles was very natural and, moreover, it followed well known patterns in other fields of physics. For example, in 1930–ties Wigner asked the question: there is known the spin–symmetry unitary group $SU_S(2)$ and also the isospin group $SU_T(2)$, then, is there of any physical meaning the wider symmetry group which would comprise, as a subgroup, the direct product of $SU_S(2) \times SU_T(2)$? The answer was yes — it is the group $SU(4)$ which up to now has an established value in atomic, nuclear, and elementary particle physics. The bases of irreducible representations of the group $SO(4)$ are known as "Wigner supermultiplets" (the prefix "super" has nothing to do with the present meaning of supersymmetry groups and their supermultiplets). The second example came from 1960–ties when Gell–Mann posed the similar question: there is the charge independence of nuclear interactions, the $SU_T(2)$ symmetry, and there is also a conservation of the strange scalar observable — the strangeness connected with the symmetry $U(1)$, then is of any physical meaning the wider symmetry group which has as its subgroup the direct product $SU_T(2) \times U(1)$? The answer was also yes, and the symmetry group is the famous unitary (flavour) group $SU(3)$ which the fundamental representation is the three member representation known to–day as up (u), down (d) and strange (s) quarks. Confirmation of quarks some ten years later has been one of the greatest triumph of the symmetry considerations in physics.

Similarly, the beginning of supersymmetry considerations in particle physics has followed the same pattern starting with the similar question. There is a fundamental relativistic symmetry of space–time degree of freedom described by the invariance transformations of the Poincarè group (P). There is also known the internal symmetry of elementary particle interactions, at present it is the standard model of the (local) symmetry of electroweak and strong (quark) interactions $SU(3) \times SU(2) \times U(1) \equiv I$. Then: is there any influence of both symmetries: the external (P) and internal (I) ones? In the group theoretical language it is a question of finding a wider symmetry group G, with the physical meaning, which has, as its subgroup, the direct product of the Poincarè and internal symmetry group

$$G \supset P \times I \tag{1}$$

Astronishingly and dramatically the answer was "no" and it was given by the so called "No–Go" theorem of Coleman and Mandula. However, there has been one narrow path to escape from the No–Go theorem by changing one of its fundamental assumptions. Up to this point the symmetry considerations have been applied either to the fermion system or to the boson system. In each representation the transformation operators could change fermion- to fermion states and boson- to boson states. If we allow to include the transformations changing fermion states to boson states or vice versa, then the No–Go theorem looses its validity. However, in such a case there are immediate two problems: physical and mathematical. In physics the question arises of the fermion conservation law which is broken by the additional transformation. In mathematics, the wider set of transformation operators span an algebra but with two different "multiplications": commutators characterizing of the Lie algebra and anticommutators. Such an algebraic structure has opened a new mathematics branch, the Lie superalgebra and the supersymmetry group theory. In physics there is an answer to the problem going back to the first moments of our Universe where there had been no difference between fermions and bosons which had been separated each other after one of the phase transitions in the Universe evolution. Coming back to the present time where there is a distinct difference between fermions and bosons, the supersymmetry assumption must be furnitured with the further condition of construction of operators for the physical observables in such a way as to fulfil the fermion conservation law. In the last 30 years this new branch of mathematics, the supergroup theory and the super–Lie algebras has been widely developed both in its fundamental mathematical research and from the point of view of its application to physics.

The supersymmetry groups has been classified in analogy to the group theory. Besides of some "exceptional" supergroups which have found, so far, no applications in physics, there are only two supergroup chains, namely the unitary–unitary chain $U(n|m)$ and the orthosymplectic chain $Osp(N|M)$. The main feature of supersymmetry chains is, that they have as their subgroups either the direct product of two unitary groups or the direct product of the orthogonal and symplectic groups

$$U(n|m) \supset U(n) \times U(m) \tag{2}$$

$$Osp(N|M) \supset SO(N) \times Sp(M) \tag{3}$$

The second dramatic turning point is that the Poincarè group is not a simple group and it could not be considered as a subgroup of any group (2, 3). The solution of the problem has been provided by the anti–de Sitter group $SO(3,2)$. It is almost trivial in formal understanding, the orthogonal transformation group in five–dimensional space of two like–time dimesions and three space–dimensions. The Poincarè group can be then recovered from the anti–de Sitter group by the so called contraction process. It also means that the supersymmetry group for the elementary particle physics has been, if any, in the chain (3).

The third dramatic point follows the impossibility, for some subtle reason, to take the anti–de Sitter group $SO(3,2)$ as the orthogonal part of the chain (3). Fortunately, there is an accidental isomorphism of two groups

$$SO(5) \approx Sp(4)$$

both in their compact and non–compact forms. Then, the anti–de Sitter group $SO(3,2)$ can be considered, now without objections, as a member of the chain (3) in its symplectic part. The supersymmetry group is then

$$Osp(N|4) \supset SO(N) \times Sp(4) \tag{4}$$

where the symplectic group $Sp(4)$ is in its non–compact version. Then, it is only an unique way to comprise the internal symmetry group in the orghogonal part $SO(N)$ of the chain (4). However, the supermultiplet contents of the supergroup (4), which are obtained with the help of Clifford algebra are only those, with physical meaning, which depend on the signature N mod 8. It means that in the supergroups (4) we can take only $N = 1, 2, \ldots, 8$. Taking the maximum $N = 8$, i.e. the group $SO(8)$ is not enough to comprise the standard model symmetry group $SU(3) \times SU(2) \times U(1)$. And that is the most dramatic point in the supersymmetry discussion in elementary particle physics. It means that the main goal to provide the mutual dependence of the space–time and internal symmetry in the frame of a proper supergroup can not be achieved. However, not everything is lost, because the supersymmetry structures have other features which we are looking for in elementary particle considerations. Because we could not take as big N as to comprise the standard model symmetry, we in the simplest way can take the supersymmetric model with $N = 1$ i.e. $Osp(1|4)$ which is called the $N = 1$ model. Exploiting the Clifford algebra one can find the supermultiplets of the group $Osp(1|4)$ which are dublets of one fermion and one boson. If we construct the direct product of the supersymmetry group and the standard model symmetry group

$$Osp(1|4) \times \{SU(3) \times SO(2) \times U(1)\} \tag{5}$$

then for each particle of the standard model there should be a supersymmetric partner of different statistic. However, neither in the known elementary particles there is supersymmetric dublet found nor there is experimental evidence of finding the predicted supersymmetric partners of known elementary fermions or bosons. And this is the final dramatic present status of supersymmetry in elementary particles. We should add that the supersymmetry notion has been applied also to the supergravity theory with rather more promising results.

SUPERSYMMETRY IN NUCLEAR PHYSICS

From the above elementary particle discussion we can simply say that the supersymmetry is the symmetry of the system with fermions and bosons. In what follows we will consider ground and low excited states of nuclei for which neither meson or quark degrees of freedom are needed. Instead we adopt the Interacting Boson Model (IBM) of Arima and Iachello [1, 2] which the main assumption says that the valence nucleons are coupled in pairs with the angular momenta $l = 0$ or 2 and that there are also unpaired protons and neutrons. The properly correlated pairs of nucleons are treated phenomenologically and approximately as ideal bosons independent, by assumption, of unpaired nucleons. In such a way we get the phenomenological system with fermions and bosons and the system is the background for testing the supersymmetry.

As an example of supersymmetry considerations in nuclei we take into account the $s - d$ nuclear shell with the oxygen non–interacting core. Because of the same single–particle energy levels for protons and neutrons we apply in our model the isospin formalism both to bosons and fermions (nucleons). Let us consider at first the boson single–particle states

$$b_i^+ |0\rangle \tag{6}$$

where "i" stands for angular momenta $(l = 0; 2)$ spin $(\sigma = 1)$ and isospin numbers $(\tau = 1)$. It follows that the vectors (6) span the 36–dimension space. The unitary transformation form the unitary group $U(36)$ with the generators

$$b_i^+ b_j \qquad\qquad i, j = 1, 2, \ldots, 36 \tag{7}$$

Similarly, the single particle $s - d$ fermion states are of the form

$$a_k^+ |0\rangle \tag{8}$$

where "k" represents the angular momenta $(l = 0; 2)$ spin $(s = 1/2)$ and isospin $(t = 1/2)$ quantum numbers. Then the operators

$$a_k^+ a_m \qquad\qquad k, m = 1, 2, \ldots, 24 \tag{9}$$

generate the unitary transformations of the group $U(24)$. To form the unitary–unitary supergroup

$$U(36|24) \supset U^B(36) \times U^F(24) \tag{10}$$

we need also the operators which transform the bosons into fermions and vice–versa. In the creation and annihilation operator representation they simply are

$$b_i^+ a_k \qquad\qquad \text{and} \qquad\qquad a_k^+ b_i \tag{11}$$

The operator representation of the physical observables are constructed with generators (7), (9) and (11) in such a way as to obey the fermion conservation law. Moreover, these generators do not change the total number of particles (fermions + bosons). The last observation provides a definition for the supermultiplets as a system with constant total number of particles

$$\nu = \nu_f + \nu_b = \text{const.}$$

Suppose we start with even–even nucleus $^{30}_{14}Si_{16}$. It belongs to the second–half of the $s - d$ shell and there are 10 nucleon holes (up to $A = 40$). In the IBM model we assume

$\nu = 5$ boson system (pair of nucleons). The member of the same supermultiplet is also the odd–odd nucleus $^{30}_{15}P_{15}$ in the isospin formalism. To the same supermultiplet belong also the nuclei even–odd $^{31}_{16}S_{15}$ and odd–even $^{31}_{15}P_{16}$ with nine nucleon–holes which form four bosons (holes) and one fermion (hole). Due to the multiplet theory four nuclei should be treated as four different states of the same object. It follows that fenomenological parameters of the supersymmetry model which have the meaning of two–body matrix elements have to be the same in the eigenvalues or mean–values calculations.

Let us begin with the energy of the multiplet. The most general two body Hamiltonian has the formal shape

$$\hat{H} = \sum_{ijkm} \langle \quad \rangle_{ijkm}\, c_i^+ c_j^+ c_k c_m \tag{12}$$

where $c^+(c)$ stands for either fermion or boson operators. If we denote

$$c^+ c \equiv G \tag{13}$$

then after commutation we get

$$\hat{H} = \sum_{\alpha\beta} \langle \quad \rangle_{\alpha\beta}\, G_\alpha G_\beta \quad (+\ \text{single particle term}) \tag{14}$$

The second order Casimir invariants of the groups under consideration have also the same shape as in (14) with special coefficients taken instead of two body matrix elements $\langle \ \rangle$. The so called dynamical approximation in the frame of IBM means to take linear combination of Casimir invariants of the assumed chain of groups instead of (14). We adopt here the so called transitional chain within the IBM model which is characterized by the group $SO(6)$. The beginning of the chain reads, following (10):

$$U^B(36) \supset SU_L^B(6) \times SU_{ST}^B(6) \supset SU_L^B(6) \times SU_{ST}^B(4)$$

$$\tag{15}$$

$$U^F(24) \supset \qquad\qquad\qquad \supset SU_L^F(6) \times SU_{ST}^F(4)$$

It is of great importance to obtain the same groups in boson and fermion representations, because we can easily construct the boson–fermion groups of transformations

$$
\begin{array}{ccc}
SU_L^{BF}(6) & \times & SU_{ST}^{BF}(4) \\
\downarrow & & \downarrow \\
SO^{BF}(6) & & SU_S^{BF}(2) \times SU_T^{BF}(2) \\
\downarrow & & \nearrow \\
SO^{BF}(5) & & \\
\downarrow & & SU_J^{BF}(2) \\
SO_L^{BF}(3) & & \swarrow
\end{array}
\tag{16}
$$

The Hamiltonian (14) is then constructed as the linear combination of the relevant Casimir invariants of the chain (16)

$$H = H_o \;+\; P\hat{C}[SU(6)] + W\hat{C}[SO(6)] + K\hat{C}[SO(5)] + D\hat{C}[SO_L(3)]$$
$$+\; B\hat{C}[SU_S(2) + F\hat{C}[SO_J(2)] + A\hat{C}[SU_T(2)] \tag{17}$$

where P, W, K, D, B, F, A are phenomenological parameters and H_o is a constant term including the rest of Casimir invariants as well as the core contribution. The parameters have to be the same, following the above discusions for all of the multiplet members which in our example are even–even, odd–odd, even–odd and odd–even nuclei with

$\nu = \nu_b + \nu_f = 5$. This is the very strong criterion of the supersymmetry assumption. The Hamiltonian (17) can be algebraically diagonalized and the energy reads

$$
\begin{aligned}
E = E_o \;+\;& P[f_1(f_2 + 5) + f_2(f_2 + 3)] + W[\sigma_1(\sigma_1 + 4) + \sigma_2(\sigma_2 + 2)] \\
+\;& K[\tau_1(\tau_1 + 3)] + \tau_2(\tau(2 + 1)] + DL(L + 1) + BS(S + 1) \qquad (18) \\
+\;& FJ(J + 1) + AT(T + 1)
\end{aligned}
$$

The numbers in parenthesis factorize the irreducible representations of the relevant groups and their values are not free but follow the strict group theory considerations for members of the supermultiplet.

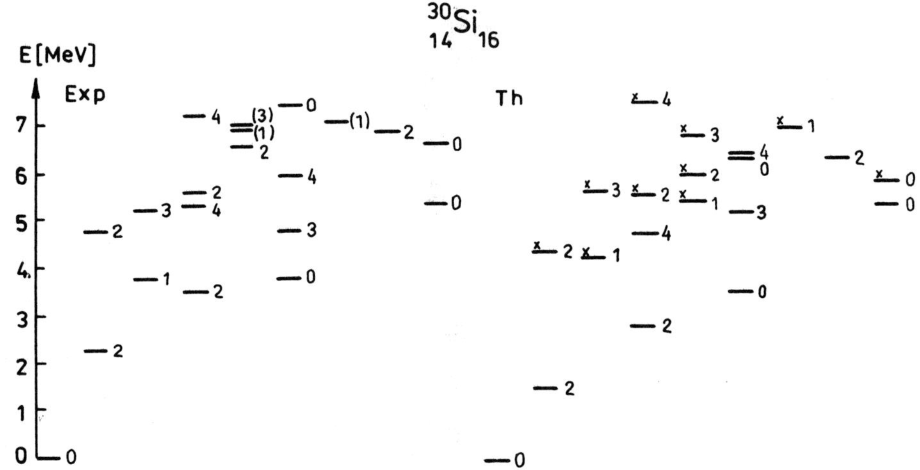

Figure 1a

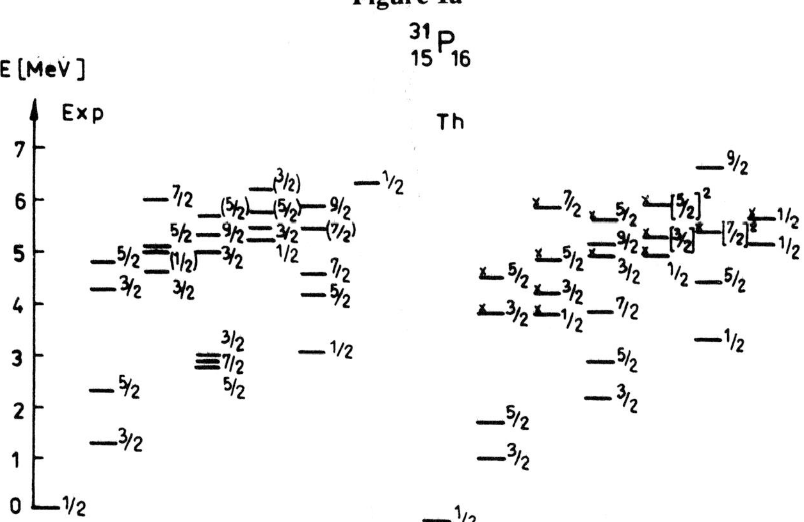

Figure 1b

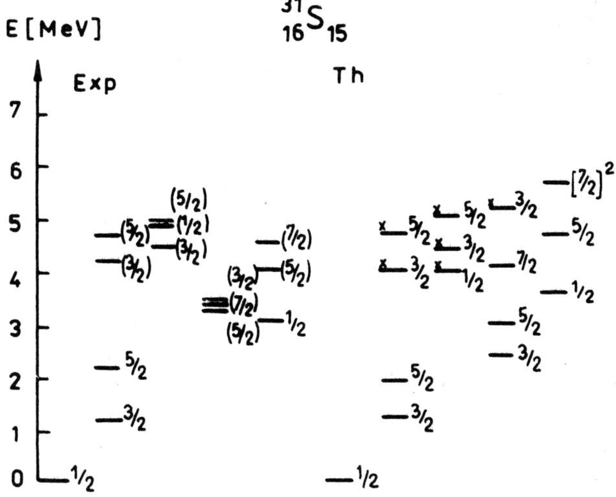

Figure 1c

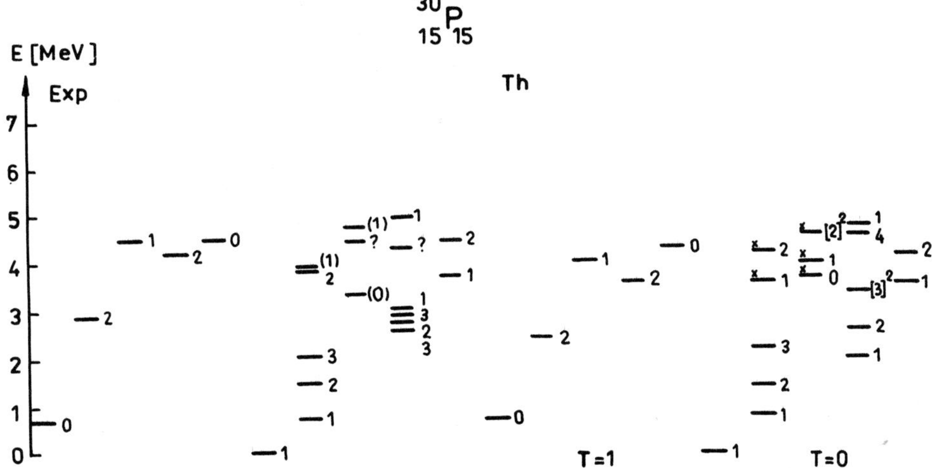

Figure 1d

Fig. 1a–d. Comparison of theoretical [5] and experimental [8] levels for four nuclei belonging to the supermultiplet $N = 5$ (holes). The Hamiltonian (17) which describes those even–even, even–odd and odd–odd–nuclei has been taken with the same parameters (two–body matrix elements) according to a supersymmetry model. The levels are organized in multiplets $SO_L^{BF}(5)$ marked by quantum numbers (h_1, h_2) of irreducible representations. The levels belong to the totally symmetric representation $[N]$ of the group $U_L^{BF}(6)$ except of those marked by stars which are of a mixed symmetry $[N-1, 1]$.

Fig. 1a–d give the comparison of experimental and theoretical spins and energies for the supermultiplet $N = 5$ (holes). It should be stressed that in Fig. 1 all of the experimental and also all of the calculated energy levels have been taken into account up to 4–6 MeV. Astonishingly, there is one to one correspondence of the experimental and theoretical spins of ground and excited levels. Calculated energies are not exactly the same but follow the same pattern. The phenomenological parameters were fixed for the nucleus $^{30}_{15}P_{15}$ and the same parameters have been then taken in the calculation of three other members of the supermultiplet. It means that the supersymmetry of the nuclear system under consideration is approximately confirmed. The approximation may have either numerical nature following the several approximations of the model or may also have the approximate behaviour of the nuclei within the supersymmetry. From physical point of view it also means that the ground and excited states of nuclei can be treated as comprised of pairs of nucleons plus some unpaired ones and that this system of fermions and approximate bosons obey the supersymmetry assumption. The similar supersymmetries have been found as well in heavy nuclei [3, 4] as in light nuclei [5, 6, 7].

4. CONCLUSIONS

In Tab. I we present the comparison of supersymmetries in elementary particle physics and in nuclear physics in the formal and physical aspects. The general conclusion is the following: the rigorous and beautiful theory of supersymmetry in elementary particle physics after several dramatic turning points in its development has led, in its simplest version, to supermultiplet fermion–boson dublets, for example quark–squark; gluon–gluino; foton–fotino; lepton–slepton and so on. However, there are, up to now, no experimental evidence for those supermultiplet partners. There is a dramatic question whether or not Nature likes supersymmetries. On the contrary, the phenomenological and approximate treatment of the nuclei–supersymmetry whose building blocks are pair of nucleons (very approximate bosons) and non–paired nucleons has shown the supersymmetry behaviour of such a system of bosons and fermions. There is no contradiction between the two conclusions, because the superesymmetry in nuclear physics exists on the level of pair of nucleons and nucleonic degrees of freedom, while discussed supersymmetry in elementary particle physics is related to the most fundamental level of degrees of freedom.

Table 1

Supersymmetry in	
elementary particles	nuclear physics
rigorous	approximate (i) $(a^\dagger a^\dagger) \approx b^\dagger$ (ii) $[b^\dagger, a] \approx 0$ (iii) $m_b \approx m_f$
relativistic (Poincarè group)	non-rel. static (no time in)
exact	broken
$Osp(N|M)$	$U(n|m)$
no experimental evidence	supersymmetry evidence in heavy and light nuclei

References

[1] F. Iachello, A. Arima, *Phys. Lett.* **53B**, 309 (1974).

[2] A. Arima, F. Iachello, *Ann. Phys. (N.Y.)*, **99**, 253 (1976); *Ann. Phys. (N.Y.)* **123**, 468 (1978).

[3] F. Iachello, *Phys. Rev. Lett.*, **44**, 772 (1980).

[4] A. B. Balantekin, I. Bars, F. Iachello, *Nucl. Phys.* **A370**, 284 (1981).

[5] S. Szpikowski, P. Kłosowski, L. Próchniak, *Nucl. Phys.* **A487**, 301 (1988).

[6] S. Szpikowski, P. Kłosowski, L. Próchniak, *Z. Phys.* **A335**, 289 (1990).

[7] L. Próchniak, S. Szpikowski, *Acta Phys. Pol.* **B24**, 557 (1993).

[8] P. M. Endt and van der Leun, *Nucl. Phys.* **A310**, *1 (1978)*.

Index